Collected Papers of

SRINIVASA RAMANUJAN

Collected Papers of
SRINIVASA RAMANUJAN

Edited by

G. H. HARDY
P. V. SESHU AIYAR
and
B. M. WILSON

Cambridge
AT THE UNIVERSITY PRESS
1927

CAMBRIDGE
UNIVERSITY PRESS

University Printing House, Cambridge CB2 8BS, United Kingdom

Cambridge University Press is part of the University of Cambridge.

It furthers the University's mission by disseminating knowledge in the pursuit of education, learning and research at the highest international levels of excellence.

www.cambridge.org
Information on this title: www.cambridge.org/9781107536517

First published 1927
First paperback edition 2015

A catalogue record for this publication is available from the British Library

ISBN 978-1-107-53651-7 Paperback

CONTENTS

CORRIGENDA

P. 37, line 14: *for* $\frac{1.2}{3^3}k^2$ *read* $\frac{1.2}{3^2}k^2$.

P. 45, equation (11): insert dots after the right-hand member.

P. 57, line 11: *for* $\Pi\{1+x^2(a+n)^2\}$ *read* $\Pi\{1+x^2/(a+n)^2\}$.

PREFACE

THIS volume contains everything published by Ramanujan except a few solutions of questions by other mathematicians printed in the *Journal of the Indian Mathematical Society*, and a certain amount of additional matter. Its publication has been made possible by the liberality of the University of Madras, the Royal Society, and Trinity College, Cambridge, each of which bodies has guaranteed a proportion of the expense of printing.

The editorial comments in Appendix I do not profess to be in any way systematic or exhaustive. We have merely put down such comments and references to the literature as occurred to us or were suggested to us by other mathematicians. In particular we are indebted to Prof. L. J. Mordell for a number of valuable suggestions.

We have also printed in Appendix II those parts of Ramanujan's letters from India which have not been printed before. It may seem that it would have been more natural to incorporate these in their proper places in the second *Notice*, but to do this would have expanded it unduly and destroyed its proportion, and the letters consist so largely of an enumeration of isolated theorems that they hardly suffer by division.

There is still a large mass of unpublished material. None of the contents of Ramanujan's notebooks has been printed, unless incorporated in later papers, except that one chapter, on generalised hypergeometric series, was analysed by Hardy* in the *Proceedings of the Cambridge Philosophical Society*. This chapter is sufficient to show that, while the notebooks are naturally unequal in quality, they contain much which should certainly be published. It would be a very formidable task to work through them systematically, select particular passages, and edit these with adequate comment, and it is impossible to print the notebooks as they stand without further monetary assistance. The singular quality of Ramanujan's work, and the romance which surrounds his career, encourage us to hope that this volume may enjoy sufficient success to make possible the publication of another.

* G. H. Hardy, "A chapter from Ramanujan's notebook", *Proc. Camb. Phil. Soc.*, XXI (1923), pp. 492–503.

SRINIVASA RAMANUJAN (1887—1920)

By P. V. Seshu Aiyar and R. Ramachandra Rao

Srinivasa Ramanujan Aiyangar, the remarkable mathematical genius who is the subject of this biographical sketch, was a member of a Brahmin family in somewhat poor circumstances in the Tanjore District of the Madras Presidency. There is nothing specially noteworthy about his ancestry to account for his great gifts. His father and paternal grandfather were *gumastas* (petty accountants) to cloth merchants in Kumbakonam, an important town in the Tanjore District. His mother, a woman of strong common-sense who still survives to mourn the loss of her distinguished son, was the daughter of a Brahmin petty official who held the position of *amin* (bailiff) in the Munsiff's court at Erode in the neighbouring district of Coimbatore. For some time after her marriage she had no children, but her father prayed to the famous goddess Namagiri, in the neighbouring town of Namakkal, to bless his daughter with children. Shortly afterwards, her eldest child, the mathematician Ramanujan, was born on the ninth day of Margasirsha in the Samvath Sarvajit, answering to the English date of 22nd December 1887.

Ramanujan was born in Erode, in the house of his maternal grandfather, to which in accordance with custom his mother had gone for the birth of her first child. In 1892, when in his fifth year, he was, as is usual with Brahmin boys, sent to a *pial* school, i.e. an indigenous elementary school conducted on very simple lines. Two years later he was admitted into the Town High School at Kumbakonam, in which he spent the rest of his school career.

During the first ten years of his life the only indication that he gave of special ability was that in 1897 he stood first amongst the successful candidates of the Tanjore District in the Primary Examination. This success secured for him the concession of being permitted to pay half-fees in his school.

Even in these early days he was remarkably quiet and meditative. It is remembered that he used to ask questions about the distances of the stars. As he held a high place in his class his class fellows used often to go to his house, but as he knew that his parents did not care for him to go out he used only to talk to them from a window which overlooked the street.

While he was in the second form he had, it appears, a great curiosity to know the "highest truth" in Mathematics, and asked some of his friends in the higher classes about it. It seems that some mentioned the Theorem of Pythagoras as the highest truth, and that some others gave the highest place to "Stocks and Shares". While in the third form, when his teacher was explaining to the class that any quantity divided by itself was equal to unity, he is said to have stood up and asked if zero divided by zero also was equal to unity. It was at about this time that he mastered the properties of the three progressions. While in the fourth form, he took to the study of Trigonometry. He is said to have borrowed a copy of the second part of Loney's *Trigonometry* from a student of the B.A. class, who was his neighbour. This student was struck with wonder to learn that this young lad of the fourth form had not only finished reading the book but could do every problem in it without any aid whatever; and not infrequently this B.A. student used to go to Ramanujan for the solution of difficult problems. While in the fifth form, he obtained unaided Euler's Theorems for the sine and the cosine and, when he found out later that the theorems had been already proved, he kept the paper containing the results secreted in the roofing of his house.

It was in 1903, while he was in the sixth form, on a momentous day for Ramanujan, that a friend of his secured for him the loan of a copy of Carr's *Synopsis of Pure Mathematics* from the library of the local Government College. Through the new world thus opened to him, Ramanujan went ranging with delight. It was this book that awakened his genius. He set himself to establish the formulæ given therein. As he was without the aid of other books, each solution was a piece of research so far as he was concerned. He first devised some methods for constructing magic squares. Then, he branched off to Geometry, where he took up the squaring of the circle and succeeded so far as to get a result for the length of the equatorial circumference of the earth which differed from the true length only by a few feet. Finding the scope of geometry limited, he turned his attention to Algebra and obtained several new series. Ramanujan used to say that the goddess of Namakkal inspired him with the formulæ in dreams. It is a remarkable fact that frequently, on rising from bed, he would note down results and rapidly verify them, though he was not always able to supply a rigorous proof. These results were embodied in a notebook which he afterwards used to show to mathematicians interested in his work.

In December 1903 he passed the Matriculation Examination of the University of Madras, and in the January of the succeeding year he joined the Junior First in Arts class of the Government College, Kumbakonam, and won the Subrahmanyam scholarship, which is generally awarded for proficiency in English and Mathematics. By this time, he was so much absorbed in the study of Mathematics that in all lecture hours—whether devoted to English, History or Physiology—he used to engage himself in some mathematical investigation, unmindful of what was happening in the class. This excessive devotion to Mathematics and his consequent neglect of the other subjects resulted in his failure to secure promotion to the senior class and in the consequent discontinuance of the scholarship. Partly owing to disappointment and partly owing to the influence of a friend, he ran away northwards into the Telugu country, but returned to Kumbakonam after some wandering and rejoined the college. As owing to his absence he failed to make sufficient attendances to obtain his term certificate in 1905, he entered Pachaiyappa's College, Madras, in 1906, but falling ill returned to Kumbakonam. He appeared as a private student for the F.A. Examination of December 1907 and failed. Afterwards he had no very definite occupation till 1909, but continued working at Mathematics in his own way and jotting down his results in another notebook.

In the summer of 1909 he married and wanted to settle down in life. Belonging to a poor and humble family, with an unfortunate college career, and without influence, he was hard put to it to secure some means of livelihood. In the hope of finding some employment he went, in 1910, to Tirukoilur, a small sub-division town in the South Arcot District, to see Mr V. Ramaswami Aiyar, M.A., the founder of the Indian Mathematical Society, who was then Deputy Collector of that place, and asked him for a clerical post in a municipal or *taluq* office of his division. This gentleman, being himself a mathematician of no mean order, and finding that the results contained in Ramanujan's notebook were remarkable, thought rightly that this unusual genius would be wasted if consigned to the dull routine of a *taluq* office, and helped Ramanujan on to Madras with a letter of introduction to Mr P. V. Seshu Aiyar, now Principal of Government College, Kumbakonam. Mr Seshu Aiyar had already known Ramanujan while the latter was at Kumbakonam, as he was the mathematical lecturer there while Ramanujan was in the F.A. class. Through him Ramanujan secured for a few months an acting post in the Madras Accountant-General's office and, when this arrangement ceased, he lived for a few months earning what little he could by giving private tuition. Not satisfied with such make-shift arrangements, Mr Seshu Aiyar sent him with a note of recommendation to Diwan Bahadur R. Ramachandra Rao, who was then Collector at Nellore, a small town 80 miles north of Madras, and who had already been introduced to Ramanujan and seen his notebook. His first interview with Ramanujan in December 1910 is better described in his own words:

"Several years ago, a nephew of mine perfectly innocent of mathematical knowledge said to me, 'Uncle, I have a visitor who talks of mathematics; I do not understand him; can you see if there is anything in his talk?' And in the plenitude of my mathematical wisdom, I condescended to permit Ramanujan to walk into my presence. A short uncouth figure, stout, unshaved, not overclean, with one conspicuous feature—shining eyes—walked in with a frayed notebook under his arm. He was miserably poor. He had run away from Kumbakonam to get leisure in Madras to pursue his studies. He never craved for any distinction. He wanted leisure; in other words, that simple food should be provided for him without exertion on his part and that he should be allowed to dream on.

"He opened his book and began to explain some of his discoveries. I saw quite at once that there was something out of the way; but my knowledge did not permit me to judge whether he talked sense or nonsense. Suspending judgment, I asked him to come over again, and he did. And then he had gauged my ignorance and shewed me some of his simpler results. These transcended existing books and I had no doubt that he was a remarkable man. Then, step by step, he led me to elliptic integrals and hypergeometric series and at last his theory of divergent series not yet announced to the world converted me. I asked him what he wanted. He said he wanted a pittance to live on so that he might pursue his researches."

Mr Ramachandra Rao sent him back to Madras, saying that it was cruel to make an intellectual giant like Ramanujan rot in a *mofussil* station like Nellore, and recommended that he should stay at Madras, undertaking to pay his expenses for a time. After a while, other attempts to obtain for him a scholarship having failed and Ramanujan being unwilling to be a burden on anybody for any length of time, he took up a small appointment on Rs 30 *per mensem* in the Madras Port Trust office, on the 9th February 1912.

He did not slacken his work at Mathematics in the meantime. His earliest contribution to the *Journal of the Indian Mathematical Society* was in the form of questions communicated by Mr Seshu Aiyar and published in the February number of Volume III (1911). His first long article was on "Some Properties of Bernoulli's Numbers" and was published in the December number of the same volume. In 1912 he contributed two more notes to the fourth volume of the same Journal, and also several questions for solution.

By this time, Mr Ramachandra Rao had induced Mr Griffith of the Madras Engineering College to take an interest in Ramanujan, and Mr Griffith spoke to Sir Francis Spring, the Chairman of the Madras Port Trust, in which Ramanujan was then employed; and from that time onwards it became easy to secure recognition of his work. Fortunately also the then manager of the Port Trust office was Mr S. Narayana Aiyar, M.A., a very keen and devoted student of Mathematics. He gave every encouragement to Ramanujan and very frequently worked with him during this period.

On the suggestion of Mr Seshu Aiyar and others, Ramanujan began a correspondence with Mr G. H. Hardy, then Fellow of Trinity College, Cambridge, on the 16th January 1913. In that letter he wrote:

"I had no University education but I have undergone the ordinary school course. After leaving school I have been employing the spare time at my disposal to work at Mathematics....I have made a special investigation of divergent series....Very recently I came across a tract published by you, styled *Orders of Infinity*, in page 36 of which I find a statement that no definite expression has been as yet found for the number of prime numbers less than any given number. I have found an expression which very nearly approximates to the real result, the error being negligible. I would request you to go through the enclosed papers. Being poor, if you are convinced that there is anything of value, I would like to have my theorems published. I have not given the actual investigations nor the expressions that I get; but I have indicated the lines on which I proceed. Being inexperienced, I would very highly value any advice you give me...."

The papers enclosed contained the enunciations of a hundred or more mathematical theorems.

In his second letter of date 27th February 1913, he wrote:

"...I have found a friend in you who views my labours sympathetically. This is already some encouragement to me to proceed....To preserve my brains, I want food and this is now my first consideration. Any sympathetic letter from you will be helpful to me here to get a scholarship either from the University or from the Government...."

But in the meantime Mr Hardy had written to the Secretary for Indian Students in London, saying that Ramanujan might prove to be a mathematician of the very highest class, and asking him to enquire whether some means could not be found for getting him a Cambridge education. This question was transmitted to the Secretary of the Students' Advisory Committee in Madras, who, in his turn, asked Ramanujan if he would go to England. But since his caste prejudices were very strong, he definitely declined to go. Upon the receipt of this unfavourable reply, the Secretary wrote, early in March 1913, to the Registrar of the University of Madras, explaining the circumstances of the case.

By this time Ramanujan's case had been brought to the notice of the University of Madras in another way. Early in February, Dr G. T. Walker, F.R.S., Director-General of Observatories, Simla, and formerly Fellow of Trinity College, Cambridge, happened to visit Madras on one of his official tours; and Sir Francis Spring took this opportunity to bring some of Ramanujan's work to Dr Walker's notice. As a result, Dr Walker addressed, on the 26th February 1913, the following letter to the Registrar of the University of Madras:

"...I have the honour to draw your attention to the case of S. Ramanujan, a clerk in the Accounts Department of the Madras Port Trust. I have not seen him, but was yesterday shewn some of his work in the presence of Sir Francis Spring. He is, I am told, 22 years of age and the character of the work that I saw impressed me as comparable in originality with that of a mathematical fellow in a Cambridge college....It was perfectly clear to me that the University would be justified in enabling S. Ramanujan for a few years at least to spend the whole of his time on Mathematics, without any anxiety as to his livelihood...."

As a result of this momentous letter and on the recommendation of the Board of Studies in Mathematics, the University granted to Ramanujan, with the previous approval of Government, a special scholarship of Rs 75 *per mensem* tenable for two years. The Syndicate took a special interest in getting this scholarship sanctioned, as may be seen from the following extract from the letter of the Registrar to the Government in this connection:

"The regulations of the University do not at present provide for such a special scholarship. But the Syndicate assumes that Section XV of the Act of Incorporation and Section 3 of the Indian Universities Act, 1904, allow of the grant of such a scholarship, subject to the express consent of the Governor of Fort St George in Council."

He was accordingly relieved from his clerical post in the Madras Port Trust office on the 1st of May 1913, and from that time he became and remained for the rest of his life a professional mathematician.

In accordance with the conditions of award of the scholarship, he submitted to the Board of Studies in Mathematics three quarterly reports on his researches on the 5th August 1913, 7th November 1913 and 9th March 1914 respectively.

But Mr Hardy was very much disappointed at Ramanujan's refusal to go to Cambridge. He had been at frequent intervals writing persuasive letters pointing out the advantages of a short stay in Cambridge, and when, early in 1914, the University of Madras invited Mr E. H. Neville, M.A., Fellow of Trinity College, Cambridge, to deliver a course of lectures at Madras, Mr Hardy used this opportunity and entrusted to Mr Neville the mission of persuading Ramanujan to give up his caste prejudices and come to Cambridge. In the meantime, many Indian friends also had been influencing him and, by the time Mr Neville approached him, Ramanujan himself had almost made up his mind; but his chief difficulty was to obtain his mother's consent. This consent was at last got very easily in an unexpected manner. For one morning his mother announced that she had had a dream on the previous night, in which she saw her son seated in a big hall amidst a group of Europeans, and that the goddess Namagiri had commanded her not to stand in the way of her son fulfilling his life's purpose. This was a very agreeable surprise to all concerned.

As soon as Ramanujan's consent was obtained, Mr Neville sent a memorandum to the authorities of the University of Madras on 28th January 1914. The memorandum ran as follows:

> "The discovery of the genius of S. Ramanujan of Madras promises to be the most interesting event of our time in the mathematical world....The importance of securing to Ramanujan a training in the refinements of modern methods and a contact with men who know what ranges of ideas have been explored and what have not cannot be overestimated....
>
> "I see no reason to doubt that Ramanujan himself will respond fully to the stimulus which contact with western mathematicians of the highest class will afford him. In that case his name will become one of the greatest in the history of mathematics and the University and the City of Madras will be proud to have assisted in his passage from obscurity to fame."

The next day, Mr R. Littlehailes, M.A., who was then Professor of Mathematics in the Presidency College, Madras, and now is the Director of Public Instruction, Madras, wrote another long letter to the Registrar of the University and made definite proposals regarding the scholarship to be granted.

The authorities of the University readily seized the opportunity and within a week decided, with the approval of the Government of Madras, to grant Ramanujan a scholarship of £250 a year, tenable in England for a period of two years, with free passage and a reasonable sum for outfit. This scholarship was subsequently extended up to 1st April 1919. Having arranged that the University should forward Rs 60 *per mensem* out of his scholarship to his mother at Kumbakonam, Ramanujan sailed for England on the 17th March

1914. He reached Cambridge in April and was admitted into Trinity College, which supplemented his scholarship by the award of an exhibition of £60.

He was now for the first time in his life in a really comfortable position and could devote himself to his researches without anxiety. Mr Hardy and Mr Littlewood helped him in publishing his papers in the English periodicals and under their guidance he developed rapidly.

On the 11th November 1915, Mr Hardy wrote to the Registrar of the Madras University:

"Ramanujan has been much handicapped by the war. Mr Littlewood, who would naturally have shared his teaching with me, has been away, and one teacher is not enough for so fertile a pupil......He is beyond question the best Indian mathematician of modern times...He will always be rather eccentric in his choice of subjects and methods of dealing with them......But of his extraordinary gifts there can be no question; in some ways he is the most remarkable mathematician I have ever known."

Mr Hardy's official report of date 16th June 1916 to the University of Madras was also in terms of very high praise*. Ramanujan had already published about a dozen papers in European journals. Everything went on well till the spring of 1917.

About May 1917, Mr Hardy wrote that it was suspected that Ramanujan had contracted an incurable disease. Since sea voyages were then risky on account of submarines and since the war had depleted India of good medical men, it was decided that he should stay in England for some time more. Hence he went into a nursing home at Cambridge in the early summer, and he was never out of bed for any length of time again. He was in sanatoria at Wells, at Matlock and in London, and it was not until the autumn of 1918 that he shewed any decided symptom of improvement.

On the 28th February 1918, he was elected a Fellow of the Royal Society. He was the first Indian on whom this high honour was conferred, and his election at the early age of thirty, and on the first occasion that his name was proposed, is a remarkable tribute to his distinguished genius. Stimulated perhaps by this election, he resumed active work, in spite of his ill-health, and some of his most beautiful theorems were discovered about this time. On the 13th October of the same year, he was elected a Fellow of Trinity College, Cambridge—a prize fellowship worth about £250 a year for six years, with no duties or conditions. In announcing this election, Mr Hardy wrote to the Registrar of the University of Madras, "He will return to India with a scientific standing and reputation such as no Indian has enjoyed before, and I am confident that India will regard him as the treasure he is", and urged the authorities of the University to make permanent provision for him in a way which could leave him free for research. This time also the University of Madras rose to the occasion and, in recognition of Ramanujan's services to the science of Mathematics, it granted him an allowance of £250 a year for

* Cf. *Journal of the Indian Mathematical Society*, 9 (1917), pp. 30–45.

five years from 1st April 1919, the date of the expiry of his scholarship, together with the actual expenses incurred by him in returning from England to India and on such passages from India to Europe and back as the Syndicate might approve of during the five years. At the suggestion of Mr Littlehailes the University of Madras also contemplated creating a University Professorship of Mathematics and offering it to him.

By this time his health shewed some signs of improvement. Although he shewed a tubercular tendency, the doctors said that he had never been gravely affected. Since the climate of England was suspected of retarding his recovery, it was decided to send him back to India. Accordingly, he left England on 27th February 1919, landed in Bombay on 27th March and arrived at Madras on the 2nd April.

When he returned he was in a precarious state of health. His friends grew very anxious. The best medical attendance was arranged for. He stayed three months in Madras and then spent two months in Kodumudi, a village on the banks of the Cauvery, not far from the place of his birth. He was a difficult patient, always inclined to revolt against medical treatment, and after a time he declined to be treated further. On the 3rd September he went to Kumbakonam, and since it was reported by many medical friends that he was getting worse, he was with great difficulty induced to come to Madras for treatment in January 1920 and was put under the best available medical care. Several philanthropic gentlemen assisted him during this period, notably Mr S. Srinivasa Aiyangar, who found all his expenses, and Rao Bahadur T. Numberumal Chetty, who gave his house free. The members of the Syndicate of the University of Madras also made a contribution towards his expenses in their individual capacity. But all this was of no avail. He died on the 26th April 1920, at Chetput, a suburb of Madras. He had no children but was survived by his parents and his wife.

We must refer to Mr Hardy's notice for an account of his mathematical work, but we add a few words about his appearance and personality. Before his illness he was inclined to stoutness; he was of moderate height (5 feet 5 inches); and had a big head with a large forehead and long wavy dark hair. His most remarkable feature was his sharp and bright dark eyes. A fairly faithful representation of him adorns the walls of the Madras University Library. On his return from England, he was very thin and emaciated and had grown very pale. He looked as if racked with pain. But his intellect was undimmed, and till about four days before he died he was engaged in work. All his work on "mock theta functions", of which only rough indications survive, was done on his death bed.

Ramanujan had definite religious views. He had a special veneration for the Namakkal goddess. Fond of the *Puranas*, he used to attend popular lectures on the Great Epics of Ramayana and Nahabharata, and to enter into discussions with learned pundits. He believed in the existence of a Supreme

Being and in the attainment of Godhood by men by proper methods of service and realisation of oneness with the Deity. He had settled convictions about the problem of life and after, and even the certain approach of death did not unsettle his faculties or spirits.

In manners he was very simple and he had absolutely no conceit. In a letter of date 26th November 1918, i.e. after Ramanujan had been honoured by being elected a Fellow of the Royal Society and a Fellow of Trinity, Mr Hardy wrote: "His natural simplicity has never been affected in the least by his success; indeed all that is wanted is to get him to realise that he really is a success." He was much distressed, when he had so little money for his own expenses, about his inability to help his poor parents; and when he received his scholarship, his first act was to devote a part of it to them. Ramanujan's simplicity and largeness of heart are further revealed in the following letter that he sent to the Registrar of the University of Madras:

2 COLINETTE ROAD, PUTNEY, S.W. 15.
11th January 1919.

To The Registrar,
University of Madras.

SIR,

I beg to acknowledge the receipt of your letter of 9th December 1918, and gratefully accept the very generous help which the University offers me.

I feel, however, that, after my return to India, which I expect to happen as soon as arrangements can be made, the total amount of money to which I shall be entitled will be much more than I shall require. I should hope that, after my expenses in England have been paid, £50 a year will be paid to my parents and that the surplus, after my necessary expenses are met, should be used for some educational purpose, such in particular as the reduction of school-fees for poor boys and orphans and provision of books in schools. No doubt it will be possible to make an arrangement about this after my return.

I feel very sorry that, as I have not been well, I have not been able to do so much mathematics during the last two years as before. I hope that I shall soon be able to do more and will certainly do my best to deserve the help that has been given me.

I beg to remain, Sir,
Your most obedient servant,
S. RAMANUJAN.

SRINIVASA RAMANUJAN (1887—1920)

BY G. H. HARDY*

I

SRINIVASA RAMANUJAN, who died at Kumbakonam on April 26th, 1920, had been a member of the Society since 1917. He was not a man who talked much about himself, and until recently I knew very little of his early life. Two notices, by P. V. Seshu Aiyar and R. Ramachandra Rao, two of the most devoted of Ramanujan's Indian friends, have been published recently in the *Journal of the Indian Mathematical Society*; and Sir Francis Spring has very kindly placed at my disposal an article which appeared in the *Madras Times* of April 5th, 1919. From these sources of information I can now supply a good many details with which I was previously unacquainted. Ramanujan (Srinivasa Iyengar Ramanuja Iyengar, to give him for once his proper name) was born on December 22nd, 1887, at Erode in southern India. His father was an accountant (*gumasta*) to a cloth merchant at Kumbakonam, while his maternal grandfather had served as *amin* in the *Munsiff's* (or local judge's) Court at Erode. He first went to school at five, and was transferred before he was seven to the Town High School at Kumbakonam, where he held a "free scholarship", and where his extraordinary powers appear to have been recognised immediately. "He used", so writes an old schoolfellow to Mr Seshu Aiyar, "to borrow Carr's *Synopsis of Pure Mathematics* from the College library, and delight in verifying some of the formulæ given there....He used to entertain his friends with his theorems and formulæ, even in those early days.... He had an extraordinary memory and could easily repeat the complete lists of Sanscrit roots (*atmanepada* and *parasmepada*); he could give the values of $\surd 2$, π, e, ... to any number of decimal places....In manners, he was simplicity itself...."

He passed his matriculation examination to the Government College at Kumbakonam in 1904; and secured the "Junior Subrahmanyam Scholarship". Owing to weakness in English, he failed in his next examination and lost his scholarship; and left Kumbakonam, first for Vizagapatam and then for Madras. Here he presented himself for the "First Examination in Arts' in December 1906, but failed and never tried again. For the next few years he continued his independent work in mathematics, "jotting down his results in two good-sized notebooks": I have one of these notebooks in my possession still. In 1909 he married, and it became necessary for him to find some permanent employment. I quote Mr Seshu Aiyar:

To this end, he went to Tirukoilur, a small sub-division town in South Arcot District, to see Mr V. Ramaswami Aiyar, the founder of the Indian Mathematical Society, but

* Obituary notice in the *Proceedings of the London Mathematical Society* (2), XIX (1921), pp. xl—lviii. The same notice was printed, with slight changes, in the *Proceedings of the Royal Society* (A), XCIX (1921), pp. xiii—xxix.

Mr Aiyar, seeing his wonderful gifts, persuaded him to go to Madras. It was then after some four years' interval that Ramanujan met me at Madras, with his two good-sized notebooks referred to above. I sent Ramanujan with a note of recommendation to that true lover of Mathematics, Diwan Bahadur R. Ramachandra Rao, who was then District Collector at Nellore, a small town some eighty miles north of Madras. Mr Rao sent him back to me, saying it was cruel to make an intellectual giant like Ramanujan rot at a *mofussil* station like Nellore, and recommended his stay at Madras, generously undertaking to pay Ramanujan's expenses for a time. This was in December 1910. After a while, other attempts to obtain for him a scholarship having failed, and Ramanujan himself being unwilling to be a burden on anybody for any length of time, he decided to take up a small appointment under the Madras Port Trust in 1912.

But he never slackened his work at Mathematics. His earliest contribution to the *Journal of the Indian Mathematical Society* was in the form of questions communicated by me in Vol. III (1911). His first long article on "Some Properties of Bernoulli's Numbers" was published in the December number of the same volume. Ramanujan's methods were so terse and novel and his presentation was so lacking in clearness and precision, that the ordinary reader, unaccustomed to such intellectual gymnastics, could hardly follow him. This particular article was returned more than once by the Editor before it took a form suitable for publication. It was during this period that he came to me one day with some theorems on Prime Numbers, and when I referred him to Hardy's Tract on *Orders of Infinity*, he observed that Hardy had said on p. 36 of his Tract "the exact order of $\rho(x)$ [defined by the equation

$$\rho(x) = \pi(x) - \int_2^x \frac{dt}{\log t},$$

where $\pi(x)$ denotes the number of primes less than x], has not yet been determined", and that he himself had discovered a result which gave the order of $\rho(x)$. On this I suggested that he might communicate his result to Mr Hardy, together with some more of his results.

This passage brings me to the beginning of my own acquaintance with Ramanujan. But before I say anything about the letters which I received from him, and which resulted ultimately in his journey to England, I must add a little more about his Indian career. Dr G. T. Walker, F.R.S., Head of the Meteorological Department, and formerly Fellow and Mathematical Lecturer of Trinity College, Cambridge, visited Madras for some official purpose some time in 1912; and Sir Francis Spring, K.C.I.E., the Chairman of the Madras Port Authority, called his attention to Ramanujan's work. Dr Walker was far too good a mathematician not to recognise its quality, little as it had in common with his own. He brought Ramanujan's case to the notice of the Government and the University of Madras. A research studentship, "Rs. 75 *per mensem* for a period of two years", was awarded him; and he became, and remained for the rest of his life, a professional mathematician.

II

Ramanujan wrote to me first on January 16th, 1913, and at fairly regular intervals until he sailed for England in 1914. I do not believe that his letters were entirely his own. His knowledge of English, at that stage of his life, could scarcely have been sufficient, and there is an occasional phrase which is hardly characteristic. Indeed I seem to remember his telling me that his friends had given him some assistance. However, it was the mathematics that mattered, and that was very emphatically his.

MADRAS, *16th January* 1913.

DEAR SIR,

I beg to introduce myself to you as a clerk in the Accounts Department of the Port Trust Office at Madras on a salary of only £20 per annum. I am now about 23 years of age. I have had no University education but I have undergone the ordinary school course. After leaving school I have been employing the spare time at my disposal to work at Mathematics. I have not trodden through the conventional regular course which is followed in a University course, but I am striking out a new path for myself. I have made a special investigation of divergent series in general and the results I get are termed by the local mathematicians as "startling".

Just as in elementary mathematics you give a meaning to a^n when n is negative and fractional to conform to the law which holds when n is a positive integer, similarly the whole of my investigations proceed on giving a meaning to Eulerian Second Integral for all values of n. My friends who have gone through the regular course of University education tell me that $\int_0^\infty x^{n-1} e^{-x}\,dx = \Gamma(n)$ is true only when n is positive. They say that this integral relation is not true when n is negative. Supposing this is true only for positive values of n and also supposing the definition $n\Gamma(n) = \Gamma(n+1)$ to be universally true, I have given meanings to these integrals and under the conditions I state the integral is true for all values of n negative and fractional. My whole investigations are based upon this and I have been developing this to a remarkable extent so much so that the local mathematicians are not able to understand me in my higher flights.

Very recently I came across a tract published by you styled *Orders of Infinity* in page 36 of which I find a statement that no definite expression has been as yet found for the number of prime numbers less than any given number. I have found an expression which very nearly approximates to the real result, the error being negligible. I would request you to go through the enclosed papers. Being poor, if you are convinced that there is anything of value I would like to have my theorems published. I have not given the actual investigations nor the expressions that I get but I have indicated the lines on which I proceed. Being inexperienced I would very highly value any advice you give me. Requesting to be excused for the trouble I give you.

I remain, Dear Sir, Yours truly,

S. RAMANUJAN.

P.S. My address is S. Ramanujan, Clerk Accounts Department, Port Trust, Madras, India.

I quote now from the "papers enclosed", and from later letters*:

In page 36 it is stated that "the number of prime numbers less than

$$x = \int_2^x \frac{dt}{\log t} + \rho(x)$$

where the precise order of $\rho(x)$ has not been determined...."

I have observed that $\rho(e^{2\pi x})$ is of such a nature that its value is very small when x lies between 0 and 3 (its value is less than a few hundreds when $x=3$) and rapidly increases when x is greater than 3....

The difference between the number of prime numbers of the form $4n-1$ and which are less than x and those of the form $4n+1$ less than x is infinite when x becomes infinite....

* [See Appendix II for parts of the letters not printed here.]

The following are a few examples from my theorems:

(1) The numbers of the form $2^p 3^q$ less than $n = \frac{1}{2}\frac{\log(2n)\log(3n)}{\log 2 \log 3}$ where p and q may have any positive integral value including 0.

(2) Let us take all numbers containing an odd number of dissimilar prime divisors, viz.

2, 3, 5, 7, 11, 13, 17, 19, 23, 29, 30, 31, 37, 41, 42, 43, 47, ...

(*a*) The number of such numbers less than $n = \frac{3n}{\pi^2}$.

(*b*) $\frac{1}{2^2} + \frac{1}{3^2} + \frac{1}{5^2} + \frac{1}{7^2} + \ldots + \frac{1}{30^2} + \frac{1}{31^2} + \ldots = \frac{9}{2\pi^2}$.

(*c*) $\frac{1}{2^4} + \frac{1}{3^4} + \frac{1}{5^4} + \frac{1}{7^4} + \ldots = \frac{15}{2\pi^4}$.

(3) Let us take the number of divisors of natural numbers, viz.

1, 2, 2, 3, 2, 4, 2, 4, 3, 4, 2, ... (1 having 1 divisor, 2 having 2, 3 having 2, 4 having 3, 5 having 2, ...).

The sum of such numbers to n terms

$$= n(2\gamma - 1 + \log n) + \tfrac{1}{2} \text{ of the number of divisors of } n$$

where $\gamma = \cdot 5772156649\ldots$, the Eulerian Constant.

(4) 1, 2, 4, 5, 8, 9, 10, 13, 16, 17, 18, ... are numbers which are either themselves squares or which can be expressed as the sum of two squares.

The number of such numbers greater than A and less than B

$$= K\int_A^B \frac{dx}{\sqrt{\log x}} + \theta(x)^* \quad \text{where} \quad K = \cdot 764\ldots$$

and $\theta(x)$ is very small when compared with the previous integral. K and $\theta(x)$ have been exactly found though complicated....

Ramanujan's theory of primes was vitiated by his ignorance of the theory of functions of a complex variable. It was (so to say) what the theory might be if the Zeta-function had no complex zeros. His methods of proof depended upon a wholesale use of divergent series. He disregarded entirely all the difficulties which are involved in the interchange of double limit operations; he did not distinguish, for example, between the sum of a series Σa_n and the value of the Abelian limit

$$\lim_{x \to 1} \Sigma a_n x^n,$$

or that of any other limit which might be used for similar purposes by a modern analyst. There are regions of mathematics in which the precepts of modern rigour may be disregarded with comparative safety, but the Analytic Theory of Numbers is not one of them, and Ramanujan's Indian work on primes, and on all the allied problems of the theory, was definitely wrong. That his proofs should have been invalid was only to be expected. But the mistakes went deeper than that, and many of the actual results were false. He had obtained the dominant terms of the classical formulæ, although by invalid methods; but none of them is such a close approximation as he supposed.

* This should presumably be $\theta(B)$.

This may be said to have been Ramanujan's one great failure. And yet I am not sure that, in some ways, his failure was not more wonderful than any of his triumphs. Consider, for example, problem (4). The dominant term, viz. $KB(\log B)^{-\frac{1}{2}}$, in Ramanujan's notation, was first obtained by Landau in 1908. Ramanujan had none of Landau's weapons at his command; he had never seen a French or German book; his knowledge even of English was insufficient to enable him to qualify for a degree. It is sufficiently marvellous that he should have even dreamt of problems such as these, problems which it has taken the finest mathematicians in Europe a hundred years to solve, and of which the solution is incomplete to the present day.

...IV. Theorems on integrals. The following are a few examples:

(1) $$\int_0^\infty \frac{1+\left(\frac{x}{b+1}\right)^2}{1+\left(\frac{x}{a}\right)^2} \cdot \frac{1+\left(\frac{x}{b+2}\right)^2}{1+\left(\frac{x}{a+1}\right)^2} \dots dx = \frac{\sqrt{\pi}}{2} \frac{\Gamma(a+\frac{1}{2})}{\Gamma(a)} \frac{\Gamma(b+1)}{\Gamma(b+\frac{1}{2})} \frac{\Gamma(b-a+\frac{1}{2})}{\Gamma(b-a+1)}.$$

......

(3) If $$\int_0^\infty \frac{\cos nx}{e^{2\pi\sqrt{x}}-1} dx = \phi(n),$$

then $$\int_0^\infty \frac{\sin nx}{e^{2\pi\sqrt{x}}-1} dx = \phi(n) - \frac{1}{2n} + \phi\left(\frac{\pi^2}{n}\right)\sqrt{\frac{2\pi^3}{n^3}}.$$

$\phi(n)$ is a complicated function. The following are certain special values:

$$\phi(0)=\frac{1}{12};\quad \phi\left(\frac{\pi}{2}\right)=\frac{1}{4\pi};\quad \phi(\pi)=\frac{2-\sqrt{2}}{8};\quad \phi(2\pi)=\frac{1}{16};$$

$$\phi\left(\frac{2\pi}{5}\right)=\frac{8-8\sqrt{5}}{16};\quad \phi\left(\frac{\pi}{5}\right)=\frac{6+\sqrt{5}}{4}-\frac{5\sqrt{10}}{8};\quad \phi(\infty)=0;$$

$$\phi\left(\frac{2\pi}{3}\right)=\frac{1}{3}-\sqrt{3}\left(\frac{3}{16}-\frac{1}{8\pi}\right).$$

(4) $$\int_0^\infty \frac{dx}{(1+x^2)(1+r^2x^2)(1+r^4x^2)\dots} = \frac{\pi}{2(1+r+r^3+r^6+r^{10}+\dots)}$$

where 1, 3, 6, 10... are sums of natural numbers.

(5) $$\int_0^\infty \frac{\sin 2nx}{x(\cosh \pi x + \cos \pi x)} dx = \frac{\pi}{4} - 2\left(\frac{e^{-n}\cos n}{\cosh\frac{\pi}{2}} - \frac{e^{-3n}\cos 3n}{3\cosh\frac{3\pi}{2}}\dots\right)$$

......

V. Theorems on summation of series*; e.g.

(1) $$\frac{1}{1^3}\cdot\frac{1}{2}+\frac{1}{2^3}\cdot\frac{1}{2^2}+\frac{1}{3^3}\cdot\frac{1}{2^3}+\frac{1}{4^3}\cdot\frac{1}{2^4}+\dots=\frac{1}{6}(\log 2)^3-\frac{\pi^2}{12}\log 2+\left(\frac{1}{1^3}+\frac{1}{3^3}+\frac{1}{5^3}+\dots\right).$$

(2) $$1+9\cdot(\tfrac{1}{4})^4+17\cdot\left(\frac{1\cdot 5}{4\cdot 8}\right)^4+25\cdot\left(\frac{1\cdot 5\cdot 9}{4\cdot 8\cdot 12}\right)^4+\dots=\frac{2\sqrt{2}}{\sqrt{\pi}\,\{\Gamma(\frac{3}{4})\}^2}.$$

* There is always more in one of Ramanujan's formulæ than meets the eye, as anyone who sets to work to verify those which look the easiest will soon discover. In some the interest lies very deep, in others comparatively near the surface; but there is not one which is not curious and entertaining.

(3) $1-5\left(\frac{1}{2}\right)^3+9\left(\frac{1\,.\,3}{2\,.\,4}\right)^3-\ldots=\frac{2}{\pi}$.

(4) $\frac{1^{13}}{e^{2\pi}-1}+\frac{2^{13}}{e^{4\pi}-1}+\frac{3^{13}}{e^{6\pi}-1}+\ldots=\frac{1}{24}$.

(5) $\frac{\coth \pi}{1^7}+\frac{\coth 2\pi}{2^7}+\frac{\coth 3\pi}{3^7}+\ldots=\frac{19\pi^7}{56700}$.

(6) $\frac{1}{1^5 \cosh \frac{\pi}{2}}-\frac{1}{3^5 \cosh \frac{3\pi}{2}}+\frac{1}{5^5 \cosh \frac{5\pi}{2}}-\ldots=\frac{\pi^5}{768}$.

......

VI. Theorems on transformation of series and integrals, e.g.

(1) $\pi\left(\frac{1}{2}-\frac{1}{\sqrt{1}+\sqrt{3}}+\frac{1}{\sqrt{3}+\sqrt{5}}-\frac{1}{\sqrt{5}+\sqrt{7}}+\ \ldots\right)=\frac{1}{1\sqrt{1}}-\frac{1}{3\sqrt{3}}+\frac{1}{5\sqrt{5}}-\ldots$.

......

(3) $$1-\frac{x^2\,3\,!}{(1\,!\,2\,!)^3}+\frac{x^4\,6\,!}{(2\,!\,4\,!)^3}-\frac{x^6\,9\,!}{(3\,!\,6\,!)^3}+\ldots$$
$$=\left\{1+\frac{x}{(1\,!)^3}+\frac{x^2}{(2\,!)^3}+\ \ldots\right\}\left\{1-\frac{x}{(1\,!)^3}+\frac{x^2}{(2\,!)^3}-\ \ldots\right\}.$$

......

(6) If $\alpha\beta=\pi^2$, then
$$\frac{1}{\sqrt[4]{\alpha}}\left\{1+4\alpha\int_0^\infty \frac{xe^{-\alpha x^2}}{e^{2\pi x}-1}\,dx\right\}=\frac{1}{\sqrt[4]{\beta}}\left\{1+4\beta\int_0^\infty \frac{xe^{-\beta x^2}}{e^{2\pi x}-1}\,dx\right\}.$$

(7) $n\left(e^{-n^2}-\frac{e^{-\frac{1}{3}n^2}}{3\sqrt{3}}+\frac{e^{-\frac{1}{5}n^2}}{5\sqrt{5}}-\ \ldots\right)=\sqrt{\pi}\,(e^{-n\sqrt{\pi}}\sin n\sqrt{\pi}-e^{-n\sqrt{3\pi}}\sin n\sqrt{3\pi}+\ldots)$.

(8) If n is any positive integer excluding 0,
$$\frac{1^{4n}}{(e^{\pi}-e^{-\pi})^2}+\frac{2^{4n}}{(e^{2\pi}-e^{-2\pi})^2}+\ldots=\frac{n}{\pi}\left\{\frac{B_{4n}}{8n}+\frac{1^{4n-1}}{e^{2\pi}-1}+\frac{2^{4n-1}}{e^{4\pi}-1}+\ldots\right\}$$
where $B_2=\frac{1}{6}$, $B_4=\frac{1}{30}$, ...

VII. Theorems on approximate integration and summation of series.

......

(2) $1+\frac{x}{1\,!}+\frac{x^2}{2\,!}+\frac{x^3}{3\,!}+\ldots+\frac{x^x}{x\,!}\,\theta=\frac{e^x}{2}$

where $\theta=\frac{1}{3}+\frac{4}{135\,(x+k)}$ where k lies between $\frac{8}{45}$ and $\frac{2}{21}$.

(3) $1+\left(\frac{x}{1\,!}\right)^5+\left(\frac{x^2}{2\,!}\right)^5+\left(\frac{x^3}{3\,!}\right)^5+\ldots=\frac{\sqrt{5}}{4\pi^2}\cdot\frac{e^{5x}}{5x^2-x+\theta}$

where θ vanishes when $x=\infty$.

(4) $$\frac{1^2}{e^x-1}+\frac{2^2}{e^{2x}-1}+\frac{3^2}{e^{3x}-1}+\frac{4^2}{e^{4x}-1}+\ldots=\frac{2}{x^3}\left(\frac{1}{1^3}+\frac{1}{2^3}+\frac{1}{3^3}+\ \ldots\right)-\frac{1}{12x}-\frac{x}{1440}+\frac{x^3}{181440}$$
$$+\frac{x^5}{7257600}+\frac{x^7}{159667200}+\ldots \text{ when } x \text{ is small.}$$

(*Note*: x may be given values from 0 to 2.)

(5) $\frac{1}{1001}+\frac{1}{1002^2}+\frac{3}{1003^3}+\frac{4^2}{1004^4}+\frac{5^3}{1005^5}+\ldots=\frac{1}{1000}-10^{-440}\times 1{\cdot}0125$ nearly.

(6) $\int_0^a e^{-x^2}\,dx=\frac{\sqrt{\pi}}{2}-\frac{e^{-a^2}}{2a}\,\frac{1}{+\,a}\,\frac{2}{+\,2a}\,\frac{3}{+\,a}\,\frac{4}{+\,2a\,+\ldots}$.

(7) The coefficient of x^n in $\frac{1}{1-2x+2x^4-2x^9+2x^{16}-\ldots}$

$$=\text{the nearest integer to } \frac{1}{4n}\left\{\cosh(\pi\sqrt{n})-\frac{\sinh(\pi\sqrt{n})}{\pi\sqrt{n}}\right\}*.$$

......

IX. Theorems on continued fractions, a few examples are:

(1) $\frac{4}{x+}\,\frac{1^2}{2x+}\,\frac{3^2}{2x+}\,\frac{5^2}{2x+}\,\frac{7^2}{2x+\ldots}=\left\{\frac{\Gamma\left(\frac{x+1}{4}\right)}{\Gamma\left(\frac{x+3}{4}\right)}\right\}^2$.

......

(4) If $u=\frac{x}{1+}\,\frac{x^5}{1+}\,\frac{x^{10}}{1+}\,\frac{x^{15}}{1+}\,\frac{x^{20}}{1+\ldots}$

and $v=\frac{\sqrt[5]{x}}{1+}\,\frac{x}{1+}\,\frac{x^2}{1+}\,\frac{x^3}{1+\ldots}$,

then $v^5=u\,.\,\frac{1-2u+4u^2-3u^3+u^4}{1+3u+4u^2+2u^3+u^4}$.

(5) $\frac{1}{1+}\,\frac{e^{-2\pi}}{1+}\,\frac{e^{-4\pi}}{1+}\,\frac{e^{-6\pi}}{1+\ldots}=\left(\sqrt{\frac{5+\sqrt{5}}{2}}-\frac{\sqrt{5}+1}{2}\right)\sqrt[5]{e^{2\pi}}$.

(6) $\frac{1}{1-}\,\frac{e^{-\pi}}{1+}\,\frac{e^{-2\pi}}{1-}\,\frac{e^{-3\pi}}{1+\ldots}=\left(\sqrt{\frac{5-\sqrt{5}}{3}}-\frac{\sqrt{5}-1}{2}\right)\sqrt[5]{e^{\pi}}$.

(7) $\frac{1}{1+}\,\frac{e^{-\pi\sqrt{n}}}{1+}\,\frac{e^{-2\pi\sqrt{n}}}{1+}\,\frac{e^{-3\pi\sqrt{n}}}{1+\ldots}$ can be exactly found if n be any positive rational quantity....

27 *February* 1913.

...I have found a friend in you who views my labours sympathetically. This is already some encouragement to me to proceed....I find in many a place in your letter rigorous proofs are required and you ask me to communicate the methods of proof.... I told him † that the sum of an infinite number of terms of the series $1+2+3+4+\ldots=-\frac{1}{12}$ under my theory. If I tell you this you will at once point out to me the lunatic asylum as my goal....What I tell you is this. Verify the results I give and if they agree with your results...you should at least grant that there may be some truths in my fundamental basis....

To preserve my brains I want food and this is now my first consideration. Any sympathetic letter from you will be helpful to me here to get a scholarship either from the University or from Government....

1. The number of prime numbers less than $e^a=\int_0^{\infty}\frac{a^x\,dx}{xS_{x+1}\Gamma(x+1)}$,

where $S_{x+1}=\frac{1}{1^{x+1}}+\frac{1}{2^{x+1}}+\ldots$.

2. The number of prime numbers less than n

$$=\frac{2}{\pi}\left\{\frac{2}{B_2}\left(\frac{\log n}{2\pi}\right)+\frac{4}{3B_4}\left(\frac{\log n}{2\pi}\right)^3+\frac{6}{5B_6}\left(\frac{\log n}{2\pi}\right)^5+\ldots\right\},$$

where $B_2=\frac{1}{6}$, $B_4=\frac{1}{30}$, ..., the Bernoullian numbers....

* This is quite untrue. But the formula is extremely interesting for a variety of reasons.

† Referring to a previous correspondence.

The order of $\theta(x)$ which you asked in your letter is $\sqrt{\left(\frac{x}{\log x}\right)}$.

......

(1) If
$$F(x)=\frac{1}{1}+\frac{x}{1}+\frac{x^2}{1}+\frac{x^3}{1}+\frac{x^4}{1}+\frac{x^5}{1}+\ldots,$$

then
$$\left\{\frac{\sqrt{5}+1}{2}+e^{-\frac{2\alpha}{5}}F(e^{-2\alpha})\right\}\left\{\frac{\sqrt{5}+1}{2}+e^{-\frac{2\beta}{5}}F(e^{-2\beta})\right\}=\frac{5+\sqrt{5}}{2}$$

with the conditions $\alpha\beta=\pi^2\ldots$,

e.g.
$$\frac{1}{1}+\frac{e^{-2\pi\sqrt{5}}}{1}+\frac{e^{-4\pi\sqrt{5}}}{1}+\ldots=e^{\frac{2\pi}{\sqrt{5}}}\left\{\frac{\sqrt{5}}{1+\sqrt[5]{5^{\frac{3}{4}}\left(\frac{\sqrt{5}-1}{2}\right)^{\frac{5}{2}}-1}}-\frac{\sqrt{5}+1}{2}\right\}.$$

The above theorem is a particular case of a theorem on the continued fraction
$$\frac{1}{1}+\frac{ax}{1}+\frac{ax^2}{1}+\frac{ax^3}{1}+\frac{ax^4}{1}+\frac{ax^5}{1}+\ldots,$$
which is a particular case of the continued fraction
$$\frac{1}{1}+\frac{ax}{1+bx}+\frac{ax^2}{1+bx^2}+\frac{ax^3}{1+bx^3}+\ldots,$$
which is a particular case of a general theorem on continued fractions.

(2) (i) $$4\int_0^\infty \frac{xe^{-x\sqrt{5}}}{\cosh x}\,dx=\frac{1}{1}+\frac{1^2}{1}+\frac{1^2}{1}+\frac{2^2}{1}+\frac{2^2}{1}+\frac{3^2}{1}+\frac{3^2}{1}+\ldots.$$

(ii) $$4\int_0^\infty \frac{x^2e^{-x\sqrt{3}}}{\cosh x}\,dx=\frac{1}{1}+\frac{1^3}{1}+\frac{1^3}{3}+\frac{2^3}{1}+\frac{2^3}{5}+\frac{3^3}{1}+\frac{3^3}{7}+\ldots.$$

(3) $$1-5.\left(\frac{1}{2}\right)^5+9.\left(\frac{1.3}{2.4}\right)^5-13.\left(\frac{1.3.5}{2.4.6}\right)^5+\ldots=\frac{2}{\{\Gamma(\frac{3}{4})\}^4}.$$

......

(6) If
$$v=\frac{x}{1}+\frac{x^3+x^6}{1}+\frac{x^6+x^{12}}{1}+\frac{x^9+x^{18}}{1}+\ldots,$$

then
(i) $$x\left(1+\frac{1}{v}\right)=\frac{1+x+x^3+x^6+x^{10}+\ldots}{1+x^9+x^{27}+x^{54}+x^{90}+\ldots},$$

(ii) $$x^3\left(1+\frac{1}{v^3}\right)=\left(\frac{1+x+x^3+x^6+x^{10}+\ldots}{1+x^3+x^9+x^{18}+x^{30}+\ldots}\right)^4.$$

(7) If n is any odd integer,
$$\frac{1}{\cosh\frac{\pi}{2n}+\cos\frac{\pi}{2n}}-\frac{1}{3\left(\cosh\frac{3\pi}{2n}+\cos\frac{3\pi}{2n}\right)}+\frac{1}{5\left(\cosh\frac{5\pi}{2n}+\cos\frac{5\pi}{2n}\right)}-\ldots=\frac{\pi}{8}.$$

......

(10) If $$F(\alpha,\beta,\gamma,\delta,\epsilon)=1+\frac{\alpha}{1!}.\frac{\beta}{\delta}.\frac{\gamma}{\epsilon}+\frac{\alpha(\alpha+1)}{2!}.\frac{\beta(\beta+1)}{\delta(\delta+1)}.\frac{\gamma(\gamma+1)}{\epsilon(\epsilon+1)}+\ldots,$$

then
$$F(\alpha,\beta,\gamma,\delta,\epsilon)=\frac{\Gamma(\delta)\Gamma(\delta-\alpha-\beta)}{\Gamma(\delta-\alpha)\Gamma(\delta-\beta)}.F(\alpha,\beta,\epsilon-\gamma,\alpha+\beta-\delta+1,\epsilon)$$
$$+\frac{\Gamma(\delta)\Gamma(\epsilon)\Gamma(\alpha+\beta-\delta)\Gamma(\delta+\epsilon-\alpha-\beta-\gamma)}{\Gamma(\alpha)\Gamma(\beta)\Gamma(\epsilon-\gamma)\Gamma(\delta+\epsilon-\alpha-\beta)}$$
$$\times F(\delta-\alpha,\delta-\beta,\delta+\epsilon-\alpha-\beta-\gamma,\delta-\alpha-\beta+1,\delta+\epsilon-\alpha-\beta).$$

(13) $$\frac{a}{1+n+}\,\frac{a^2}{3+n+}\,\frac{(2a)^2}{5+n+}\,\frac{(3a)^2}{7+n+}\dots$$
$$=2a\int_0^1 z^{\frac{n}{\sqrt{(1+a^2)}}}\,\frac{dz}{\{\sqrt{(1+a^2)}+1\}+z^2\{\sqrt{(1+a^2)}-1\}}.$$

(14) If $$F(\alpha,\beta)=\alpha+\frac{(1+\beta)^2+k}{2\alpha\;+}\,\frac{(3+\beta)^2+k}{2\alpha\;+}\,\frac{(5+\beta)^2+k}{2\alpha\;+\dots},$$
then $$F(\alpha,\beta)=F(\beta,\alpha).$$

(15) If $$F(\alpha,\beta)=\frac{\alpha}{n+}\,\frac{\beta^2}{n+}\,\frac{(2\alpha)^2}{n\;+}\,\frac{(3\beta)^2}{n\;+\dots},$$
then $$F(\alpha,\beta)+F(\beta,\alpha)=2F\{\tfrac{1}{2}(\alpha+\beta),\ \sqrt{(\alpha\beta)}\}.$$

......

(17) If $F(k)=1+\left(\frac{1}{2}\right)^2k+\left(\frac{1\,.\,3}{2\,.\,4}\right)^2k^2+\dots$ and $F(1-k)=\sqrt{(210)}\,F(k)$,
then
$$k=(\sqrt{2}-1)^4(2-\sqrt{3})^2(\sqrt{7}-\sqrt{6})^4(8-3\sqrt{7})^2(\sqrt{10}-3)^4(4-\sqrt{15})^4(\sqrt{15}-\sqrt{14})^2(6-\sqrt{35})^2.$$

......

(20) If $$F(\alpha)=\int_0^{\frac{1}{2}\pi}\frac{d\phi}{\sqrt{\{1-(1-\alpha)\sin^2\phi\}}}\Bigg/\int_0^{\frac{1}{2}\pi}\frac{d\phi}{\sqrt{\{1-\alpha\sin^2\phi\}}}$$
and $$F(\alpha)=3F(\beta)=5F(\gamma)=15F(\delta),$$
then (i) $$[(\alpha\delta)^{\frac{1}{8}}+\{(1-\alpha)(1-\delta)\}^{\frac{1}{8}}][(\beta\gamma)^{\frac{1}{8}}+\{(1-\beta)(1-\gamma)\}^{\frac{1}{8}}]=1.$$

......

(v) $$(\alpha\beta\gamma\delta)^{\frac{1}{8}}+\{(1-\alpha)(1-\beta)(1-\gamma)(1-\delta)\}^{\frac{1}{8}}$$
$$+\{16\alpha\beta\gamma\delta(1-\alpha)(1-\beta)(1-\gamma)(1-\delta)\}^{\frac{1}{24}}=1.$$

......

(21) If $$F(\alpha)=3F(\beta)=13F(\gamma)=39F(\delta)$$
or $$F(\alpha)=5F(\beta)=11F(\gamma)=55F(\delta)$$
or $$F(\alpha)=7F(\beta)=\ 9F(\gamma)=63F(\delta),$$
then $$\frac{\{(1-\alpha)(1-\delta)\}^{\frac{1}{8}}-(\alpha\delta)^{\frac{1}{8}}}{\{(1-\beta)(1-\gamma)\}^{\frac{1}{8}}-(\beta\gamma)^{\frac{1}{8}}}=\frac{1+\{(1-\alpha)(1-\delta)\}^{\frac{1}{4}}+(\alpha\delta)^{\frac{1}{4}}}{1+\{(1-\beta)(1-\gamma)\}^{\frac{1}{4}}+(\beta\gamma)^{\frac{1}{4}}}.$$

......

(23) $$(1+e^{-\pi\sqrt{1353}})(1+e^{-3\pi\sqrt{1353}})(1+e^{-5\pi\sqrt{1353}})\dots$$
$$=\sqrt[4]{2}\,e^{-\frac{1}{24}\pi\sqrt{1353}}\times\sqrt{\left\{\sqrt{\left(\frac{569+99\sqrt{33}}{8}\right)}+\sqrt{\left(\frac{561+99\sqrt{33}}{8}\right)}\right\}}$$
$$\times\sqrt{\left\{\sqrt{\left(\frac{25+3\sqrt{33}}{8}\right)}+\sqrt{\left(\frac{17+3\sqrt{33}}{8}\right)}\right\}}\times\sqrt[4]{\left(\frac{123+11}{\sqrt{2}}\right)}$$
$$\times\sqrt[8]{(10+3\sqrt{11})}\times\sqrt[8]{(26+15\sqrt{3})}\times\sqrt[12]{\left(\frac{6817+321\sqrt{451}}{\sqrt{2}}\right)}.$$

......

17 *April* 1913.

...I am a little pained to see what you have written*....I am not in the least apprehensive of my method being utilized by others. On the contrary my method has been in my possession for the last eight years and I have not found anyone to appreciate the method. As I wrote in my last letter I have found a sympathetic friend in you and

* Ramanujan might very reasonably have been reluctant to give away his secrets to an English mathematician, and I had tried to reassure him on this point as well as I could.

I am willing to place unreservedly in your hands what little I have. It was on account of the novelty of the method I have used that I am a little diffident even now to communicate my own way of arriving at the expressions I have already given....

...I am glad to inform you that the local University has been pleased to grant me a scholarship of £60 per annum for two years and this was at the instance of Dr Walker, F.R.S., Head of the Meteorological Department in India, to whom my thanks are due.... I request you to convey my thanks also to Mr Littlewood, Dr Barnes, Mr Berry and others who take an interest in me....

III

It is unnecessary to repeat the story of how Ramanujan was brought to England. There were serious difficulties; and the credit for overcoming them is due primarily to Prof. E. H. Neville, in whose company Ramanujan arrived in April 1914. He had a scholarship from Madras of £250, of which £50 was allotted to the support of his family in India, and an exhibition of £60 from Trinity. For a man of his almost ludicrously simple tastes, this was an ample income; and he was able to save a good deal of money which was badly wanted later. He had no duties and could do as he pleased; he wished indeed to qualify for a Cambridge degree as a research student, but this was a formality. He was now, for the first time in his life, in a really comfortable position, and could devote himself to his researches without anxiety.

There was one great puzzle. What was to be done in the way of teaching him modern mathematics? The limitations of his knowledge were as startling as its profundity. Here was a man who could work out modular equations, and theorems of complex multiplication, to orders unheard of, whose mastery of continued fractions was, on the formal side at any rate, beyond that of any mathematician in the world, who had found for himself the functional equation of the Zeta-function, and the dominant terms of many of the most famous problems in the analytic theory of numbers; and he had never heard of a doubly periodic function or of Cauchy's theorem, and had indeed but the vaguest idea of what a function of a complex variable was. His ideas as to what constituted a mathematical proof were of the most shadowy description. All his results, new or old, right or wrong, had been arrived at by a process of mingled argument, intuition, and induction, of which he was entirely unable to give any coherent account.

It was impossible to ask such a man to submit to systematic instruction, to try to learn mathematics from the beginning once more. I was afraid too that, if I insisted unduly on matters which Ramanujan found irksome, I might destroy his confidence or break the spell of his inspiration. On the other hand there were things of which it was impossible that he should remain in ignorance. Some of his results were wrong, and in particular those which concerned the distribution of primes, to which he attached the greatest importance. It was impossible to allow him to go through life supposing that all the zeros of the Zeta-function were real. So I had to try to teach him, and in a measure I succeeded, though obviously I learnt from him much more than he learnt from me. In a few years' time he had a very tolerable knowledge of the theory of functions and the analytic theory of numbers. He was never a mathematician of the modern school, and it was hardly desirable that he should become one; but he knew when he had proved a theorem and when he had not. And his flow of original ideas shewed no symptom of abatement.

I should add a word here about Ramanujan's interests outside mathematics. Like his mathematics, they shewed the strangest contrasts. He had very little interest, I should say, in literature as such, or in art, though he could tell good literature from bad. On the other hand, he was a keen philosopher, of what appeared, to followers of the modern Cambridge school, a rather nebulous kind, and an ardent politician, of a pacifist and ultra-radical type. He adhered, with a severity most unusual in Indians resident in England, to the religious observances of his caste; but his religion was a matter of observance and not of intellectual conviction, and I remember well his telling me (much to my surprise) that all religions seemed to him more or less equally true. Alike in literature, philosophy, and mathematics, he had a passion for what was unexpected, strange, and odd; he had quite a small library of books by circle-squarers and other cranks.

It was in the spring of 1917 that Ramanujan first appeared to be unwell. He went into a Nursing Home at Cambridge in the early summer, and was never out of bed for any length of time again. He was in sanatoria at Wells, at Matlock, and in London, and it was not until the autumn of 1918 that he shewed any decided symptom of improvement. He had then resumed active work, stimulated perhaps by his election to the Royal Society, and some of his most beautiful theorems were discovered about this time. His election to a Trinity Fellowship was a further encouragement; and each of those famous societies may well congratulate themselves that they recognised his claims before it was too late. Early in 1919 he had recovered, it seemed, sufficiently for the voyage home to India, and the best medical opinion held out hopes of a permanent restoration. I was rather alarmed by not hearing from him for a considerable time; but a letter reached me in February 1920, from which it appeared that he was still active in research.

UNIVERSITY OF MADRAS.
12th January 1920.

I am extremely sorry for not writing you a single letter up to now....I discovered very interesting functions recently which I call "Mock" ϑ-functions. Unlike the "False" ϑ-functions (studied partially by Prof. Rogers in his interesting paper) they enter into mathematics as beautifully as the ordinary ϑ-functions. I am sending you with this letter some examples....

Mock ϑ-functions

$$\phi(q)=1+\frac{q}{1+q^2}+\frac{q^4}{(1+q^2)(1+q^4)}+\dots,$$

$$\psi(q)=\frac{q}{1-q}+\frac{q^4}{(1-q)(1-q^3)}+\frac{q^9}{(1-q)(1-q^3)(1-q^5)}+\dots.$$

......

Mock ϑ-functions (of 5th order)

$$f(q)=1+\frac{q}{1+q}+\frac{q^4}{(1+q)(1+q^2)}+\frac{q^9}{(1+q)(1+q^2)(1+q^3)}+\ldots.$$

......

Mock ϑ-functions (of 7th order)

$$\text{(i)}\quad 1+\frac{q}{1-q^2}+\frac{q^4}{(1-q^3)(1-q^4)}+\frac{q^9}{(1-q^4)(1-q^5)(1-q^6)}+\ldots.$$

......

He said little about his health, and what he said was not particularly discouraging; and I was quite unprepared for the news of his death.

IV

Ramanujan published the following papers in Europe:

(1) "Some definite integrals", *Messenger of Mathematics*, Vol. 44 (1914), pp. 10—18.

(2) "Some definite integrals connected with Gauss's sums", *ibid.*, pp. 75—85.

(3) "Modular equations and approximations to π", *Quarterly Journal of Mathematics*, Vol. 45 (1914), pp. 350—372.

(4) "New expressions for Riemann's functions $\xi(s)$ and $\Xi(t)$", *ibid.*, Vol. 46 (1915), pp. 253—260.

(5) "On certain infinite series", *Messenger of Mathematics*, Vol. 45 (1916), pp. 11—15.

(6) "Summation of a certain series", *ibid.*, pp. 157—160.

(7) "Highly composite numbers", *Proc. London Math. Soc.*, Ser. 2, Vol. 14 (1915), pp. 347—409.

(8) "Some formulæ in the analytic theory of numbers", *Messenger of Mathematics*, Vol. 45 (1916), pp. 81—84.

(9) "On certain arithmetical functions", *Trans. Cambridge Phil. Soc.*, Vol. 22 (1916), No. 9, pp. 159—184.

(10) "A series for Euler's constant γ", *Messenger of Mathematics*, Vol. 46 (1917), pp. 73—80.

(11) "On the expression of a number in the form $ax^2+by^2+cz^2+dt^2$", *Proc. Cambridge Phil. Soc.*, Vol. 19 (1917), pp. 11—21.

*(12) "Une formule asymptotique pour le nombre des partitions de n", *Comptes Rendus*, 2 Jan. 1917.

*(13) "Asymptotic formulæ for the distribution of integers of various types", *Proc. London Math. Soc.*, Ser. 2, Vol. 16 (1917), pp. 112—132.

*(14) "The normal number of prime factors of a number n", *Quarterly Journal of Mathematics*, Vol. 48 (1917), pp. 76—92.

*(15) "Asymptotic formulæ in Combinatory Analysis", *Proc. London Math. Soc.*, Ser. 2, Vol. 17 (1918), pp. 75—115.

*(16) "On the coefficients in the expansions of certain modular functions", *Proc. Roy. Soc.* (A), Vol. 95 (1918), pp. 144—155.

(17) "On certain trigonometrical sums and their applications in the theory of numbers", *Trans. Cambridge Phil. Soc.*, Vol. 22, No. 13 (1918), pp. 259—276.

(18) "Some properties of $p(n)$, the number of partitions of n", *Proc. Cambridge Phil. Soc.*, Vol. 19 (1919), pp. 207—210.

(19) "Proof of certain identities in Combinatory Analysis", *ibid.*, pp. 214—216.

(20) "A class of definite integrals", *Quarterly Journal of Mathematics*, Vol. 48 (1920), pp. 294—310.

(21) "Congruence properties of partitions", *Math. Zeitschrift*, Vol. 9 (1921), pp. 147—153.

Of these, those marked with an asterisk were written in collaboration with me, and (21) is a posthumous extract from a much larger unpublished manuscript in my possession.* He also published a number of short notes in the *Records of Proceedings* at our meetings, and in the *Journal of the Indian Mathematical Society*. The complete list of these is as follows:

Records of Proceedings at Meetings.

*(22) "Proof that almost all numbers n are composed of about $\log\log n$ prime factors", 14 Dec. 1916.

*(23) "Asymptotic formulæ in Combinatory Analysis", 1 March 1917.

(24) "Some definite integrals", 17 Jan. 1918.

(25) "Congruence properties of partitions", 13 March 1919.

(26) "Algebraic relations between certain infinite products", 13 March 1919.

Journal of the Indian Mathematical Society.

(A) Articles and Notes.

(27) "Some properties of Bernoulli's numbers", Vol. 3 (1911), pp. 219—234.

(28) "On Q. 330 of Prof. Sanjana", Vol. 4 (1912), pp. 59—61.

(29) "A set of equations", Vol. 4 (1912), pp. 94—96.

(30) "Irregular numbers", Vol. 5 (1913), pp. 105—106.

(31) "Squaring the circle", Vol. 5 (1913), p. 132.

(32) "On the integral $\int_0^x \frac{\tan^{-1} t}{t}\, dt$", Vol. 7 (1915), pp. 93—96.

(33) "On the divisors of a number", Vol. 7 (1915), pp. 131—133.

(34) "The sum of the square roots of the first n natural numbers", Vol. 7 (1915), pp. 173—175.

(35) "On the product $\Pi\left[1+\frac{x^2}{(a+nd)^2}\right]$", Vol. 7 (1915), pp. 209—211.

(36) "Some definite integrals", Vol. 11 (1919), pp. 81—87.

(37) "A proof of Bertrand's postulate", Vol. 11 (1919), pp. 181—182.

(38) (Communicated by S. Narayana Aiyar), Vol. 3 (1911), p. 60.

(B) Questions proposed and solved.

Nos. 260, 261, 283, 284, 289, 294, 295, 308, 353, 358, 359, 386, 427, 441, 464, 489, 507, 524, 525, 541, 546, 571, 605, 606, 629, 642, 666, 682, 700, 723, 724, 739, 740, 753, 768, 769, 783, 785, 1070.

(C) Questions proposed but not solved.

Nos. 327, 352, 387, 441, 463, 469, 526, 584, 661, 662, 681, 699, 722, 738, 754, 755, 770, 784, 1049, and 1076.

Finally, I may mention the following writings by other authors, concerned with Ramanujan's work.†

"Proof of a formula of Mr Ramanujan", by G. H. Hardy (*Messenger of Mathematics*, Vol. 44, 1915, pp. 18—21).

* All of Ramanujan's manuscripts passed through my hands, and I edited them very carefully for publication. The earlier ones I rewrote completely. I had no share of any kind in the results, except of course when I was actually a collaborator, or when explicit acknowledgment is made. Ramanujan was almost absurdly scrupulous in his desire to acknowledge the slightest help.

† [Further references will be found in Appendix I.]

"Mr S. Ramanujan's mathematical work in England", by G. H. Hardy (Report to the University of Madras, 1916, privately printed).
"On Mr Ramanujan's empirical expansions of modular functions", by L. J. Mordell (*Proc. Cambridge Phil. Soc.*, Vol. 19, 1917, pp. 117—124).
"Life sketch of Ramanujan" (editorial in the *Journal of the Indian Math. Soc.*, Vol. 11, 1919, p. 122).
"Note on the parity of the number which enumerates the partitions of a number", by P. A. MacMahon (*Proc. Cambridge Phil. Soc.*, Vol. 20, 1921, pp. 281—283).
"Proof of certain identities and congruences enunciated by S. Ramanujan", by H. B. C. Darling (*Proc. London Math. Soc.*, Ser. 2, Vol. 19, 1921, pp. 350—372).
"On a type of modular relation", by L. J. Rogers (*ibid.*, pp. 387—397).

It is plainly impossible for me, within the limits of a notice such as this, to attempt a reasoned estimate of Ramanujan's work. Some of it is very intimately connected with my own, and my verdict could not be impartial; there is much too that I am hardly competent to judge; and there is a mass of unpublished material, in part new and in part anticipated, in part proved and in part only conjectured, that still awaits analysis. But it may be useful if I state, shortly and dogmatically, what seems to me Ramanujan's finest, most independent, and most characteristic work.

His most remarkable papers appear to me to be (3), (7), (9), (17), (18), (19), and (21). The first of these is mainly Indian work, done before he came to England; and much of it had been anticipated. But there is much that is new, and in particular a very striking series of algebraic approximations to π. I may mention only the formulæ

$$\pi = \frac{63}{25}\,\frac{17+15\sqrt{5}}{7+15\sqrt{5}}, \quad \frac{1}{2\pi\sqrt{2}} = \frac{1103}{99^2},$$

correct to 9 and 8 places of decimals respectively.

The long memoir (7) represents work, perhaps, in a backwater of mathematics, and is somewhat overloaded with detail; but the elementary analysis of "highly composite" numbers—numbers which have more divisors than any preceding number—is most remarkable, and shews very clearly Ramanujan's extraordinary mastery over the algebra of inequalities. Papers (9) and (17) should be read together, and in connection with Mr Mordell's paper mentioned above; for Mr Mordell afterwards proved a great deal that Ramanujan conjectured. They contain, in particular, very original and important contributions to the theory of the representation of numbers by sums of squares. But I am inclined to think that it was in the theory of partitions, and the allied parts of the theories of elliptic functions and continued fractions, that Ramanujan shews at his very best. It is in papers (18), (19), and (21), and in the papers of Prof. Rogers and Mr Darling that I have quoted, that this side of his work (so far as it has been published) is to be found. It would be difficult to find more beautiful formulæ than the "Rogers-Ramanujan" identities, proved in (19); but here Ramanujan must take second place to

Prof. Rogers; and, if I had to select one formula from all Ramanujan's work, I would agree with Major MacMahon in selecting a formula from (18), viz.

$$p(4)+p(9)x+p(14)x^2+\ldots=5\frac{\{(1-x^5)(1-x^{10})(1-x^{15})\ldots\}^5}{\{(1-x)(1-x^2)(1-x^3)\ldots\}^6},$$

where $p(n)$ is the number of partitions of n.

I have often been asked whether Ramanujan had any special secret; whether his methods differed in kind from those of other mathematicians; whether there was anything really abnormal in his mode of thought. I cannot answer these questions with any confidence or conviction; but I do not believe it. My belief is that all mathematicians think, at bottom, in the same kind of way, and that Ramanujan was no exception. He had, of course, an extraordinary memory. He could remember the idiosyncrasies of numbers in an almost uncanny way. It was Mr Littlewood (I believe) who remarked that "every positive integer was one of his personal friends." I remember once going to see him when he was lying ill at Putney. I had ridden in taxi-cab No. 1729, and remarked that the number (7 . 13 . 19) seemed to me rather a dull one, and that I hoped it was not an unfavourable omen. "No," he replied, "it is a very interesting number; it is the smallest number expressible as a sum of two cubes in two different ways." I asked him, naturally, whether he knew the answer to the corresponding problem for fourth powers; and he replied, after a moment's thought, that he could see no obvious example, and thought that the first such number must be very large.* His memory, and his powers of calculation, were very unusual, but they could not reasonably be called "abnormal". If he had to multiply two large numbers, he multiplied them in the ordinary way; he would do it with unusual rapidity and accuracy, but not more rapidly or more accurately than any mathematician who is naturally quick and has the habit of computation. There is a table of partitions at the end of our paper (15). This was, for the most part, calculated independently by Ramanujan and Major MacMahon; and Major MacMahon was, in general, slightly the quicker and more accurate of the two.

It was his insight into algebraical formulæ, transformations of infinite series, and so forth, that was most amazing. On this side most certainly I have never met his equal, and I can compare him only with Euler or Jacobi. He worked, far more than the majority of modern mathematicians, by induction from numerical examples; all of his congruence properties of partitions, for example, were discovered in this way. But with his memory, his patience, and his power of calculation, he combined a power of generalisation, a feeling for form, and a capacity for rapid modification of his hypotheses, that were often really startling, and made him, in his own peculiar field, without a rival in his day.

* Euler gave $158^4+59^4=134^4+133^4$ as an example. For other solutions see L. E. Dickson, *History of the Theory of Numbers*, Vol. 2, pp. 644—647.

It is often said that it is much more difficult now for a mathematician to be original than it was in the great days when the foundations of modern analysis were laid; and no doubt in a measure it is true. Opinions may differ as to the importance of Ramanujan's work, the kind of standard by which it should be judged, and the influence which it is likely to have on the mathematics of the future. It has not the simplicity and the inevitableness of the very greatest work; it would be greater if it were less strange. One gift it has which no one can deny, profound and invincible originality. He would probably have been a greater mathematician if he had been caught and tamed a little in his youth; he would have discovered more that was new, and that, no doubt, of greater importance. On the other hand he would have been less of a Ramanujan, and more of a European professor, and the loss might have been greater than the gain.

1

SOME PROPERTIES OF BERNOULLI'S NUMBERS

(*Journal of the Indian Mathematical Society*, III, 1911, 219—234)

1. Let the well-known expansion of $x \cot x$ (*vide* Edwards' *Differential Calculus*, § 149) be written in the form

$$x \cot x = 1 - \frac{B_2}{2!}(2x)^2 - \frac{B_4}{4!}(2x)^4 - \frac{B_6}{6!}(2x)^6 - \dots, \quad\dots\dots\dots\dots(1)$$

from which we infer that B_0 may be supposed to be -1.

Now
$$\cot x = \frac{\cos x}{\sin x} = \frac{1 - \frac{x^2}{2!} + \frac{x^4}{4!} - \frac{x^6}{6!} + \dots}{x - \frac{x^3}{3!} + \frac{x^5}{5!} - \frac{x^7}{7!} + \dots}$$

$$= \frac{\sin 2x}{1 - \cos 2x} = \frac{\frac{2x}{1!} - \frac{(2x)^3}{3!} + \frac{(2x)^5}{5!} - \dots}{\frac{(2x)^2}{2!} - \frac{(2x)^4}{4!} + \frac{(2x)^6}{6!} - \dots}$$

$$= \frac{1 + \cos 2x}{\sin 2x} = \frac{2 - \frac{(2x)^2}{2!} + \frac{(2x)^4}{4!} - \frac{(2x)^6}{6!} + \dots}{2x - \frac{(2x)^3}{3!} + \frac{(2x)^5}{5!} - \frac{(2x)^7}{7!} + \dots}.$$

Multiplying both sides in each of the above three relations by the denominator of the right-hand side and equating the coefficients of x^n on both sides, we can write the results thus:

$$c_1 \frac{B_{n-1}}{2} - c_3 \frac{B_{n-3}}{2^3} + c_5 \frac{B_{n-5}}{2^5} - \dots + \frac{(-1)^{\frac{1}{2}(n-1)}}{2^n} B_0 + \frac{n}{2^n}(-1)^{\frac{1}{2}(n-1)} = 0, \quad\dots\dots(2)$$

where n is any odd integer;

$$c_2 B_{n-2} - c_4 B_{n-4} + c_6 B_{n-6} - \dots + (-1)^{\frac{1}{2}(n-2)} B_0 + \frac{n}{2}(-1)^{\frac{1}{2}(n-2)} = 0, \quad\dots\dots(3)$$

where n is any even integer;

$$c_1 B_{n-1} - c_3 B_{n-3} + c_5 B_{n-5} - \dots + (-1)^{\frac{1}{2}(n-1)} B_0 + \frac{n}{2}(-1)^{\frac{1}{2}(n-1)} = 0, \quad\dots\dots(4)$$

where n is any odd integer greater than unity.

From any one of (2), (3), (4) we can calculate the B's. But as n becomes greater and greater the calculation will get tedious. So we shall try to find simpler methods.

2. We know $$(x\cot x)^2 = -x^2\left(1+\frac{d\cot x}{dx}\right).$$

Using (1) and equating the coefficients of x^n on both sides, and simplifying, we have

$$\tfrac{1}{2}(n+1)B_n = c_2 B_2 B_{n-2} + c_4 B_4 B_{n-4} + c_6 B_6 B_{n-6} + \ldots,$$

the last term being $c_{\frac{1}{2}n-1} B_{\frac{1}{2}n-1} B_{\frac{1}{2}n+1}$ or $\frac{1}{2}c_{\frac{1}{2}n}(B_{\frac{1}{2}n})^2$ according as $\frac{1}{2}n$ is odd or even. ....(5)

A similar result can be obtained by equating the coefficients of x^n in the identity

$$\frac{d\tan x}{dx} = 1 + \tan^2 x.$$

3. Again

$$-\tfrac{1}{2}x(\cot\tfrac{1}{2}x + \coth\tfrac{1}{2}x) = -\tfrac{1}{2}x(\cot\tfrac{1}{2}x + i\cot\tfrac{1}{2}ix)$$
$$= 2\left\{B_0 + B_4\frac{x^4}{4!} + B_8\frac{x^8}{8!} + \ldots\right\},$$

by using (1). The expression may also be written

$$-\tfrac{1}{2}x\frac{(\cos\frac{1}{2}x\sin\frac{1}{2}ix + i\sin\frac{1}{2}x\cos\frac{1}{2}ix)}{\sin\frac{1}{2}x\sin\frac{1}{2}ix}$$
$$= -\tfrac{1}{2}x\frac{(1+i)\sin\frac{1}{2}x(1+i) - (1-i)\sin\frac{1}{2}x(1-i)}{\cos\frac{1}{2}x(1-i) - \cos\frac{1}{2}x(1+i)}$$
$$= -x\frac{\dfrac{x}{1!} - \dfrac{x^5}{2^2.5!} + \dfrac{x^9}{2^4.9!} - \ldots}{\dfrac{x^2}{2!} - \dfrac{x^6}{2^2.6!} + \dfrac{x^{10}}{2^4.10!} - \ldots},$$

by expanding the numerator and the denominator, and simplifying by De Moivre's theorem.

Hence $$2\left(B_0 + B_4\frac{x^4}{4!} + B_8\frac{x^8}{8!} + \ldots\right) = -x\frac{\dfrac{x}{1!} - \dfrac{x^5}{2^2.5!} + \ldots}{\dfrac{x^2}{2!} - \dfrac{x^6}{2^2.6!} + \ldots}. \qquad \text{.........(6)}$$

Similarly

$$-\tfrac{1}{2}x(\cot\tfrac{1}{2}x - \coth\tfrac{1}{2}x) = 2\left(B_2\frac{x^2}{2!} + B_6\frac{x^6}{6!} + B_{10}\frac{x^{10}}{10!} + \ldots\right)$$
$$= x\frac{\frac{1}{2}(1-i)\sin\frac{1}{2}x(1+i) - \frac{1}{2}(1+i)\sin\frac{1}{2}x(1-i)}{\cos\frac{1}{2}x(1+i) - \cos\frac{1}{2}x(1-i)}$$
$$= x\frac{\dfrac{x^3}{2.3!} - \dfrac{x^7}{2^3.7!} + \dfrac{x^{11}}{2^5.11!} + \ldots}{\dfrac{x^2}{2!} - \dfrac{x^6}{2^2.6!} + \dfrac{x^{10}}{2^4.10!} - \ldots}. \qquad \text{........................(7)}$$

Proceeding as in § 1 we have, if n is an even integer greater than 2,

$$c_2 \frac{B_{n-2}}{2} - c_6 \frac{B_{n-6}}{2^3} + c_{10} \frac{B_{n-10}}{2^5} - \ldots + \frac{n}{2^{\frac{1}{2}(n+2)}}(-1)^{\frac{1}{4}n} \text{ or } \frac{n}{2^{\frac{1}{2}(n+2)}}(-1)^{\frac{1}{4}(n-2)} = 0, \quad (8)$$

according as n or $n-2$ is a multiple of 4.

Analogous results can be obtained from $\tan \frac{1}{2}x \pm \tanh \frac{1}{2}x$.

In (2), (3) and (4) there are $\frac{1}{2}n$ terms, while in (5) and (8) there are $\frac{1}{4}n$ or $\frac{1}{4}(n-2)$ terms. Thus B_n can be found from only half of the previous B's.

4. A still simpler method can be deduced from the following identities.

If 1, ω, ω^2 be the three cube roots of unity, then

$$4 \sin x \sin x\omega \sin x\omega^2 = -(\sin 2x + \sin 2x\omega + \sin 2x\omega^2),$$

as may easily be verified.

By logarithmic differentiation, we have

$$\cot x + \omega \cot x\omega + \omega^2 \cot x\omega^2 = 2\frac{\cos 2x + \omega \cos 2x\omega + \omega^2 \cos 2x\omega^2}{\sin 2x + \sin 2x\omega + \sin 2x\omega^2}.$$

Writing $\frac{1}{2}x$ for x,

$$-\tfrac{1}{2}x(\cot \tfrac{1}{2}x + \omega \cot \tfrac{1}{2}x\omega + \omega^2 \cot \tfrac{1}{2}x\omega^2) = -x\frac{\cos x + \omega \cos x\omega + \omega^2 \cos x\omega^2}{\sin x + \sin x\omega + \sin x\omega^2},$$

and, proceeding as in § 3, we get

$$3\left(B_0 + B_6 \frac{x^6}{6!} + B_{12}\frac{x^{12}}{12!} + \ldots\right) = -x \frac{\dfrac{x^2}{2!} - \dfrac{x^8}{8!} + \dfrac{x^{14}}{14!} - \ldots}{\dfrac{x^3}{3!} - \dfrac{x^9}{9!} + \dfrac{x^{15}}{15!} - \ldots}. \quad \ldots\ldots(9)$$

Again

$$\cot \tfrac{1}{2}x\omega - \cot \tfrac{1}{2}x\omega^2 = \frac{\cos x\omega^2 - \cos x\omega}{2 \sin \frac{1}{2}x \sin \frac{1}{2}x\omega \sin \frac{1}{2}x\omega^2} = \frac{2(\cos x\omega - \cos x\omega^2)}{\sin x + \sin x\omega + \sin x\omega^2}.$$

Multiplying both sides by $-\frac{1}{2}x(\omega^2 - \omega)$ and adding to the corresponding sides of the previous result, we have

$$-\tfrac{1}{2}x(\cot \tfrac{1}{2}x + \omega^2 \cot \tfrac{1}{2}x\omega + \omega \cot \tfrac{1}{2}x\omega^2) = -x\frac{\cos x + \omega^2 \cos x\omega + \omega \cos x\omega^2}{\sin x + \sin x\omega + \sin x\omega^2}.$$

Hence, as before,

$$3\left(B_2 \frac{x^2}{2!} + B_8 \frac{x^8}{8!} + B_{14}\frac{x^{14}}{14!} + \ldots\right) = x \frac{\dfrac{x^4}{4!} - \dfrac{x^{10}}{10!} + \dfrac{x^{16}}{16!} - \ldots}{\dfrac{x^3}{3!} - \dfrac{x^9}{9!} + \dfrac{x^{15}}{15!} - \ldots}. \quad \ldots\ldots(10)$$

Similarly

$$-x(\cot \tfrac{1}{2}x + \cot \tfrac{1}{2}x\omega + \cot \tfrac{1}{2}x\omega^2) = x\frac{\cos x + \cos x\omega + \cos x\omega^2 - 3}{\sin x + \sin x\omega + \sin x\omega^2},$$

and therefore

$$6\left(B_4 \frac{x^4}{4!} + B_{10} \frac{x^{10}}{10!} + B_{16}\frac{x^{16}}{16!} + \ldots\right) = x \frac{\dfrac{x^6}{6!} - \dfrac{x^{12}}{12!} + \dfrac{x^{18}}{18!} - \ldots}{\dfrac{x^3}{3!} - \dfrac{x^9}{9!} + \dfrac{x^{15}}{15!} - \ldots}. \quad \ldots(11)$$

Multiplying up and equating coefficients in (9), (10) and (11) as usual, we have,

$$c_3 B_{n-3} - c_9 B_{n-9} + c_{15} B_{n-15} - \ldots = 0, \quad \ldots\ldots\ldots\ldots\ldots\ldots (12)$$

the last term being $\frac{1}{6}n(-1)^{\frac{1}{6}(n-1)}$, $\frac{1}{3}n(-1)^{\frac{1}{6}(n+1)}$, or $\frac{1}{2}n(-1)^{\frac{1}{6}(n-3)}$.

Again, dividing both sides in (10) by x and differentiating, we have

$$3\left(B_2\frac{1}{2!} + 7B_8\frac{x^6}{8!} + 13B_{14}\frac{x^{12}}{14!} + \ldots\right) = \frac{d}{dx}\left(\frac{\frac{x^4}{4!} - \frac{x^{10}}{10!} + \frac{x^{16}}{16!} - \ldots}{\frac{x^3}{3!} - \frac{x^9}{9!} + \frac{x^{15}}{15!} - \ldots}\right)$$

$$= 1 - \frac{\frac{x^2}{2!} - \frac{x^8}{8!} + \frac{x^{14}}{14!} - \ldots}{\frac{x^3}{3!} - \frac{x^9}{9!} + \frac{x^{15}}{15!} - \ldots} \cdot \frac{\frac{x^4}{4!} - \frac{x^{10}}{10!} + \frac{x^{16}}{16!} - \ldots}{\frac{x^3}{3!} - \frac{x^9}{9!} + \frac{x^{15}}{15!} - \ldots}.$$

Hence by (9) and (10),

$$3\left(B_2\frac{x^2}{2!} + 7B_8\frac{x^8}{8!} + 13B_{14}\frac{x^{14}}{14!} + \ldots\right)$$

$$= x^2 + 9\left(B_0 + B_6\frac{x^6}{6!} + B_{12}\frac{x^{12}}{12!} + \ldots\right)\left(B_2\frac{x^2}{2!} + B_8\frac{x^8}{8!} + B_{14}\frac{x^{14}}{14!} + \ldots\right).$$

Equating the coefficients of x^n we have, if $n > 2$ and $n-2$ is a multiple of 6,

$$\tfrac{1}{3}(n+2)B_n = c_6 B_{n-6} B_6 + c_{12} B_{n-12} B_{12} + c_{18} B_{n-18} B_{18} + \ldots \quad \ldots\ldots (13)$$

From (12) the B's can be calculated very quickly and (13) may prove useful in checking the calculations. The number of terms is one-third of that in (4); thus B_{24} is found from B_{18}, B_{12} and B_6.

5. We shall see later on how the B's can be obtained from their properties only. But to know these properties, it will be convenient to calculate a few B's by substituting 3, 5, 7, 9, ..., for n in succession in (12). Thus

$$B_0 = -1;\ B_2 = \frac{1}{6};\ B_4 = \frac{1}{30};\ B_6 = \frac{1}{42};\ B_8 - \tfrac{1}{3}B_2 = -\frac{1}{45};$$

$$B_{10} - \frac{5}{2}B_4 = -\frac{1}{132};\ B_{12} - 11B_6 = -\frac{4}{455};\ B_{14} - \frac{143}{4}B_8 + \frac{B_2}{5} = \frac{1}{120};$$

$$B_{16} - \frac{286}{3}B_{10} + 4B_4 = \frac{1}{306};\ B_{18} - 221B_{12} + \frac{204}{5}B_6 = \frac{3}{665};$$

$$B_{20} - \frac{3230}{7}B_{14} + \frac{1938}{7}B_8 - \frac{B_2}{7} = -\frac{1}{231};$$

$$B_{22} - \frac{3553}{4}B_{16} + \frac{7106}{5}B_{10} - \frac{11}{2}B_4 = -\frac{1}{552};$$

and so on. Hence we have finally the following values:

$$B_2 = \frac{1}{6};\ B_4 = \frac{1}{30};\ B_6 = \frac{1}{42};\ B_8 = \frac{1}{30};\ B_{10} = \frac{5}{66};\ B_{12} = \frac{691}{2730};$$

$$B_{14} = \frac{7}{6};\ B_{16} = \frac{3617}{510};\ B_{18} = \frac{43867}{798};\ B_{20} = \frac{174611}{330};\ B_{22} = \frac{854513}{138};$$

$$B_{24} = \frac{236364091}{2730};\ B_{26} = \frac{8553103}{6};\ B_{28} = \frac{23749461029}{870};$$

$$B_{30} = \frac{8615841276005}{14322};\ B_{32} = \frac{7709321041217}{510};\ B_{34} = \frac{2577687858367}{6};$$

$$B_{36} = \frac{26315271553053477373}{1919190};\ B_{38} = \frac{2929993913841559}{6};$$

$$B_{40} = \frac{261082718496449122051}{13530};\ \ldots,\ B_\infty = \infty.$$

6. It will be observed* that, if n is even but not equal to zero,

(i) B_n is a fraction and the numerator of B_n/n in its lowest terms is a prime number, ...(14)

(ii) the denominator of B_n contains each of the factors 2 and 3 once and only once, ...(15)

(iii) $2^n(2^n-1)B_n/n$ is an integer and consequently $2(2^n-1)B_n$ is an *odd* integer. ....(16)

From (16) it can easily be shewn that the denominator of $2(2^n-1)B_n/n$ in its lowest terms is the greatest power of 2 which divides n; and consequently, if n is not a multiple of 4, then $4(2^n-1)B_n/n$ is an odd integer. ....(17)

It follows from (14) that the numerator of B_n in its lowest terms is divisible by the greatest measure of n prime to the denominator, and the quotient is a prime number. ....(18)

Examples: (*a*) 2 and 3 are the only prime factors of 12, 24 and 36, and they are found in the denominators of B_{12}, B_{24} and B_{36} and their numerators are prime numbers.

(*b*) 11 is not found in the denominator of B_{22}, and hence its numerator is divisible by 11; similarly, the numerators of B_{26}, B_{34}, B_{38} are divisible by 13, 17, 19, respectively and the quotients in all cases are prime numbers.

(*c*) 5 is found in the denominator of B_{20} and not in that of B_{30}, and consequently the numerator of B_{30} is divisible by 5 while that of B_{20} is a prime number. Thus we may say that if a prime number appearing in n is not found in the denominator it will appear in the numerator, and *vice versa*.

* See § 12 below.

7. Next, let us consider the denominators.

All the denominators are divisible by 6; those of $B_4, B_8, B_{12}, \ldots$ by 5; those of $B_6, B_{12}, B_{18}, \ldots$ by 7; those of $B_{10}, B_{20}, B_{30}, \ldots$ by 11; but those of $B_8, B_{16}, B_{24}, \ldots$ are *not* divisible by 9; and those of $B_{14}, B_{28}, \ldots$ are *not* divisible by 15. Hence we may infer that:

the denominator of B_n is the continued product of prime numbers which are the next numbers (in the natural order) to the factors of n (including unity and the number itself). ....(19)

As an example take the denominator of B_{24}. Write all the factors of 24, viz. 1, 2, 3, 4, 6, 8, 12, 24. The next numbers to these are 2, 3, 4, 5, 7, 9, 13, 25. Strike out the *composite* numbers and we have the prime numbers 2, 3, 5, 7, 13. And the denominator of B_{24} is the product of 2, 3, 5, 7, 13, i.e. 2730.

It is unnecessary to write the *odd* factors of n except unity, as the next numbers to these are even and hence composite.

The following are some further examples:

Even factors of n and unity	Denominator of B_n
B_2 ... 1, 2	2.3=6
B_6 ... 1, 2, 6	2.3.7=42
B_{12} ... 1, 2, 4, 6, 12	2.3.5.7.13=2730
B_{20} ... 1, 2, 4, 10, 20	2.3.5.11=330
B_{30} ... 1, 2, 6, 10, 30	2.3.7.11.31=14322
B_{42} ... 1, 2, 6, 14, 42	2.3.7.43=1806
B_{56} ... 1, 2, 4, 8, 14, 28, 56	2.3.5.29=870
B_{72} ... 1, 2, 4, 6, 8, 12, 18, 24, 36, 72 ...	2.3.5.7.13.19.37.73=140100870
B_{90} ... 1, 2, 6, 10, 18, 30, 90	2.3.7.11.19.31=272118
B_{110} ... 1, 2, 10, 22, 110	2.3.11.23=1518

8. Again taking the fractional part of any B and splitting it into partial fractions, we see that:

the fractional part of $B_n = (-1)^{\frac{1}{2}n}$ {the sum of the reciprocals of the prime factors of the denominator of B_n} $-(-1)^{\frac{1}{2}n}$. ....(20)

Thus the fractional part of $B_{16} = \frac{1}{2} + \frac{1}{3} + \frac{1}{5} + \frac{1}{17} - 1 = \frac{47}{510}$;

that of $B_{22} = 1 - \frac{1}{2} - \frac{1}{3} - \frac{1}{23} = \frac{17}{138}$;

that of $B_{28} = \frac{1}{2} + \frac{1}{3} + \frac{1}{5} + \frac{1}{29} - 1 = \frac{59}{870}$;

and so on.

9. It can be inferred from (20) that:

if G be the G.C.M. and L the L.C.M. of the denominators of B_m and B_n, then L/G is the denominator of $B_m - (-1)^{\frac{1}{2}(m-n)} B_n$, and hence, if the denominators of B_m and B_n are equal, then $B_m - (-1)^{\frac{1}{2}(m-n)} B_n$ is an integer. ....(21)

Example: $B_{24}-B_{12}$ and $B_{32}-B_{16}$ are integers, while the denominator of $B_{10}+B_{20}$ is 5.

It will be observed that:

(1) if n is a multiple of 4, then the numerator of $B_n-\frac{1}{30}$ in its lowest terms is divisible by 20; but if n is not a multiple of 4 then that of $\frac{B_n}{n}-\frac{1}{12}$ in its lowest terms is divisible by 5; ....(22)

(2) if n is any integer, then

$$2\left(2^{4n+2}-1\right)\frac{B_{4n+2}}{2n+1},\quad 2\left(2^{8n+4}-1\right)\frac{B_{8n+4}}{2n+1},\quad 2\left(2^{8n+4}-1\right)\frac{B_{16n+8}}{2n+1}$$

are integers of the form $30p+1$. ....(23)

10. If a B is known to lie between certain limits, then it is possible to find its exact value from the above properties.

Suppose we know that B_{22} lies between 6084 and 6244; its exact value can be found as follows.

The fractional part of $B_{22}=\frac{17}{138}$ by (20), also B_{22} is divisible by 11 by (18). And by (22) $B_{22}-\frac{11}{6}$ must be divisible by 5. To satisfy these conditions B_{23} must be either $6137\frac{17}{138}$ or $6192\frac{17}{138}$.

But according to (18) the numerator of B_{22} should be a *prime* number after it is divided by 11; and consequently B_{22} must be equal to $6192\frac{17}{138}$ or $\frac{854513}{138}$, since the numerator of $6137\frac{17}{138}$ is divisible not only by 11 but also by 7 and 17.

11. It is known (Edwards' *Differential Calculus*, Ch. v, Ex. 29) that

$$B_n=\frac{2\,.\,n\,!}{(2\pi)^n}\left(\frac{1}{1^n}+\frac{1}{2^n}+\frac{1}{3^n}+\ldots\right),$$

or

$$\frac{2\,.\,n\,!}{(2\pi)^n}=B_n\left(1-\frac{1}{2^n}\right)\left(1-\frac{1}{3^n}\right)\left(1-\frac{1}{5^n}\right)\ldots,\quad \text{...........(24)}$$

where 2, 3, 5, ... are prime numbers.

Also

$$\frac{B_n}{2n}=\int_0^\infty \frac{x^{n-1}}{e^{2\pi x}-1}\,dx. \quad \text{..........................(25)}$$

For

$$\int_0^\infty \frac{x^{n-1}}{e^{2\pi x}-1}\,dx=\int_0^\infty x^{n-1}\left(e^{-2\pi x}+e^{-4\pi x}+\ldots\right)dx$$

$$=\frac{(n-1)!}{(2\pi)^n}\left(\frac{1}{1^n}+\frac{1}{2^n}+\frac{1}{3^n}+\ldots\right)=\frac{B_n}{2n}$$

by (24). In a similar manner

$$\int_0^\infty \frac{x^n}{\left(e^{\pi x}-e^{-\pi x}\right)^2}\,dx=\frac{B_n}{4\pi},$$

and

$$\int^\infty x^{n-2}\log\left(1-e^{-2\pi x}\right)dx=-\frac{\pi B_n}{n(n-1)}. \quad \text{...........(26)}$$

Take logarithms of both sides in (24) and write for $\log_e n!$ the well-known expansion of $\log_e \Gamma(n+1)$, as in Carr's *Synopsis*, viz.

$$(n+\tfrac{1}{2})\log n - n + \tfrac{1}{2}\log 2\pi + \frac{B_2}{1.2n} - \frac{B_4}{3.4n^3} + \frac{B_6}{5.6n^5} - \dots$$
$$- (-1)^p \frac{B_{2p}\,\theta}{(2p-1)\,2pn^{2p-1}}, \qquad \text{..............(27)}$$

where $0 < \theta < 1$, and where

$$\frac{B_{2p}\,\theta}{(2p-1)\,2pn^{2p-1}} = \frac{B_{2p}}{(2p-1)\,2pn^{2p-1}} - \frac{B_{2p+2}}{(2p+1)(2p+2)\,n^{2p+1}} + \dots$$
$$= -\frac{1}{\pi}\int_0^\infty \frac{x^{2p-2}}{n^{2p-1}} \log(1-e^{-2\pi x})\,dx + \frac{1}{\pi}\int_0^\infty \frac{x^{2p}}{n^{2p+1}} \log(1-e^{-2\pi x})\,dx - \dots$$
$$= -\frac{1}{\pi}\int_0^\infty \left(\frac{x^{2p-2}}{n^{2p-1}} - \frac{x^{2p}}{n^{2p+1}} + \dots\right) \log(1-e^{-2\pi x})\,dx$$
$$= -\frac{1}{\pi}\int_0^\infty \frac{x^{2p-2}}{n^{2p-3}(n^2+x^2)} \log(1-e^{-2\pi x})\,dx$$
$$= -\int_0^\infty \frac{x^{2p-2}\log(1-e^{-2\pi n x})}{\pi(1+x^2)}\,dx.$$

We can find the integral part of B_n, and since the fractional part can be found, as shewn in § 8, the exact value of B_n is known. Unless the calculation is made to depend upon the values of $\log_{10} e$, $\log_e 10$, π, ..., which are known to a great number of decimal places, we should have to find the logarithms of certain numbers whose values are not found in the tables to as many places of decimals as we require. Such difficulties are removed by the method given in § 13.

12. Results (14) to (17), (20) and (21) can be obtained as follows. We have

$$\frac{1}{2x^2} + \frac{1}{(x+1)^2} + \frac{1}{(x+2)^2} + \frac{1}{(x+3)^2} + \frac{1}{(x+4)^2} + \dots$$
$$-\frac{1}{x} - \frac{1}{6(x^3-x)} + \frac{1}{5(x^5-x)} + \frac{1}{7(x^7-x)} + \frac{1}{11(x^{11}-x)} + \dots$$
$$= \frac{1}{x^{15}} - \frac{7}{x^{17}} + \frac{55}{x^{19}} - \frac{529}{x^{21}} + \dots, \qquad \text{..(28)}$$

where 5, 7, 11, 13, are prime numbers above 3. If we can prove that the left-hand side of (28) can be expanded in ascending powers of $1/x$ with integral coefficients, then (20) and (21) are at once deduced as follows.

From (27) we have

$$\frac{d^2 \log \Gamma(n+1)}{dn^2} = \frac{1}{(n+1)^2} + \frac{1}{(n+2)^2} + \frac{1}{(n+3)^2} + \dots$$

$$= \frac{1}{n} - \frac{1}{2n^2} + \frac{B_2}{n^3} - \frac{B_4}{n^5} + \frac{B_6}{n^7} - \frac{B_8}{n^9} + \dots - (-)^p \frac{B_{2p}\theta}{n^{2p+1}}, \quad \dots\dots\dots\dots\dots(29)$$

where

$$\frac{B_{2p}\theta}{n^{2p+1}} = \frac{B_{2p}}{n^{2p+1}} - \frac{B_{2p+2}}{n^{2p+3}} + \dots$$

$$= 4\pi \int_0^\infty \frac{x^{2p}}{n^{2p+1}(e^{\pi x} - e^{-\pi x})^2} dx - 4\pi \int_0^\infty \frac{x^{2p+2}}{n^{2p+3}(e^{\pi x} - e^{-\pi x})^2} dx + \dots$$

$$= \pi \int_0^\infty \left(\frac{x^{2p}}{n^{2p+1}} - \frac{x^{2p+2}}{n^{2p+3}} + \dots \right) \frac{dx}{\sinh^2 \pi x}$$

$$= \pi \int_0^\infty \frac{x^{2p}}{n^{2p-1}(n^2 + x^2)} \frac{dx}{\sinh^2 \pi x} = \int_0^\infty \frac{\pi x^{2p}}{(1 + x^2)\sinh^2 \pi n x} dx.$$

Substituting the result of (29) in (28) we see that

$$\frac{B_2}{x^3} - \frac{B_4}{x^5} + \frac{B_6}{x^7} - \dots - \frac{1}{6(x^3 - x)} + \frac{1}{5(x^5 - x)} + \frac{1}{7(x^7 - x)} + \frac{1}{11(x^{11} - x)} + \dots,$$

where 5, 7, 11, ... are prime numbers, can be expanded in ascending powers of $1/x$ with integral coefficients.

Therefore $B_2 - \frac{1}{6}$, $-B_4 - \frac{1}{6} + \frac{1}{5}$, $B_6 - \frac{1}{6} + \frac{1}{7}$, $-B_8 - \frac{1}{6} + \frac{1}{5}$, $B_{10} - \frac{1}{6} + \frac{1}{11}$, ..., which are the coefficients of $1/x^3$, $1/x^5$, $1/x^7$, ..., are integers.

Writing $\frac{1}{2} + \frac{1}{3} - 1$ for $-\frac{1}{6}$ we get the results of (20) and (21).

Again changing n to $\frac{1}{2}n$ in (29), and subtracting half of the result from (29), we have

$$\frac{1}{(n+1)^2} - \frac{1}{(n+2)^2} + \frac{1}{(n+3)^2} - \dots = \frac{1}{2n^2} - \frac{(2^2-1)B_2}{n^3} + \frac{(2^4-1)B_4}{n^5}$$

$$- \frac{(2^6-1)B_6}{n^7} + \dots + (-1)^p (2^{2p} - 1) \frac{B_{2p}\theta}{n^{2p+1}}, \quad \dots\dots(30)$$

where $0 < \theta < 1$, and also, by (29),

$$(2^{2p} - 1) \frac{B_{2p}\theta}{n^{2p+1}} = \int_0^\infty \frac{\pi x^{2p} \cosh \pi n x}{(1 + x^2)\sinh^2 \pi n x} dx.$$

Thus we see that, if we can prove that twice the left-hand side of (30) can be expanded in ascending powers of $1/n$ with integral coefficients, then the second part of (16) is at once proved.

Again from (27) we have

$$\frac{d\log\Gamma(n+1)}{dn}=1+\tfrac{1}{2}+\tfrac{1}{3}+\tfrac{1}{4}+\ldots+\frac{1}{n}-\gamma$$

$$=\log n+\frac{1}{2n}-\frac{B_2}{2n^2}+\frac{B_4}{4n^4}-\frac{B_6}{6n^6}+\frac{B_8}{8n^8}-\ldots+(-1)^p\frac{B_{2p}\,\theta}{2pn^{2p}},\quad\ldots(31)$$

where $0<\theta<1$; and also, by (25),

$$\frac{B_{2p}\,\theta}{2pn^{2p}}=\int_0^\infty\frac{2x^{2p-1}}{(1+x^2)(e^{2\pi nx}-1)}dx,$$

from which it can easily be shewn that

$$\frac{1}{n+2}-\frac{1}{n+4}+\frac{1}{n+6}-\frac{1}{n+8}+\frac{1}{n+10}-\ldots$$

$$=\frac{1}{2n}-2(2^2-1)\frac{B_2}{2n^2}+2^3(2^4-1)\frac{B_4}{4n^4}-2^5(2^6-1)\frac{B_6}{6n^6}+\ldots$$

$$+(-1)^p\,2^{2p-1}(2^{2p}-1)\frac{B_{2p}\,\theta}{2pn^{2p}}-\ldots,\quad\ldots\ldots\ldots\ldots(32)$$

where $0<\theta<1$; and also, by (31),

$$2^{2p-1}(2^{2p}-1)\frac{B_{2p}\,\theta}{2pn^{2p}}=\int_0^\infty\frac{x^{2p-1}}{2(1+x^2)\sinh\frac{1}{2}(\pi nx)}dx.$$

From the above theorem we see that, if we can prove that

$$2\left(\frac{1}{n+2}-\frac{1}{n+4}+\frac{1}{n+6}-\ldots\right),$$

can be expanded in ascending powers of $1/n$ with integral coefficients, then the first part of (16) at once follows.

13. The first few digits, and the number of digits in the integral part as well as in the numerator of B_n, can be found from the approximate formula:

$$\log_{10}B_n=(n+\tfrac{1}{2})\log_{10}n-1{\cdot}2324743503n+0{\cdot}700120,$$

the true value being greater by about $0{\cdot}0362/n$ when n is great. ......(33)

This formula is proved as follows: taking logarithms of both sides in (24),

$$\log_e B_n=(n+\tfrac{1}{2})\log_e n-n(1+\log_e 2\pi)+\tfrac{1}{2}\log_e 8\pi$$

nearly. Multiplying both sides by $\log_{10}e$ or $\cdot4342944819$, and reducing, we can get the result.

14. Changing n to $n-2$ in (24) and taking the ratio of the two results, we have

$$B_n=\frac{n(n-1)}{4\pi^2}B_{n-2}\left(1-\frac{2^2-1}{2^n-1}\right)\left(1-\frac{3^2-1}{3^n-1}\right)\left(1-\frac{5^2-1}{5^n-1}\right)\ldots,\quad(34)$$

where 2, 3, 5, ... are prime numbers.

Hence we see that $\dfrac{B_n}{B_{n-2}}$ approaches $\dfrac{n(n-1)}{4\pi^2}$ very rapidly as n becomes greater and greater. ....(35)

From the value of π, viz. 3·14159, 26535, 89793, 23846, 26433, 83279, 50288, 41971, 69399, ..., the integral part of any B can be found from the previous B; and from the integral part the exact value can at once be written by help of (20) as follows:

Approximate ratio of any B to previous B	lies between *	Hence the exact value is
B_2	0 and 1 ...	$1-\frac{1}{2}-\frac{1}{3} = \frac{1}{6}$
$B_4 = \frac{3.4}{4\pi^2} B_2$ ...	0 and 1 ...	$\frac{1}{2}+\frac{1}{3}+\frac{1}{5}-1 = \frac{1}{30}$
$B_6 = \frac{5.6}{4\pi^2} B_4$ ...	0 and 1 ...	$1-\frac{1}{2}-\frac{1}{3}-\frac{1}{7} = \frac{1}{42}$
$B_8 = \frac{7.8}{4\pi^2} B_6$ ...	0 and 1 ...	$\frac{1}{2}+\frac{1}{3}+\frac{1}{5}-1 = \frac{1}{30}$
$B_{10} = \frac{9.10}{4\pi^2} B_8$ ...	0 and 1 ...	$1-\frac{1}{2}-\frac{1}{3}-\frac{1}{11} = \frac{5}{66}$
$B_{12} = \frac{11.12}{4\pi^2} B_{10}$ ...	0 and 1 ...	$\frac{1}{2}+\frac{1}{3}+\frac{1}{5}+\frac{1}{7}+\frac{1}{13}-1 = \frac{691}{2730}$
$B_{14} = \frac{13.14}{4\pi^2} B_{12}$ ...	0 and 2 ...	$2-\frac{1}{2}-\frac{1}{3} = \frac{7}{6}$
$B_{16} = \frac{15.16}{4\pi^2} B_{14}$ ...	7 and 8 ...	$6+\frac{1}{2}+\frac{1}{3}+\frac{1}{5}+\frac{1}{17} = \frac{3617}{510}$
$B_{18} = \frac{17.18}{4\pi^2} B_{16}$ ...	54 and 55 ...	$56-\frac{1}{2}-\frac{1}{3}-\frac{1}{7}-\frac{1}{19} = \frac{43867}{798}$
$B_{20} = \frac{19.20}{4\pi^2} B_{18}$ ...	529 and 530 ...	$528+\frac{1}{2}+\frac{1}{3}+\frac{1}{5}+\frac{1}{11} = \frac{174611}{330}$
... ...		

15. From the preceding theorems we know some of the properties of B_n for all positive even values of n. As an example let us take $B_{444} = N/D$.

The fractional part of B_{444} is $\frac{23975417}{90709710}$ by (20). The numerator of B_{444} is divisible by 37 and the quotient is a prime number by (18). Again $\log_{10} B_{444} = 630{\cdot}2433$, nearly, by (33). Therefore the integral part of B_{444} contains 631 digits, the first 4 digits being 1751. Again

$$\log_{10} N = \log_{10} B_{444} + \log_{10} D = 630{\cdot}2433 + \log_{10} 90709710 = 638{\cdot}2010$$

nearly. Therefore N contains 639 digits, the first four digits being 1588.

* These integral limits are got from a rough calculation of any B from the preceding B by the formula given in the first column.

Again the numerator of $B_{444} - \frac{1}{30}$ is divisible by 20; that is to say, if $[B_{444}]$ is the integral part of B_{444}, $[B_{444}] + \frac{23975417}{90709710} - \frac{1}{30} = [B_{444}] + \frac{698392}{30236657}$ has a numerator divisible by 20. Therefore the integral part of B_{444} ends with 4 and also the figure next to the last is even.

Hence N ends with 57 and also the third figure from the last is even.

16. Instead of starting with $\cot x$ as in §§ 2 and 3, we may start with $\tan x$ or $\operatorname{cosec} x$ and get other similar results.

Thus:

$$\text{(i)}\quad \tfrac{4}{3}B_n(2^n - 1) = c_6 B_{n-6}(2^{n-6} - 1) - c_{12} B_{n-12}(2^{n-12} - 1) + \dots$$
$$+ \tfrac{1}{3}n(-1)^{\frac{1}{6}(n-2)} \text{ or } \tfrac{1}{6}n(-1)^{\frac{1}{6}(n-4)} \text{ or } \tfrac{1}{3}n(-1)^{\frac{1}{6}(n-6)} \dots, \quad \dots\dots(36)$$

$$\text{(ii)}\quad c_3\left(1 - \frac{1}{2^{n-4}}\right)B_{n-3} - c_9\left(1 - \frac{1}{2^{n-10}}\right)B_{n-9} + c_{15}\left(1 - \frac{1}{2^{n-16}}\right)B_{n-15} - \dots$$
$$= \tfrac{2}{3}\cdot\frac{n}{2^n}\{3^{\frac{1}{2}(n-1)} + (-1)^{\frac{1}{6}(n-3)} \text{ or } (-1)^{\frac{1}{6}(n+1)} \text{ or } (-1)^{\frac{1}{6}(n+5)}\}. \quad \dots(37)$$

17. The formulæ obtained in §§ 1, 3, 4 may be called the one interval, two interval and three interval formulæ respectively. The p interval formulæ can be got by taking the pth roots of unity or of i according as p is odd or even.

For example, let us take the fifth roots of unity $(1, \alpha, \alpha^2, \alpha^3, \alpha^4)$, and find the 5 interval formulæ.

Let
$$\phi(x) = \sin x + \sin x\alpha + \sin x\alpha^2 + \sin x\alpha^3 + \sin x\alpha^4$$
$$= 5\left(\frac{x^5}{5!} - \frac{x^{15}}{15!} + \frac{x^{25}}{25!} - \dots\right).$$

Then it can easily be shewn that

$$16 \sin x \sin x\alpha \sin x\alpha^2 \sin x\alpha^3 \sin x\alpha^4$$
$$= \phi(2x) - \phi\{2x(\alpha + \alpha^4)\} - \phi\{2x(\alpha^2 + \alpha^3)\}$$
$$= \phi(2x) + \phi\{x(1 + \sqrt{5})\} + \phi\{x(1 - \sqrt{5})\}.$$

Taking logarithms and differentiating both sides, we have

$$5\left(B_0 + B_{10}\frac{x^{10}}{10!} + B_{20}\frac{x^{20}}{20!} + B_{30}\frac{x^{30}}{30!} + \dots\right)$$
$$= -x\,\frac{\dfrac{x^4}{4!}(1 + \alpha_5) - \dfrac{x^{14}}{14!}(1 + \alpha_{15}) + \dots}{\dfrac{x^5}{5!}(1 + \alpha_5) - \dfrac{x^{15}}{15!}(1 + \alpha_{15}) + \dots}, \quad \dots\dots\dots\dots\dots\dots(38)$$

where $\alpha_n = \left(\frac{1 + \sqrt{5}}{2}\right)^n + \left(\frac{1 - \sqrt{5}}{2}\right)^n$, so that $\alpha_n\alpha_m = \alpha_{n+m} + (-1)^n\alpha_{m-n}$.

Similarly,

(i) $$5\left(B_2\frac{x^2}{2!}+B_{12}\frac{x^{12}}{12!}+B_{22}\frac{x^{22}}{22!}+\ldots\right)$$
$$=x\,\frac{\frac{x^6}{6!}(1+\alpha_7)-\frac{x^{16}}{16!}(1+\alpha_{17})+\ldots}{\frac{x^5}{5!}(1+\alpha_5)-\frac{x^{15}}{15!}(1+\alpha_{15})+\ldots}, \quad\ldots\ldots(39)$$

(ii) $$5\left(B_4\frac{x^4}{4!}+B_{14}\frac{x^{14}}{14!}+B_{24}\frac{x^{24}}{24!}+\ldots\right)$$
$$=x\,\frac{\frac{x^8}{8!}(\alpha_7-1)-\frac{x^{18}}{18!}(\alpha_{17}-1)+\ldots}{\frac{x^5}{5!}(1+\alpha_5)-\frac{x^{15}}{15!}(1+\alpha_{15})+\ldots}, \quad\ldots\ldots(40)$$

(iii) $$10\left(B_6\frac{x^6}{6!}+B_{16}\frac{x^{16}}{16!}+B_{26}\frac{x^{26}}{26!}+\ldots\right)$$
$$=x\,\frac{\frac{x^{10}}{10!}(\alpha_{10}-3)-\frac{x^{20}}{20!}(\alpha_{20}-3)+\frac{x^{30}}{30!}(\alpha_{30}-3)-\ldots}{\frac{x^5}{5!}(1+\alpha_5)-\frac{x^{15}}{15!}(1+\alpha_{15})+\frac{x^{25}}{25!}(1+\alpha_{25})-\ldots}, \quad(41)$$

(iv) $$5\left(B_8\frac{x^8}{8!}+B_{18}\frac{x^{18}}{18!}+B_{28}\frac{x^{28}}{28!}+\ldots\right)$$
$$=x\,\frac{\frac{x^{12}}{12!}(\alpha_{11}-1)-\frac{x^{22}}{22!}(\alpha_{21}-1)+\frac{x^{32}}{32!}(\alpha_{31}-1)-\ldots}{\frac{x^5}{5!}(1+\alpha_5)-\frac{x^{15}}{15!}(1+\alpha_{15})+\frac{x^{25}}{25!}(1+\alpha_{25})-\ldots}. \quad(42)$$

Again from

$$16\cos x\cos x\alpha\cos x\alpha^2\cos x\alpha^3\cos x\alpha^4$$
$$=1+\psi[2x]+\psi[2x(\alpha+\alpha^4)]+\psi[2x(\alpha^2+\alpha^3)],$$

where $$\psi(x)=\cos x+\cos x\alpha+\cos x\alpha^2+\cos x\alpha^3+\cos x\alpha^4$$
$$=5\left(1-\frac{x^{10}}{10!}+\frac{x^{20}}{20!}-\frac{x^{30}}{30!}+\ldots\right),$$

and similar relations, we can get many other identities.

18. The four interval formulæ can be got from the following identities:

If $a_n=\left(1+\frac{1}{\sqrt{2}}\right)^n+\left(1-\frac{1}{\sqrt{2}}\right)^n$ and $b_n=\left(1+\frac{1}{\sqrt{2}}\right)^n-\left(1-\frac{1}{\sqrt{2}}\right)^n$,

so that $a_{m+n}=a_ma_n-a_{m-n}/2^n$; and $b_{m+n}=a_mb_n+b_{m-n}/2^n$; then:

(i) $$4\left\{B_0+B_8\frac{x^8}{8!}+B_{16}\frac{x^{16}}{16!}+B_{24}\frac{x^{24}}{24!}+\ldots\right\}$$
$$=-x\,\frac{\frac{x^3}{3!}a_2-\frac{x^{11}}{11!}a_6+\frac{x^{19}}{19!}a_{10}-\ldots}{\frac{x^4}{4!}a_2-\frac{x^{12}}{12!}a_6+\frac{x^{20}}{20!}a_{10}-\ldots}, \quad\ldots\ldots(43)$$

(ii) $$4\left\{B_2\frac{x^2}{2!}+B_{10}\frac{x^{10}}{10!}+B_{18}\frac{x^{18}}{18!}+\dots\right\}$$

$$=-x\,\frac{\dfrac{x^5}{5!}a_3-\dfrac{x^{13}}{13!}a_7+\dfrac{x^{21}}{21!}a_{11}-\dots}{\dfrac{x^4}{4!}a_2-\dfrac{x^{12}}{12!}a_6+\dfrac{x^{20}}{20!}a_{10}-\dots}, \qquad \text{.....(44)}$$

(iii) $$4\sqrt{2}\left\{B_4\frac{x^4}{4!}+B_{12}\frac{x^{12}}{12!}+B_{20}\frac{x^{20}}{20!}+\dots\right\}$$

$$=x\,\frac{\dfrac{x^7}{7!}b_3-\dfrac{x^{15}}{15!}b_7+\dfrac{x^{23}}{23!}b_{11}-\dots}{\dfrac{x^4}{4!}a_2-\dfrac{x^{12}}{12!}a_6+\dfrac{x^{20}}{20!}a_{10}-\dots}, \qquad \text{.....(45)}$$

(iv) $$4\sqrt{2}\left\{B_6\frac{x^6}{6!}+B_{14}\frac{x^{14}}{14!}+B_{22}\frac{x^{22}}{22!}+\dots\right\}$$

$$=x\,\frac{\dfrac{x^9}{9!}b_4-\dfrac{x^{17}}{17!}b_8+\dfrac{x^{25}}{25!}b_{12}-\dots}{\dfrac{x^4}{4!}a_2-\dfrac{x^{12}}{12!}a_6+\dfrac{x^{20}}{20!}a_{10}-\dots}. \qquad \text{.....(46)}$$

2

ON QUESTION 330 OF PROFESSOR SANJANA

(*Journal of the Indian Mathematical Society*, IV, 1912, 59—61)

1. Prof. Sanjana remarks that it is not easy to evaluate the series

$$\frac{1}{1^n}+\frac{1}{2}\frac{1}{3^n}+\frac{1.3}{2.4}\frac{1}{5^n}+\frac{1.3.5}{2.4.6}\frac{1}{7^n}+\dots \text{ ad inf.},$$

if $n>3$. In attempting to sum the series for *all* values of n, I have arrived at the following results:

Let
$$f(p)=\frac{1}{1+p}+\frac{1}{2}\frac{1}{3+p}+\frac{1.3}{2.4}\frac{1}{5+p}+\dots$$
$$=\int_0^1 x^p\left(1+\frac{1}{2}x^2+\frac{1.3}{2.4}x^4+\dots\right)dx$$
$$=\int_0^1\frac{x^p}{\sqrt{1-x^2}}\,dx=\tfrac{1}{2}\int_0^1 x^{\frac{1}{2}(p-1)}(1-x)^{-\frac{1}{2}}\,dx$$
$$=\tfrac{1}{2}\frac{\Gamma\left(\frac{p+1}{2}\right)\Gamma(\tfrac{1}{2})}{\Gamma\left(\frac{p+2}{2}\right)}=\frac{\pi^{\frac{1}{2}}}{2}\frac{\Gamma\left(\frac{p+1}{2}\right)}{\Gamma\left(\frac{p+2}{2}\right)}.$$

But
$$\Gamma\left(\frac{p+1}{2}\right)=\frac{\pi^{\frac{1}{2}}}{2^p}\frac{\Gamma(p+1)}{\Gamma\left(\frac{p+2}{2}\right)}$$

(*vide* Williamson, *Integral Calculus*, p. 164).

Therefore
$$f(p)=\frac{\pi}{2^{p+1}}\frac{\Gamma(p+1)}{\left\{\Gamma\left(\frac{p+2}{2}\right)\right\}^2}.$$

Therefore
$$\log\{f(p)\}=\log(\tfrac{1}{2}\pi)-p\log 2$$
$$+\frac{p^2}{2}\left(1-\frac{1}{2}\right)S_2-\frac{p^3}{3}\left(1-\frac{1}{2^2}\right)S_3+\dots, \qquad \dots\dots(1)$$

where $S_n\equiv\frac{1}{1^n}+\frac{1}{2^n}+\frac{1}{3^n}+\dots$ ad inf. (*vide* Carr's *Synopsis*, 2295).

Again, by expanding $f(p)$ in ascending powers of p, we have

$$f(p)=\left(1+\frac{1}{2}\frac{1}{3}+\frac{1.3}{2.4}\frac{1}{5}+\dots\right)-p\left(1+\frac{1}{2}\frac{1}{3^2}+\frac{1.3}{2.4}\frac{1}{5^2}+\dots\right)$$
$$+p^2\left(1+\frac{1}{2}\frac{1}{3^3}+\frac{1.3}{2.4}\frac{1}{5^3}+\dots\right)-\dots$$
$$=\frac{\pi}{2}\{\phi(0)-p\phi(1)+p^2\phi(2)-p^3\phi(3)+\dots\},$$

where $$\frac{1}{1^{n+1}}+\frac{1}{2}\frac{1}{3^{n+1}}+\frac{1.3}{2.4}\frac{1}{5^{n+1}}+\ldots\equiv\frac{\pi}{2}\phi(n).$$

Hence (1) may be written

$$\log\tfrac{1}{2}\pi+\log\{\phi(0)-p.\phi(1)+p^2.\phi(2)-\ldots\}$$

$$=\log(\tfrac{1}{2}\pi)-p\log 2+\frac{p^2}{2}\left(1-\frac{1}{2}\right)S_2-\frac{p^3}{3}\left(1-\frac{1}{2^2}\right)S_3+\ldots$$

$$=\log(\tfrac{1}{2}\pi)-p\sigma_1+\frac{p^2}{2}\sigma_2-\frac{p^3}{3}\sigma_3+\ldots,$$

where $$\sigma_n\equiv 1-\frac{1}{2^n}+\frac{1}{3^n}-\frac{1}{4^n}+\ldots.$$

Differentiating with respect to p, and equating the coefficients of p^{n-1}, we have

$$n\phi(n)\equiv\sigma_1\phi(n-1)+\sigma_2\phi(n-2)+\sigma_3\phi(n-3)+\ldots \text{ to } n \text{ terms.}$$

Thus we see that

$$\frac{\pi}{2}\phi(0)\equiv 1+\frac{1}{2}\frac{1}{3}+\frac{1.3}{2.4}\frac{1}{5}+\ldots=\frac{\pi}{2},$$

$$\frac{\pi}{2}\phi(1)\equiv 1+\frac{1}{2}\frac{1}{3^2}+\frac{1.3}{2.4}\frac{1}{5^2}+\ldots=\frac{\pi}{2}\log 2,$$

$$\frac{\pi}{2}\phi(2)\equiv 1+\frac{1}{2}\frac{1}{3^3}+\frac{1.3}{2.4}\frac{1}{5^3}+\ldots=\frac{\pi^3}{48}+\frac{\pi}{4}(\log 2)^2,$$

$$\frac{\pi}{2}\phi(3)\equiv 1+\frac{1}{2}\frac{1}{3^4}+\frac{1.3}{2.4}\frac{1}{5^4}+\ldots=\frac{\pi^3}{48}\log 2+\frac{\pi^3}{12}(\log 2)^3+\frac{\pi}{6}\sigma_3$$

$$=\frac{\pi^3}{48}\log 2+\frac{\pi}{12}(\log 2)^3+\frac{\pi}{8}S_3,$$

and so on.

2. More generally, consider the series

$$\frac{1}{b^n}-\frac{a}{1!}\frac{1}{(b+1)^n}+\frac{a(a-1)}{2!}\frac{1}{(b+2)^n}-\ldots.$$

Writing $\dfrac{\Gamma(b)\Gamma(a+1)}{\Gamma(a+b+1)}\phi(n-1)$ for this, and taking the identity

$$\frac{1}{b+p}-\frac{a}{1!}\frac{1}{b+1+p}+\frac{a(a-1)}{2!}\frac{1}{b+2+p}-\ldots$$

$$=\int_0^1 x^{b+p-1}(1-x)^a\,dx=\frac{\Gamma(b+p)\Gamma(a+1)}{\Gamma(a+b+p+1)},$$

we find

$$n\phi(n)=\sigma_1\phi(n-1)+\sigma_2\phi(n-2)+\sigma_3\phi(n-3)+\ldots \text{ to } n \text{ terms,}$$

where $$\sigma_n\equiv\frac{1}{b^n}-\frac{1}{(a+b+1)^n}+\frac{1}{(b+1)^n}-\frac{1}{(a+b+2)^n}+\ldots.$$

Examples: Put $a = -\frac{1}{2}$, $b = \frac{1}{4}$. Then we see that

(i) $$1 + \frac{1}{2}\frac{1}{5} + \frac{1.3}{2.4}\frac{1}{9} + \frac{1.3.5}{2.4.6}\frac{1}{13} + \dots = \frac{\{\Gamma(\frac{1}{4})\}^2}{4\sqrt{(2\pi)}},$$

(ii) $$1 + \frac{1}{2}\frac{1}{5^2} + \frac{1.3}{2.4}\frac{1}{9^2} + \frac{1.3.5}{2.4.6}\frac{1}{13^2} + \dots = \frac{\{\Gamma(\frac{1}{4})\}^2}{4\sqrt{(2\pi)}}\frac{\pi}{4},$$

(iii) $$1 + \frac{1}{2}\frac{1}{5^3} + \frac{1.3}{2.4}\frac{1}{9^3} + \frac{1.3.5}{2.4.6}\frac{1}{13^3} + \dots = \frac{\{\Gamma(\frac{1}{4})\}^2}{4\sqrt{(2\pi)}}\left\{\frac{\pi^2}{32} + \frac{1}{2}S_2'\right\},$$

(iv) $$1 + \frac{1}{2}\frac{1}{5^4} + \frac{1.3}{2.4}\frac{1}{9^4} + \frac{1.3.5}{2.4.6}\frac{1}{13^4} + \dots = \frac{\{\Gamma(\frac{1}{4})\}^2}{4\sqrt{(2\pi)}}\left\{\frac{5\pi^3}{384} + \frac{\pi}{8}S_2' + \frac{1}{3}S_3'\right\},$$

where $S_r' = \frac{1}{1^r} - \frac{1}{3^r} + \frac{1}{5^r} - \frac{1}{7^r} + \dots$.

3

NOTE ON A SET OF SIMULTANEOUS EQUATIONS*

(*Journal of the Indian Mathematical Society*, IV, 1912, 94—96)

1. Consider the equations

$$x_1 + x_2 + x_3 + \dots + x_n = a_1,$$
$$x_1 y_1 + x_2 y_2 + x_3 y_3 + \dots + x_n y_n = a_2,$$
$$x_1 y_1^2 + x_2 y_2^2 + x_3 y_3^2 + \dots + x_n y_n^2 = a_3,$$
$$x_1 y_1^3 + x_2 y_2^3 + x_3 y_3^3 + \dots + x_n y_n^3 = a_4,$$
$$\dots\dots\dots\dots\dots\dots\dots\dots\dots\dots\dots\dots$$
$$x_1 y_1^{2n-1} + x_2 y_2^{2n-1} + x_3 y_3^{2n-1} + \dots + x_n y_n^{2n-1} = a_{2n},$$

where $x_1, x_2, x_3, \dots x_n$ and $y_1, y_2, y_3, \dots y_n$ are $2n$ unknown quantities.

Now, let us take the expression

$$\phi(\theta) \equiv \frac{x_1}{1-\theta y_1} + \frac{x_2}{1-\theta y_2} + \frac{x_3}{1-\theta y_3} + \dots + \frac{x_n}{1-\theta y_n} \quad \dots\dots\dots\dots(1)$$

and expand it in ascending powers of θ. Then we see that the expression is equal to

$$a_1 + a_2\theta + a_3\theta^2 + \dots + a_{2n}\theta^{2n-1} + \dots . \quad \dots\dots\dots\dots\dots\dots(2)$$

But (1), when simplified, will have for its numerator an expression of the $(n-1)$th degree in θ, and for its denominator an expression of the nth degree in θ.

Thus we may suppose that

$$\phi(\theta) = \frac{A_1 + A_2\theta + A_3\theta^2 + \dots + A_n\theta^{n-1}}{1 + B_1\theta + B_2\theta^2 + B_3\theta^3 + \dots + B_n\theta^n} \quad \dots\dots\dots\dots\dots(3)$$
$$= a_1 + a_2\theta + a_3\theta^2 + \dots + a_{2n}\theta^{2n-1} + \dots;$$

and so $$(1 + B_1\theta + \dots)(a_1 + a_2\theta + \dots) = A_1 + A_2\theta + \dots .$$

Equating the coefficients of like powers of θ, we have

$$A_1 = a_1,$$
$$A_2 = a_2 + a_1 B_1,$$
$$A_3 = a_3 + a_2 B_1 + a_1 B_2,$$
$$A_n = a_n + a_{n-1}B_1 + a_{n-2}B_2 + \dots + a_1 B_{n-1},$$
$$0 = a_{n+1} + a_n B_1 + \dots + a_1 B_n,$$
$$0 = a_{n+2} + a_{n+1} B_1 + \dots + a_2 B_n,$$
$$0 = a_{n+3} + a_{n+2} B_1 + \dots + a_3 B_n,$$
$$\dots\dots\dots\dots\dots\dots\dots\dots\dots\dots$$
$$0 = a_{2n} + a_{2n-1} B_1 + \dots + a_n B_n.$$

* For a solution, by determinants, of a similar set of equations, see Burnside and Panton, *Theory of Equations*, Vol. II, p. 106, Ex. 3. [Editor, *J. Indian Math. Soc.*]

From these $B_1, B_2, \ldots B_n$ can easily be found, and since $A_1, A_2, \ldots A_n$ depend upon these values they can also be found.

Now, splitting (3) into partial fractions in the form

$$\frac{p_1}{1-q_1\theta}+\frac{p_2}{1-q_2\theta}+\frac{p_3}{1-q_3\theta}+\ldots+\frac{p_n}{1-q_n\theta},$$

and comparing with (1), we see that

$$x_1=p_1,\ y_1=q_1;$$
$$x_2=p_2,\ y_2=q_2;$$
$$x_3=p_3,\ y_3=q_3;$$
$$\ldots\ldots\ldots\ldots\ldots\ldots\ldots$$

2. As an example we may solve the equations:

$$\begin{aligned}
x+y+z+u+v&=2,\\
px+qy+rz+su+tv&=3,\\
p^2x+q^2y+r^2z+s^2u+t^2v&=16,\\
p^3x+q^3y+r^3z+s^3u+t^3v&=31,\\
p^4x+q^4y+r^4z+s^4u+t^4v&=103,\\
p^5x+q^5y+r^5z+s^5u+t^5v&=235,\\
p^6x+q^6y+r^6z+s^6u+t^6v&=674,\\
p^7x+q^7y+r^7z+s^7u+t^7v&=1669,\\
p^8x+q^8y+r^8z+s^8u+t^8v&=4526,\\
p^9x+q^9y+r^9z+s^9u+t^9v&=11595,
\end{aligned}$$

where $x, y, z, u, v, p, q, r, s, t$ are the unknowns. Proceeding as before, we have

$$\frac{x}{1-\theta p}+\frac{y}{1-\theta q}+\frac{z}{1-\theta r}+\frac{u}{1-\theta s}+\frac{v}{1-\theta t}$$
$$=2+3\theta+16\theta^2+31\theta^3+103\theta^4+235\theta^5+674\theta^6+1669\theta^7+4526\theta^8+11595\theta^9+\ldots.$$

By the method of indeterminate coefficients, this can be shewn to be equal to

$$\frac{2+\theta+3\theta^2+2\theta^3+\theta^4}{1-\theta-5\theta^2+\theta^3+3\theta^4-\theta^5}.$$

Splitting this into partial fractions, we get the values of the unknowns, as follows:

$$x=-\frac{3}{5},\qquad p=-1,$$
$$y=\frac{18+\sqrt{5}}{10},\qquad q=\frac{3+\sqrt{5}}{2},$$
$$z=\frac{18-\sqrt{5}}{10},\qquad r=\frac{3-\sqrt{5}}{2},$$
$$u=-\frac{8+\sqrt{5}}{2\sqrt{5}},\qquad s=\frac{\sqrt{5}-1}{2},$$
$$v=\frac{8-\sqrt{5}}{2\sqrt{5}};\qquad t=-\frac{\sqrt{5}+1}{2}.$$

4

IRREGULAR NUMBERS

(*Journal of the Indian Mathematical Society*, v, 1913, 105—106)

1. Let $a_2, a_3, a_5, a_7, \ldots$ denote numbers less than unity, where the subscripts 2, 3, 5, 7, ... are the series of prime numbers. Then

$$\frac{1}{1-a_2}\cdot\frac{1}{1-a_3}\cdot\frac{1}{1-a_5}\ldots = 1 + a_2 + a_3 + a_2 . a_2 + a_5$$
$$+ a_2 . a_3 + a_7 + a_2 . a_2 . a_2 + a_3 . a_3 + \ldots, \qquad \ldots\ldots(1)$$

the terms being so arranged that the products obtained by multiplying the subscripts are the series of natural numbers 2, 3, 4, 5, 6, 7, 8, 9, ...

The above result is easily got if we remember that the natural numbers are formed by multiplying primes and their powers.

2. Similarly, we have

$$\frac{1}{1+a_2}\cdot\frac{1}{1+a_3}\cdot\frac{1}{1+a_5}\ldots = 1 - a_2 - a_3 + a_2 . a_2 - a_5$$
$$+ a_2 . a_3 - a_7 - a_2 . a_2 . a_2 + a_3 . a_3 + \ldots, \qquad \ldots\ldots(2)$$

where the sign is negative whenever a term contains an *odd* number of prime subscripts.

3. Put $a_2 = 1/2^n$, $a_3 = 1/3^n$, $a_5 = 1/5^n$, ... in (1), and we get

$$\left(1-\frac{1}{2^n}\right)\left(1-\frac{1}{3^n}\right)\left(1-\frac{1}{5^n}\right)\left(1-\frac{1}{7^n}\right)\ldots = \frac{1}{S_n}, \qquad \ldots\ldots\ldots\ldots(3)$$

where S_n denotes $1/1^n + 1/2^n + 1/3^n + 1/4^n + \ldots$.

Changing n into $2n$ in (3) and dividing by the original, we obtain

$$\left(1+\frac{1}{2^n}\right)\left(1+\frac{1}{3^n}\right)\left(1+\frac{1}{5^n}\right)\left(1+\frac{1}{7^n}\right)\ldots = \frac{S_n}{S_{2n}}. \qquad \ldots\ldots\ldots\ldots(4)$$

Examples:

$$\text{(i)}\quad \left(1+\frac{1}{2^2}\right)\left(1+\frac{1}{3^2}\right)\left(1+\frac{1}{5^2}\right)\ldots = \frac{15}{\pi^2}, \qquad \ldots\ldots\ldots\ldots(5)$$

$$\text{(ii)}\quad \left(1+\frac{1}{2^4}\right)\left(1+\frac{1}{3^4}\right)\left(1+\frac{1}{5^4}\right)\ldots = \frac{105}{\pi^4}, \qquad \ldots\ldots\ldots(6)$$

since
$$S_2 = \pi^2/6,\quad S_4 = \pi^4/90,\quad S_8 = \pi^8/9450.$$

4. Subtract (2) from (1) and put $a_2 = 2^{-n}\ldots$; then

$$\frac{1}{2^n}+\frac{1}{3^n}+\frac{1}{5^n}+\frac{1}{7^n}+\frac{1}{8^n}+\frac{1}{11^n}+\frac{1}{12^n}+\ldots = \frac{S_n^2 - S_{2n}}{2S_n},$$

where the numbers 2, 3, 5, 7, 8, ... contain an *odd* number of prime divisors.

Examples: (i) $$\frac{1}{2^2}+\frac{1}{3^2}+\frac{1}{5^2}+\frac{1}{7^2}+\frac{1}{8^2}+\ldots=\frac{\pi^2}{20},\ \ldots\ldots(7)$$

(ii) $$\frac{1}{2^4}+\frac{1}{3^4}+\frac{1}{5^4}+\frac{1}{7^4}+\frac{1}{8^4}+\ldots=\frac{\pi^4}{1260}.\ \ldots\ldots(8)$$

5. Again (2, 3, 5, 7, ... being the prime numbers)

$$(1+a_2)(1+a_3)(1+a_5)(1+a_7)\ldots=1+a_2+a_3+a_5$$
$$+a_2.a_3+a_7+a_2.a_5+a_{11}+a_{13}+\ldots,\ \ldots(9)$$

where the product of the subscripts in any term is a natural number containing *dissimilar* prime divisors; and

$$(1-a_2)(1-a_3)(1-a_5)(1-a_7)\ldots=1-a_2-a_3-a_5+a_2.a_3-a_7,\ldots(10)$$

where the signs are negative whenever the number of factors is odd.

6. Replacing as before a_2, a_3, a_5, ... by the values given in § 3 and using (4), we deduce that

$$1+\frac{1}{2^n}+\frac{1}{3^n}+\frac{1}{5^n}+\frac{1}{6^n}+\ldots=\frac{S_n}{S_{2n}},\ \ldots\ldots(11)$$

where 2, 3, 5, 6, 7, ... are the numbers containing *dissimilar* prime divisors.

7. Also taking half the difference between (3) and (4),

$$\frac{1}{2^n}+\frac{1}{3^n}+\frac{1}{5^n}+\frac{1}{7^n}+\frac{1}{11^n}+\frac{1}{13^n}+\frac{1}{17^n}+\frac{1}{19^n}+\frac{1}{23^n}+\frac{1}{29^n}$$
$$+\frac{1}{30^n}+\frac{1}{31^n}+\ldots=\frac{S_n^2-S_{2n}}{2S_nS_{2n}},\ \ldots\ldots(12)$$

where 2, 3, 5, ... are numbers containing an *odd* number of *dissimilar* prime divisors.

Examples: (i) $$\frac{1}{2^2}+\frac{1}{3^2}+\frac{1}{5^2}+\ldots=\frac{9}{2\pi^2},\ \ldots\ldots(13)$$

(ii) $$\frac{1}{2^4}+\frac{1}{3^4}+\frac{1}{5^4}+\ldots=\frac{15}{2\pi^4}.\ \ldots\ldots(14)$$

8. Subtracting (11) from S_n, we have

$$\frac{1}{4^n}+\frac{1}{8^n}+\frac{1}{9^n}+\frac{1}{12^n}+\ldots=\frac{S_n(S_{2n}-1)}{S_{2n}},\ \ldots\ldots(15)$$

where 4, 8, 9, ... are composite numbers having *at least* two *equal* prime divisors.

5

SQUARING THE CIRCLE

(*Journal of the Indian Mathematical Society*, v, 1913, 132)

Let PQR be a circle with centre O, of which a diameter is PR. Bisect PO at H and let T be the point of trisection of OR nearer R. Draw TQ perpendicular to PR and place the chord $RS = TQ$.

Join PS, and draw OM and TN parallel to RS. Place a chord $PK = PM$, and draw the tangent $PL = MN$. Join RL, RK and KL. Cut off $RC = RH$. Draw CD parallel to KL, meeting RL at D.

Then the square on RD will be equal to the circle PQR approximately.

For
$$RS^2 = \tfrac{5}{36}d^2,$$
where d is the diameter of the circle.

Therefore
$$PS^2 = \tfrac{31}{36}d^2.$$

But PL and PK are equal to MN and PM respectively.

Therefore
$$PK^2 = \tfrac{31}{144}d^2, \text{ and } PL^2 = \tfrac{31}{324}d^2.$$

Hence
$$RK^2 = PR^2 - PK^2 = \tfrac{113}{144}d^2,$$
and
$$RL^2 = PR^2 + PL^2 = \tfrac{355}{324}d^2.$$

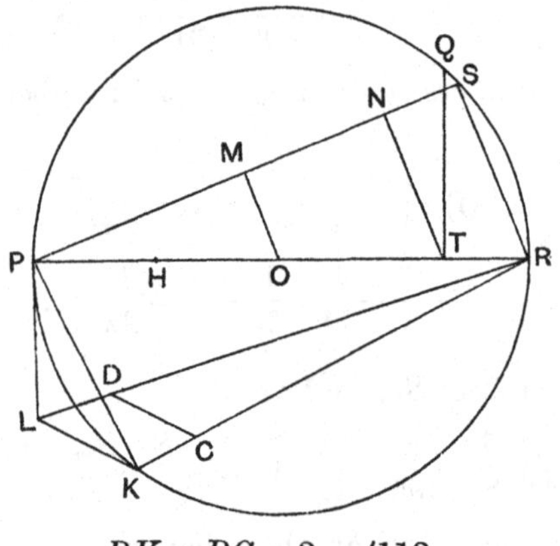

But
$$\frac{RK}{RL} = \frac{RC}{RD} = \frac{3}{2}\sqrt{\frac{113}{355}},$$
and
$$RC = \tfrac{3}{4}d.$$

Therefore
$$RD = \frac{d}{2}\sqrt{\frac{355}{113}} = r\sqrt{\pi}, \text{ very nearly.}$$

Note.—If the area of the circle be 140,000 square miles, then RD is greater than the true length by about an inch.

6

MODULAR EQUATIONS AND APPROXIMATIONS TO π

(*Quarterly Journal of Mathematics*, XLV, 1914, 350—372)

1. If we suppose that

$$(1+e^{-\pi\sqrt{n}})(1+e^{-3\pi\sqrt{n}})(1+e^{-5\pi\sqrt{n}})\ldots = 2^{\frac{1}{4}}\, e^{-\pi\sqrt{n}/24}\, G_n \quad \text{.........(1)}$$

and

$$(1-e^{-\pi\sqrt{n}})(1-e^{-3\pi\sqrt{n}})(1-e^{-5\pi\sqrt{n}})\ldots = 2^{\frac{1}{4}}\, e^{-\pi\sqrt{n}/24}\, g_n, \quad \text{.........(2)}$$

then G_n and g_n can always be expressed as roots of algebraical equations when n is any rational number. For we know that

$$(1+q)(1+q^3)(1+q^5)\ldots = 2^{\frac{1}{6}} q^{\frac{1}{24}} (kk')^{-\frac{1}{12}} \quad \text{..............(3)}$$

and

$$(1-q)(1-q^3)(1-q^5)\ldots = 2^{\frac{1}{6}} q^{\frac{1}{24}} k^{-\frac{1}{12}} k'^{\frac{1}{6}}. \quad \text{..............(4)}$$

Now the relation between the moduli k and l, which makes

$$n\frac{K'}{K} = \frac{L'}{L},$$

where $n = r/s$, r and s being positive integers, is expressed by the modular equation of the rsth degree. If we suppose that $k = l'$, $k' = l$, so that $K = L'$, $K' = L$, then

$$q = e^{-\pi L'/L} = e^{-\pi\sqrt{n}},$$

and the corresponding value of k may be found by the solution of an algebraical equation.

From (1), (2), (3), and (4) it may easily be deduced that

$$g_{4n} = 2^{\frac{1}{4}}\, g_n G_n, \quad \text{................................(5)}$$

$$G_n = G_{1/n}, \quad 1/g_n = g_{4/n}, \quad \text{..........................(6)}$$

$$(g_n G_n)^8 (G_n{}^8 - g_n{}^8) = \tfrac{1}{4}. \quad \text{..........................(7)}$$

I shall consider only integral values of n. It follows from (7) that we need consider only one of G_n or g_n for any given value of n; and from (5) that we may suppose n not divisible by 4. It is most convenient to consider g_n when n is even, and G_n when n is odd.

2. Suppose then that n is odd. The values of G_n and g_{2n} are got from the same modular equation. For example, let us take the modular equation of the 5th degree, viz.

$$\left(\frac{u}{v}\right)^3 + \left(\frac{v}{u}\right)^3 = 2\left(u^2v^2 - \frac{1}{u^2v^2}\right), \quad \text{.......................(8)}$$

where

$$2^{\frac{1}{4}} q^{\frac{1}{24}} u = (1+q)(1+q^3)(1+q^5)\ldots$$

and

$$2^{\frac{1}{4}} q^{\frac{5}{24}} v = (1+q^5)(1+q^{15})(1+q^{25})\ldots.$$

By changing q to $-q$ the above equation may also be written as

$$\left(\frac{v}{u}\right)^3 - \left(\frac{u}{v}\right)^3 = 2\left(u^2v^2 + \frac{1}{u^2v^2}\right), \quad \text{.......................(9)}$$

where $2^{\frac{1}{4}} q^{\frac{1}{24}} u = (1-q)(1-q^3)(1-q^5)\ldots$

and $2^{\frac{1}{4}} q^{\frac{5}{24}} v = (1-q^5)(1-q^{15})(1-q^{25})\ldots.$

If we put $q = e^{-\pi/\sqrt{5}}$ in (8), so that $u = G_{\frac{1}{5}}$ and $v = G_5$, and hence $u = v$, we see that

$$v^4 - v^{-4} = 1.$$

Hence $$v^4 = \frac{1+\sqrt{5}}{2}, \quad G_5 = \left(\frac{1+\sqrt{5}}{2}\right)^{\frac{1}{4}}.$$

Similarly, by putting $q = e^{-\pi\sqrt{\frac{2}{5}}}$, so that $u = g_{\frac{2}{5}}$ and $v = g_{10}$, and hence $u = 1/v$, we see that

$$v^6 - v^{-6} = 4.$$

Hence $$v^2 = \frac{1+\sqrt{5}}{2}, \quad g_{10} = \sqrt{\left(\frac{1+\sqrt{5}}{2}\right)}.$$

Similarly it can be shewn that

$$G_9 = \left(\frac{1+\sqrt{3}}{\sqrt{2}}\right)^{\frac{1}{3}}, \quad g_{18} = (\sqrt{2}+\sqrt{3})^{\frac{1}{3}},$$

$$G_{17} = \sqrt{\left(\frac{5+\sqrt{17}}{8}\right)} + \sqrt{\left(\frac{\sqrt{17}-3}{8}\right)},$$

$$g_{34} = \sqrt{\left(\frac{7+\sqrt{17}}{8}\right)} + \sqrt{\left(\frac{\sqrt{17}-1}{8}\right)},$$

and so on.

3. In order to obtain approximations for π we take logarithms of (1) and (2). Thus

$$\left.\begin{aligned} \pi &= \frac{24}{\sqrt{n}} \log(2^{\frac{1}{4}} G_n) \\ \pi &= \frac{24}{\sqrt{n}} \log(2^{\frac{1}{4}} g_n) \end{aligned}\right\}, \quad \text{.......................(10)}$$

approximately, the error being nearly $\frac{24}{\sqrt{n}} e^{-\pi\sqrt{n}}$ in both cases. These equations may also be written as

$$e^{\pi\sqrt{n}/24} = 2^{\frac{1}{4}} G_n, \quad e^{\pi\sqrt{n}/24} = 2^{\frac{1}{4}} g_n. \quad \text{.....................(11)}$$

In those cases in which G_n^{12} and g_n^{12} are simple quadratic surds we may use the forms

$$(G_n^{12} + G_n^{-12})^{\frac{1}{12}}, \quad (g_n^{12} + g_n^{-12})^{\frac{1}{12}},$$

instead of G_n and g_n, for we have

$$g_n^{12} = \tfrac{1}{8}e^{\frac{1}{2}\pi\sqrt{n}} - \tfrac{3}{2}e^{-\frac{1}{2}\pi\sqrt{n}},$$

approximately, and so

$$g_n^{12} + g_n^{-12} = \tfrac{1}{8}e^{\frac{1}{2}\pi\sqrt{n}} + \tfrac{13}{2}e^{-\frac{1}{2}\pi\sqrt{n}},$$

approximately, so that

$$\pi = \frac{2}{\sqrt{n}} \log \{8 (g_n^{12} + g_n^{-12})\}, \quad \text{.....................(12)}$$

the error being about $\frac{104}{\sqrt{n}} e^{-\pi\sqrt{n}}$, which is of the same order as the error in the formulæ (10). The formula (12) often leads to simpler results. Thus the second of formulae (10) gives

$$e^{\pi\sqrt{18}/24} = 2^{\frac{1}{4}} g_{18}$$

or

$$e^{\frac{1}{4}\pi\sqrt{18}} = 10\sqrt{2} + 8\sqrt{3}.$$

But if we use the formula (12), or

$$e^{\pi\sqrt{n}/24} = 2^{\frac{1}{4}} (g_n^{12} + g_n^{-12})^{\frac{1}{12}},$$

we get a simpler form, viz.

$$e^{\frac{1}{8}\pi\sqrt{18}} = 2\sqrt{7}.$$

4. The values of g_{2n} and G_n are obtained from the same equation. The approximation by means of g_{2n} is preferable to that by G_n for the following reasons.

(*a*) It is more accurate. Thus the error when we use G_{65} contains a factor $e^{-\pi\sqrt{65}}$, whereas that when we use g_{130} contains a factor $e^{-\pi\sqrt{130}}$.

(*b*) For many values of n, g_{2n} is simpler in form than G_n; thus

$$g_{130} = \sqrt{\left\{(2+\sqrt{5})\left(\frac{3+\sqrt{13}}{2}\right)\right\}},$$

while

$$G_{65} = \left\{\left(\frac{1+\sqrt{5}}{2}\right)\left(\frac{3+\sqrt{13}}{2}\right)\right\}^{\frac{1}{4}} \sqrt{\left\{\sqrt{\left(\frac{9+\sqrt{65}}{8}\right)} + \sqrt{\left(\frac{1+\sqrt{65}}{8}\right)}\right\}}.$$

(*c*) For many values of n, g_{2n} involves quadratic surds only, even when G_n is a root of an equation of higher order. Thus G_{23}, G_{29}, G_{31} are roots of cubic equations, G_{47}, G_{79} are those of quintic equations, and G_{71} is that of a septic equation, while g_{46}, g_{58}, g_{62}, g_{94}, g_{142}, and g_{158} are all expressible by quadratic surds.

5. Since G_n and g_n can be expressed as roots of algebraical equations with rational coefficients, the same is true of G_n^{24} or g_n^{24}. So let us suppose that

$$1 = a g_n^{-24} - b g_n^{-48} + \ldots,$$

or

$$g_n^{24} = a - b g_n^{-24} + \ldots.$$

But we know that

$$64 e^{-\pi\sqrt{n}} g_n^{24} = 1 - 24 e^{-\pi\sqrt{n}} + 276 e^{-2\pi\sqrt{n}} - \ldots,$$

$$64 g_n^{24} = e^{\pi\sqrt{n}} - 24 + 276 e^{-\pi\sqrt{n}} - \ldots,$$

$$64a - 64b g_n^{-24} + \ldots = e^{\pi\sqrt{n}} - 24 + 276 e^{-\pi\sqrt{n}} - \ldots,$$

$$64a - 4096b e^{-\pi\sqrt{n}} + \ldots = e^{\pi\sqrt{n}} - 24 + 276 e^{-\pi\sqrt{n}} - \ldots,$$

that is

$$e^{\pi\sqrt{n}} = (64a + 24) - (4096b + 276) e^{-\pi\sqrt{n}} + \ldots. \quad \text{.........(13)}$$

Similarly, if $$1 = aG_n^{-24} - bG_n^{-48} + \ldots,$$
then $$e^{\pi\sqrt{n}} = (64a - 24) - (4096b + 276)\, e^{-\pi\sqrt{n}} + \ldots. \qquad \ldots\ldots\ldots(14)$$

From (13) and (14) we can find whether $e^{\pi\sqrt{n}}$ is very nearly an integer for given values of n, and ascertain also the number of 9's or 0's in the decimal part. But if G_n and g_n be simple quadratic surds we may work independently as follows. We have, for example,

$$g_{22} = \sqrt{(1+\sqrt{2})}.$$

Hence $$64 g_{22}^{24} = e^{\pi\sqrt{22}} - 24 + 276 e^{-\pi\sqrt{22}} - \ldots,$$
$$64 g_{22}^{-24} = 4096 e^{-\pi\sqrt{22}} + \ldots,$$
so that
$$64\,(g_{22}^{24} + g_{22}^{-24}) = e^{\pi\sqrt{22}} - 24 + 4372 e^{-\pi\sqrt{22}} + \ldots = 64\,\{(1+\sqrt{2})^{12} + (1-\sqrt{2})^{12}\}.$$
Hence $$e^{\pi\sqrt{22}} = 2508951 \cdot 9982\ldots.$$

Again $$G_{37} = (6+\sqrt{37})^{\frac{1}{4}},$$
$$64 G_{37}^{24} = e^{\pi\sqrt{37}} + 24 + 276 e^{-\pi\sqrt{37}} + \ldots,$$
$$64 G_{37}^{-24} = 4096 e^{-\pi\sqrt{37}} - \ldots,$$
so that
$$64\,(G_{37}^{24} + G_{37}^{-24}) = e^{\pi\sqrt{37}} + 24 + 4372 e^{-\pi\sqrt{37}} - \ldots = 64\,\{(6+\sqrt{37})^{6} + (6-\sqrt{37})^{6}\}.$$
Hence $$e^{\pi\sqrt{37}} = 199148647 \cdot 999978\ldots.$$

Similarly, from $$g_{58} = \sqrt{\left(\frac{5+\sqrt{29}}{2}\right)},$$
we obtain
$$64\,(g_{58}^{24} + g_{58}^{-24}) = e^{\pi\sqrt{58}} - 24 + 4372 e^{-\pi\sqrt{58}} + \ldots$$
$$= 64\left\{\left(\frac{5+\sqrt{29}}{2}\right)^{12} + \left(\frac{5-\sqrt{29}}{2}\right)^{12}\right\}.$$
Hence $$e^{\pi\sqrt{58}} = 24591257751 \cdot 99999982\ldots.$$

6. I have calculated the values of G_n and g_n for a large number of values of n. Many of these results are equivalent to results given by Weber; for example,

$$G_{13}^4 = \frac{3+\sqrt{13}}{2}, \qquad G_{25} = \frac{1+\sqrt{5}}{2},$$
$$g_{30}^6 = (2+\sqrt{5})(3+\sqrt{10}), \qquad G_{37}^4 = 6+\sqrt{37},$$
$$G_{49} = \frac{7^{\frac{1}{4}} + \sqrt{(4+\sqrt{7})}}{2}, \qquad g_{58}^2 = \frac{5+\sqrt{29}}{2},$$
$$g_{70}^2 = \frac{(3+\sqrt{5})(1+\sqrt{2})}{2},$$
$$G_{73} = \sqrt{\left(\frac{9+\sqrt{73}}{8}\right)} + \sqrt{\left(\frac{1+\sqrt{73}}{8}\right)},$$
$$G_{85} = \left(\frac{1+\sqrt{5}}{2}\right)\left(\frac{9+\sqrt{85}}{2}\right)^{\frac{1}{4}},$$

$$G_{97} = \sqrt{\left(\frac{13+\sqrt{97}}{8}\right)} + \sqrt{\left(\frac{5+\sqrt{97}}{8}\right)},$$

$$g_{190}{}^2 = (2+\sqrt{5})(3+\sqrt{10}),$$

$$G_{385}{}^2 = \tfrac{1}{8}(3+\sqrt{11})(\sqrt{5}+\sqrt{7})(\sqrt{7}+\sqrt{11})(3+\sqrt{5}),$$

and so on. I have also many results not given by Weber. I give a complete table of new results. In Weber's notation, $G_n = 2^{-\frac{1}{4}} f\{\sqrt{(-n)}\}$ and $g_n = 2^{-\frac{1}{4}} f_1\{\sqrt{(-n)}\}$.

TABLE I.

$$g_{62} + \frac{1}{g_{62}} = \tfrac{1}{2}\{\sqrt{(1+\sqrt{2})} + \sqrt{(9+5\sqrt{2})}\},$$

$$G_{65}{}^2 = \sqrt{\left\{\left(\frac{1+\sqrt{5}}{2}\right)\left(\frac{3+\sqrt{13}}{2}\right)\right\}}\left\{\sqrt{\left(\frac{1+\sqrt{65}}{8}\right)} + \sqrt{\left(\frac{9+\sqrt{65}}{8}\right)}\right\},$$

$$g_{66}{}^2 = \sqrt{(\sqrt{2}+\sqrt{3})}(7\sqrt{2}+3\sqrt{11})^{\frac{1}{6}}\left\{\sqrt{\left(\frac{7+\sqrt{33}}{8}\right)} + \sqrt{\left(\frac{\sqrt{33}-1}{8}\right)}\right\},$$

$$G_{69}{}^2 = (3\sqrt{3}+\sqrt{23})^{\frac{1}{4}}\left(\frac{5+\sqrt{23}}{4}\right)^{\frac{1}{6}}\left\{\sqrt{\left(\frac{6+3\sqrt{3}}{4}\right)} + \sqrt{\left(\frac{2+3\sqrt{3}}{4}\right)}\right\},$$

$$G_{77}{}^2 = \{\tfrac{1}{2}(\sqrt{7}+\sqrt{11})(8+3\sqrt{7})\}^{\frac{1}{4}}\left\{\sqrt{\left(\frac{6+\sqrt{11}}{4}\right)} + \sqrt{\left(\frac{2+\sqrt{11}}{4}\right)}\right\},$$

$$G_{81}{}^3 = \frac{(2\sqrt{3}+2)^{\frac{1}{3}}+1}{(2\sqrt{3}-2)^{\frac{1}{3}}-1},$$

$$g_{90} = \{(2+\sqrt{5})(\sqrt{5}+\sqrt{6})\}^{\frac{1}{6}}\left\{\sqrt{\left(\frac{3+\sqrt{6}}{4}\right)} + \sqrt{\left(\frac{\sqrt{6}-1}{4}\right)}\right\},$$

$$g_{94} + \frac{1}{g_{94}} = \tfrac{1}{2}\{\sqrt{(7+\sqrt{2})} + \sqrt{(7+5\sqrt{2})}\},$$

$$g_{98} + \frac{1}{g_{98}} = \tfrac{1}{2}\{\sqrt{2} + \sqrt{(14+4\sqrt{14})}\},$$

$$g_{114}{}^2 = \sqrt{(\sqrt{2}+\sqrt{3})}(3\sqrt{2}+\sqrt{19})^{\frac{1}{6}}\left\{\sqrt{\left(\frac{23+3\sqrt{57}}{8}\right)} + \sqrt{\left(\frac{15+3\sqrt{57}}{8}\right)}\right\},$$

$$G_{117} = \tfrac{1}{2}\left(\frac{3+\sqrt{13}}{2}\right)^{\frac{1}{4}}(2\sqrt{3}+\sqrt{13})^{\frac{1}{6}}\{3^{\frac{1}{4}} + \sqrt{(4+\sqrt{3})}\},$$

$$\left.\begin{aligned} G_{121} + \frac{1}{G_{121}} &= \left(\frac{11}{2}\right)^{\frac{1}{6}}\left\{\left(3+\frac{1}{3\sqrt{3}}\right)^{\frac{1}{3}} + \left(3-\frac{1}{3\sqrt{3}}\right)^{\frac{1}{3}}\right\} \\ \frac{1}{G_{121}} &= \frac{1}{3\sqrt{2}}\left[(11-3\sqrt{11})^{\frac{1}{3}}\{(3\sqrt{11}+3\sqrt{3}-4)^{\frac{1}{3}}\right. \\ &\qquad \left.+(3\sqrt{11}-3\sqrt{3}-4)^{\frac{1}{3}}\}-2\right] \end{aligned}\right\},$$

$$g_{126} = \sqrt{\left(\frac{\sqrt{3}+\sqrt{7}}{2}\right)}(\sqrt{6}+\sqrt{7})^{\frac{1}{6}}\left\{\sqrt{\left(\frac{3+\sqrt{2}}{4}\right)} + \sqrt{\left(\frac{\sqrt{2}-1}{4}\right)}\right\}^2,$$

$$g_{138}^{2} = \sqrt{\left(\frac{3\sqrt{3}+\sqrt{23}}{2}\right)}\,(78\sqrt{2}+23\sqrt{23})^{\frac{1}{6}} \times \left\{\sqrt{\left(\frac{5+2\sqrt{6}}{4}\right)} + \sqrt{\left(\frac{1+2\sqrt{6}}{4}\right)}\right\},$$

$$G_{141}^{2} = (4\sqrt{3}+\sqrt{47})^{\frac{1}{4}}\left(\frac{7+\sqrt{47}}{\sqrt{2}}\right)^{\frac{1}{6}}\left\{\sqrt{\left(\frac{18+9\sqrt{3}}{4}\right)} + \sqrt{\left(\frac{14+9\sqrt{3}}{4}\right)}\right\},$$

$$G_{145}^{2} = \sqrt{\left\{\frac{(2+\sqrt{5})(5+\sqrt{29})}{2}\right\}}\left\{\sqrt{\left(\frac{17+\sqrt{145}}{8}\right)} + \sqrt{\left(\frac{9+\sqrt{145}}{8}\right)}\right\},$$

$$\frac{1}{G_{147}} = 2^{-\frac{1}{12}}\left[\frac{1}{2} + \frac{1}{\sqrt{3}}\left\{\sqrt{\left(\frac{7}{4}\right)} - (28)^{\frac{1}{6}}\right\}\right],$$

$$G_{153} = \left\{\sqrt{\left(\frac{5+\sqrt{17}}{8}\right)} + \sqrt{\left(\frac{\sqrt{17}-3}{8}\right)}\right\}^{2} \times \left\{\sqrt{\left(\frac{37+9\sqrt{17}}{4}\right)} + \sqrt{\left(\frac{33+9\sqrt{17}}{4}\right)}\right\}^{\frac{1}{3}},$$

$$g_{154}^{2} = \sqrt{\left\{(2\sqrt{2}+\sqrt{7})\left(\frac{\sqrt{7}+\sqrt{11}}{2}\right)\right\}} \times \left\{\sqrt{\left(\frac{13+2\sqrt{22}}{4}\right)} + \sqrt{\left(\frac{9+2\sqrt{22}}{4}\right)}\right\},$$

$$g_{158} + \frac{1}{g_{158}} = \tfrac{1}{2}\left\{\sqrt{(9+\sqrt{2})} + \sqrt{(17+13\sqrt{2})}\right\},$$

$$\left.\begin{aligned} G_{169} + \frac{1}{G_{169}} &= \left(\frac{13}{4}\right)^{\frac{1}{6}}\left\{\left(1+\frac{1}{3\sqrt{3}}\right)^{\frac{1}{3}} + \left(1-\frac{1}{3\sqrt{3}}\right)^{\frac{1}{3}}\right\}^{2} \\ \frac{1}{G_{169}} &= \frac{1}{3}\left[(\sqrt{13}-2) + \left(\frac{13-3\sqrt{13}}{2}\right)^{\frac{1}{3}}\right. \\ &\quad \left.\times\left\{\left(3\sqrt{3}-\frac{11-\sqrt{13}}{2}\right)^{\frac{1}{3}} - \left(3\sqrt{3}+\frac{11-\sqrt{13}}{2}\right)^{\frac{1}{3}}\right\}\right] \end{aligned}\right\},$$

$$g_{198} = \sqrt{(1+\sqrt{2})}\,(4\sqrt{2}+\sqrt{33})^{\frac{1}{6}}\left\{\sqrt{\left(\frac{9+\sqrt{33}}{8}\right)} + \sqrt{\left(\frac{1+\sqrt{33}}{8}\right)}\right\},$$

$$G_{205} = \left(\frac{1+\sqrt{5}}{2}\right)\left(\frac{3\sqrt{5}+\sqrt{41}}{2}\right)^{\frac{1}{4}}\left\{\sqrt{\left(\frac{7+\sqrt{41}}{8}\right)} + \sqrt{\left(\frac{\sqrt{41}-1}{8}\right)}\right\},$$

$$G_{213}^{2} = (5\sqrt{3}+\sqrt{71})^{\frac{1}{4}}\left(\frac{59+7\sqrt{71}}{4}\right)^{\frac{1}{6}} \times \left\{\sqrt{\left(\frac{21+12\sqrt{3}}{2}\right)} + \sqrt{\left(\frac{19+12\sqrt{3}}{2}\right)}\right\},$$

$$G_{217}^{2} = \left\{\sqrt{\left(\frac{9+4\sqrt{7}}{2}\right)} + \sqrt{\left(\frac{11+4\sqrt{7}}{2}\right)}\right\} \times \left\{\sqrt{\left(\frac{12+5\sqrt{7}}{4}\right)} + \sqrt{\left(\frac{16+5\sqrt{7}}{4}\right)}\right\},$$

$$G_{225} = \left(\frac{1+\sqrt{5}}{4}\right)(2+\sqrt{3})^{\frac{1}{3}}\left\{\sqrt{(4+\sqrt{15})} + 15^{\frac{1}{4}}\right\},$$

$$g_{238} = \left\{\sqrt{\left(\frac{1+2\sqrt{2}}{4}\right)} + \sqrt{\left(\frac{5+2\sqrt{2}}{4}\right)}\right\} \times \left\{\sqrt{\left(\frac{1+3\sqrt{2}}{4}\right)} + \sqrt{\left(\frac{5+3\sqrt{2}}{4}\right)}\right\},$$

$$G_{265}^2 = \sqrt{\left\{(2+\sqrt{5})\left(\frac{7+\sqrt{53}}{2}\right)\right\}}\left\{\sqrt{\left(\frac{89+5\sqrt{265}}{8}\right)} + \sqrt{\left(\frac{81+5\sqrt{265}}{8}\right)}\right\},$$

$$G_{289} = \left[\sqrt{\left\{\frac{17+\sqrt{17}+17^{\frac{1}{4}}(5+\sqrt{17})}{16}\right\}} + \sqrt{\left\{\frac{1+\sqrt{17}+17^{\frac{1}{4}}(5+\sqrt{17})}{16}\right\}}\right]^2,$$

$$G_{301}^2 = \left\{(8+3\sqrt{7})\left(\frac{23\sqrt{43}+57\sqrt{7}}{2}\right)\right\}^{\frac{1}{4}} \times \left\{\sqrt{\left(\frac{46+7\sqrt{43}}{4}\right)} + \sqrt{\left(\frac{42+7\sqrt{43}}{4}\right)}\right\},$$

$$g_{310} = \left(\frac{1+\sqrt{5}}{2}\right)\sqrt{(1+\sqrt{2})}\left\{\sqrt{\left(\frac{7+2\sqrt{10}}{4}\right)} + \sqrt{\left(\frac{3+2\sqrt{10}}{4}\right)}\right\},$$

$$\left.\begin{aligned} G_{325} &= \left(\frac{3+\sqrt{13}}{2}\right)^{\frac{1}{4}} t, \text{ where} \\ & t^3 + t^2\left(\frac{1-\sqrt{13}}{2}\right)^2 + t\left(\frac{1+\sqrt{13}}{2}\right)^2 + 1 \\ & \qquad = \sqrt{5}\left\{t^3 - t^2\left(\frac{1+\sqrt{13}}{2}\right) + t\left(\frac{1-\sqrt{13}}{2}\right) - 1\right\} \end{aligned}\right\},$$

$$G_{333} = \tfrac{1}{2}(6+\sqrt{37})^{\frac{1}{4}}(7\sqrt{3}+2\sqrt{37})^{\frac{1}{6}}\{\sqrt{(7+2\sqrt{3})} + \sqrt{(3+2\sqrt{3})}\},$$

$$\left.\begin{aligned} G_{363} &= 2^{\frac{5}{12}} t, \text{ where} \\ & 2t^3 - t^2\{(4+\sqrt{33}) + \sqrt{(11+2\sqrt{33})}\} \\ & \qquad - t\{1+\sqrt{(11+2\sqrt{33})}\} - 1 = 0 \end{aligned}\right\},$$

$$G_{441}^2 = \left(\frac{\sqrt{3}+\sqrt{7}}{2}\right)(2+\sqrt{3})^{\frac{1}{3}}\left\{\frac{2+\sqrt{7}+\sqrt{(7+4\sqrt{7})}}{2}\right\}\left\{\frac{\sqrt{(3+\sqrt{7})}+(6\sqrt{7})^{\frac{1}{4}}}{\sqrt{(3+\sqrt{7})}-(6\sqrt{7})^{\frac{1}{4}}}\right\},$$

$$G_{445} = \sqrt{(2+\sqrt{5})}\left(\frac{21+\sqrt{445}}{2}\right)^{\frac{1}{4}}\sqrt{\left\{\left(\frac{13+\sqrt{89}}{8}\right) + \sqrt{\left(\frac{5+\sqrt{89}}{8}\right)}\right\}},$$

$$G_{465}^2 = \sqrt{\left\{(2+\sqrt{3})\left(\frac{1+\sqrt{5}}{2}\right)\left(\frac{3\sqrt{3}+\sqrt{31}}{2}\right)\right\}}(5\sqrt{5}+2\sqrt{31})^{\frac{1}{6}} \times \left\{\sqrt{\left(\frac{2+\sqrt{31}}{4}\right)} + \sqrt{\left(\frac{6+\sqrt{31}}{4}\right)}\right\} \times \left\{\sqrt{\left(\frac{11+2\sqrt{31}}{2}\right)} + \sqrt{\left(\frac{13+2\sqrt{31}}{2}\right)}\right\},$$

$$G_{505}^2 = (2+\sqrt{5})\sqrt{\left\{\left(\frac{1+\sqrt{5}}{2}\right)(10+\sqrt{101})\right\}} \times \left\{\left(\frac{5\sqrt{5}+\sqrt{101}}{4}\right) + \sqrt{\left(\frac{105+\sqrt{505}}{8}\right)}\right\},$$

$$g_{522} = \sqrt{\left(\frac{5+\sqrt{29}}{2}\right)}(5\sqrt{29}+11\sqrt{6})^{\frac{1}{6}}\left\{\sqrt{\left(\frac{9+3\sqrt{6}}{4}\right)} + \sqrt{\left(\frac{5+3\sqrt{6}}{4}\right)}\right\},$$

$$G_{553}{}^2 = \left\{\sqrt{\left(\frac{96 + 11\sqrt{79}}{4}\right)} + \sqrt{\left(\frac{100 + 11\sqrt{79}}{4}\right)}\right\}$$
$$\times \left\{\sqrt{\left(\frac{141 + 16\sqrt{79}}{2}\right)} + \sqrt{\left(\frac{143 + 16\sqrt{79}}{2}\right)}\right\},$$

$$g_{630} = (\sqrt{14} + \sqrt{15})^{\frac{1}{6}} \sqrt{\left\{(1 + \sqrt{2})\left(\frac{3 + \sqrt{5}}{2}\right)\left(\frac{\sqrt{3} + \sqrt{7}}{2}\right)\right\}}$$
$$\times \left\{\sqrt{\left(\frac{\sqrt{15} + \sqrt{7} + 2}{4}\right)} + \sqrt{\left(\frac{\sqrt{15} + \sqrt{7} - 2}{4}\right)}\right\}$$
$$\times \left\{\sqrt{\left(\frac{\sqrt{15} + \sqrt{7} + 4}{8}\right)} + \sqrt{\left(\frac{\sqrt{15} + \sqrt{7} - 4}{8}\right)}\right\},$$

$$G_{765}{}^2 = \left(\frac{3 + \sqrt{5}}{2}\right)(16 + \sqrt{255})^{\frac{1}{6}} \sqrt{\left\{(4 + \sqrt{15})\left(\frac{9 + \sqrt{85}}{2}\right)\right\}}$$
$$\times \left\{\sqrt{\left(\frac{6 + \sqrt{51}}{4}\right)} + \sqrt{\left(\frac{10 + \sqrt{51}}{4}\right)}\right\}$$
$$\times \left\{\sqrt{\left(\frac{18 + 3\sqrt{51}}{4}\right)} + \sqrt{\left(\frac{22 + 3\sqrt{51}}{4}\right)}\right\}$$

$$G_{777}{}^2 = \sqrt{\left\{(2 + \sqrt{3})(6 + \sqrt{37})\left(\frac{\sqrt{3} + \sqrt{7}}{2}\right)\right\}}(246\sqrt{7} + 107\sqrt{37})^{\frac{1}{6}}$$
$$\times \left\{\sqrt{\left(\frac{6 + 3\sqrt{7}}{4}\right)} + \sqrt{\left(\frac{10 + 3\sqrt{7}}{4}\right)}\right\}$$
$$\times \left\{\sqrt{\left(\frac{15 + 6\sqrt{7}}{2}\right)} + \sqrt{\left(\frac{17 + 6\sqrt{7}}{2}\right)}\right\},$$

$$G_{1225} = \left(\frac{1 + \sqrt{5}}{2}\right)(6 + \sqrt{35})^{\frac{1}{4}} \left\{\frac{7^{\frac{1}{4}} + \sqrt{(4 + \sqrt{7})}}{2}\right\}^{\frac{3}{2}}$$
$$\times \left[\sqrt{\left\{\frac{43 + 15\sqrt{7} + (8 + 3\sqrt{7})\sqrt{(10\sqrt{7})}}{8}\right\}}\right.$$
$$\left. + \sqrt{\left\{\frac{35 + 15\sqrt{7} + (8 + 3\sqrt{7})\sqrt{(10\sqrt{7})}}{8}\right\}}\right],$$

$$G_{1353}{}^2 = \sqrt{\left\{(3 + \sqrt{11})(5 + 3\sqrt{3})\left(\frac{11 + \sqrt{123}}{2}\right)\right\}}$$
$$\times \left(\frac{6817 + 321\sqrt{451}}{4}\right)^{\frac{1}{6}}$$
$$\times \left\{\sqrt{\left(\frac{17 + 3\sqrt{33}}{8}\right)} + \sqrt{\left(\frac{25 + 3\sqrt{33}}{8}\right)}\right\}$$
$$\times \left\{\sqrt{\left(\frac{561 + 99\sqrt{33}}{8}\right)} + \sqrt{\left(\frac{569 + 99\sqrt{33}}{8}\right)}\right\},$$

$$G_{1645}{}^2 = (2 + \sqrt{5})\sqrt{\left\{(3 + \sqrt{7})\left(\frac{7 + \sqrt{47}}{2}\right)\right\}}\left(\frac{73\sqrt{5} + 9\sqrt{329}}{2}\right)^{\frac{1}{4}}$$
$$\times \left\{\sqrt{\left(\frac{119 + 7\sqrt{329}}{8}\right)} + \sqrt{\left(\frac{127 + 7\sqrt{329}}{8}\right)}\right\}$$
$$\times \left\{\sqrt{\left(\frac{743 + 41\sqrt{329}}{8}\right)} + \sqrt{\left(\frac{751 + 41\sqrt{329}}{8}\right)}\right\}.$$

7. Hence we deduce the following approximate formulæ.

TABLE II.

$$e^{\frac{1}{8}\pi\sqrt{18}} = 2\sqrt{7}, \quad e^{\pi\sqrt{22}/12} = 2 + \sqrt{2}, \quad e^{\frac{1}{4}\pi\sqrt{30}} = 20\sqrt{3} + 16\sqrt{6},$$

$$e^{\frac{1}{4}\pi\sqrt{34}} = 12(4 + \sqrt{17}), \quad e^{\frac{1}{2}\pi\sqrt{46}} = 144(147 + 104\sqrt{2}),$$

$$e^{\frac{1}{4}\pi\sqrt{42}} = 84 + 32\sqrt{6}, \quad e^{\pi\sqrt{58}/12} = \frac{5 + \sqrt{29}}{\sqrt{2}},$$

$$e^{\frac{1}{4}\pi\sqrt{70}} = 60\sqrt{35} + 96\sqrt{14}, \quad e^{\frac{1}{4}\pi\sqrt{78}} = 300\sqrt{3} + 208\sqrt{6},$$

$$e^{\pi\sqrt{55}/24} = \frac{1 + \sqrt{(3 + 2\sqrt{5})}}{\sqrt{2}}, \quad e^{\frac{1}{4}\pi\sqrt{102}} = 800\sqrt{3} + 196\sqrt{51},$$

$$e^{\frac{1}{4}\pi\sqrt{130}} = 12(323 + 40\sqrt{65}), \quad e^{\pi\sqrt{190}/12} = (2\sqrt{2} + \sqrt{10})(3 + \sqrt{10}),$$

$$\pi = \frac{12}{\sqrt{130}} \log\left\{\frac{(2 + \sqrt{5})(3 + \sqrt{13})}{\sqrt{2}}\right\},$$

$$\pi = \frac{24}{\sqrt{142}} \log\left\{\sqrt{\left(\frac{10 + 11\sqrt{2}}{4}\right)} + \sqrt{\left(\frac{10 + 7\sqrt{2}}{4}\right)}\right\},$$

$$\pi = \frac{12}{\sqrt{190}} \log\{(2\sqrt{2} + \sqrt{10})(3 + \sqrt{10})\},$$

$$\pi = \frac{12}{\sqrt{310}} \log\left[\tfrac{1}{4}(3 + \sqrt{5})(2 + \sqrt{2})\{(5 + 2\sqrt{10}) + \sqrt{(61 + 20\sqrt{10})}\}\right],$$

$$\pi = \frac{4}{\sqrt{522}} \log\left[\left(\frac{5 + \sqrt{29}}{\sqrt{2}}\right)^3 (5\sqrt{29} + 11\sqrt{6}) \times \left\{\sqrt{\left(\frac{9 + 3\sqrt{6}}{4}\right)} + \sqrt{\left(\frac{5 + 3\sqrt{6}}{4}\right)}\right\}^6\right].$$

The last five formulæ are correct to 15, 16, 18, 22, and 31 places of decimals respectively.

8. Thus we have seen how to approximate to π by means of logarithms of surds. I shall now shew how to obtain approximations in terms of surds only. If

$$n\frac{K'}{K} = \frac{L'}{L},$$

we have

$$\frac{n\,dk}{kk'^2 K^2} = \frac{dl}{ll'^2 L^2}.$$

But, by means of the modular equation connecting k and l, we can express dk/dl as an algebraic function of k, a function moreover in which all coefficients which occur are algebraic numbers. Again,

$$q = e^{-\pi K'/K}, \quad q^n = e^{-\pi L'/L},$$

$$\frac{q^{\frac{1}{12}}(1 - q^2)(1 - q^4)(1 - q^6)\ldots}{q^{\frac{1}{12}n}(1 - q^{2n})(1 - q^{4n})(1 - q^{6n})\ldots} = \left(\frac{kk'}{ll'}\right)^{\frac{1}{6}} \sqrt{\left(\frac{K}{L}\right)}. \quad \text{......(15)}$$

Differentiating this equation logarithmically, and using the formula

$$\frac{dq}{dk}=\frac{\pi^2 q}{2kk'^2K^2},$$

we see that

$$n\left\{1-24\left(\frac{q^{2n}}{1-q^{2n}}+\frac{2q^{4n}}{1-q^{4n}}+\dots\right)\right\}-\left\{1-24\left(\frac{q^2}{1-q^2}+\frac{2q^4}{1-q^4}+\dots\right)\right\}=\frac{KL}{\pi^2}A(k),\quad\dots(16)$$

where $A(k)$ denotes an algebraic function of the special class described above. I shall use the letter A generally to denote a function of this type.

Now, if we put $k=l'$ and $k'=l$ in (16), we have

$$n\left\{1-24\left(\frac{1}{e^{2\pi\sqrt{n}}-1}+\frac{2}{e^{4\pi\sqrt{n}}-1}+\dots\right)\right\}-\left\{1-24\left(\frac{1}{e^{2\pi/\sqrt{n}}-1}+\frac{2}{e^{4\pi/\sqrt{n}}-1}+\dots\right)\right\}=\left(\frac{K}{\pi}\right)^2 A(k).\quad\dots(17)$$

The algebraic function $A(k)$ of course assumes a purely numerical form when we substitute the value of k deduced from the modular equation. But by substituting $k=l'$ and $k'=l$ in (15) we have

$$n^{\frac{1}{4}}e^{-\pi\sqrt{n}/12}(1-e^{-2\pi\sqrt{n}})(1-e^{-4\pi\sqrt{n}})(1-e^{-6\pi\sqrt{n}})\dots$$
$$=e^{-\pi/(12\sqrt{n})}(1-e^{-2\pi/\sqrt{n}})(1-e^{-4\pi/\sqrt{n}})(1-e^{-6\pi/\sqrt{n}})\dots.$$

Differentiating the above equation logarithmically we have

$$n\left\{1-24\left(\frac{1}{e^{2\pi\sqrt{n}}-1}+\frac{2}{e^{4\pi\sqrt{n}}-1}+\dots\right)\right\}+\left\{1-24\left(\frac{1}{e^{2\pi/\sqrt{n}}-1}+\frac{2}{e^{4\pi/\sqrt{n}}-1}+\dots\right)\right\}=\frac{6\sqrt{n}}{\pi}.\quad\dots\dots\dots(18)$$

Now, adding (17) and (18), we have

$$1-\frac{3}{\pi\sqrt{n}}-24\left(\frac{1}{e^{2\pi\sqrt{n}}-1}+\frac{2}{e^{4\pi\sqrt{n}}-1}+\dots\right)=\left(\frac{K}{\pi}\right)^2 A(k).\quad\dots\dots(19)$$

But it is known that

$$1-24\left(\frac{q}{1+q}+\frac{3q^3}{1+q^3}+\frac{5q^5}{1+q^5}+\dots\right)=\left(\frac{2K}{\pi}\right)^2(1-2k^2),$$

so that

$$1-24\left(\frac{1}{e^{\pi\sqrt{n}}+1}+\frac{3}{e^{3\pi\sqrt{n}}+1}+\dots\right)=\left(\frac{K}{\pi}\right)^2 A(k).\quad\dots\dots\dots(20)$$

Hence, dividing (19) by (20), we have

$$\frac{1-\dfrac{3}{\pi\sqrt{n}}-24\left(\dfrac{1}{e^{2\pi\sqrt{n}}-1}+\dfrac{2}{e^{4\pi\sqrt{n}}-1}+\dots\right)}{1-24\left(\dfrac{1}{e^{\pi\sqrt{n}}+1}+\dfrac{3}{e^{3\pi\sqrt{n}}+1}+\dots\right)}=R,\quad\dots\dots\dots(21)$$

where R can always be expressed in radicals if n is any rational number. Hence we have

$$\pi = \frac{3}{(1-R)\sqrt{n}}, \quad \text{.........................(22)}$$

nearly, the error being about $8\pi e^{-\pi\sqrt{n}}(\pi\sqrt{n}-3)$.

9. We may get a still closer approximation from the following results. It is known that

$$1 + 240 \sum_{r=1}^{r=\infty} \frac{r^3 q^{2r}}{1-q^{2r}} = \left(\frac{2K}{\pi}\right)^4 (1 - k^2 k'^2),$$

and also that

$$1 - 504 \sum_{r=1}^{r=\infty} \frac{r^5 q^{2r}}{1-q^{2r}} = \left(\frac{2K}{\pi}\right)^6 (1-2k^2)(1+\tfrac{1}{2}k^2 k'^2).$$

Hence, from (19), we see that

$$\left\{1 - \frac{3}{\pi\sqrt{n}} - 24 \sum_{r=1}^{r=\infty} \frac{r}{e^{2\pi r\sqrt{n}}-1}\right\}\left\{1 + 240 \sum_{r=1}^{r=\infty} \frac{r^3}{e^{2\pi r\sqrt{n}}-1}\right\}$$

$$= R'\left\{1 - 504 \sum_{r=1}^{r=\infty} \frac{r^5}{e^{2\pi r\sqrt{n}}-1}\right\}, \quad \text{......(23)}$$

where R' can always be expressed in radicals for any rational value of n. Hence

$$\pi = \frac{3}{(1-R')\sqrt{n}}, \quad \text{............................(24)}$$

nearly, the error being about $24\pi(10\pi\sqrt{n}-31)e^{-2\pi\sqrt{n}}$.

It will be seen that the error in (24) is much less than that in (22), if n is at all large.

10. In order to find R and R' the series in (16) must be calculated in finite terms. I shall give the final results for a few values of n.

TABLE III.

$$q = e^{-\pi K'/K}, \quad q^n = e^{-\pi L'/L},$$

$$f(q) = n\left(1 - 24 \sum_1^\infty \frac{q^{2mn}}{1-q^{2mn}}\right) - \left(1 - 24 \sum_1^\infty \frac{q^{2m}}{1-q^{2m}}\right),$$

$$f(2) = \frac{4KL}{\pi^2}(k' + l),$$

$$f(3) = \frac{4KL}{\pi^2}(1 + kl + k'l'),$$

$$f(4) = \frac{4KL}{\pi^2}(\sqrt{k'} + \sqrt{l})^2,$$

$$f(5) = \frac{4KL}{\pi^2}(3 + kl + k'l')\sqrt{\left(\frac{1+kl+k'l'}{2}\right)},$$

$$f(7) = \frac{12KL}{\pi^2}(1 + kl + k'l'),$$

$$f(11)=\frac{8KL}{\pi^2}\{2(1+kl+k'l')+\surd(kl)+\surd(k'l')-\surd(kk'll')\},$$

$$f(15)=\frac{4KL}{\pi^2}[\{1+(kl)^{\frac{1}{4}}+(k'l')^{\frac{1}{4}}\}^4-\{1+kl+k'l'\}],$$

$$f(17)=\frac{4KL}{\pi^2}\surd\{44(1+k^2l^2+k'^2l'^2)+168(kl+k'l'-kk'll')$$
$$-102(1-kl-k'l')(4kk'll')^{\frac{1}{3}}-192(4kk'll')^{\frac{2}{3}}\},$$

$$f(19)=\frac{24KL}{\pi^2}\{(1+kl+k'l')+\surd(kl)+\surd(k'l')-\surd(kk'll')\},$$

$$f(23)=\frac{4KL}{\pi^2}[11(1+kl+k'l')-16(4kk'll')^{\frac{1}{6}}\{1+\surd(kl)+\surd(k'l')\}$$
$$-20(4kk'll')^{\frac{1}{3}}],$$

$$f(31)=\frac{12KL}{\pi^2}[3(1+kl+k'l')+4\{\surd(kl)+\surd(k'l')+\surd(kk'll')\}$$
$$-4(kk'll')^{\frac{1}{4}}\{1+(kl)^{\frac{1}{4}}+(k'l')^{\frac{1}{4}}\}],$$

$$f(35)=\frac{4KL}{\pi^2}[2\{\surd(kl)+\surd(k'l')-\surd(kk'll')\}$$
$$+(4kk'll')^{-\frac{1}{6}}\{1-\surd(kl)-\surd(k'l')\}^3].$$

Thus the sum of the series (19) can be found in finite terms, when $n=2, 3, 4, 5, \ldots$, from the equations in Table III. We can use the same table to find the sum of (19) when $n=9, 25, 49, \ldots$; but then we have also to use the equation

$$\frac{3}{\pi}=1-24\left(\frac{1}{e^{2\pi}-1}+\frac{2}{e^{4\pi}-1}+\frac{3}{e^{6\pi}-1}+\ldots\right),$$

which is got by putting $k=k'=1/\surd 2$ and $n=1$ in (18).

Similarly we can find the sum of (19) when $n=21, 33, 57, 93, \ldots$, by combining the values of $f(3)$ and $f(7)$, $f(3)$ and $f(11)$, and so on, obtained from Table III.

11. The errors in (22) and (24) being about

$$8\pi e^{-\pi\surd n}(\pi\surd n-3), \quad 24\pi(10\pi\surd n-31)e^{-2\pi\surd n},$$

we cannot expect a high degree of approximation for small values of n. Thus, if we put $n=7$, 9, 16, and 25 in (24), we get

$$\frac{19}{16}\surd 7=3\cdot 14180\ldots,$$

$$\frac{7}{3}\left(1+\frac{\surd 3}{5}\right)=3\cdot 14162\ldots,$$

$$\frac{99}{80}\left(\frac{7}{7-3\surd 2}\right)=3\cdot 14159274\ldots,$$

$$\frac{63}{25}\left(\frac{17+15\surd 5}{7+15\surd 5}\right)=3\cdot 14159265380\ldots,$$

while $$\pi = 3{\cdot}14159265358\ldots.$$
But if we put $n = 25$ in (22), we get only
$$\frac{9}{5} + \sqrt{\left(\frac{9}{5}\right)} = 3{\cdot}14164\ldots.$$

12. Another curious approximation to π is
$$\left(9^2 + \frac{19^2}{22}\right)^{\frac{1}{4}} = 3{\cdot}14159265262\ldots.$$
This value was obtained empirically, and it has no connection with the preceding theory.

The actual value of π, which I have used for purposes of calculation, is
$$\frac{355}{113}\left(1 - \frac{{\cdot}0003}{3533}\right) = 3{\cdot}1415926535897943\ldots,$$
which is greater than π by about 10^{-15}. This is obtained by simply taking the reciprocal of $1 - (113\pi/355)$.

In this connection it may be interesting to note the following simple geometrical constructions for π. The first merely gives the ordinary value 355/113. The second gives the value $(9^2 + 19^2/22)^{\frac{1}{4}}$ mentioned above.

(1) Let AB (Fig. 1) be a diameter of a circle whose centre is O.

Bisect AO at M and trisect OB at T.

Draw TP perpendicular to AB and meeting the circumference at P.

Draw a chord BQ equal to PT and join AQ.

Draw OS and TR parallel to BQ and meeting AQ at S and R respectively.

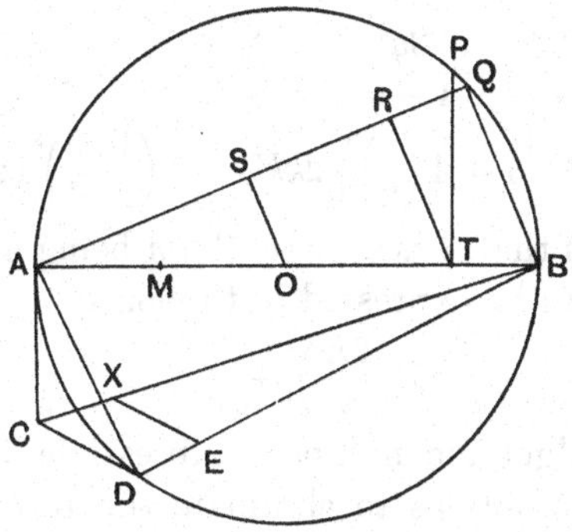

Fig. 1.

Draw a chord AD equal to AS and a tangent $AC = RS$.

Join BC, BD, and CD; cut off $BE = BM$, and draw EX, parallel to CD, meeting BC at X.

Then the square on BX is very nearly equal to the area of the circle, the error being less than a tenth of an inch when the diameter is 40 miles long.

(2) Let AB (Fig. 2) be a diameter of a circle whose centre is O. Bisect the arc ACB at C and trisect AO at T.

Join BC and cut off from it CM and MN equal to AT.

Join AM and AN and cut off from the latter AP equal to AM.

Through P draw PQ parallel to MN and meeting AM at Q.

Join OQ and through T draw TR, parallel to OQ, and meeting AQ at R.

Draw AS perpendicular to AO and equal to AR, and join OS.

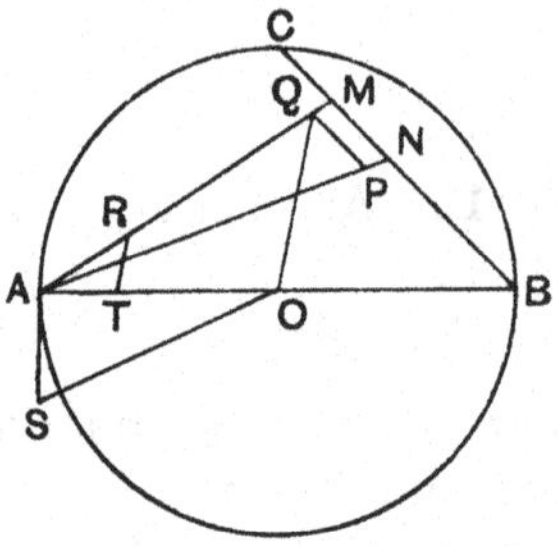

Fig. 2.

Then the mean proportional between OS and OB will be very nearly equal to a sixth of the circumference, the error being less than a twelfth of an inch when the diameter is 8000 miles long.

13. I shall conclude this paper by giving a few series for $1/\pi$.

It is known that, when $k \leqslant 1/\sqrt{2}$,

$$\left(\frac{2K}{\pi}\right)^2 = 1 + \left(\frac{1}{2}\right)^3 (2kk')^2 + \left(\frac{1\,.\,3}{2\,.\,4}\right)^3 (2kk')^4 + \dots \quad \dots\dots\dots(25)$$

Hence we have

$$q^{\frac{1}{3}}(1-q^2)^4(1-q^4)^4(1-q^6)^4 \dots$$
$$= \left(\frac{1}{4}kk'\right)^{\frac{2}{3}} \left\{1 + \left(\frac{1}{2}\right)^3 (2kk')^2 + \left(\frac{1\,.\,3}{2\,.\,4}\right)^3 (2kk')^4 + \dots\right\}. \quad \dots\dots(26)$$

Differentiating both sides in (26) logarithmically with respect to k, we can easily shew that

$$1 - 24\left(\frac{q^2}{1-q^2} + \frac{2q^4}{1-q^4} + \frac{3q^6}{1-q^6} + \dots\right)$$
$$= (1-2k^2)\left\{1 + 4\left(\frac{1}{2}\right)^3 (2kk')^2 + 7\left(\frac{1\,.\,3}{2\,.\,4}\right)^3 (2kk')^4 + \dots\right\}. \dots(27)$$

But it follows from (19) that, when $q = e^{-\pi\sqrt{n}}$, n being a rational number, the left-hand side of (27) can be expressed in the form

$$A\left(\frac{2K}{\pi}\right)^2 + \frac{B}{\pi},$$

where A and B are algebraic numbers expressible by surds. Combining (25) and (27) in such a way as to eliminate the term $(2K/\pi)^2$, we are left with a series for $1/\pi$. Thus, for example,

$$\frac{4}{\pi} = 1 + \frac{7}{4}\left(\frac{1}{2}\right)^3 + \frac{13}{4^2}\left(\frac{1\,.\,3}{2\,.\,4}\right)^3 + \frac{19}{4^3}\left(\frac{1\,.\,3\,.\,5}{2\,.\,4\,.\,6}\right)^3 + \dots,$$
$$(q = e^{-\pi\sqrt{3}},\ 2kk' = \tfrac{1}{2}), \quad \dots\dots\dots\dots\dots(28)$$

$$\frac{16}{\pi} = 5 + \frac{47}{64}\left(\frac{1}{2}\right)^3 + \frac{89}{64^2}\left(\frac{1\,.\,3}{2\,.\,4}\right)^3 + \frac{131}{64^3}\left(\frac{1\,.\,3\,.\,5}{2\,.\,4\,.\,6}\right)^3 + \dots,$$
$$(q = e^{-\pi\sqrt{7}},\ 2kk' = \tfrac{1}{8}), \quad \dots\dots\dots\dots\dots(29)$$

$$\frac{32}{\pi}=(5\sqrt{5}-1)+\frac{47\sqrt{5}+29}{64}\left(\frac{1}{2}\right)^3\left(\frac{\sqrt{5}-1}{2}\right)^8$$
$$+\frac{89\sqrt{5}+59}{64^2}\left(\frac{1.3}{2.4}\right)^3\left(\frac{\sqrt{5}-1}{2}\right)^{16}+\ldots,$$
$$\left[q=e^{-\pi\sqrt{15}},\ 2kk'=\frac{1}{8}\left(\frac{\sqrt{5}-1}{2}\right)\right]; \quad\text{.........(30)}$$

here $5\sqrt{5}-1$, $47\sqrt{5}+29$, $89\sqrt{5}+59$, ... are in arithmetical progression.

14. The ordinary modular equations express the relations which hold between k and l when $nK'/K=L'/L$, or $q^n=Q$, where

$$q=e^{-\pi K'/K},\quad Q=e^{-\pi L'/L},$$
$$K=1+\left(\frac{1}{2}\right)^2 k^2+\left(\frac{1.3}{2.4}\right)^2 k^4+\ldots.$$

There are corresponding theories in which q is replaced by one or other of the functions

$$q_1=e^{-\pi K_1'\sqrt{2}/K_1},\quad q_2=e^{-2\pi K_2'/(K_2\sqrt{3})},\quad q_3=e^{-2\pi K_3'/K_3},$$

where

$$K_1=1+\frac{1.3}{4^2}k^2+\frac{1.3.5.7}{4^2.8^2}k^4+\frac{1.3.5.7.9.11}{4^2.8^2.12^2}k^6+\ldots,$$
$$K_2=1+\frac{1.2}{3^3}k^2+\frac{1.2.4.5}{3^2.6^2}k^4+\frac{1.2.4.5.7.8}{3^2.6^2.9^2}k^6+\ldots,$$
$$K_3=1+\frac{1.5}{6^2}k^2+\frac{1.5.7.11}{6^2.12^2}k^4+\frac{1.5.7.11.13.17}{6^2.12^2.18^2}k^6+\ldots.$$

From these theories we can deduce further series for $1/\pi$, such as

$$\frac{27}{4\pi}=2+17\frac{1}{2}\frac{1}{3}\frac{2}{3}\left(\frac{2}{27}\right)$$
$$+32\frac{1.3}{2.4}\frac{1.4}{3.6}\frac{2.5}{3.6}\left(\frac{2}{27}\right)^2+\ldots,\quad\text{......(31)}$$

$$\frac{15\sqrt{3}}{2\pi}=4+37\frac{1}{2}\frac{1}{3}\frac{2}{3}\left(\frac{4}{125}\right)$$
$$+70\frac{1.3}{2.4}\frac{1.4}{3.6}\frac{2.5}{3.6}\left(\frac{4}{125}\right)^2+\ldots,\quad\text{......(32)}$$

$$\frac{5\sqrt{5}}{2\pi\sqrt{3}}=1+12\frac{1}{2}\frac{1}{6}\frac{5}{6}\left(\frac{4}{125}\right)$$
$$+23\frac{1.3}{2.4}\frac{1.7}{6.12}\frac{5.11}{6.12}\left(\frac{4}{125}\right)^2+\ldots,\quad\text{...(33)}$$

$$\frac{85\sqrt{85}}{18\pi\sqrt{3}}=8+141\frac{1}{2}\frac{1}{6}\frac{5}{6}\left(\frac{4}{85}\right)^3$$
$$+274\frac{1.3}{2.4}\frac{1.7}{6.12}\frac{5.11}{6.12}\left(\frac{4}{85}\right)^6+\ldots,\quad\text{...(34)}$$

$$\frac{4}{\pi}=\frac{3}{2}-\frac{23}{2^3}\frac{1}{2}\frac{1.3}{4^2}+\frac{43}{2^5}\frac{1.3}{2.4}\frac{1.3.5.7}{4^2.8^2}-\ldots, \quad\text{.............(35)}$$

$$\frac{4}{\pi\sqrt{3}}=\frac{3}{4}-\frac{31}{3.4^3}\frac{1}{2}\frac{1.3}{4^2}+\frac{59}{3^2.4^5}\frac{1.3}{2.4}\frac{1.3.5.7}{4^2.8^2}-\ldots, \quad\text{.........(36)}$$

$$\frac{4}{\pi}=\frac{23}{18}-\frac{283}{18^3}\frac{1}{2}\frac{1.3}{4^2}+\frac{543}{18^5}\frac{1.3}{2.4}\frac{1.3.5.7}{4^2.8^2}-\ldots, \quad\text{.........(37)}$$

$$\frac{4}{\pi\sqrt{5}}=\frac{41}{72}-\frac{685}{5.72^3}\frac{1}{2}\frac{1.3}{4^2}+\frac{1329}{5^2.72^5}\frac{1.3}{2.4}\frac{1.3.5.7}{4^2.8^2}-\ldots, \quad\text{......(38)}$$

$$\frac{4}{\pi}=\frac{1123}{882}-\frac{22583}{882^3}\frac{1}{2}\frac{1.3}{4^2}+\frac{44043}{882^5}\frac{1.3}{2.4}\frac{1.3.5.7}{4^2.8^2}-\ldots, \quad\text{...(39)}$$

$$\frac{2\sqrt{3}}{\pi}=1+\frac{9}{9}\frac{1}{2}\frac{1.3}{4^2}+\frac{17}{9^2}\frac{1.3}{2.4}\frac{1.3.5.7}{4^2.8^2}+\ldots, \quad\text{.................(40)}$$

$$\frac{1}{2\pi\sqrt{2}}=\frac{1}{9}+\frac{11}{9^3}\frac{1}{2}\frac{1.3}{4^2}+\frac{21}{9^5}\frac{1.3}{2.4}\frac{1.3.5.7}{4^2.8^2}+\ldots, \quad\text{..............(41)}$$

$$\frac{1}{3\pi\sqrt{3}}=\frac{3}{49}+\frac{43}{49^3}\frac{1}{2}\frac{1.3}{4^2}+\frac{83}{49^5}\frac{1.3}{2.4}\frac{1.3.5.7}{4^2.8^2}+\ldots, \quad\text{...........(42)}$$

$$\frac{2}{\pi\sqrt{11}}=\frac{19}{99}+\frac{299}{99^3}\frac{1}{2}\frac{1.3}{4^2}+\frac{579}{99^5}\frac{1.3}{2.4}\frac{1.3.5.7}{4^2.8^2}+\ldots, \quad\text{...........(43)}$$

$$\frac{1}{2\pi\sqrt{2}}=\frac{1103}{99^2}+\frac{27493}{99^6}\frac{1}{2}\frac{1.3}{4^2}+\frac{53883}{99^{10}}\frac{1.3}{2.4}\frac{1.3.5.7}{4^2.8^2}+\ldots \quad\text{...(44)}$$

In all these series the first factors in each term form an arithmetical progression; e.g. 2, 17, 32, 47, ..., in (31), and 4, 37, 70, 103, ..., in (32). The first two series belong to the theory of q_2, the next two to that of q_3, and the rest to that of q_1.

The last series (44) is extremely rapidly convergent. Thus, taking only the first term, we see that

$$\frac{1103}{99^2}=\cdot 11253953678\ldots,$$

$$\frac{1}{2\pi\sqrt{2}}=\cdot 11253953951\ldots.$$

15. In concluding this paper I have to remark that the series

$$1-24\left(\frac{q^2}{1-q^2}+\frac{2q^4}{1-q^4}+\frac{3q^6}{1-q^6}+\ldots\right),$$

which has been discussed in §§ 8—13, is very closely connected with the perimeter of an ellipse whose eccentricity is k. For, if a and b be the semi-major and the semi-minor axes, it is known that

$$p=2\pi a\left\{1-\frac{1}{2^2}k^2-\frac{1^2.3}{2^2.4^2}k^4-\frac{1^2.3^2.5}{2^2.4^4.6^2}k^6-\ldots\right\}, \quad\text{.........(45)}$$

where p is the perimeter and k the eccentricity. It can easily be seen from (45) that

$$p = 4ak'^2\left\{K + k\frac{dK}{dk}\right\}. \quad\ldots\ldots\ldots\ldots\ldots\ldots\ldots(46)$$

But, taking the equation

$$q^{\frac{1}{12}}(1-q^2)(1-q^4)(1-q^6)\ldots = (2kk')^{\frac{1}{6}}\sqrt{(K/\pi)},$$

and differentiating both sides logarithmically with respect to k, and combining the result with (46) in such a way as to eliminate dK/dk, we can shew that

$$p = \frac{4a}{3K}\left[K^2(1+k'^2) + (\tfrac{1}{2}\pi)^2\left\{1 - 24\left(\frac{q^2}{1-q^2} + \frac{2q^4}{1-q^4} + \ldots\right)\right\}\right]. \quad (47)$$

But we have shewn already that the right-hand side of (47) can be expressed in terms of K if $q = e^{-\pi\sqrt{n}}$, where n is any rational number. It can also be shewn that K can be expressed in terms of Γ-functions if q be of the forms $e^{-\pi n}$, $e^{-\pi n\sqrt{2}}$ and $e^{-\pi n\sqrt{3}}$, where n is rational. Thus, for example, we have

$$\left.\begin{aligned}
&k = \sin\frac{\pi}{4}, \quad q = e^{-\pi},\\
&\qquad p = a\sqrt{\left(\frac{\pi}{2}\right)}\left\{\frac{\Gamma(\frac{1}{4})}{\Gamma(\frac{3}{4})} + \frac{\Gamma(\frac{3}{4})}{\Gamma(\frac{5}{4})}\right\},\\
&k = \tan\frac{\pi}{8}, \quad q = e^{-\pi\sqrt{2}},\\
&\qquad p = a\sqrt{\left(\frac{\pi}{4}\right)}\left\{\frac{\Gamma(\frac{1}{8})}{\Gamma(\frac{5}{8})} + \frac{\Gamma(\frac{5}{8})}{\Gamma(\frac{9}{8})}\right\},\\
&k = \sin\frac{\pi}{12}, \quad q = e^{-\pi\sqrt{3}},\\
&\qquad p = a\sqrt{\left(\frac{\pi}{\sqrt{3}}\right)}\left\{\left(1 + \frac{1}{\sqrt{3}}\right)\frac{\Gamma(\frac{1}{3})}{\Gamma(\frac{5}{6})} + 2\frac{\Gamma(\frac{5}{6})}{\Gamma(\frac{1}{3})}\right\},\\
&\frac{b}{a} = \tan^2\frac{\pi}{8}, \quad q = e^{-2\pi},\\
&\qquad p = (a+b)\sqrt{\left(\frac{\pi}{2}\right)}\left\{\frac{1}{2}\frac{\Gamma(\frac{1}{4})}{\Gamma(\frac{3}{4})} + \frac{\Gamma(\frac{3}{4})}{\Gamma(\frac{5}{4})}\right\},
\end{aligned}\right\} \ldots\ldots(48)$$

and so on.

16. The following approximations for p were obtained empirically:

$$p = \pi\left[3(a+b) - \sqrt{\{(a+3b)(3a+b)\}} + \epsilon\right], \ldots\ldots\ldots\ldots\ldots(49)$$

where ϵ is about $ak^{12}/1048576$;

$$p = \pi\left\{(a+b) + \frac{3(a-b)^2}{10(a+b) + \sqrt{(a^2+14ab+b^2)}} + \epsilon\right\}, \ldots\ldots\ldots(50)$$

where ϵ is about $3ak^{20}/68719476736$.

7

ON THE INTEGRAL $\int_0^x \frac{\tan^{-1} t}{t} dt$

(*Journal of the Indian Mathematical Society*, VII, 1915, 93—96)

1. Let
$$\phi(x) = \int_0^x \frac{\tan^{-1} t}{t} dt. \quad \text{.........................(1)}$$
Then it is easy to see that
$$\phi(x) + \phi(-x) = 0; \quad \text{.........................(2)}$$
and that
$$\phi(x) = \frac{x}{1^2} - \frac{x^3}{3^2} + \frac{x^5}{5^2} - \frac{x^7}{7^2} + \dots \quad \text{.....................(3)}$$
provided that $|x| \leqslant 1$.

Changing t into $1/t$ in (1), we obtain
$$\phi(x) - \phi\left(\frac{1}{x}\right) = \tfrac{1}{2}\pi \log x, \quad \text{.......................(4)}$$
provided that the real part of x is positive.

The results in the following two sections can be very easily proved by differentiating both sides with respect to x.

2. If $0 < x < \frac{1}{2}\pi$, then
$$\frac{\sin 2x}{1^2} + \frac{\sin 6x}{3^2} + \frac{\sin 10x}{5^2} + \dots = \phi(\tan x) - x \log(\tan x). \quad \text{......(5)}$$

If, in particular, we put $x = \frac{1}{8}\pi$ and $\frac{1}{12}\pi$ in (5), we obtain
$$\frac{1}{1^2} - \frac{1}{5^2} + \frac{1}{9^2} - \frac{1}{13^2} + \dots = \frac{1}{\sqrt{2}}\phi(\sqrt{2}-1) + \frac{\pi}{8\sqrt{2}}\log(1+\sqrt{2}) + \frac{\pi^2}{16}; \quad \text{...(6)*}$$
and
$$2\phi(1) = 3\phi(2-\sqrt{3}) + \tfrac{1}{4}\pi \log(2+\sqrt{3}). \quad \text{..............(7)}$$

If $-\frac{1}{2}\pi < x < \frac{1}{2}\pi$, then
$$2\phi\left(\tan\frac{x}{2}\right) = \sin x + \frac{2}{3}\frac{\sin^3 x}{3} + \frac{2.4}{3.5}\frac{\sin^5 x}{5} + \dots. \quad \text{.........(8)}$$

If $0 < x < \frac{1}{2}\pi$, then
$$\frac{\sin x}{1^2}\cos x + \frac{\sin 2x}{2^2}\cos^2 x + \frac{\sin 3x}{3^2}\cos^3 x + \dots$$
$$= \phi(\tan x) + \tfrac{1}{2}\pi \log \cos x - x \log \sin x; \quad \text{...........(9)}$$
and
$$\frac{\cos x + \sin x}{1^2} + \frac{1}{2}\frac{\cos^3 x + \sin^3 x}{3^2} + \frac{1.3}{2.4}\frac{\cos^5 x + \sin^5 x}{5^2} + \dots$$
$$= \phi(\tan x) + \tfrac{1}{2}\pi \log(2\cos x). \quad \text{......................(10)}$$

* This equation is incorrect: see Appendix, p. 337.

If $-\frac{1}{2}\pi < x < \frac{1}{2}\pi$ and a be any number such that

$$|(1-a)\sin x| \leqslant 1, \quad \left|\left(1-\frac{1}{a}\right)\cos x\right| \leqslant 1,$$

then
$$\frac{\sin x}{1^2}\left(1-\frac{1}{a}\right)\cos x + \frac{\sin 2x}{2^2}\left(1-\frac{1}{a}\right)^2\cos^2 x + \frac{\sin 3x}{3^2}\left(1-\frac{1}{a}\right)^3\cos^3 x + \ldots$$
$$+\frac{\sin(x+\frac{1}{2}\pi)}{1^2}(1-a)\sin x - \frac{\sin 2(x+\frac{1}{2}\pi)}{2^2}(1-a)^2\sin^2 x + \ldots$$
$$= \phi(\tan x) - \phi(a\tan x) + x\log a. \qquad \text{.........(11)}$$

3. Let $R(x)$ and $I(x)$ denote the real and the imaginary parts of x respectively. Then, if $-1 < R(x) < 1$,

$$\log\left(1-\frac{x^2}{1^2}\right) - 3\log\left(1-\frac{x^2}{3^2}\right) + 5\log\left(1-\frac{x^2}{5^2}\right) - \ldots$$
$$= \frac{4}{\pi}[\phi(1) - \phi\{\tan\tfrac{1}{4}\pi(1-x)\}] + \log\tan\tfrac{1}{4}\pi(1-x). \qquad \text{...(12)}$$

Putting $x = \frac{2}{3}$ in (12) and using (7), we obtain

$$\left(1-\frac{4}{3^2}\right)\left(1-\frac{4}{9^2}\right)^{-3}\left(1-\frac{4}{15^2}\right)^5\left(1-\frac{4}{21^2}\right)^{-7}\left(1-\frac{4}{27^2}\right)^9 \ldots = (2-\sqrt{3})^{\frac{2}{3}}e^n,$$

where
$$n = \frac{4}{3\pi}\phi(1). \qquad \text{.........(13)}$$

Again, subtracting $\log(1-x)$ from both sides in (12) and making $x \to 1$, we obtain

$$\left(1-\frac{1}{3^2}\right)^{-3}\left(1-\frac{1}{5^2}\right)^5\left(1-\frac{1}{7^2}\right)^{-7}\left(1-\frac{1}{9^2}\right)^9 \ldots = \frac{\pi}{8}e^{3n}. \qquad \text{.........(14)}$$

If $-1 < I(x) < 1$, then

$$\log\left(1+\frac{x^2}{1^2}\right) - 3\log\left(1+\frac{x^2}{3^2}\right) + 5\log\left(1+\frac{x^2}{5^2}\right) - \ldots$$
$$= \frac{4}{\pi}\{\phi(1) - \phi(e^{-\frac{1}{2}\pi x})\} - 2x\tan^{-1}e^{-\frac{1}{2}\pi x}. \qquad \text{.........(15)}$$

From this and (7) we see that, if $\frac{1}{2}\pi x = \log(2+\sqrt{3})$, then

$$\left(1+\frac{x^2}{1^2}\right)\left(1+\frac{x^2}{3^2}\right)^{-3}\left(1+\frac{x^2}{5^2}\right)^5\left(1+\frac{x^2}{7^2}\right)^{-7} \ldots = e^n, \qquad \text{.........(16)}$$

where n is the same as in (13).

It follows at once from (12) and (15) that, if

$$-1 < R(\beta) < 1, \quad -1 < I(\alpha) < 1,$$

then
$$e^{\frac{1}{2}\pi\alpha\beta} = \left(\frac{1^2+\alpha^2}{1^2-\beta^2}\right)\left(\frac{3^2-\beta^2}{3^2+\alpha^2}\right)^3\left(\frac{5^2+\alpha^2}{5^2-\beta^2}\right)^5\left(\frac{7^2-\beta^2}{7^2+\alpha^2}\right)^7 \ldots, \qquad \text{......(17)}$$

provided that $\cosh\frac{1}{2}\pi\alpha = \sec\frac{1}{2}\pi\beta$.

4. Now changing x into $2x(1+i)$ in (15), we have

$$\log\left(1+\frac{8ix^2}{1^2}\right)-3\log\left(1+\frac{8ix^2}{3^2}\right)+5\log\left(1+\frac{8ix^2}{5^2}\right)-\dots$$

$$=\frac{4}{\pi}\phi(1)-4x(1+i)\tan^{-1}e^{-\pi x(1+i)}-\frac{4}{\pi}\left\{\frac{1}{1^2}e^{-\pi x(1+i)}-\frac{1}{3^2}e^{-3\pi x(1+i)}+\dots\right\}.$$

Equating real and imaginary parts we see that, if x is positive, then

$$\log\left(1+\frac{64x^4}{1^4}\right)-3\log\left(1+\frac{64x^4}{3^4}\right)+5\log\left(1+\frac{64x^4}{5^4}\right)-\dots$$

$$=\frac{8}{\pi}\phi(1)-2x\log\left(\frac{\cosh\pi x+\sin\pi x}{\cosh\pi x-\sin\pi x}\right)-4x\tan^{-1}\left(\frac{\cos\pi x}{\sinh\pi x}\right)$$

$$-\frac{8}{\pi}\left\{\frac{\cos\pi x}{1^2}e^{-\pi x}-\frac{\cos3\pi x}{3^2}e^{-3\pi x}+\frac{\cos5\pi x}{5^2}e^{-5\pi x}-\dots\right\};\quad\dots(18)$$

and
$$\tan^{-1}\frac{8x^2}{1^2}-3\tan^{-1}\frac{8x^2}{3^2}+5\tan^{-1}\frac{8x^2}{5^2}-\dots$$

$$=x\log\left(\frac{\cosh\pi x+\sin\pi x}{\cosh\pi x-\sin\pi x}\right)-2x\tan^{-1}\left(\frac{\cos\pi x}{\sinh\pi x}\right)$$

$$+\frac{4}{\pi}\left\{\frac{\sin\pi x}{1^2}e^{-\pi x}-\frac{\sin3\pi x}{3^2}e^{-3\pi x}+\frac{\sin5\pi x}{5^2}e^{-5\pi x}-\dots\right\}.\quad\dots(19)$$

It follows from (18) that, if n is a positive odd integer, then

$$\left(1+\frac{4n^4}{1^4}\right)\left(1+\frac{4n^4}{3^4}\right)^{-3}\left(1+\frac{4n^4}{5^4}\right)^{5}\left(1+\frac{4n^4}{7^4}\right)^{-7}\dots$$

$$=e^{\frac{8}{\pi}\phi(1)}\left(\frac{1-e^{-\frac{1}{2}\pi n}}{1+e^{-\frac{1}{2}\pi n}}\right)^{2n\cos\frac{1}{2}(n-1)\pi},\quad\dots\dots\dots(20)$$

and, if n is any even integer, then

$$\left(1+\frac{4n^4}{1^4}\right)\left(1+\frac{4n^4}{3^4}\right)^{-3}\left(1+\frac{4n^4}{5^4}\right)^{5}\left(1+\frac{4n^4}{7^4}\right)^{-7}\dots$$

$$=\exp\left\{\frac{8}{\pi}\phi(1)-\frac{8}{\pi}(-1)^{\frac{1}{2}n}\left[\phi(e^{-\frac{1}{2}\pi n})+\tfrac{1}{2}\pi n\tan^{-1}e^{-\frac{1}{2}\pi n}\right]\right\}.\quad\dots(21)$$

Similarly from (19) we see that, if n is any positive odd integer, then

$$\tan^{-1}\frac{2n^2}{1^2}-3\tan^{-1}\frac{2n^2}{3^2}+5\tan^{-1}\frac{2n^2}{5^2}-\dots$$

$$=\frac{4}{\pi}(-1)^{\frac{1}{2}(n-1)}\left\{\frac{\pi n}{4}\log\left(\frac{1+e^{-\frac{1}{2}\pi n}}{1-e^{-\frac{1}{2}\pi n}}\right)+\frac{1}{1^2}e^{-\frac{1}{2}\pi n}+\frac{1}{3^2}e^{-\frac{3}{2}\pi n}+\frac{1}{5^2}e^{-\frac{5}{2}\pi n}+\dots\right\};\quad(22)$$

and, if n is a positive even integer, then

$$\tan^{-1}\frac{2n^2}{1^2}-3\tan^{-1}\frac{2n^2}{3^2}+5\tan^{-1}\frac{2n^2}{5^2}-\dots=2n(-1)^{\frac{1}{2}n-1}\tan^{-1}e^{-\frac{1}{2}\pi n}.\quad\dots(23)$$

In this connection it may be interesting to note that

$$\tan^{-1}e^{-\frac{1}{2}\pi n}=\frac{\pi}{4}-\left(\tan^{-1}\frac{n}{1}-\tan^{-1}\frac{n}{3}+\tan^{-1}\frac{n}{5}-\dots\right)\quad\dots\dots\dots(24)$$

for all real values of n.

5. Remembering that

$$\frac{\pi}{4\cosh \pi x} = \frac{1}{1^2+4x^2} - \frac{3}{3^2+4x^2} + \frac{5}{5^2+4x^2} - \dots$$

we have

$$\frac{\pi}{4}\sum_{1}^{\infty}\frac{1}{n^2\cosh \pi n x} = \sum_{n=1}^{n=\infty}\left\{\frac{1}{n^2(1^2+4n^2x^2)} - \frac{3}{n^2(3^2+4n^2x^2)} + \dots\right\}$$

$$= \frac{\pi^3}{8}\left(\frac{1}{3}+\frac{x^2}{2}\right) - \pi x\left(\frac{\coth\frac{\pi}{2x}}{1^2} - \frac{\coth\frac{3\pi}{2x}}{3^2} + \frac{\coth\frac{5\pi}{2x}}{5^2} - \dots\right). \quad \dots(25)$$

That is to say, if α and β are real and $\alpha\beta = \pi^2$, then

$$\phi(1) + 2\phi(e^{-\alpha}) + 2\phi(e^{-2\alpha}) + 2\phi(e^{-3\alpha}) + \dots$$

$$= \frac{\pi}{8}\left(\frac{\alpha}{3}+\frac{\beta}{2}\right) - \frac{\pi}{4\beta}\left\{\frac{1}{1^2\cosh\beta} + \frac{1}{2^2\cosh 2\beta} + \dots\right\}. \quad \dots\dots(26)$$

If, in particular, we put $\alpha = \beta = \pi$ in (26), we obtain

$$\phi(1) = \frac{5\pi^2}{48} - 2\left\{\frac{1}{1^2(e^{\pi}-1)} - \frac{1}{3^2(e^{3\pi}-1)} + \frac{1}{5^2(e^{5\pi}-1)}\dots\right\}$$

$$-\frac{1}{2}\left\{\frac{1}{1^2(e^{\pi}+e^{-\pi})} + \frac{1}{2^2(e^{2\pi}+e^{-2\pi})} + \frac{1}{3^2(e^{3\pi}+e^{-3\pi})} + \dots\right\}$$

$$= \cdot 9159655942, \quad \dots\dots(27)$$

approximately.

8

ON THE NUMBER OF DIVISORS OF A NUMBER

(*Journal of the Indian Mathematical Society*, VII, 1915, 131—133)

1. If δ be a divisor of N, then there is a conjugate divisor δ' such that $\delta\delta' = N$. Thus we see that

the number of divisors from 1 to $\sqrt{N}$ is equal to the number of divisors from $\sqrt{N}$ to N. ....(1)

From this it evidently follows that

$$d(N) < 2\sqrt{N}, \qquad \text{.........(2)}$$

where $d(N)$ denotes the number of divisors of N (including unity and the number itself). This is only a trivial result, as all the numbers from 1 to $\sqrt{N}$ cannot be divisors of N. So let us try to find the best possible superior limit for $d(N)$ by using purely elementary reasoning.

2. First let us consider the case in which all the prime divisors of N are known. Let

$$N = p_1^{a_1} . p_2^{a_2} . p_3^{a_3} \dots p_n^{a_n},$$

where $p_1, p_2, p_3 \dots p_n$ are a given set of n primes. Then it is easy to see that

$$d(N) = (1 + a_1)(1 + a_2)(1 + a_3) \dots (1 + a_n). \qquad \text{.........(3)}$$

But $$\frac{1}{n}\{(1 + a_1)\log p_1 + (1 + a_2)\log p_2 + \dots + (1 + a_n)\log p_n\}$$

$$> \{(1 + a_1)(1 + a_2) \dots (1 + a_n)\log p_1 \log p_2 \dots \log p_n\}^{\frac{1}{n}}, \qquad \text{......(4)}$$

since the arithmetic mean of unequal positive numbers is always greater than their geometric mean. Hence

$$\frac{1}{n}\{\log p_1 + \log p_2 + \dots + \log p_n + \log N\} > \{\log p_1 \log p_2 \dots \log p_n . d(N)\}^{\frac{1}{n}}.$$

In other words

$$d(N) < \frac{\left\{\frac{1}{n}\log(p_1 p_2 p_3 \dots p_n N)\right\}^n}{\log p_1 \log p_2 \log p_3 \dots \log p_n}, \qquad \text{.........(5)}$$

for all values of N whose prime divisors are $p_1, p_2, p_3 \dots p_n$.

3. Next let us consider the case in which only the number of prime divisors of N is known. Let

$$N = p_1^{a_1} p_2^{a_2} p_3^{a_3} \dots p_n^{a_n},$$

where n is a given number; and let

$$N' = 2^{a_1} . 3^{a_2} . 5^{a_3} \dots p^{a_n},$$

where p is the natural nth prime. Then it is evident that

$$N' \leqslant N; \qquad \text{.....(6)}$$

and

$$d(N') = d(N). \qquad \text{.....(7)}$$

But

$$d(N') < \frac{\left\{\frac{1}{n}\log(2.3.5\ldots p.N')\right\}^n}{\log 2 \log 3 \log 5 \ldots \log p}, \qquad \text{.....(8)}$$

by virtue of (5). It follows from (6) to (8) that, if p be the natural nth prime, then

$$d(N) < \frac{\left\{\frac{1}{n}\log(2.3.5\ldots p.N)\right\}^n}{\log 2 \log 3 \log 5 \ldots \log p}, \qquad \text{.....(9)}$$

for all values of N having n prime divisors.

4. Finally, let us consider the case in which nothing is known about N. Any integer N can be written in the form

$$2^{a_2}.\,3^{a_3}.\,5^{a_5}\ldots,$$

where $a_\lambda \geqslant 0$. Now let

$$x^h = 2, \qquad \text{.....(10)}$$

where h is any positive number. Then we have

$$\frac{d(N)}{N^h} = \frac{1+a_2}{2^{ha_2}} \cdot \frac{1+a_3}{3^{ha_3}} \cdot \frac{1+a_5}{5^{ha_5}} \cdot \qquad \text{.....(11)}$$

But from (10) we see that, if q be any prime greater than x, then

$$\frac{1+a_q}{q^{ha_q}} \leqslant \frac{1+a_q}{x^{ha_q}} = \frac{1+a_q}{2^{a_q}} \leqslant 1. \qquad \text{.....(12)}$$

It follows from (11) and (12) that, if p be the largest prime not exceeding x, then

$$\frac{d(N)}{N^h} \leqslant \frac{1+a_2}{2^{ha_2}} \cdot \frac{1+a_3}{3^{ha_3}} \cdot \frac{1+a_5}{5^{ha_5}} \cdots \frac{1+a_p}{p^{ha_p}}$$
$$\leqslant \frac{1+a_2}{2^{ha_2}} \cdot \frac{1+a_3}{2^{ha_3}} \cdot \frac{1+a_5}{2^{ha_5}} \cdots \frac{1+a_p}{2^{ha_p}} \cdot \qquad \text{.....(13)}$$

But it is easy to shew that the maximum value of $(1+a)\,2^{-ha}$ for the variable a is $\dfrac{2^h}{he \log 2}$. Hence

$$\frac{d(N)}{N^h} \leqslant \left(\frac{2^h}{he \log 2}\right)^{\varpi(x)}, \qquad \text{.....(14)}$$

where $\varpi(x)$ denotes the number of primes not exceeding x. But from (10) we have

$$h = \frac{\log 2}{\log x}.$$

Substituting this in (14), we obtain

$$d(N) \leqslant N^{\frac{\log 2}{\log x}} \left\{ \frac{2^{\frac{\log 2}{\log x}} \log x}{e(\log 2)^2} \right\}^{\omega(x)} . \quad \text{.................}(15)$$

But it is easy to verify that, if $x \geqslant 6{\cdot}05$, then

$$2^h < e(\log 2)^2.$$

From this and (15) it follows that, if $x \geqslant 6{\cdot}05$, then

$$d(N) < 2^{(\log N)/(\log x)} (\log x)^{\omega(x)} \quad \text{.....................}(16)$$

for all values of N, $\omega(x)$ being the number of primes not exceeding x.

5. The symbol "O" is used in the following sense:

$$\phi(x) = O\{\psi(x)\}$$

means that there is a positive constant K such that

$$\left| \frac{\phi(x)}{\psi(x)} \right| \leqslant K$$

for all sufficiently large values of x (see Hardy, *Orders of Infinity*, pp. 5 *et seq.*). For example:

$$5x = O(x); \quad \tfrac{1}{2}x = O(x); \quad x \sin x = O(x); \quad \sqrt{x} = O(x); \quad \log x = O(x);$$

but

$$x^2 \neq O(x); \quad x \log x \neq O(x).$$

Hence it is obvious that

$$\omega(x) = O(x). \quad \text{...............................}(17)$$

Now, let us suppose that

$$x = \frac{\log N}{(\log\log N)^2}$$

in (16). Then we have

$$\log x = \log\log N + O(\log\log\log N);$$

and so

$$\frac{\log N}{\log x} = \frac{\log N}{\log\log N} + O\left\{\frac{\log N \log\log\log N}{(\log\log N)^2}\right\}. \quad \text{.........}(18)$$

Again

$$\omega(x) \log\log x = O(x \log\log x) = O\left\{\frac{\log N \log\log\log N}{(\log\log N)^2}\right\}. \quad \text{...}(19)$$

It follows from (16), (18) and (19), that

$$\log d(N) < \frac{\log 2 \log N}{\log\log N} + O\left\{\frac{\log N \log\log\log N}{(\log\log N)^2}\right\} \text{...........}(20)$$

for all sufficiently large values of N.

9

ON THE SUM OF THE SQUARE ROOTS OF THE FIRST n NATURAL NUMBERS

(*Journal of the Indian Mathematical Society*, VII, 1915, 173—175)

1. Let

$$\phi_1(n) = \sqrt{1} + \sqrt{2} + \sqrt{3} + \ldots + \sqrt{n} - (C_1 + \tfrac{2}{3}n\sqrt{n} + \tfrac{1}{2}\sqrt{n}) - \tfrac{1}{6}\sum_{\nu=0}^{\nu=\infty}\{\sqrt{(n+\nu)} + \sqrt{(n+\nu+1)}\}^{-3},$$

where C_1 is a constant such that $\phi_1(1) = 0$. Then we see that

$$\phi_1(n) - \phi_1(n+1) = -\sqrt{(n+1)} + [\tfrac{2}{3}(n+1)\sqrt{(n+1)} + \tfrac{1}{2}\sqrt{(n+1)}] - (\tfrac{2}{3}n\sqrt{n} + \tfrac{1}{2}\sqrt{n}) + \tfrac{1}{6}\{\sqrt{n} - \sqrt{(n+1)}\}^3 = 0.$$

But $\phi_1(1) = 0$. Hence $\phi_1(n) = 0$ for all values of n. That is to say

$$\sqrt{1} + \sqrt{2} + \sqrt{3} + \sqrt{4} + \ldots + \sqrt{n} = C_1 + \tfrac{2}{3}n\sqrt{n} + \tfrac{1}{2}\sqrt{n} + \tfrac{1}{6}[\{\sqrt{n} + \sqrt{(n+1)}\}^{-3} + \{\sqrt{(n+1)} + \sqrt{(n+2)}\}^{-3} + \{\sqrt{(n+2)} + \sqrt{(n+3)}\}^{-3} + \ldots]. \quad \ldots\ldots\ldots(1)$$

But it is known that

$$C_1 = -\frac{1}{4\pi}\left(\frac{1}{1\sqrt{1}} + \frac{1}{2\sqrt{2}} + \frac{1}{3\sqrt{3}} + \ldots\right). \quad \ldots\ldots\ldots\ldots\ldots(2)$$

Putting $n = 1$ in (1) and using (2), we obtain

$$2\pi\left\{\frac{1}{(\sqrt{1})^3} + \frac{1}{(\sqrt{1}+\sqrt{2})^3} + \frac{1}{(\sqrt{2}+\sqrt{3})^3} + \frac{1}{(\sqrt{3}+\sqrt{4})^3} + \ldots\right\} = 3\left\{\frac{1}{(\sqrt{1})^3} + \frac{1}{(\sqrt{2})^3} + \frac{1}{(\sqrt{3})^3} + \frac{1}{(\sqrt{4})^3} + \ldots\right\}. \quad \ldots\ldots(3)$$

2. Again let

$$\phi_2(n) = 1\sqrt{1} + 2\sqrt{2}\ldots + n\sqrt{n} - (C_2 + \tfrac{2}{5}n^2\sqrt{n} + \tfrac{1}{2}n\sqrt{n} + \tfrac{1}{8}\sqrt{n}) - \tfrac{1}{40}\sum_{\nu=0}^{\nu=\infty}[\sqrt{(n+\nu)} + \sqrt{(n+\nu+1)}]^{-5},$$

where C_2 is a constant such that $\phi_2(1) = 0$. Then we have

$$\phi_2(n) - \phi_2(n+1) = -(n+1)\sqrt{(n+1)} + \{\tfrac{2}{5}(n+1)^2\sqrt{(n+1)} + \tfrac{1}{2}(n+1)\sqrt{(n+1)} + \tfrac{1}{8}\sqrt{(n+1)}\} - \{\tfrac{2}{5}n^2\sqrt{n} + \tfrac{1}{2}n\sqrt{n} + \tfrac{1}{8}\sqrt{n}\} + \tfrac{1}{40}\{\sqrt{n} - \sqrt{(n+1)}\}^5 = 0.$$

But $\phi_2(1) = 0$. Hence $\phi_2(n) = 0$. In other words

$$1\sqrt{1} + 2\sqrt{2} + 3\sqrt{3} + \ldots n\sqrt{n} = C_2 + \tfrac{2}{5}n^2\sqrt{n} + \tfrac{1}{2}n\sqrt{n} + \tfrac{1}{8}\sqrt{n} + \tfrac{1}{40}[\{\sqrt{n} + \sqrt{(n+1)}\}^{-5} + \{\sqrt{(n+1)} + \sqrt{(n+2)}\}^{-5} + \{\sqrt{(n+2)} + \sqrt{(n+3)}\}^{-5} + \ldots]. \quad \ldots\ldots\ldots\ldots\ldots(4)$$

But it is known that

$$C_2 = -\frac{3}{16\pi^2}\left(\frac{1}{1^2\sqrt{1}} + \frac{1}{2^2\sqrt{2}} + \frac{1}{3^2\sqrt{3}} + \dots\right). \qquad \dots\dots\dots\dots\dots(5)$$

It is easy to see from (4) and (5) that

$$2\pi^2\left\{\frac{1}{(\sqrt{1})^5} + \frac{1}{(\sqrt{1}+\sqrt{2})^5} + \frac{1}{(\sqrt{2}+\sqrt{3})^5} + \frac{1}{(\sqrt{3}+\sqrt{4})^5} + \dots\right\}$$
$$= 15\left\{\frac{1}{(\sqrt{1})^5} + \frac{1}{(\sqrt{2})^5} + \frac{1}{(\sqrt{3})^5} + \frac{1}{(\sqrt{4})^5} + \dots\right\}. \qquad \dots\dots(6)$$

3. The corresponding results for higher powers are not so neat as the previous ones. Thus for example

$$1^2\sqrt{1} + 2^2\sqrt{2} + 3^2\sqrt{3} + \dots + n^2\sqrt{n} = C_3 + \sqrt{n}\left(\tfrac{2}{7}n^3 + \tfrac{1}{2}n^2 + \tfrac{5}{24}n\right)$$
$$-\tfrac{1}{96}\left[\{\sqrt{n}+\sqrt{(n+1)}\}^{-3} + \{\sqrt{(n+1)}+\sqrt{(n+2)}\}^{-3} + \dots\right]$$
$$+\tfrac{1}{224}\left[\{\sqrt{n}+\sqrt{(n+1)}\}^{-7} + \{\sqrt{(n+1)}+\sqrt{(n+2)}\}^{-7}\right.$$
$$\left. + \{\sqrt{(n+2)}+\sqrt{(n+3)}\}^{-7} + \dots\right]; \qquad \dots\dots\dots(7)$$

$$1^3\sqrt{1} + 2^3\sqrt{2} + \dots + n^3\sqrt{n} = C_4 + \sqrt{n}\left(\tfrac{2}{9}n^4 + \tfrac{1}{2}n^3 + \tfrac{7}{24}n^2 - \tfrac{7}{384}\right)$$
$$-\tfrac{1}{192}\left[\{\sqrt{n}+\sqrt{(n+1)}\}^{-5} + \{\sqrt{(n+1)}+\sqrt{(n+2)}\}^{-5} + \dots\right]$$
$$+\tfrac{1}{1152}\left[\{\sqrt{n}+\sqrt{(n+1)}\}^{-9} + \{\sqrt{(n+1)}+\sqrt{(n+2)}\}^{-9} + \dots\right]; \dots(8)$$

and so on.

The constants $C_3, C_4, \dots$ can be ascertained from the well-known result that *the constant in the approximate summation of the series* $1^{r-1} + 2^{r-1} + 3^{r-1} + \dots + n^{r-1}$ *is*

$$\frac{2\Gamma(r)}{(2\pi)^r}\left(\frac{1}{1^r} + \frac{1}{2^r} + \frac{1}{3^r} + \frac{1}{4^r} + \dots\right)\cos\tfrac{1}{2}\pi r, \qquad \dots\dots\dots\dots\dots(9)$$

provided that the real part of r is greater than 1.

4. Similarly we can shew, by induction, that

$$\frac{1}{\sqrt{1}} + \frac{1}{\sqrt{2}} + \frac{1}{\sqrt{3}} + \dots + \frac{1}{\sqrt{n}} = C_0 + 2\sqrt{n} + \frac{1}{2\sqrt{n}}$$
$$-\tfrac{1}{2}\left\{\frac{\{\sqrt{n}+\sqrt{(n+1)}\}^{-3}}{\sqrt{\{n(n+1)\}}} + \frac{\{\sqrt{(n+1)}+\sqrt{(n+2)^{-3}}\}}{\sqrt{\{(n+1)(n+2)\}}} + \dots\right\}. \dots\dots(10)$$

The value of C_0 can be determined as follows: from (10) we have

$$\frac{1}{\sqrt{1}} + \frac{1}{\sqrt{2}} + \frac{1}{\sqrt{3}} + \dots + \frac{1}{\sqrt{(2n)}} - 2\sqrt{(2n)} \to C_0, \qquad \dots\dots\dots(11)$$

as $n \to \infty$. Also

$$2\left(\frac{1}{\sqrt{2}} + \frac{1}{\sqrt{4}} + \frac{1}{\sqrt{6}} + \dots + \frac{1}{\sqrt{(2n)}}\right) - 2\sqrt{(2n)} \to C_0\sqrt{2}, \dots\dots(12)$$

as $n \to \infty$.

Now subtracting (12) from (11) we see that

$$\frac{1}{\sqrt{1}}-\frac{1}{\sqrt{2}}+\frac{1}{\sqrt{3}}-\frac{1}{\sqrt{4}}+\ldots-\frac{1}{\sqrt{(2n)}}\rightarrow C_0(1-\sqrt{2}), \text{ as } n\rightarrow\infty.$$

That is to say

$$C_0=-(1+\sqrt{2})\left(\frac{1}{\sqrt{1}}-\frac{1}{\sqrt{2}}+\frac{1}{\sqrt{3}}-\frac{1}{\sqrt{4}}+\ldots\right). \qquad \ldots\ldots\ldots(13)$$

We can also shew, by induction, that

$$\sqrt{1}+\sqrt{2}+\sqrt{3}+\ldots+\sqrt{n}=C_1+\frac{2}{3}n\sqrt{n}+\frac{1}{2}\sqrt{n}+\frac{1}{24\sqrt{n}}$$
$$-\frac{1}{24}\left[\frac{\{\sqrt{n}+\sqrt{(n+1)}\}^{-5}}{\sqrt{\{n(n+1)\}}}+\frac{\{\sqrt{(n+1)}+\sqrt{(n+2)}\}^{-5}}{\sqrt{\{(n+1)(n+2)\}}}+\ldots\right]. \quad \ldots(14)$$

The asymptotic expansion of $\sqrt{1}+\sqrt{2}+\sqrt{3}+\ldots+\sqrt{n}$ for large values of n can be shewn to be

$$C_1+\frac{2}{3}n\sqrt{n}+\frac{1}{2}\sqrt{n}+\frac{1}{\sqrt{n}}\left(\frac{1}{24}-\frac{1}{1920n^2}+\frac{1}{9216n^4}-\ldots\right), \quad \ldots\ldots(15)$$

by using the Euler-Maclaurin sum formula.

10

ON THE PRODUCT $\prod_{n=0}^{n=\infty}\left[1+\left(\frac{x}{a+nd}\right)^3\right]$

(*Journal of the Indian Mathematical Society*, VII, 1915, 209—211)

1. Let
$$\phi(\alpha,\beta)=\left\{1+\left(\frac{\alpha+\beta}{1+\alpha}\right)^3\right\}\left\{1+\left(\frac{\alpha+\beta}{2+\alpha}\right)^3\right\}\dots \quad\text{.............(1)}$$

It is easy to see that

$$\left\{1+\left(\frac{\alpha+\beta}{n+\alpha}\right)^3\right\}\left\{1+\left(\frac{\alpha+\beta}{n+\beta}\right)^3\right\}$$
$$=\frac{\left(1+\frac{\alpha+2\beta}{n}\right)\left(1+\frac{\beta+2\alpha}{n}\right)}{\left(1+\frac{\alpha}{n}\right)^3\left(1+\frac{\beta}{n}\right)^3}\left[1-\left\{\frac{(\alpha-\beta)+i(\alpha+\beta)\sqrt{3}}{2n}\right\}^2\right]$$
$$\times\left[1-\left\{\frac{(\alpha-\beta)-i(\alpha+\beta)\sqrt{3}}{2n}\right\}^2\right];\quad\text{.......................(2)}$$

$$\prod_{n=1}^{n=\infty}\left\{\frac{\left(1+\frac{\alpha+2\beta}{n}\right)\left(1+\frac{\beta+2\alpha}{n}\right)}{\left(1+\frac{\alpha}{n}\right)^3\left(1+\frac{\beta}{n}\right)^3}\right\}=\frac{\{\Gamma(1+\alpha)\Gamma(1+\beta)\}^3}{\Gamma(1+\alpha+2\beta)\Gamma(1+\beta+2\alpha)};\quad\text{...(3)}$$

and

$$\prod_{n=1}^{n=\infty}\left[1-\left\{\frac{(\alpha-\beta)+i(\alpha+\beta)\sqrt{3}}{2n}\right\}^2\right]\left[1-\left\{\frac{(\alpha-\beta)-i(\alpha+\beta)\sqrt{3}}{2n}\right\}^2\right]$$
$$=\frac{\cosh\pi(\alpha+\beta)\sqrt{3}-\cos\pi(\alpha-\beta)}{2\pi^2(\alpha^2+\alpha\beta+\beta^2)}.\quad\text{..................(4)}$$

It follows from (1)—(4) that

$$\phi(\alpha,\beta)\,\phi(\beta,\alpha)$$
$$=\frac{\{\Gamma(1+\alpha)\Gamma(1+\beta)\}^3}{\Gamma(1+\alpha+2\beta)\Gamma(1+\beta+2\alpha)}\left\{\frac{\cosh\pi(\alpha+\beta)\sqrt{3}-\cos\pi(\alpha-\beta)}{2\pi^2(\alpha^2+\alpha\beta+\beta^2)}\right\}.\quad\text{...(5)}$$

But it is evident that, if $\alpha-\beta$ be any integer, then $\phi(\alpha,\beta)/\phi(\beta,\alpha)$ can be expressed in finite terms. From this and (5) it follows that $\phi(\alpha,\beta)$ can be expressed in finite terms, if $\alpha-\beta$ be any integer. That is to say

$$\left\{1+\left(\frac{x}{a}\right)^3\right\}\left\{1+\left(\frac{x}{a+d}\right)^3\right\}\left\{1+\left(\frac{x}{a+2d}\right)^3\right\}\dots$$

can be expressed in finite terms if $x-2a$ be a multiple of d.

2. Suppose now that $\alpha=\beta$ in (5). We obtain

$$\left\{1+\left(\frac{2\alpha}{1+\alpha}\right)^3\right\}\left\{1+\left(\frac{2\alpha}{2+\alpha}\right)^3\right\}\left\{1+\left(\frac{2\alpha}{3+\alpha}\right)^3\right\}\cdots$$
$$=\frac{\{\Gamma(1+\alpha)\}^3}{\Gamma(1+3\alpha)}\frac{\sinh\pi\alpha\sqrt{3}}{\pi\alpha\sqrt{3}}. \quad \text{.............(6)}$$

Similarly, putting $\beta=\alpha+1$ in (5), we obtain

$$\left\{1+\left(\frac{2\alpha+1}{1+\alpha}\right)^3\right\}\left\{1+\left(\frac{2\alpha+1}{2+\alpha}\right)^3\right\}\cdots$$
$$=\frac{\{\Gamma(1+\alpha)\}^3}{\Gamma(2+3\alpha)}\frac{\cosh\pi(\frac{1}{2}+\alpha)\sqrt{3}}{\pi}. \quad \text{......(7)}$$

Again, since

$$\left\{1+\left(\frac{\alpha}{n}\right)^3\right\}\left\{1+3\left(\frac{\alpha}{2n+\alpha}\right)^2\right\}=\frac{\left(1+\frac{\alpha}{n}\right)\left(1+\frac{\alpha^2}{n^2}+\frac{\alpha^4}{n^4}\right)}{\left(1+\frac{\alpha}{2n}\right)^2},$$

it is easy to see that

$$\left[\left(1+\frac{\alpha^3}{1^3}\right)\left(1+\frac{\alpha^3}{2^3}\right)\cdots\right]\left[\left\{1+3\left(\frac{\alpha}{2+\alpha}\right)^2\right\}\left\{1+3\left(\frac{\alpha}{4+\alpha}\right)^2\right\}\cdots\right]$$
$$=\frac{\Gamma(\frac{1}{2}\alpha)}{\Gamma\{\frac{1}{2}(1+\alpha)\}}\left(\frac{\cosh\pi\alpha\sqrt{3}-\cos\pi\alpha}{2^{\alpha+2}\pi\alpha\sqrt{\pi}}\right). \quad \text{..................(8)}$$

3. It is known that, if the real part of α is positive, then

$$\log\Gamma(\alpha)=(\alpha-\tfrac{1}{2})\log\alpha-\alpha+\tfrac{1}{2}\log 2\pi+2\int_0^\infty\frac{\tan^{-1}(x/\alpha)}{e^{2\pi x}-1}dx. \quad \text{...(9)}$$

From this we can shew that, if the real part of α is positive, then

$$\tfrac{1}{2}\log 2\pi\alpha+\frac{\pi\alpha}{\sqrt{3}}+\log\left\{\left(1+\frac{\alpha^3}{1^3}\right)\left(1+\frac{\alpha^3}{2^3}\right)\left(1+\frac{\alpha^3}{3^3}\right)\cdots\right\}$$
$$=\log\left(\frac{\cosh\pi\alpha\sqrt{3}-\cos\pi\alpha}{\pi\alpha}\right)+2\int_0^\infty\frac{\tan^{-1}(x/\alpha)^3}{e^{2\pi x}-1}dx. \quad \text{.........(10)}$$

From this and the previous section it follows that

$$\int_0^\infty\frac{\tan^{-1}x^3}{e^{2\pi nx}-1}dx$$

can be expressed in finite terms if n is a positive integer. Thus, for example,

$$\int_0^\infty\frac{\tan^{-1}x^3}{e^{2\pi x}-1}dx=\tfrac{1}{4}\log 2\pi-\frac{\pi}{4\sqrt{3}}-\tfrac{1}{2}\log(1+e^{-\pi\sqrt{3}}); \quad \text{.........(11)}$$

$$\int_0^\infty\frac{\tan^{-1}x^3}{e^{4\pi x}-1}dx=\tfrac{1}{8}\log 12\pi-\frac{\pi}{4\sqrt{3}}-\tfrac{1}{4}\log(1-e^{-2\pi\sqrt{3}}); \quad \text{......(12)}$$

and so on.

4. It is also easy to see that

$$\frac{1^2}{1^3+n^3}-\frac{2^2}{2^3+n^3}+\frac{3^2}{3^3+n^3}-\frac{4^2}{4^3+n^3}+\dots$$
$$=\frac{1}{3}\left(\frac{1}{1+n}-\frac{1}{2+n}+\frac{1}{3+n}-\frac{1}{4+n}+\dots\right)$$
$$+\frac{4}{3}\left\{\frac{2-n}{(2-n)^2+3n^2}-\frac{4-n}{(4-n)^2+3n^2}+\frac{6-n}{(6-n)^2+3n^2}-\dots\right\}. \quad \dots(13)$$

Since
$$\frac{\pi}{4\cosh\frac{1}{2}\pi x}=\frac{1}{1^2+x^2}-\frac{3}{3^2+x^2}+\frac{5}{5^2+x^2}-\dots,$$

it is clear that the left-hand side of (13) can be expressed in finite terms if n is any odd integer. For example,

$$\frac{1^2}{1^3+1}-\frac{2^2}{2^3+1}+\frac{3^2}{3^3+1}-\frac{4^2}{4^3+1}+\dots=\frac{1}{3}(1-\log 2+\pi\operatorname{sech}\tfrac{1}{2}\pi\sqrt{3}). \quad (14)$$

The corresponding integral in this case is

$$\int_0^\infty \frac{x^5}{\sinh \pi x}\frac{dx}{n^6+x^6}=\frac{2}{\pi}\int_0^\infty \left\{\frac{1}{2x^2}+\sum_{\nu=1}^{\nu=\infty}\frac{(-1)^\nu}{\nu^2+x^2}\right\}\frac{x^6\,dx}{n^6+x^6}$$
$$=\frac{1}{3}\left(\frac{1}{n}-\frac{1}{n+1}+\frac{1}{n+2}-\frac{1}{n+3}+\dots\right)$$
$$-\frac{4}{3}\left\{\frac{n+2}{(n+2)^2+3n^2}-\frac{n+4}{(n+4)^2+3n^2}+\frac{n+6}{(n+6)^2+3n^2}-\dots\right\}; \quad \dots(15)$$

and so the integral on the left-hand side of (15) can be expressed in finite terms if n is any odd integer. For example,

$$\int_0^\infty \frac{x^5}{\sinh\pi x}\frac{dx}{1+x^6}=\frac{1}{3}(\log 2-1+\pi\operatorname{sech}\tfrac{1}{2}\pi\sqrt{3}). \quad \dots\dots\dots(16)$$

11

SOME DEFINITE INTEGRALS

(*Messenger of Mathematics*, XLIV, 1915, 10—18)

1. Consider the integral

$$\int_0^\infty \frac{\cos 2mx\,dx}{\{1+x^2/a^2\}\{1+x^2/(a+1)^2\}\{1+x^2/(a+2)^2\}\ldots},$$

where m and a are positive.

It can be easily proved that

$$\left\{1-\left(\frac{t}{a}\right)^2\right\}\left\{1-\left(\frac{t}{a+1}\right)^2\right\}\left\{1-\left(\frac{t}{a+2}\right)^2\right\}\ldots\left\{1-\left(\frac{t}{a+n-1}\right)^2\right\}$$
$$=\frac{\Gamma(a+n-t)\,\Gamma(a+n+t)\,\{\Gamma(a)\}^2}{\Gamma(a-t)\,\Gamma(a+t)\,\{\Gamma(a+n)\}^2},$$

where n is any positive integer. Hence, by splitting

$$\frac{1}{\{1+x^2/a^2\}\{1+x^2/(a+1)^2\}\ldots\{1+x^2/(a+n-1)^2\}}$$

into partial fractions, we see that it is equal to

$$\frac{2\Gamma(2a)\{\Gamma(a+n)\}^2}{\{\Gamma(a)\}^2\,\Gamma(n)\,\Gamma(2a+n)}\left\{\frac{a}{a^2+x^2}-\frac{2a}{1\,!}\,\frac{n-1}{n+2a}\,\frac{a+1}{(a+1)^2+x^2}\right.$$
$$\left.+\frac{2a(2a+1)}{2\,!}\,\frac{(n-1)(n-2)}{(n+2a)(n+2a+1)}\,\frac{a+2}{(a+2)^2+x^2}-\ldots\right\}.$$

Multiplying both sides by $\cos 2mx$ and integrating from 0 to ∞ with respect to x, we have

$$\int_0^\infty \frac{\cos 2mx\,dx}{\{1+x^2/a^2\}\{1+x^2/(a+1)^2\}\ldots\{1+x^2/(a+n-1)^2\}}$$
$$=\frac{\pi\Gamma(2a)\{\Gamma(a+n)\}^2}{\{\Gamma(a)\}^2\,\Gamma(n)\,\Gamma(2a+n)}\left\{e^{-2am}-\frac{2a}{1\,!}\,\frac{n-1}{n+2a}\,e^{-2(a+1)m}+\ldots\right\}.$$

The limit of the right-hand side, as $n\to\infty$, is

$$\frac{\pi\Gamma(2a)}{\{\Gamma(a)\}^2}\left\{e^{-2am}-\frac{2a}{1\,!}\,e^{-2(a+1)m}+\frac{2a(2a+1)}{2\,!}\,e^{-2(a+2)m}-\ldots\right\}$$
$$=\tfrac{1}{2}\sqrt{\pi}\,\frac{\Gamma(a+\frac{1}{2})}{\Gamma(a)}\,\operatorname{sech}^{2a} m.$$

Hence $$\int_0^\infty \frac{\cos 2mx\,dx}{\{1+x^2/a^2\}\{1+x^2/(a+1)^2\}\ldots}=\tfrac{1}{2}\sqrt{\pi}\,\frac{\Gamma(a+\frac{1}{2})}{\Gamma(a)}\,\operatorname{sech}^{2a} m. \quad\ldots\ldots(1)$$

Since $\left\{1+\left(\frac{x}{a}\right)^2\right\}\left\{1+\left(\frac{x}{a+1}\right)^2\right\}\left\{1+\left(\frac{x}{a+2}\right)^2\right\}\dots=\frac{\{\Gamma(a)\}^2}{\Gamma(a+ix)\Gamma(a-ix)}$,

the formula (1) is equivalent to

$$\int_0^\infty |\Gamma(a+ix)|^2 \cos 2mx\,dx = \tfrac{1}{2}\sqrt{\pi}\,\Gamma(a)\,\Gamma(a+\tfrac{1}{2})\operatorname{sech}^{2a} m. \quad \dots\dots(2)$$

2. In a similar manner we can prove that

$$\int_0^\infty \left(\frac{1+x^2/b^2}{1+x^2/a^2}\right)\left(\frac{1+x^2/(b+1)^2}{1+x^2/(a+1)^2}\right)\left(\frac{1+x^2/(b+2)^2}{1+x^2/(a+2)^2}\right)\dots\cos mx\,dx$$

$$=\frac{\pi\Gamma(2a)\{\Gamma(b)\}^2}{\{\Gamma(a)\}^2\Gamma(b+a)\Gamma(b-a)}\left\{e^{-am}-\frac{2a}{1!}\,\frac{b-a-1}{b+a}\,e^{-(a+1)m}\right.$$

$$\left.+\frac{2a(2a+1)}{2!}\,\frac{(b-a-1)(b-a-2)}{(b+a)(b+a+1)}\,e^{-(a+2)m}-\dots\right\},$$

where m is positive and $0<a<b$. When $0<a<b-\frac{1}{2}$, the integral and the series remain convergent for $m=0$, and we obtain the formulæ

$$\int_0^\infty \left(\frac{1+x^2/b^2}{1+x^2/a^2}\right)\left(\frac{1+x^2/(b+1)^2}{1+x^2/(a+1)^2}\right)\left(\frac{1+x^2/(b+2)^2}{1+x^2/(a+2)^2}\right)\dots dx$$

$$=\tfrac{1}{2}\sqrt{\pi}\,\frac{\Gamma(a+\frac{1}{2})\Gamma(b)\Gamma(b-a-\frac{1}{2})}{\Gamma(a)\Gamma(b-\frac{1}{2})\Gamma(b-a)}, \quad \dots\dots(3)$$

$$\int_0^\infty \left|\frac{\Gamma(a+ix)}{\Gamma(b+ix)}\right|^2 dx=\tfrac{1}{2}\sqrt{\pi}\,\frac{\Gamma(a)\Gamma(a+\frac{1}{2})\Gamma(b-a-\frac{1}{2})}{\Gamma(b-\frac{1}{2})\Gamma(b)\Gamma(b-a)}. \quad \dots\dots(4)$$

If $a_1, a_2, a_3, \dots, a_n$ be n positive numbers in arithmetical progression, then

$$\int_0^\infty \frac{dx}{(a_1^2+x^2)(a_2^2+x^2)(a_3^2+x^2)\dots(a_n^2+x^2)}$$

is a particular case of the above integral, and its value can be written down at once. Thus, for example, by putting $a=\frac{11}{10}$ and $b=\frac{61}{10}$, we obtain

$$\int_0^\infty \frac{dx}{(x^2+11^2)(x^2+21^2)(x^2+31^2)(x^2+41^2)(x^2+51^2)}$$

$$=\frac{5\pi}{12.13.16.17.18.22.23.24.31.32.41}.$$

3. It follows at once from equation (1), by applying Fourier's theorem

$$\int_0^\infty \cos ny\,dy\int_0^\infty \phi(x)\cos xy\,dx=\tfrac{1}{2}\pi\phi(n),$$

that, when a and n are positive,

$$\int_0^\infty \operatorname{sech}^{2a} x\cos 2nx\,dx$$

$$=\tfrac{1}{2}\sqrt{\pi}\,\frac{\Gamma(a)}{\Gamma(a+\frac{1}{2})}\,\frac{1}{\{1+n^2/a^2\}\{1+n^2/(a+1)^2\}\{1+n^2/(a+2)^2\}\dots}$$

$$=\tfrac{1}{2}\sqrt{\pi}\,\frac{|\Gamma(a+in)|^2}{\Gamma(a)\Gamma(a+\frac{1}{2})}. \quad \dots\dots\dots\dots(5)$$

Hence the function

$$\phi(a)=\int_0^\infty \operatorname{sech}^a x \cos nx\, dx \quad (0<a<2)$$

satisfies the functional equation

$$\phi(a)\,\phi(2-a)=\frac{\pi \sin \pi a}{2(1-a)(\cosh \pi n-\cos \pi a)}.$$

4. Let

$$\int_a^b f(x)\,F(nx)\,dx=\psi(n),$$

and

$$\int_\alpha^\beta \phi(x)\,F(nx)\,dx=\chi(n).$$

Then, if we suppose the functions f, ϕ, and F to be such that the order of integration is indifferent, we have

$$\int_a^b f(x)\,\chi(nx)\,dx=\int_\alpha^\beta dy\int_a^b f(x)\,\phi(y)\,F(nxy)\,dx$$

$$=\int_\alpha^\beta \phi(y)\,\psi(ny)\,dy. \quad \text{.......................(6)}$$

A number of curious relations between definite integrals may be deduced from this result. We have, for example, the formulæ

$$\int_0^\infty \frac{\cos 2nx}{\cosh \pi x}\,dx=\frac{1}{2\cosh n}, \quad \text{..............................(7)}$$

$$\int_0^\infty \frac{\cos 2nx\,dx}{1+2\cosh \frac{2}{3}\pi x}=\frac{\sqrt{3}}{2(1+2\cosh 2n)}, \quad \text{..............(8)}$$

$$\int_0^\infty e^{-x^2}\cos 2nx\,dx=\tfrac{1}{2}\sqrt{\pi}\,e^{-n^2}. \quad \text{...........................(9)}$$

By applying the general result (6) to the integrals (7) and (8), we obtain

$$\sqrt{3}\int_0^\infty \frac{dx}{\cosh \pi x\,(1+2\cosh 2nx)}=\int_0^\infty \frac{dx}{\cosh nx\,(1+2\cosh \frac{2}{3}\pi x)};$$

or, in other words, if $\alpha\beta=\frac{2}{3}\pi^2$, then

$$\sqrt{\alpha}\int_0^\infty \frac{dx}{\cosh \alpha x\,(1+2\cosh \pi x)}$$

$$=\sqrt{\beta}\int_0^\infty \frac{dx}{\cosh \beta x\,(1+2\cosh \pi x)}. \quad \text{..................(10)}$$

In the same way, from (8) and (9), we obtain

$$\sqrt{\alpha}\int_0^\infty \frac{e^{-x^2}\,dx}{1+2\cosh \alpha x}=\sqrt{\beta}\int_0^\infty \frac{e^{-x^2}\,dx}{1+2\cosh \beta x}, \quad \text{...........(11)}$$

with the condition $\alpha\beta=\frac{4}{3}\pi$; and, from (7) and (9),

$$\sqrt{\alpha}\int_0^\infty \frac{e^{-x^2}}{\cosh \alpha x}\,dx=\sqrt{\beta}\int_0^\infty \frac{e^{-x^2}}{\cosh \beta x}\,dx, \quad \text{..............(12)}$$

with the condition $\alpha\beta=\pi$*.

* Formulæ equivalent to (11) and (12) were given by Hardy, *Quarterly Journal*, XXXV, p. 193.

Similarly, by taking the two integrals

$$\int_0^\infty \frac{\sin nx}{e^{2\pi x}-1}\,dx = \tfrac{1}{2}\left(\frac{1}{e^n-1}+\frac{1}{2}-\frac{1}{n}\right),$$

and

$$\int_0^\infty xe^{-x^2}\sin nx\,dx = \tfrac{1}{4}\sqrt{\pi}\,ne^{-\frac{1}{4}n^2},$$

we can prove that, if $\alpha\beta=\pi^2$, then

$$\frac{1}{\sqrt[4]{\alpha}}\left\{1+2\alpha\int_0^\infty \frac{e^{-\alpha x}}{e^{2\pi\sqrt{x}}-1}\,dx\right\}$$
$$=\frac{1}{\sqrt[4]{\beta}}\left\{1+2\beta\int_0^\infty \frac{e^{-\beta x}}{e^{2\pi\sqrt{x}}-1}\,dx\right\}; \quad \text{.....................(13)}$$

and so on.

5. Suppose now that a, b, and n are positive, and

$$\int_0^\infty \phi(a,x)\,{\cos \atop \sin}\,nx\,dx = \psi(a,n). \quad \text{.................(14)}$$

Then, if the conditions of Fourier's double integral theorem are satisfied, we have

$$\int_0^\infty \psi(b,x)\,{\cos \atop \sin}\,nx\,dx = \tfrac{1}{2}\pi\phi(b,n). \quad \text{.................(15)}$$

Applying the formula (6) to (14) and (15), we obtain

$$\tfrac{1}{2}\pi\int_0^\infty \phi(a,x)\,\phi(b,nx)\,dx = \int_0^\infty \psi(b,x)\,\psi(a,nx)\,dx. \quad \text{......(16)}$$

Thus, when $a=b$, we have the formula

$$\tfrac{1}{2}\pi\int_0^\infty \phi(x)\,\phi(nx)\,dx = \int_0^\infty \psi(x)\,\psi(nx)\,dx,$$

where

$$\psi(t) = \int_0^\infty \phi(x)\,{\cos \atop \sin}\,tx\,dx;$$

and, in particular, if $n=1$, then

$$\tfrac{1}{2}\pi\int_0^\infty \{\phi(x)\}^2\,dx = \int_0^\infty \{\psi(x)\}^2\,dx.$$

If

$$\phi(a,x) = \frac{1}{\{1+x^2/a^2\}\{1+x^2/(a+1)^2\}\dots} \quad (a>0),$$

then, by (1),

$$\psi(a,x) = \tfrac{1}{2}\sqrt{\pi}\,\frac{\Gamma(a+\frac{1}{2})}{\Gamma(a)}\,\mathrm{sech}^{2a}\,\tfrac{1}{2}x.$$

Hence, by (16),

$$\int_0^\infty \phi(a,x)\,\phi(b,x)\,dx = \frac{\Gamma(a+\frac{1}{2})\,\Gamma(b+\frac{1}{2})}{2\Gamma(a)\,\Gamma(b)}\int_0^\infty \mathrm{sech}^{2a+2b}\,\tfrac{1}{2}x\,dx;$$

and so

$$\int_0^\infty \frac{dx}{\{1+x^2/a^2\}\{1+x^2/(a+1)^2\}\ldots\{1+x^2/b^2\}\{1+x^2/(b+1)^2\}\ldots}$$

$$= \tfrac{1}{2}\sqrt{\pi}\,\frac{\Gamma(a+\frac{1}{2})\,\Gamma(b+\frac{1}{2})\,\Gamma(a+b)}{\Gamma(a)\,\Gamma(b)\,\Gamma(a+b+\frac{1}{2})}, \quad \ldots(17)$$

a and b being positive: or

$$\int_0^\infty |\,\Gamma(a+ix)\,\Gamma(b+ix)\,|^2 dx$$

$$= \tfrac{1}{2}\sqrt{\pi}\,\frac{\Gamma(a)\,\Gamma(a+\frac{1}{2})\,\Gamma(b)\,\Gamma(b+\frac{1}{2})\,\Gamma(a+b)}{\Gamma(a+b+\frac{1}{2})}. \quad \ldots(18)$$

As particular cases of the above result, we have, when $b=1$,

$$\int_0^\infty \frac{x}{\sinh \pi x}\,\frac{dx}{\{1+x^2/a^2\}\{1+x^2/(a+1)^2\}\ldots} = \frac{a}{2(1+2a)};$$

when $b=2$,

$$\int_0^\infty \frac{x^3}{\sinh \pi x}\,\frac{dx}{\{1+x^2/a^2\}\{1+x^2/(a+1)^2\}\ldots} = \frac{a^2}{2(1+2a)(3+2a)};$$

and so on. Since $\Pi\{1+x^2(a+n)^2\}$ can be expressed in finite terms by means of hyperbolic functions when $2a$ is an integer, we can deduce a large number of special formulæ from the preceding results.

6. Another curious formula is the following. If $0<r<1$, $n>0$, and $0<a<r^{n-1}$, then

$$\int_0^\infty \frac{(1+arx)(1+ar^2x)\ldots}{(1+x)(1+rx)(1+r^2x)\ldots}\,x^{n-1}dx$$

$$= \frac{\pi}{\sin n\pi}\prod_{m=1}^{m=\infty}\frac{(1-r^{m-n})(1-ar^m)}{(1-r^m)(1-ar^{m-n})}, \quad \ldots(19)$$

unless n is an integer or a is of the form r^p, where p is a positive integer.

If $a=r^p$, the formula reduces to

$$\int_0^\infty \frac{x^{n-1}dx}{(1+x)(1+rx)\ldots(1+r^px)}$$

$$= \frac{\pi}{\sin n\pi}\,\frac{(1-r^{1-n})(1-r^{2-n})\ldots(1-r^{p-n})}{(1-r)(1-r^2)\ldots(1-r^p)}. \quad \ldots(20)$$

If n is an integer, the value of the integral is in any case

$$-\frac{\log r}{1-a}\,\frac{(1-r)(1-r^2)\ldots(1-r^{n-1})}{(r-a)(r^2-a)\ldots(r^{n-1}-a)}.$$

My own proofs of the above results make use of a general formula, the truth of which depends on conditions which I have not yet investigated completely. A direct proof depending on Cauchy's theorem will be found in Mr Hardy's note which follows this paper. The final formula used in Mr Hardy's proof can be proved as follows. Let

$$f(t) = \prod_{m=0}^{m=\infty}\left(\frac{1-btx^m}{1-atx^m}\right) = A_0 + A_1t + A_2t^2 + \ldots.$$

Then it is evident that

$$(1-at)f(t)=(1-bt)f(tx).$$

That is

$$(1-at)(A_0+A_1t+A_2t^2+\ldots)=(1-bt)(A_0+A_1tx+A_2t^2x^2+\ldots).$$

Equating the coefficients of t^n, we obtain

$$A_n=A_{n-1}\frac{a-bx^{n-1}}{1-x^n};$$

and A_0 is evidently 1. Hence we have

$$f(t)=1+t\frac{a-b}{1-x}+t^2\frac{(a-b)(a-bx)}{(1-x)(1-x^2)}+\ldots. \qquad \ldots\ldots\ldots\ldots(21)$$

7. As a particular case of (19), we have, when $a=0$,

$$\int_0^\infty \frac{x^{n-1}dx}{(1+x)(1+rx)(1+r^2x)\ldots}=\frac{\pi}{\sin n\pi}\frac{1-r^{1-n}}{1-r}\frac{1-r^{2-n}}{1-r^2}\ldots. \qquad (22)$$

When n is an integer, the value of the integral reduces to

$$-r^{-\frac{1}{2}n(n-1)}(1-r)(1-r^2)\ldots(1-r^{n-1})\log r.$$

When we put $n=\frac{1}{2}$ in (19), we have

$$\int_0^\infty \frac{1}{1+x^2}\frac{1+ar^2x^2}{1+r^2x^2}\frac{1+ar^4x^2}{1+r^4x^2}\ldots dx$$

$$=\tfrac{1}{2}\pi\frac{1-ar^2}{1-r^2}\frac{1-ar^4}{1-r^4}\ldots\frac{1-r}{1-ar}\frac{1-r^3}{1-ar^3}\ldots. \qquad \ldots\ldots(23)$$

If, in particular, $n=\frac{1}{2}$ in (22), or $a=0$ in (23), then

$$\int_0^\infty \frac{dx}{(1+x^2)(1+r^2x^2)(1+r^4x^2)\ldots}$$

$$=\tfrac{1}{2}\pi\frac{1-r}{1-r^2}\frac{1-r^3}{1-r^4}\frac{1-r^5}{1-r^6}\ldots=\frac{\pi}{2(1+r+r^3+r^6+r^{10}+\ldots)}, \qquad (24)$$

the nth term in the denominator being $r^{\frac{1}{2}n(n-1)}$. Thus, for example, when $r=e^{-5\pi}$, we have

$$\int_0^\infty \frac{dx}{(1+x^2)(1+e^{-10\pi}x^2)(1+e^{-20\pi}x^2)\ldots}=\frac{\pi}{2(1+e^{-5\pi}+e^{-15\pi}+e^{-30\pi}+\ldots)}$$

$$=\pi^{\frac{3}{4}}\Gamma(\tfrac{3}{4})\sqrt{5}\,\sqrt[8]{2}\,\tfrac{1}{2}(1+\sqrt[4]{5})\{\tfrac{1}{2}(1+\sqrt{5})\}^{\frac{1}{2}}e^{-5\pi/8}.$$

Similarly

$$\int_0^\infty \frac{dx}{(1+x^2)(1+e^{-20\pi}x^2)(1+e^{-40\pi}x^2)\ldots}$$

$$=\pi^{\frac{3}{4}}\Gamma(\tfrac{3}{4})\sqrt{5}\,\sqrt[4]{2}\,\tfrac{1}{2}(1+\sqrt[4]{5})^2\{\tfrac{1}{2}(1+\sqrt{5})\}^{\frac{5}{2}}e^{-5\pi/4};$$

and

$$\int_0^\infty \frac{dx}{(1+x^2)(1+\cdot 001x^2)(1+\cdot 00001x^2)\ldots}=\tfrac{1}{2}\pi\frac{10}{11}\frac{1110}{1111}\frac{111110}{111111}\ldots$$

$$=\frac{\pi}{2\cdot 202\,002\,000\,200\,002\,000\,002\ldots}.$$

12

SOME DEFINITE INTEGRALS CONNECTED WITH GAUSS'S SUMS

(*Messenger of Mathematics*, XLIV, 1915, 75—85)

1. If n is real and positive, and $|I(t)|$, where $I(t)$ is the imaginary part of t, is less than either n or 1, we have

$$\int_0^\infty \frac{\cos \pi tx}{\cosh \pi x} e^{-i\pi nx^2} dx = 2\int_0^\infty \int_0^\infty \frac{\cos \pi tx \cos 2\pi xy}{\cosh \pi y} e^{-i\pi nx^2} dx\,dy$$

$$= \sqrt{n} \exp\left\{-\tfrac{1}{4}i\pi\left(1-\frac{t^2}{n}\right)\right\}\int_0^\infty \frac{\cos \pi tx}{\cosh \pi nx} e^{i\pi nx^2} dx. \quad \text{...(1)}$$

When $n=1$ the above formula reduces to

$$\int_0^\infty \frac{\cos \pi tx}{\cosh \pi x} \sin \pi x^2 dx = \tan\{\tfrac{1}{8}\pi(1-t^2)\}\int_0^\infty \frac{\cos \pi tx}{\cosh \pi x} \cos \pi x^2 dx. \quad \text{...(2)}$$

If $t=0$, and

$$\left.\begin{aligned} \phi(n) &= \int_0^\infty \frac{\cos \pi nx^2}{\cosh \pi x} dx, \\ \psi(n) &= \int_0^\infty \frac{\sin \pi nx^2}{\cosh \pi x} dx, \end{aligned}\right\} \quad \text{...(3)}$$

then

$$\left.\begin{aligned} \phi(n) &= \sqrt{\left(\frac{2}{n}\right)}\,\psi\left(\frac{1}{n}\right) + \psi(n), \\ \psi(n) &= \sqrt{\left(\frac{2}{n}\right)}\,\phi\left(\frac{1}{n}\right) - \phi(n). \end{aligned}\right\} \quad \text{...(3')}$$

Similarly, if $\frac{1}{2}\sqrt{3}|I(t)|$ is less than either 1 or n, we have

$$\int_0^\infty \frac{\cos \pi tx}{1+2\cosh(2\pi x/\sqrt{3})} e^{i\pi nx^2} dx$$

$$= \sqrt{n} \exp\left\{-\tfrac{1}{4}i\pi\left(1-\frac{t^2}{n}\right)\right\}\int_0^\infty \frac{\cos \pi tx}{1+2\cosh(2\pi nx/\sqrt{3})} e^{-i\pi nx^2} dx. \quad (4)$$

If in (4) we suppose $n=1$, we obtain

$$\int_0^\infty \frac{\cos \pi tx \sin \pi x^2}{1+2\cosh(2\pi x/\sqrt{3})} dx$$

$$= \tan\{\tfrac{1}{8}\pi(1-t^2)\}\int_0^\infty \frac{\cos \pi tx \cos \pi x^2}{1+2\cosh(2\pi x/\sqrt{3})} dx; \quad \text{...(5)}$$

and if $t = 0$, and

$$\left.\begin{aligned}\phi(n) &= \int_0^\infty \frac{\cos \pi n x^2}{1 + 2\cosh(2\pi x/\sqrt{3})}\,dx,\\ \psi(n) &= \int_0^\infty \frac{\sin \pi n x^2}{1 + 2\cosh(2\pi x/\sqrt{3})}\,dx,\end{aligned}\right\} \quad \text{.............(6)}$$

then

$$\left.\begin{aligned}\phi(n) &= \sqrt{\left(\frac{2}{n}\right)}\,\psi\left(\frac{1}{n}\right) + \psi(n),\\ \psi(n) &= \sqrt{\left(\frac{2}{n}\right)}\,\phi\left(\frac{1}{n}\right) - \phi(n).\end{aligned}\right\} \quad \text{....................(6')}$$

In a similar manner we can prove that

$$\int_0^\infty \frac{\sin \pi t x}{\tanh \pi x} e^{-i\pi n x^2}\,dx = -\sqrt{n}\exp\left\{\tfrac{1}{4}i\pi\left(1 + \frac{t^2}{n}\right)\right\}\int_0^\infty \frac{\sin \pi t x}{\tanh \pi n x} e^{i\pi n x^2}\,dx. \quad (7)$$

If we put $n = 1$ in (7), we obtain

$$\int_0^\infty \frac{\sin \pi t x}{\tanh \pi x} \cos \pi x^2\,dx = \tan\{\tfrac{1}{8}\pi(1 + t^2)\}\int_0^\infty \frac{\sin \pi t x}{\tanh \pi x} \sin \pi x^2\,dx. \quad \text{...(8)}$$

Now

$$\begin{aligned}\lim_{t\to 0} \frac{1}{t}\int_0^\infty \frac{\sin a t x}{\tanh b x} e^{icx^2}\,dx &= \lim_{t\to 0}\frac{1}{t}\int_0^\infty \frac{2\sin a t x}{e^{2bx} - 1} e^{icx^2}\,dx + \lim_{t\to 0}\int_0^\infty \frac{\sin a t x}{t} e^{icx^2}\,dx\\ &= \int_0^\infty \frac{a e^{icx}}{e^{2b\sqrt{x}} - 1}\,dx + \frac{ia}{2c}. \quad \text{.................(9)}\end{aligned}$$

Hence, dividing both sides of (7) by t, and making $t \to 0$, we obtain the result corresponding to (3) and (6), viz.: if

$$\left.\begin{aligned}\phi(n) &= \int_0^\infty \frac{\cos \pi n x}{e^{2\pi\sqrt{x}} - 1}\,dx,\\ \psi(n) &= \frac{1}{2\pi n} + \int_0^\infty \frac{\sin \pi n x}{e^{2\pi\sqrt{x}} - 1}\,dx,\end{aligned}\right\} \quad \text{....................(10)}$$

then

$$\left.\begin{aligned}\phi(n) &= \frac{1}{n}\sqrt{\left(\frac{2}{n}\right)}\,\psi\left(\frac{1}{n}\right) - \psi(n),\\ \psi(n) &= \frac{1}{n}\sqrt{\left(\frac{2}{n}\right)}\,\phi\left(\frac{1}{n}\right) + \phi(n).\end{aligned}\right\} \quad \text{.................(10')}$$

2. I shall now shew that the integral (1) may be expressed in finite terms for all rational values of n. Consider the integral

$$J(t) = \int_0^\infty \frac{\cos t x}{\cosh \frac{1}{2}\pi x}\,\frac{dx}{a^2 + x^2}.$$

If $R(a)$ and t are positive, we have

$$J(t) = \frac{4}{\pi}\int_0^\infty \sum_{r=0}^{r=\infty} \frac{(-1)^r(2r+1)}{x^2+(2r+1)^2}\frac{\cos tx}{a^2+x^2}dx$$

$$= 2\sum_{r=0}^{r=\infty} \frac{(-1)^r}{a^2-(2r+1)^2}\left\{e^{-(2r+1)t} - \frac{1}{a}(2r+1)e^{-at}\right\}$$

$$= \frac{\pi e^{-at}}{2a\cos\frac{1}{2}\pi a} + 2\sum_{r=0}^{r=\infty}\frac{(-1)^r e^{-(2r+1)t}}{a^2-(2r+1)^2}, \quad \text{...............(11)}$$

and it is easy to see that this last equation remains true when t is complex, provided $R(t) > 0$ and $|I(t)| \leqslant \frac{1}{2}\pi$. Thus the integral $J(t)$ can be expressed in finite terms for all rational values of a. Thus, for example, we have

$$\left.\begin{aligned}\int_0^\infty \frac{\cos tx}{\cosh\frac{1}{2}\pi x}\frac{dx}{1+x^2} &= \cosh t\log(2\cosh t) - t\sinh t,\\ \int_0^\infty \frac{\cos 2tx}{\cosh \pi x}\frac{dx}{1+x^2} &= 2\cosh t - (e^{2t}\tan^{-1}e^{-t} + e^{-2t}\tan^{-1}e^{t}),\end{aligned}\right\} \quad \text{...(12)}$$

and so on. Now let

$$F(n) = \int_0^\infty \frac{\cos 2tx}{\cosh \pi x} e^{-i\pi n x^2}dx. \quad \text{....................(13)}$$

Then, if $R(a) > 0$,

$$\int_0^\infty e^{-an}F(n)\,dn = \int_0^\infty \frac{\cos 2tx}{\cosh \pi x}\frac{dx}{a+i\pi x^2}. \quad \text{..............(14)}$$

Now let

$$f(n) = \sum_{r=0}^{r=\infty}(-1)^r \exp\{-(2r+1)t + \tfrac{1}{4}(2r+1)^2 i\pi n\}$$
$$+ \frac{1}{\sqrt{n}}\exp\left\{-i\left(\tfrac{1}{4}\pi - \frac{t^2}{\pi n}\right)\right\}\sum_{r=0}^{r=\infty}(-1)^r\exp\left\{-(2r+1)\frac{t}{n}\right.$$
$$\left. - \tfrac{1}{4}(2r+1)^2\frac{i\pi}{n}\right\}. \quad (15)$$

Then

$$\int_0^\infty e^{-an}f(n)\,dn = \sum_{r=0}^{r=\infty}\frac{(-1)^r e^{-(2r+1)t}}{a - \frac{1}{4}(2r+1)^2 i\pi}$$
$$+ \sqrt{\left(\frac{\pi}{2a}\right)}\frac{\exp\{-\sqrt{(2a/\pi)}(1-i)t\}}{(1+i)\cosh\{(1+i)\sqrt{(\frac{1}{2}\pi a)}\}} = \int_0^\infty \frac{\cos 2tx}{\cosh \pi x}\frac{dx}{a+i\pi x^2}, \quad (16)$$

in virtue of (11); and therefore

$$\int_0^\infty e^{-an}\{F(n) - f(n)\}\,dn = 0. \quad \text{....................(17)}$$

Now it is known that, if $\phi(n)$ is continuous and

$$\int_0^\infty e^{-an}\phi(n)\,dn = 0,$$

for all positive values of a (or even only for an infinity of such values in arithmetical progression), then

$$\phi(n) = 0,$$

for all positive values of n. Hence

$$F(n) = f(n). \qquad \text{.............................(18)}$$

Equating the real and imaginary parts in (13) and (15) we have

$$\int_0^\infty \frac{\cos 2tx}{\cosh \pi x} \cos \pi n x^2 dx$$

$$= \left\{e^{-t} \cos \frac{\pi n}{4} - e^{-3t} \cos \frac{9\pi n}{4} + e^{-5t} \cos \frac{25\pi n}{4} - \ldots\right\}$$

$$+ \frac{1}{\sqrt{n}} \left\{e^{-t/n} \cos\left(\frac{\pi}{4} - \frac{t^2}{\pi n} + \frac{\pi}{4n}\right) - e^{-3t/n} \cos\left(\frac{\pi}{4} - \frac{t^2}{\pi n} + \frac{9\pi}{4n}\right) + \ldots\right\}, \quad (19)$$

$$\int_0^\infty \frac{\cos 2tx}{\cosh \pi x} \sin \pi n x^2 dx$$

$$= -\left\{e^{-t} \sin \frac{\pi n}{4} - e^{-3t} \sin \frac{9\pi n}{4} + e^{-5t} \sin \frac{25\pi n}{4} - \ldots\right\}$$

$$+ \frac{1}{\sqrt{n}} \left\{e^{-t/n} \sin\left(\frac{\pi}{4} - \frac{t^2}{\pi n} + \frac{\pi}{4n}\right) - e^{-3t/n} \sin\left(\frac{\pi}{4} - \frac{t^2}{\pi n} + \frac{9\pi}{4n}\right) + \ldots\right\}. \quad (20)$$

We can verify the results (18), (19), and (20) by means of the equation (1). This equation can be expressed as a functional equation in $F(n)$, and it is easy to see that $f(n)$ satisfies the same equation.

The right-hand side of these equations can be expressed in finite terms if n is any rational number. For let $n = a/b$, where a and b are any two positive integers and one of them is odd. Then the results (19) and (20) reduce to

$$2 \cosh bt \int_0^\infty \frac{\cos 2tx}{\cosh \pi x} \cos\left(\frac{\pi a x^2}{b}\right) dx$$

$$= [\cosh\{(1-b)t\} \cos(\pi a/4b) - \cosh\{(3-b)t\} \cos(9\pi a/4b)$$

$$+ \cosh\{(5-b)t\} \cos(25\pi a/4b) - \ldots \text{ to } b \text{ terms}]$$

$$+ \sqrt{\left(\frac{b}{a}\right)} \left[\cosh\left\{\left(1 - \frac{1}{a}\right) bt\right\} \cos\left(\frac{\pi}{4} - \frac{bt^2}{\pi a} + \frac{\pi b}{4a}\right)\right.$$

$$\left. - \cosh\left\{\left(1 - \frac{3}{a}\right) bt\right\} \cos\left(\frac{\pi}{4} - \frac{bt^2}{\pi a} + \frac{9\pi b}{4a}\right) + \ldots \text{ to } a \text{ terms}\right], \quad \text{...(21)}$$

$$2 \cosh bt \int_0^\infty \frac{\cos 2tx}{\cosh \pi x} \sin\left(\frac{\pi a x^2}{b}\right) dx$$

$$= -[\cosh\{(1-b)t\} \sin(\pi a/4b) - \cosh\{(3-b)t\} \sin(9\pi a/4b)$$

$$+ \cosh\{(5-b)t\} \sin(25\pi a/4b) - \ldots \text{ to } b \text{ terms}]$$

$$+ \sqrt{\left(\frac{b}{a}\right)} \left[\cosh\left\{\left(1 - \frac{1}{a}\right) bt\right\} \sin\left(\frac{\pi}{4} - \frac{bt^2}{\pi a} + \frac{\pi b}{4a}\right)\right.$$

$$\left. - \cosh\left\{\left(1 - \frac{3}{a}\right) bt\right\} \sin\left(\frac{\pi}{4} - \frac{bt^2}{\pi a} + \frac{9\pi b}{4a}\right) + \ldots \text{ to } a \text{ terms}\right]. \quad \text{...(22)}$$

Thus, for example, we have, when $a=1$ and $b=1$,

$$\int_0^\infty \frac{\cos \pi x^2}{\cosh \pi x} \cos 2\pi tx\,dx = \frac{1+\sqrt{2}\sin \pi t^2}{2\sqrt{2}\cosh \pi t}, \quad \text{...............(23)}$$

$$\int_0^\infty \frac{\sin \pi x^2}{\cosh \pi x} \cos 2\pi tx\,dx = \frac{-1+\sqrt{2}\cos \pi t^2}{2\sqrt{2}\cosh \pi t}. \quad \text{...........(24)}$$

It is easy to verify that (23) and (24) satisfy the relation (2).

The values of the integrals

$$\int_0^\infty \frac{\cos \pi n x^2}{\cosh \pi x}\,dx, \qquad \int_0^\infty \frac{\sin \pi n x^2}{\cosh \pi x}\,dx$$

can be obtained easily from the preceding results by putting $t=0$, and need no special discussion. By successive differentiations of the results (19) and (20) with respect to t and n, we can evaluate the integrals

$$\left.\begin{aligned} &\int_0^\infty x^{2m-1} \frac{\sin tx}{\cosh \pi x} \,{}^{\cos}_{\sin}\, \pi n x^2\,dx, \\ &\int_0^\infty x^{2m} \frac{\cos tx}{\cosh \pi x} \,{}^{\cos}_{\sin}\, \pi n x^2\,dx, \end{aligned}\right\} \quad \text{...................(25)}$$

for all rational values of n and all positive integral values of m. Thus, for example, we have

$$\left.\begin{aligned} &\int_0^\infty x^2 \frac{\cos \pi x^2}{\cosh \pi x}\,dx = \frac{1}{8\sqrt{2}} - \frac{1}{4\pi}, \\ &\int_0^\infty x^2 \frac{\sin \pi x^2}{\cosh \pi x}\,dx = \frac{1}{8} - \frac{1}{8\sqrt{2}}. \end{aligned}\right\} \quad \text{.................(26)}$$

3. We can get many interesting results by applying the theory of Cauchy's reciprocal functions to the preceding results. It is known that, if

$$\int_0^\infty \phi(x) \cos knx\,dx = \psi(n), \quad \text{....................(27)}$$

then (i) $\frac{1}{2}\alpha\{\frac{1}{2}\phi(0)+\phi(\alpha)+\phi(2\alpha)+\phi(3\alpha)+\dots\}$

$$= \tfrac{1}{2}\psi(0)+\psi(\beta)+\psi(2\beta)+\psi(3\beta)+\dots, \quad \text{...(27)}$$

with the condition $\alpha\beta = 2\pi/k$;

(ii) $\alpha\sqrt{2}\{\phi(\alpha)-\phi(3\alpha)-\phi(5\alpha)+\phi(7\alpha)+\phi(9\alpha)-\dots\}$

$$= \psi(\beta)-\psi(3\beta)-\psi(5\beta)+\psi(7\beta)+\psi(9\beta)-\dots, \quad (27)$$

with the condition $\alpha\beta = \pi/4k$;

(iii) $\alpha\sqrt{3}\{\phi(\alpha)-\phi(5\alpha)-\phi(7\alpha)+\phi(11\alpha)+\phi(13\alpha)-\dots\}$

$$= \psi(\beta)-\psi(5\beta)-\psi(7\beta)+\psi(11\beta)+\psi(13\beta)-\dots, \quad (27)$$

with the condition $\alpha\beta = \pi/6k$, where 1, 5, 7, 11, 13, ... are the odd natural numbers without the multiples of 3.

There are of course corresponding results for the function

$$\int_0^\infty \phi(x) \sin knx\,dx = \psi(n), \quad \text{......................(28)}$$

such as

$$\alpha\{\phi(\alpha) - \phi(3\alpha) + \phi(5\alpha) - \ldots\} = \psi(\beta) - \psi(3\beta) + \psi(5\beta) - \ldots,$$

with the condition $\alpha\beta = \pi/2k$.

Thus from (23) and (27)(i) we obtain the following results. If

$$F(\alpha, \beta) = \sqrt{\alpha}\left\{\tfrac{1}{2} + \sum_{r=1}^{r=\infty} \frac{\cos r^2\pi\alpha^2}{\cosh r\pi\alpha}\right\} - \sqrt{\beta}\sum_{r=1}^{r=\infty} \frac{\sin r^2\pi\beta^2}{\cosh r\pi\beta}, \quad \text{......(29)}$$

then $\quad F(\alpha, \beta) = F(\beta, \alpha) = \sqrt{(2\alpha)}\,\{\tfrac{1}{2} + e^{-\pi\alpha} + e^{-4\pi\alpha} + e^{-9\pi\alpha} + \ldots\}^2,$

provided that $\alpha\beta = 1$.

4. If, instead of starting with the integral (11), we start with the corresponding sine integral, we can shew that, when $R(a)$ and $R(t)$ are positive and $|I(t)| \leqslant \pi$,

$$\int_0^\infty \frac{\sin tx}{\sinh \pi x}\frac{dx}{a^2 + x^2} = \frac{1}{2a^2} - \frac{\pi e^{-at}}{2a \sin \pi a} + \sum_{r=1}^{r=\infty} \frac{(-1)^r e^{-rt}}{a^2 - r^2}. \quad \text{......(30)}$$

Hence the above integral can be expressed in finite terms for all rational values of a. For example, we have

$$\int_0^\infty \frac{\sin tx}{\sinh \frac{1}{2}\pi x}\frac{dx}{1 + x^2} = e^t \tan^{-1} e^{-t} - e^{-t}\tan^{-1} e^t. \quad \text{............(31)}$$

From (30) we can deduce that

$$\int_0^\infty \frac{\sin 2tx}{\sinh \pi x} e^{-i\pi n x^2} dx = \tfrac{1}{2} - e^{-2t + i\pi n} + e^{-4t + 4i\pi n} - e^{-6t + 9i\pi n} + \ldots$$

$$- \frac{1}{\sqrt{n}} \exp\left\{\left(\tfrac{1}{4}\pi + \frac{t^2}{\pi n}\right) i\right\} \{e^{-(t + \frac{1}{2}i\pi)/n} + e^{-(3t + \frac{3}{2}i\pi)/n} + \ldots\}, \quad (32)$$

$R(t)$ being positive and $|I(t)| \leqslant \frac{1}{2}\pi$. The right-hand side can be expressed in finite terms for all rational values of n. Thus, for example, we have

$$\int_0^\infty \frac{\cos \pi x^2}{\sinh \pi x} \sin 2\pi t x\,dx = \frac{\cosh \pi t - \cos \pi t^2}{2 \sinh \pi t}, \quad \text{............(33)}$$

$$\int_0^\infty \frac{\sin \pi x^2}{\sinh \pi x} \sin 2\pi t x\,dx = \frac{\sin \pi t^2}{2 \sinh \pi t}, \quad \text{......................(34)}$$

and so on.

Applying the formula (28) to (33) and (34), we have, when $\alpha\beta = \frac{1}{4}$,

$$\left.\begin{aligned} &\sqrt{\alpha}\sum_{r=0}^{r=\infty} (-1)^r \frac{\cos\{(2r+1)^2\pi\alpha^2\}}{\sinh\{(2r+1)\pi\alpha\}} + \sqrt{\beta}\sum_{r=0}^{r=\infty} (-1)^r \frac{\cos\{(2r+1)^2\pi\beta^2\}}{\sinh\{(2r+1)\pi\beta\}} \\ &\qquad = 2\sqrt{\alpha}\{\tfrac{1}{2} + e^{-2\pi\alpha} + e^{-8\pi\alpha} + e^{-18\pi\alpha} + \ldots\}^2; \\ &\sqrt{\alpha}\sum_{r=0}^{r=\infty} (-1)^r \frac{\sin\{(2r+1)^2\pi\alpha^2\}}{\sinh\{(2r+1)\pi\alpha\}} = \sqrt{\beta}\sum_{r=0}^{r=\infty} (-1)^r \frac{\sin\{(2r+1)^2\pi\beta^2\}}{\sinh\{(2r+1)\pi\beta\}}. \end{aligned}\right\} \quad \text{...(35)}$$

By successive differentiation of (32) with respect to t and n we can evaluate the integrals

$$\left.\begin{aligned}&\int_0^\infty x^{2m-1}\frac{\cos tx}{\sinh \pi x}\begin{matrix}\cos\\ \sin\end{matrix}\,\pi n x^2\,dx,\\ &\int_0^\infty x^{2m}\frac{\sin tx}{\sinh \pi x}\begin{matrix}\cos\\ \sin\end{matrix}\,\pi n x^2\,dx\end{aligned}\right\}\quad\ldots\ldots\ldots\ldots\ldots\ldots(36)$$

for all rational values of n and all positive integral values of m. Thus, for example, we have

$$\left.\begin{aligned}&\int_0^\infty x\frac{\cos \pi x^2}{\sinh \pi x}dx=\frac{1}{8},\qquad \int_0^\infty x\frac{\sin \pi x^2}{\sinh \pi x}dx=\frac{1}{4\pi},\\ &\int_0^\infty x^3\frac{\cos \pi x^2}{\sinh \pi x}dx=\frac{1}{16}\left(\frac{1}{4}-\frac{3}{\pi^2}\right),\qquad \int_0^\infty x^3\frac{\sin \pi x^2}{\sinh \pi x}dx=\frac{1}{16\pi},\end{aligned}\right\}\quad(37)$$

and so on.

The denominators of the integrands in (25) and (36) are $\cosh \pi x$ and $\sinh \pi x$. Similar integrals having the denominators of their integrands equal to

$$\prod_1^r \cosh \pi a_r x \sinh \pi b_r x$$

can be evaluated, if a_r and b_r are rational, by splitting up the integrand into partial fractions.

5. The preceding formulæ may be generalised. Thus it may be shewn that, if $R(a)$ and $R(t)$ are positive, $|I(t)|\leqslant \pi$, and $-1<R(\theta)<1$, then

$$\sin \pi\theta\int_0^\infty \frac{\cos tx}{\cosh \pi x+\cos \pi\theta}\,\frac{dx}{a^2+x^2}=\frac{\pi}{2a}\,\frac{e^{-at}\sin \pi\theta}{\cos \pi a+\cos \pi\theta}$$
$$+\sum_{r=0}^{r=\infty}\left\{\frac{e^{-(2r+1-\theta)t}}{a^2-(2r+1-\theta)^2}-\frac{e^{-(2r+1+\theta)t}}{a^2-(2r+1+\theta)^2}\right\}.\quad(38)$$

From (38) it can be deduced that, if n and $R(t)$ are positive, $|I(t)|\leqslant\pi$, and $-1<\theta<1$, then

$$\sin \pi\theta\int_0^\infty \frac{\cos tx}{\cosh \pi x+\cos \pi\theta}\,e^{-i\pi n x^2}dx$$
$$=\sum_{r=0}^{r=\infty}\{e^{-(2r+1-\theta)t+(2r+1-\theta)^2 i\pi n}-e^{-(2r+1+\theta)t+(2r+1+\theta)^2 i\pi n}\}$$
$$+\frac{1}{\sqrt{n}}\exp\left\{-\frac{1}{4}i\left(\pi-\frac{t^2}{\pi n}\right)\right\}\sum_{r=1}^{r=\infty}(-1)^{r-1}\sin r\pi\theta\, e^{-(2rt+r^2 i\pi)/4n}.\quad(39)$$

The right-hand side can be expressed in finite terms if n and θ are rational. In particular, when $\theta=\frac{1}{3}$, we have

$$\int_0^\infty \frac{\cos tx}{1+2\cosh(2\pi x/\sqrt{3})}\,e^{-i\pi n x^2}dx$$
$$=\tfrac{1}{2}\{e^{-\frac{1}{3}(t\sqrt{3}-i\pi n)}-e^{-\frac{1}{3}(2t\sqrt{3}-4i\pi n)}+e^{-\frac{1}{3}(4t\sqrt{3}-16i\pi n)}-\ldots\}$$
$$+\frac{1}{2\sqrt{n}}\exp\left\{-\frac{1}{4}i\left(\pi-\frac{t^2}{\pi n}\right)\right\}\{e^{-(t\sqrt{3}+i\pi)/3n}-e^{-(2t\sqrt{3}+4i\pi)/3n}$$
$$+e^{-(4t\sqrt{3}+16i\pi)/3n}-\ldots\},\quad\ldots(40)$$

where 1, 2, 4, 5, ... are the natural numbers without the multiples of 3.

As an example, when $n=1$, we have

$$\left.\begin{aligned}\int_0^\infty \frac{\cos\pi x^2\cos\pi tx}{1+2\cosh(2\pi x/\sqrt{3})}dx &= \frac{1-2\sin\{(\pi-3\pi t^2)/12\}}{8\cosh(\pi t/\sqrt{3})-4},\\ \int_0^\infty \frac{\sin\pi x^2\cos\pi tx}{1+2\cosh(2\pi x/\sqrt{3})}dx &= \frac{-\sqrt{3}+2\cos\{(\pi-3\pi t^2)/12\}}{8\cosh(\pi t/\sqrt{3})-4}.\end{aligned}\right\} \quad \text{...(41)}$$

6. The formula (32) assumes a neat and elegant form when t is changed to $t+\frac{1}{2}i\pi$. We have then

$$\begin{aligned}&\int_0^\infty \frac{\sin tx}{\tanh\pi x}e^{-i\pi nx^2}dx \quad (n>0,\ t>0)\\ &=\{\tfrac{1}{2}+e^{-t+i\pi n}+e^{-2t+4i\pi n}+e^{-3t+9i\pi n}+\dots\}\\ &-\frac{1}{\sqrt{n}}\exp\left\{\frac{1}{4}i\left(\pi+\frac{t^2}{\pi n}\right)\right\}\{\tfrac{1}{2}+e^{-(t+i\pi)/n}+e^{-(2t+4i\pi)/n}+\dots\}. \quad \text{...(42)}\end{aligned}$$

In particular, when $n=1$, we have

$$\left.\begin{aligned}\int_0^\infty \frac{\cos\pi x^2}{\tanh\pi x}\sin 2\pi tx\,dx &= \tfrac{1}{2}\tanh\pi t\{1-\cos(\tfrac{1}{4}\pi+\pi t^2)\},\\ \int_0^\infty \frac{\sin\pi x^2}{\tanh\pi x}\sin 2\pi tx\,dx &= \tfrac{1}{2}\tanh\pi t\sin(\tfrac{1}{4}\pi+\pi t^2).\end{aligned}\right\} \quad \text{...(43)}$$

We shall now consider an important special case of (42). It can easily be seen from (9) that the left-hand side of (42), when divided by t, tends to

$$\int_0^\infty \frac{\cos\pi nx}{e^{2\pi\sqrt{x}}-1}dx - i\left\{\frac{1}{2\pi n}+\int_0^\infty \frac{\sin\pi nx}{e^{2\pi\sqrt{x}}-1}dx\right\} \quad \text{............(44)}$$

as $t\to 0$. But the limit of the right-hand side of (42) divided by t can be found when n is rational. Let then $n=a/b$, where a and b are any two positive integers, and let

$$\phi(n)=\int_0^\infty \frac{\cos\pi nx}{e^{2\pi\sqrt{x}}-1}dx, \quad \psi(n)=\frac{1}{2\pi n}+\int_0^\infty \frac{\sin\pi nx}{e^{2\pi\sqrt{x}}-1}dx.$$

The relation between $\phi(n)$ and $\psi(n)$ has been stated already in (10′). From (42) and (44) it can easily be deduced that, if a and b are both odd, then

$$\left.\begin{aligned}\phi\left(\frac{a}{b}\right) &= \tfrac{1}{4}\sum_{r=1}^{r=b}(b-2r)\cos\left(\frac{r^2\pi a}{b}\right)-\frac{b}{4a}\sqrt{\left(\frac{b}{a}\right)}\sum_{r=1}^{r=a}(a-2r)\sin\left(\tfrac{1}{4}\pi+\frac{r^2b\pi}{a}\right),\\ \psi\left(\frac{a}{b}\right) &= -\tfrac{1}{4}\sum_{r=1}^{r=b}(b-2r)\sin\left(\frac{r^2\pi a}{b}\right)+\frac{b}{4a}\sqrt{\left(\frac{b}{a}\right)}\sum_{r=1}^{r=a}(a-2r)\cos\left(\tfrac{1}{4}\pi+\frac{r^2\pi b}{a}\right).\end{aligned}\right\}$$

......(45)

It can easily be seen that these satisfy the relation (10′). Similarly, when one of a and b is odd and the other even, it can be shewn that

$$\left.\begin{aligned}\phi\left(\frac{a}{b}\right) &= \frac{\sigma}{4\pi a\sqrt{a}} - \tfrac{1}{2}\sum_{r=1}^{r=b} r\left(1-\frac{r}{b}\right)\cos\left(\frac{r^2\pi a}{b}\right)\\ &\quad + \frac{b}{2a}\sqrt{\left(\frac{b}{a}\right)}\sum_{r=1}^{r=a} r\left(1-\frac{r}{a}\right)\sin\left(\tfrac{1}{4}\pi + \frac{r^2\pi b}{a}\right),\\ \psi\left(\frac{a}{b}\right) &= \frac{\sigma'}{4\pi a\sqrt{a}} + \tfrac{1}{2}\sum_{r=1}^{r=b} r\left(1-\frac{r}{b}\right)\sin\left(\frac{r^2\pi a}{b}\right)\\ &\quad - \frac{b}{2a}\sqrt{\left(\frac{b}{a}\right)}\sum_{r=1}^{r=a} r\left(1-\frac{r}{a}\right)\cos\left(\tfrac{1}{4}\pi + \frac{r^2\pi b}{a}\right),\end{aligned}\right\} \quad \text{.........(46)}$$

where

$$\left.\begin{aligned}\sigma &= \sqrt{b}\sum_1^a \cos\left(\tfrac{1}{4}\pi + \frac{r^2\pi b}{a}\right) = \sqrt{a}\sum_1^b \sin\left(\frac{r^2\pi a}{b}\right),\\ \sigma' &= \sqrt{b}\sum_1^a \sin\left(\tfrac{1}{4}\pi + \frac{r^2\pi b}{a}\right) = \sqrt{a}\sum_1^b \cos\left(\frac{r^2\pi a}{b}\right).\end{aligned}\right\} \quad \text{.........(47)}$$

Thus, for example, we have

$$\left.\begin{aligned}&\phi(0) = \frac{1}{12},\ \phi(1) = \frac{2-\sqrt{2}}{8},\ \phi(2) = \frac{1}{16},\ \phi(4) = \frac{3-\sqrt{2}}{32},\\ &\phi(6) = \frac{13-4\sqrt{3}}{144},\ \phi\left(\frac{1}{2}\right) = \frac{1}{4\pi},\ \phi\left(\frac{2}{5}\right) = \frac{8-3\sqrt{5}}{16},\end{aligned}\right\} \quad (48)$$

and so on.

By differentiating (42) with respect to n, we can evaluate the integrals

$$\int_0^\infty \frac{x^m}{e^{2\pi\sqrt{x}}-1} \begin{matrix}\cos\\ \sin\end{matrix}\, \pi n x\, dx \quad \text{.....................(49)}$$

for all rational values of n and positive integral values of m. Thus, for example, we have

$$\left.\begin{aligned}\int_0^\infty \frac{x\cos\frac{1}{2}\pi x}{e^{2\pi\sqrt{x}}-1}\,dx &= \frac{13-4\pi}{8\pi^2},\\ \int_0^\infty \frac{x\cos 2\pi x}{e^{2\pi\sqrt{x}}-1}\,dx &= \frac{1}{64}\left(\frac{1}{2}-\frac{3}{\pi}+\frac{5}{\pi^2}\right),\\ \int_0^\infty \frac{x^2\cos 2\pi x}{e^{2\pi\sqrt{x}}-1}\,dx &= \frac{1}{256}\left(1-\frac{5}{\pi}+\frac{5}{\pi^2}\right),\end{aligned}\right\} \quad \text{...............(50)}$$

and so on.

13

SUMMATION OF A CERTAIN SERIES

(*Messenger of Mathematics*, XLIV, 1915, 157—160)

1. Let
$$\Phi(s, x) = \sum_{n=0}^{n=\infty} \{\sqrt{(x+n)} + \sqrt{(x+n+1)}\}^{-s}$$
$$= \sum_{n=0}^{n=\infty} \{\sqrt{(x+n+1)} - \sqrt{(x+n)}\}^{s}.$$

The object of this paper is to give a finite expression of $\Phi(s, 0)$ in terms of Riemann ζ-functions, when s is an odd integer greater than 1.

Let $\zeta(s, x)$, where $x > 0$, denote the function expressed by the series
$$x^{-s} + (x+1)^{-s} + (x+2)^{-s} + \ldots,$$
and its analytical continuations. Then
$$\zeta(s, 1) = \zeta(s), \quad \zeta(s, \tfrac{1}{2}) = (2^s - 1)\,\zeta(s), \qquad \text{.............(1)}$$
where $\zeta(s)$ is the Riemann ζ-function;
$$\zeta(s, x) - \zeta(s, x+1) = x^{-s}; \qquad \text{.....................(2)}$$
$$\left.\begin{array}{l} 1^s + 2^s + 3^s + \ldots + n^s = \zeta(-s) - \zeta(-s, n+1), \\ 1^s + 3^s + 5^s + \ldots + (2n-1)^s = (1-2^s)\,\zeta(-s) - 2^s\,\zeta(-s, n+\tfrac{1}{2}), \end{array}\right\} \ldots(3)$$
if n is a positive integer; and
$$\lim_{x\to\infty} \left\{\zeta(s, x) - \tfrac{1}{2}x^{-s} + \left(\frac{x^{1-s}}{1-s} - B_2 \frac{s}{2!} x^{-s-1} + B_4 \frac{s(s+1)(s+2)}{4!} x^{-s-3} \right.\right.$$
$$\left.\left. - B_6 \frac{s(s+1)(s+2)(s+3)(s+4)}{6!} x^{-s-5} + \ldots \text{ to } n \text{ terms}\right)\right\} = 0, \ldots(4)$$
if n is a positive integer, $-(2n-1) < s < 1$, and $B_2 = \frac{1}{6}$, $B_4 = \frac{1}{30}$, $B_6 = \frac{1}{42}$, $B_8 = \frac{1}{30}$, ..., are Bernoulli's numbers.

Suppose now that
$$\psi(x) = 6\zeta(-\tfrac{1}{2}, x) + (4x-3)\sqrt{x} + \Phi(3, x).$$
Then from (2) we see that
$$\psi(x) - \psi(x+1) = 6\sqrt{x} + (4x-3)\sqrt{x} - (4x+1)\sqrt{(x+1)}$$
$$+ \{\sqrt{(x+1)} - \sqrt{x}\}^3 = 0;$$

and from (4) that $\psi(x) \to 0$ as $x \to \infty$. It follows that $\psi(x) = 0$. That is to say,

$$6\zeta(-\tfrac{1}{2}, x) + (4x-3)\sqrt{x} + \Phi(3, x) = 0. \quad \text{...............(5)}$$

Similarly, we can shew that

$$40\zeta(-\tfrac{3}{2}, x) + (16x^2 - 20x + 5)\sqrt{x} + \Phi(5, x) = 0. \quad \text{............(6)}$$

2. Remembering the functional equation satisfied by $\zeta(s)$, viz.,

$$\zeta(1-s) = 2(2\pi)^{-s}\,\Gamma(s)\,\zeta(s)\cos\tfrac{1}{2}\pi s, \quad \text{..................(7)}$$

we see from (3) and (5) that

$$\sqrt{1} + \sqrt{2} + \sqrt{3} + \ldots + \sqrt{n} = \tfrac{2}{3}n^{\frac{3}{2}} + \tfrac{1}{2}\sqrt{n} - \frac{1}{4\pi}\zeta(\tfrac{3}{2}) + \tfrac{1}{6}\Phi(3, n); \quad \text{...(8)}$$

and $\sqrt{1} + \sqrt{3} + \sqrt{5} + \ldots + \sqrt{(2n-1)}$

$$= \tfrac{1}{3}(2n-1)^{\frac{3}{2}} + \tfrac{1}{2}\sqrt{(2n-1)} + \frac{\sqrt{2}-1}{4\pi}\zeta(\tfrac{3}{2}) + \frac{1}{3\sqrt{2}}\Phi(3, n-\tfrac{1}{2}). \quad \text{...(9)}$$

Similarly from (6), we have

$1\sqrt{1} + 2\sqrt{2} + 3\sqrt{3} + \ldots + n\sqrt{n}$

$$= \tfrac{2}{5}n^{\frac{5}{2}} + \tfrac{1}{2}n^{\frac{3}{2}} + \tfrac{1}{8}\sqrt{n} - \frac{3}{16\pi^2}\zeta(\tfrac{5}{2}) + \tfrac{1}{40}\Phi(5, n); \quad \text{......(10)}$$

and $1\sqrt{1} + 3\sqrt{3} + 5\sqrt{5} + \ldots + (2n-1)\sqrt{(2n-1)}$

$$= \tfrac{1}{5}(2n-1)^{\frac{5}{2}} + \tfrac{1}{2}(2n-1)^{\frac{3}{2}} + \tfrac{1}{4}\sqrt{(2n-1)}$$

$$+ \frac{3(2\sqrt{2}-1)}{16\pi^2}\zeta(\tfrac{5}{2}) + \frac{1}{10\sqrt{2}}\Phi(5, n-\tfrac{1}{2}). \quad \text{.........(11)}$$

It also follows from (5) and (6) that

$\sqrt{(a+d)} + \sqrt{(a+2d)} + \sqrt{(a+3d)} + \ldots + \sqrt{(a+nd)}$

$$= C + \frac{2}{3d}(a+nd)^{\frac{3}{2}} + \tfrac{1}{2}\sqrt{(a+nd)} + \tfrac{1}{6}\sqrt{d}\,\Phi(3, n+a/d); \quad \text{...(12)}$$

and $(a+d)^{\frac{3}{2}} + (a+2d)^{\frac{3}{2}} + (a+3d)^{\frac{3}{2}} + \ldots + (a+nd)^{\frac{3}{2}}$

$$= C' + \frac{2}{5d}(a+nd)^{\frac{5}{2}} + \tfrac{1}{2}(a+nd)^{\frac{3}{2}}$$

$$+ \tfrac{1}{8}d\sqrt{(a+nd)} + \tfrac{1}{40}d\sqrt{d}\,\Phi(5, n+a/d), \quad \text{......(13)}$$

where C and C' are independent of n.

Putting $n = 1$ in (8) and (10), we obtain

$$\Phi(3, 0) = \frac{3}{2\pi}\zeta(\tfrac{3}{2}), \quad \Phi(5, 0) = \frac{15}{2\pi^2}\zeta(\tfrac{5}{2}). \quad \text{...............(14)}$$

3. The preceding results may be generalised as follows. If s be an odd integer greater than 1, then

$$\Phi(s, x) + \tfrac{1}{2}\{\sqrt{x} + \sqrt{(x-1)}\}^s + \tfrac{1}{2}\{\sqrt{x} - \sqrt{(x-1)}\}^s$$
$$+ \frac{s}{1!} 2^{s-2} \zeta(1 - \tfrac{1}{2}s, x) + \frac{s(s-4)(s-5)}{3!} 2^{s-6} \zeta(3 - \tfrac{1}{2}s, x)$$
$$+ \frac{s(s-6)(s-7)(s-8)(s-9)}{5!} 2^{s-10} \zeta(5 - \tfrac{1}{2}s, x)$$
$$+ \frac{s(s-8)(s-9)(s-10)(s-11)(s-12)(s-13)}{7!} 2^{s-14}$$
$$\times \zeta(7 - \tfrac{1}{2}s, x) + \dots \text{ to } [\tfrac{1}{4}(s+1)] \text{ terms} = 0, \quad \dots\dots(15)$$

where $[x]$ denotes, as usual, the integral part of x. This can be proved by induction, using the formula

$$\{\sqrt{x} + \sqrt{(x \pm 1)}\}^s + \{\sqrt{x} - \sqrt{(x \pm 1)}\}^s$$
$$= (2\sqrt{x})^s \pm \frac{s}{1!}(2\sqrt{x})^{s-2} + \frac{s(s-3)}{2!}(2\sqrt{x})^{s-4}$$
$$\pm \frac{s(s-4)(s-5)}{3!}(2\sqrt{x})^{s-6} + \dots \text{ to } [1 + \tfrac{1}{2}s] \text{ terms}, \quad \dots\dots(16)$$

which is true for all positive integral values of s.

Similarly, we can shew that if s is a positive even integer, then

$$\frac{s}{1!} 2^{s-2} \{\zeta(1 - \tfrac{1}{2}s) - \zeta(1 - \tfrac{1}{2}s, x)\}$$
$$+ \frac{s(s-4)(s-5)}{3!} 2^{s-6} \{\zeta(3 - \tfrac{1}{2}s) - \zeta(3 - \tfrac{1}{2}s, x)\}$$
$$+ \frac{s(s-6)(s-7)(s-8)(s-9)}{5!} 2^{s-10} \{\zeta(5 - \tfrac{1}{2}s) - \zeta(5 - \tfrac{1}{2}s, x)\}$$
$$+ \dots \text{ to } [\tfrac{1}{4}(s+2)] \text{ terms}$$
$$= \tfrac{1}{2}\{\sqrt{x} + \sqrt{(x-1)}\}^s + \tfrac{1}{2}\{\sqrt{x} - \sqrt{(x-1)}\}^s - 1. \quad \dots\dots(17)$$

Now, remembering (7) and putting $x = 1$ in (15), we obtain

$$\Phi(s, 0) = -\frac{s}{\sqrt{2}} \pi^{-\frac{1}{2}(1+s)} \cos \tfrac{1}{4}\pi s \, \{1.3.5 \dots (s-2) \pi \zeta(\tfrac{1}{2}s)$$
$$- 3.5.7 \dots (s-4) \tfrac{1}{2}(s-5) \tfrac{1}{3}\pi^3 \zeta(\tfrac{1}{2}s - 2)$$
$$+ 5.7.9 \dots (s-6) \tfrac{1}{2}(s-7) \tfrac{1}{4}(s-9) \tfrac{1}{5}\pi^5 \zeta(\tfrac{1}{2}s - 4)$$
$$- 7.9.11 \dots (s-8) \tfrac{1}{2}(s-9) \tfrac{1}{4}(s-11) \tfrac{1}{6}(s-13) \tfrac{1}{7}\pi^7 \zeta(\tfrac{1}{2}s - 6)$$
$$+ 9.11.13 \dots (s-10) \tfrac{1}{2}(s-11) \tfrac{1}{4}(s-13) \tfrac{1}{6}(s-15) \tfrac{1}{8}(s-17)$$
$$\times \tfrac{1}{9}\pi^9 \zeta(\tfrac{1}{2}s - 8) - \dots \text{ to } [\tfrac{1}{4}(s+1)] \text{ terms}\}, \quad \dots\dots(18)$$

if s is an odd integer greater than 1. Similarly, putting $x = \tfrac{1}{2}$ in (15), we can express $\Phi(s, \tfrac{1}{2})$ in terms of ζ-functions, if s is an odd integer greater than 1.

4. It is also easy to shew that, if

$$\Psi(s, x) = \sum_{n=0}^{n=\infty} \frac{\{\sqrt{(x+n)} + \sqrt{(x+n+1)}\}^{-s}}{\sqrt{\{(x+n)(x+n+1)\}}},$$

then

$$\Psi(s, x) - \tfrac{1}{2}\frac{\{\sqrt{x} + \sqrt{(x-1)}\}^{s} - \{\sqrt{x} - \sqrt{(x-1)}\}^{s}}{\sqrt{\{x(x-1)\}}}$$

$$= \frac{s-2}{1!} 2^{s-2} \zeta(2 - \tfrac{1}{2}s, x) + \frac{(s-4)(s-5)(s-6)}{3!} 2^{s-6} \zeta(4 - \tfrac{1}{2}s, x)$$

$$+ \frac{(s-6)(s-7)(s-8)(s-9)(s-10)}{5!} 2^{s-10} \zeta(6 - \tfrac{1}{2}s, x)$$

$$+ \dots \text{ to } [\tfrac{1}{4}(s+1)] \text{ terms}, \qquad \qquad (19)$$

provided that s is a positive odd integer. For example,

$$\left.\begin{aligned} \Psi(1, x) &= \frac{1}{\sqrt{x}}, \\ \Psi(3, x) &= 4\sqrt{x} - \frac{1}{\sqrt{x}} + 2\zeta(\tfrac{1}{2}, x), \\ \Psi(5, x) &= 16x\sqrt{x} - 12\sqrt{x} + \frac{1}{\sqrt{x}} + 24\zeta(-\tfrac{1}{2}, x), \end{aligned}\right\} \qquad (20)$$

and so on.

14

NEW EXPRESSIONS FOR RIEMANN'S FUNCTIONS $\xi(s)$ AND $\Xi(t)$

(*Quarterly Journal of Mathematics*, XLVI, 1915, 253—260)

1. The principal object of this paper is to prove that if the real parts of α and β are positive, and $\alpha\beta = \pi^2$, and t is real, then

$$\alpha^{-\frac{1}{4}}\left\{\frac{1}{1+t^2}+4\alpha\int_0^\infty\left(\frac{1}{3^2+t^2}-\frac{\alpha}{1\,!}\frac{x^2}{7^2+t^2}+\frac{\alpha^2}{2\,!}\frac{x^4}{11^2+t^2}-\ldots\right)\frac{x\,dx}{e^{2\pi x}-1}\right\}$$
$$-\beta^{-\frac{1}{4}}\left\{\frac{1}{1+t^2}+4\beta\int_0^\infty\left(\frac{1}{3^2+t^2}-\frac{\beta}{1\,!}\frac{x^2}{7^2+t^2}+\frac{\beta^2}{2\,!}\frac{x^4}{11^2+t^2}-\ldots\right)\frac{x\,dx}{e^{2\pi x}-1}\right\}$$
$$=\frac{\pi^{-\frac{3}{2}}}{4t}\Gamma\left(\frac{-1+it}{4}\right)\Gamma\left(\frac{-1-it}{4}\right)\Xi\left(\frac{t}{2}\right)\sin\left(\frac{t}{8}\log\frac{\beta}{\alpha}\right). \quad \ldots\ldots\ldots\ldots\ldots(1)$$

Consider the integral

$$J(u)=\int_0^\infty\frac{xe^{-\pi ux^2}}{e^{2\pi x}-1}\,dx,$$

where the real part of u is positive. Since

$$\int_0^\infty\frac{\sin\pi nx}{e^{\pi x}-1}\,dx=\frac{1}{e^{2\pi n}-1}+\frac{1}{2}-\frac{1}{2\pi n},$$

we have $$J(u)+\frac{1}{4\pi u}-\frac{1}{4\pi\sqrt{u}}=\int_0^\infty xe^{-\pi ux^2}\left(\frac{1}{e^{2\pi x}-1}+\frac{1}{2}-\frac{1}{2\pi x}\right)dx$$
$$=\int_0^\infty\int_0^\infty xe^{-\pi ux^2}\frac{\sin\pi xy}{e^{\pi y}-1}\,dx\,dy=u^{-\frac{3}{2}}\int_0^\infty\frac{xe^{-\pi x^2/u}}{e^{2\pi x}-1}\,dx; \quad \ldots\ldots(2)$$

and so $$J(u)-\frac{1}{4\pi\sqrt{u}}=u^{-\frac{3}{2}}\int_0^\infty xe^{-\pi x^2/u}\left(\frac{1}{e^{2\pi x}-1}-\frac{1}{2\pi x}\right)dx. \quad \ldots\ldots\ldots(3)$$

Suppose now that $s=\sigma+it$, where $0<\sigma<1$. Then, from (3), we have

$$\int_0^1 u^{\frac{1}{2}(s-1)}\left\{J(nu)-\frac{1}{4\pi\sqrt{(nu)}}\right\}du$$
$$=n^{-\frac{3}{2}}\int_0^1 u^{\frac{1}{2}(s-4)}\,du\int_0^\infty xe^{-\pi x^2/nu}\left(\frac{1}{e^{2\pi x}-1}-\frac{1}{2\pi x}\right)dx. \ \ldots\ldots(4)$$

Changing u into $1/v$, we obtain

$$n^{-\frac{3}{2}}\int_1^\infty v^{-\frac{1}{2}s}\,dv\int_0^\infty xe^{-\pi vx^2/n}\left(\frac{1}{e^{2\pi x}-1}-\frac{1}{2\pi x}\right)dx$$
$$=n^{-\frac{3}{2}}\left\{\int_0^\infty v^{-\frac{1}{2}s}\,dv\int_0^\infty xe^{-\pi vx^2/n}\left(\frac{1}{e^{2\pi x}-1}-\frac{1}{2\pi x}\right)dx\right.$$
$$\left.-\int_0^1 v^{-\frac{1}{2}s}\,dv\int_0^\infty xe^{-\pi vx^2/n}\left(\frac{1}{e^{2\pi x}-1}-\frac{1}{2\pi x}\right)dx\right\}.$$

But

$$\int_0^1 u^{\frac{1}{2}(s-1)}\left\{J(nu)-\frac{1}{4\pi\sqrt{(nu)}}\right\}du=\int_0^1 u^{\frac{1}{2}(s-1)}J(nu)\,du-\frac{1}{2\pi s\sqrt{n}}$$

$$=-\frac{1}{2\pi s\sqrt{n}}+\int_0^1 u^{\frac{1}{2}(s-1)}\,du\int_0^\infty \frac{xe^{-\pi nux^2}}{e^{2\pi x}-1}\,dx$$

$$=-\frac{1}{2\pi s\sqrt{n}}+\int_0^\infty \frac{x\,dx}{e^{2\pi x}-1}\int_0^1 u^{\frac{1}{2}(s-1)}e^{-\pi nux^2}\,du$$

$$=-\frac{1}{2\pi s\sqrt{n}}+2\int_0^\infty\left\{\frac{1}{1+s}-\frac{\pi nx^2}{1!(3+s)}+\frac{(\pi nx^2)^2}{2!(5+s)}-\dots\right\}\frac{x\,dx}{e^{2\pi x}-1}.\quad\dots(5)$$

Also

$$n^{-\frac{3}{2}}\int_0^\infty v^{-\frac{1}{2}s}\,dv\int_0^\infty xe^{-\pi vx^2/n}\left(\frac{1}{e^{2\pi x}-1}-\frac{1}{2\pi x}\right)dx$$

$$=n^{-\frac{3}{2}}\int_0^\infty x\left(\frac{1}{e^{2\pi x}-1}-\frac{1}{2\pi x}\right)dx\int_0^\infty v^{-\frac{1}{2}s}e^{-\pi vx^2/n}\,dv$$

$$=\pi^{\frac{1}{2}(s-2)}n^{-\frac{1}{2}(s+1)}\Gamma(1-\tfrac{1}{2}s)\int_0^\infty x^{s-1}\left(\frac{1}{e^{2\pi x}-1}-\frac{1}{2\pi x}\right)dx$$

$$=-\frac{n^{-\frac{1}{2}(s+1)}}{4\pi\sqrt{\pi}}\Gamma\left(-\frac{s}{2}\right)\Gamma\left(\frac{s-1}{2}\right)\xi(s),\quad\dots\dots\dots\dots(6)$$

where $\xi(s)=(s-1)\Gamma(1+\tfrac{1}{2}s)\pi^{-\frac{1}{2}s}\zeta(s)$.

Finally

$$n^{-\frac{3}{2}}\int_0^1 v^{-\frac{1}{2}s}\,dv\int_0^\infty xe^{-\pi vx^2/n}\left(\frac{1}{e^{2\pi x}-1}-\frac{1}{2\pi x}\right)dx$$

$$=-\frac{n^{-\frac{3}{2}}}{2\pi}\int_0^1 v^{-\frac{1}{2}s}\,dv\int_0^\infty e^{-\pi vx^2/n}\,dx+n^{-\frac{3}{2}}\int_0^1 v^{-\frac{1}{2}s}\,dv\int_0^\infty\frac{xe^{-\pi vx^2/n}}{e^{2\pi x}-1}\,dx$$

$$=-\frac{1}{4\pi n}\int_0^1 v^{-\frac{1}{2}(1+s)}\,dv+n^{-\frac{3}{2}}\int_0^\infty\frac{x\,dx}{e^{2\pi x}-1}\int_0^1 v^{-\frac{1}{2}s}e^{-\pi vx^2/n}\,dv$$

$$=-\frac{1}{2\pi n(1-s)}+2n^{-\frac{3}{2}}\int_0^\infty\left\{\frac{1}{2-s}-\frac{\pi x^2/n}{1!(4-s)}+\frac{(\pi x^2/n)^2}{2!(6-s)}-\dots\right\}\frac{x\,dx}{e^{2\pi x}-1}.$$

$$\dots\dots(7)$$

All the inversions of the order of integration, effected in the preceding argument, are easily justified, since every integral remains convergent when the subject of integration is replaced by its modulus.

It follows from (4)—(7) that, if the real parts of α and β are positive, and $\alpha\beta=\pi^2$, then

$$\alpha^{-\frac{1}{4}}\left\{\frac{1}{1-s}-4\alpha\int_0^\infty\left(\frac{1}{1+s}-\frac{\alpha}{1!}\frac{x^2}{3+s}+\frac{\alpha^2}{2!}\frac{x^4}{5+s}-\dots\right)\frac{x\,dx}{e^{2\pi x}-1}\right\}$$

$$+\beta^{-\frac{1}{4}}\left\{\frac{1}{s}-4\beta\int_0^\infty\left(\frac{1}{2-s}-\frac{\beta}{1!}\frac{x^2}{4-s}+\frac{\beta^2}{2!}\frac{x^4}{6-s}-\dots\right)\frac{x\,dx}{e^{2\pi x}-1}\right\}$$

$$=\tfrac{1}{2}\pi^{-\frac{3}{2}}\left(\frac{\alpha}{\beta}\right)^{\frac{1}{4}-\frac{1}{2}s}\Gamma\left(-\frac{s}{2}\right)\Gamma\left(\frac{s-1}{2}\right)\xi(s).\quad\dots\dots\dots(8)$$

Changing s to $\frac{1}{2}(1+it)$ in (8), and writing as usual

$$\xi(\tfrac{1}{2}+\tfrac{1}{2}it)=\Xi(\tfrac{1}{2}t),$$

and equating the real and imaginary parts, we obtain the formula (1), and also the formula

$$\alpha^{-\frac{1}{4}}\left\{\frac{1}{1+t^2}-4\alpha\int_0^\infty\left(\frac{3}{3^2+t^2}-\frac{\alpha}{1!}\frac{7x^2}{7^2+t^2}+\frac{\alpha^2}{2!}\frac{11x^4}{11^2+t^2}-\dots\right)\frac{x\,dx}{e^{2\pi x}-1}\right\}$$
$$+\beta^{-\frac{1}{4}}\left\{\frac{1}{1+t^2}-4\beta\int_0^\infty\left(\frac{3}{3^2+t^2}-\frac{\beta}{1!}\frac{7x^2}{7^2+t^2}+\frac{\beta^2}{2!}\frac{11x^4}{11^2+t^2}-\dots\right)\frac{x\,dx}{e^{2\pi x}-1}\right\}$$
$$=\tfrac{1}{4}\pi^{-\frac{3}{4}}\Gamma\left(\frac{-1+it}{4}\right)\Gamma\left(\frac{-1-it}{4}\right)\Xi(\tfrac{1}{2}t)\cos\left(\frac{t}{8}\log\frac{\alpha}{\beta}\right). \quad\text{.........(9)}$$

2. We have proved (8) on the assumption that $0<\sigma<1$. But it can be shewn that the formula is true for all values of s other than integral values.

Suppose first that $-1<\sigma<0$. The formula (3) is equivalent to

$$J(u)=u^{-\frac{3}{2}}\int_0^\infty xe^{-\pi x^2/u}\left(\frac{1}{e^{2\pi x}-1}-\frac{1}{2\pi x}+\frac{1}{2}\right)dx. \quad\text{.........(10)}$$

Using this formula as we used (3) in the previous section, we can shew that (8) is true in the strip $-1<\sigma<0$ also. In the right-hand side of (3), the first term in the expansion of $1/(e^{2\pi x}-1)$, viz. $1/(2\pi x)$, is removed, and in that of (10) two terms are removed. By considering the corresponding formulæ in which more and more terms in the expansion of $1/(e^{2\pi x}-1)$, viz.

$$\frac{1}{2\pi x}-\frac{1}{2}+\frac{\pi x}{6}-\frac{\pi^3x^3}{90}+\frac{\pi^5x^5}{945}-\frac{\pi^7x^7}{9450}+\frac{\pi^9x^9}{93555}-\dots,$$

are removed, we can shew that the formula (8) is true in the strips $-2<\sigma<1$, $-3<\sigma<-2$, and so on. That it is also true in the strips $1<\sigma<2$, $2<\sigma<3,\dots$ is easily deduced from the functional equation $\xi(s)=\xi(1-s)$.

The formula also holds on the lines which divide the strips, except at the special points $s=k$, where k is an integer. This follows at once from the continuity of $\xi(s)$ and the uniform convergence of the integrals in question.

3. As a particular case of (9) we have, when $\alpha=\beta=\pi$,

$$\frac{1}{1+t^2}-4\pi\int_0^\infty\left(\frac{3}{3^2+t^2}-\frac{\pi}{1!}\frac{7x^2}{7^2+t^2}+\frac{\pi^2}{2!}\frac{11x^4}{11^2+t^2}-\dots\right)\frac{x\,dx}{e^{2\pi x}-1}$$
$$=\frac{1}{8\sqrt{\pi}}\Gamma\left(\frac{-1+it}{4}\right)\Gamma\left(\frac{-1-it}{4}\right)\Xi(\tfrac{1}{2}t). \quad\text{............(11)}$$

But the left-hand side of (11) is equal to

$$\int_0^\infty\left\{e^{-z}-4\pi\int_0^\infty\left(e^{-3z}-\frac{\pi x^2}{1!}e^{-7z}+\frac{\pi^2x^4}{2!}e^{-11z}-\dots\right)\frac{x\,dx}{e^{2\pi x}-1}\right\}\cos tz\,dz.$$

Hence,
$$\int_0^\infty \left\{e^{-z} - 4\pi \int_0^\infty \frac{xe^{-3z-\pi x^2 e^{-4z}}}{e^{2\pi x}-1}\,dx\right\}\cos tz\,dz$$
$$= \frac{1}{8\sqrt{\pi}}\,\Gamma\left(\frac{-1+it}{4}\right)\Gamma\left(\frac{-1-it}{4}\right)\Xi\left(\tfrac{1}{2}t\right). \qquad (12)$$

It follows from this and Fourier's theorem that
$$e^{-n} - 4\pi e^{-3n}\int_0^\infty \frac{xe^{-\pi x^2 e^{-4n}}}{e^{2\pi x}-1}\,dx$$
$$= \frac{1}{4\pi\sqrt{\pi}}\int_0^\infty \Gamma\left(\frac{-1+it}{4}\right)\Gamma\left(\frac{-1-it}{4}\right)\Xi\left(\tfrac{1}{2}t\right)\cos nt\,dt. \qquad (13)$$

But it is easily seen from (2) that, if α and β are positive and $\alpha\beta = \pi^2$, then
$$\alpha^{-\frac{1}{4}}\left\{1 + 4\alpha\int_0^\infty \frac{xe^{-\alpha x^2}}{e^{2\pi x}-1}\,dx\right\} = \beta^{-\frac{1}{4}}\left\{1 + 4\beta\int_0^\infty \frac{xe^{-\beta x^2}}{e^{2\pi x}-1}\,dx\right\}. \qquad (14)$$

From this it follows that the left-hand side of (13) is an even function of n, and so the formula (13) is true for all real values of n.

4. It can easily be shewn that, if $\alpha\beta = 4\pi^2$ and $R(s)$, where $R(s)$ is the real part of s, is greater than -1, then
$$\frac{\zeta(1-s)}{4\cos\frac{1}{2}\pi s}\alpha^{\frac{1}{2}(s-1)} + \frac{\zeta(-s)}{8\sin\frac{1}{2}\pi s}\alpha^{\frac{1}{2}(s+1)}$$
$$+ \alpha^{\frac{1}{2}(s+1)}\int_0^\infty\int_0^\infty \frac{x^s \sin\alpha xy}{(e^{2\pi x}-1)(e^{2\pi y}-1)}\,dx\,dy$$
$$= \frac{\zeta(1-s)}{4\cos\frac{1}{2}\pi s}\beta^{\frac{1}{2}(s-1)} + \frac{\zeta(-s)}{8\sin\frac{1}{2}\pi s}\beta^{\frac{1}{2}(s+1)}$$
$$+ \beta^{\frac{1}{2}(s+1)}\int_0^\infty\int_0^\infty \frac{x^s \sin\beta xy}{(e^{2\pi x}-1)(e^{2\pi y}-1)}\,dx\,dy. \qquad (15)$$

From this we can shew, by arguments similar to those of §§ 1—2, that if $\alpha\beta = 4\pi^2$ and $R(s) > -1$ then
$$\frac{\zeta(1-s)}{4\cos\frac{1}{2}\pi s}\,\frac{\alpha^{\frac{1}{2}(s-1)}}{s-1-t} + \frac{\zeta(-s)}{8\sin\frac{1}{2}\pi s}\,\frac{\alpha^{\frac{1}{2}(s+1)}}{s+1-t}$$
$$+ \alpha^{\frac{1}{2}(s+1)}\int_0^\infty\int_0^\infty \left\{\frac{\alpha xy}{1!\,(s+3-t)} - \frac{(\alpha xy)^3}{3!\,(s+7-t)}\right.$$
$$\left. + \frac{(\alpha xy)^5}{5!\,(s+11-t)} - \dots\right\}\frac{x^s\,dx\,dy}{(e^{2\pi x}-1)(e^{2\pi y}-1)}$$
$$+ \frac{\zeta(1-s)}{4\cos\frac{1}{2}\pi s}\,\frac{\beta^{\frac{1}{2}(s-1)}}{s-1+t} + \frac{\zeta(-s)}{8\sin\frac{1}{2}\pi s}\,\frac{\beta^{\frac{1}{2}(s+1)}}{s+1+t}$$
$$+ \beta^{\frac{1}{2}(s+1)}\int_0^\infty\int_0^\infty \left\{\frac{\beta xy}{1!\,(s+3+t)} - \frac{(\beta xy)^3}{3!\,(s+7+t)}\right.$$
$$\left. + \frac{(\beta xy)^5}{5!\,(s+11+t)} - \dots\right\}\frac{x^s\,dx\,dy}{(e^{2\pi x}-1)(e^{2\pi y}-1)}$$
$$= \left(\frac{\alpha}{\beta}\right)^{\frac{1}{4}t}\frac{2^{\frac{1}{2}(s-3)}}{\pi}\,\frac{\Gamma\{\frac{1}{4}(s-1+t)\}\,\Gamma\{\frac{1}{4}(s-1-t)\}}{(s+1)^2-t^2}$$
$$\times\,\xi\left(\frac{1+s+t}{2}\right)\xi\left(\frac{1+s-t}{2}\right). \qquad (16)$$

From this we deduce that, if $\alpha\beta = 4\pi^2$ and $R(s) > -1$, then

$$\frac{\zeta(1-s)}{4\cos\frac{1}{2}\pi s}\frac{s-1}{(s-1)^2+t^2}\{\alpha^{\frac{1}{2}(s-1)}+\beta^{\frac{1}{2}(s-1)}\}$$

$$+\frac{\zeta(-s)}{8\sin\frac{1}{2}\pi s}\frac{s+1}{(s+1)^2+t^2}\{\alpha^{\frac{1}{2}(s+1)}+\beta^{\frac{1}{2}(s+1)}\}$$

$$+\alpha^{\frac{1}{2}(s+1)}\int_0^\infty\int_0^\infty\left\{\frac{\alpha xy}{1!}\frac{s+3}{(s+3)^2+t^2}-\frac{(\alpha xy)^3}{3!}\frac{s+7}{(s+7)^2+t^2}+\ldots\right\}\frac{x^s\,dx\,dy}{(e^{2\pi x}-1)(e^{2\pi y}-1)}$$

$$+\beta^{\frac{1}{2}(s+1)}\int_0^\infty\int_0^\infty\left\{\frac{\beta xy}{1!}\frac{s+3}{(s+3)^2+t^2}-\frac{(\beta xy)^3}{3!}\frac{s+7}{(s+7)^2+t^2}+\ldots\right\}\frac{x^s\,dx\,dy}{(e^{2\pi x}-1)(e^{2\pi y}-1)}$$

$$=\frac{2^{\frac{1}{2}(s-3)}}{\pi}\frac{\Gamma\{\frac{1}{4}(s-1+it)\}\Gamma\{\frac{1}{4}(s-1-it)\}}{(s+1)^2+t^2}$$

$$\times\Xi\left(\frac{t+is}{2}\right)\Xi\left(\frac{t-is}{2}\right)\cos\left(\tfrac{1}{4}t\log\frac{\alpha}{\beta}\right);\quad\ldots\ldots(17)$$

and

$$\frac{\zeta(1-s)}{4\cos\frac{1}{2}\pi s}\frac{1}{(s-1)^2+t^2}\{\alpha^{\frac{1}{2}(s-1)}-\beta^{\frac{1}{2}(s-1)}\}$$

$$+\frac{\zeta(-s)}{8\sin\frac{1}{2}\pi s}\frac{1}{(s+1)^2+t^2}\{\alpha^{\frac{1}{2}(s+1)}-\beta^{\frac{1}{2}(s+1)}\}$$

$$+\alpha^{\frac{1}{2}(s+1)}\int_0^\infty\int_0^\infty\left\{\frac{\alpha xy}{1!}\frac{1}{(s+3)^2+t^2}-\frac{(\alpha xy)^3}{3!}\frac{1}{(s+7)^2+t^2}+\ldots\right\}\frac{x^s\,dx\,dy}{(e^{2\pi x}-1)(e^{2\pi y}-1)}$$

$$-\beta^{\frac{1}{2}(s+1)}\int_0^\infty\int_0^\infty\left\{\frac{\beta xy}{1!}\frac{1}{(s+3)^2+t^2}-\frac{(\beta xy)^3}{3!}\frac{1}{(s+7)^2+t^2}+\ldots\right\}\frac{x^s\,dx\,dy}{(e^{2\pi x}-1)(e^{2\pi y}-1)}$$

$$=\frac{2^{\frac{1}{2}(s-3)}}{\pi}\frac{\Gamma\{\frac{1}{4}(s-1+it)\}\Gamma\{\frac{1}{4}(s-1-it)\}}{(s+1)^2+t^2}$$

$$\times\Xi\left(\frac{t+is}{2}\right)\Xi\left(\frac{t-is}{2}\right)\sin\left(\tfrac{1}{4}t\log\frac{\alpha}{\beta}\right).\quad\ldots\ldots\ldots(18)$$

5. Proceeding as in § 3 we can shew that, if n is real, and

$$F(n)=\int_0^\infty\frac{\Gamma\{\frac{1}{4}(s-1+it)\}\Gamma\{\frac{1}{4}(s-1-it)\}}{(s+1)^2+t^2}\Xi\left(\frac{t+is}{2}\right)\Xi\left(\frac{t-is}{2}\right)\cos nt\,dt,$$

then, if $R(s) > 1$,

$$F(n) = \tfrac{1}{8}(4\pi)^{-\frac{1}{2}(s-3)} \left\{ \int_0^\infty \frac{x^s\,dx}{(e^{xe^n}-1)(e^{xe^{-n}}-1)} - 2\Gamma(s)\zeta(s)\cosh n(1-s) \right\}; \quad \text{.........(19)}$$

if $-1 < R(s) < 1$,

$$F(n) = \tfrac{1}{8}(4\pi)^{-\frac{1}{2}(s-3)} \int_0^\infty x^s \left(\frac{1}{e^{xe^n}-1} - \frac{1}{xe^n}\right)\left(\frac{1}{e^{xe^{-n}}-1} - \frac{1}{xe^{-n}}\right) dx; \quad \text{...(20)}$$

if $-3 < R(s) < -1$,

$$F(n) = \tfrac{1}{8}(4\pi)^{-\frac{1}{2}(s-3)} \left\{ \int_0^\infty x^s \left(\frac{1}{e^{xe^n}-1} - \frac{1}{xe^n} + \frac{1}{2}\right) \right.$$

$$\left. \times \left(\frac{1}{e^{xe^{-n}}-1} - \frac{1}{xe^{-n}} + \frac{1}{2}\right) dx - \Gamma(1+s)\zeta(1+s)\cosh n(1+s) \right\}; \quad \text{...(21)}$$

and so on. If, in particular, we put $s = 0$ in (20), we obtain

$$\int_0^\infty \Gamma\left(\frac{-1+it}{4}\right)\Gamma\left(\frac{-1-it}{4}\right)\{\Xi(\tfrac{1}{2}t)\}^2 \frac{\cos nt}{1+t^2}\,dt$$

$$= \pi\sqrt{\pi} \int_0^\infty \left(\frac{1}{e^{xe^n}-1} - \frac{1}{xe^n}\right)\left(\frac{1}{e^{xe^{-n}}-1} - \frac{1}{xe^{-n}}\right) dx. \quad \text{......(22)}$$

15

HIGHLY COMPOSITE NUMBERS

(*Proceedings of the London Mathematical Society*, 2, XIV, 1915, 347—409)

CONTENTS

I.

Introduction and Summary of Results.

1. The number $d(N)$ of divisors of N varies with extreme irregularity as N tends to infinity, tending itself to infinity or remaining small according to the form of N. In this paper I prove a large number of results which add a good deal to our knowledge of the behaviour of $d(N)$.

It was proved by Dirichlet* that

$$\frac{d(1)+d(2)+d(3)+\ldots+d(N)}{N}=\log N+2\gamma-1+O\left(\frac{1}{\sqrt{N}}\right)\dagger,$$

where γ is the Eulerian constant. Voronöi‡ and Landau§ have shewn that the error term may be replaced by $O(N^{-\frac{2}{3}+\epsilon})$, or indeed $O(N^{-\frac{2}{3}}\log N)$. It seems not unlikely that the real value of the error is of the form $O(N^{-\frac{3}{4}+\epsilon})$, but this is as yet unproved. Mr Hardy has, however, shewn recently|| that the equation

$$\frac{d(1)+d(2)+d(3)+\ldots+d(N)}{N}=\log N+2\gamma-1+o(N^{-\frac{3}{4}})$$

is certainly false. He has also proved that

$$d(1)+d(2)+\ldots+d(N-1)+\tfrac{1}{2}d(N)-N\log N-(2\gamma-1)N-\tfrac{1}{4}$$
$$=\sqrt{N}\sum_{1}^{\infty}\frac{d(n)}{\sqrt{n}}[H_1\{4\pi\sqrt{(nN)}\}-Y_1\{4\pi\sqrt{(nN)}\}],$$

where Y_n is the ordinary second solution of Bessel's equation, and

$$H_1(x)=\frac{2}{\pi}\int_1^{\infty}\frac{we^{-xw}\,dw}{\sqrt{(w^2-1)}};$$

and that the series on the right-hand side is the sum of the series

$$\frac{N^{\frac{1}{4}}}{\pi\sqrt{2}}\sum_{1}^{\infty}\frac{d(n)}{n^{\frac{3}{4}}}\cos\{4\pi\sqrt{(nN)}-\tfrac{1}{4}\pi\},$$

and an absolutely and uniformly convergent series.

* *Werke*, Vol. 2, p. 49.

† $f=O(\phi)$ means that a constant exists such that $|f|<K\phi$: $f=o(\phi)$ means that $f/\phi\to 0$.

‡ *Crelle's Journal*, Vol. 126, p. 241. § *Göttinger Nachrichten*, 1912.

|| *Comptes Rendus*, May 10, 1915. See Appendix, p. 338.

The "average" order of $d(N)$ is thus known with considerable accuracy. In this paper I consider, not the average order of $d(N)$, but its maximum order. This problem has been much less studied. It is obvious that

$$d(N) < 2\sqrt{N}.$$

It was shewn by Wigert* that

$$d(N) < 2^{\frac{\log N}{\log\log N}(1+\epsilon)} \qquad \text{(i)}$$

for all positive values of ϵ and all sufficiently large values of N, and that

$$d(N) > 2^{\frac{\log N}{\log\log N}(1-\epsilon)} \qquad \text{(ii)}$$

for an infinity of values of N. From (i) it follows in particular that

$$d(N) < N^{\delta}$$

for all positive values of δ and all sufficiently large values of N.

Wigert proves (i) by purely elementary reasoning, but uses the "Prime Number Theorem†" to prove (ii). This is, however, unnecessary, the inequality (ii) being also capable of elementary proof. In §5 I shew, by elementary reasoning, that

$$d(N) < 2^{\frac{\log N}{\log\log N} + O\frac{\log N}{(\log\log N)^2}}$$

for all values of N, and

$$d(N) > 2^{\frac{\log N}{\log\log N} + O\frac{\log N}{(\log\log N)^2}}$$

for an infinity of values of N. I also shew later on that, if we assume all known results concerning the distribution of primes, then

$$d(N) < 2^{Li(\log N) + O[\log N e^{-a\sqrt{(\log\log N)}}]}$$

for all values of N, and

$$d(N) > 2^{Li(\log N) + O[\log N e^{-a\sqrt{(\log\log N)}}]}$$

for an infinity of values of N, where a is a positive constant.

I then adopt a different point of view. I define a highly composite number as a number whose number of divisors exceeds that of all its predecessors. Writing such a number in the form

$$N = 2^{a_2} \cdot 3^{a_3} \cdot 5^{a_5} \dots p^{a_p},$$

I prove that $\qquad a_2 \geqslant a_3 \geqslant a_5 \geqslant \dots \geqslant a_p,$

and that $\qquad a_p = 1,$

for all highly composite values of N except 4 and 36.

* *Arkiv för Matematik*, Vol. 3, No. 18.

† The theorem that $\qquad \pi(x) \sim \dfrac{x}{\log x},$

$\pi(x)$ being the number of primes not exceeding x.

I then go on to prove that the indices near the beginning form a decreasing sequence in the stricter sense, i.e., that

$$a_2 > a_3 > a_5 > \dots > a_\lambda,$$

where λ is a certain function of p.

Near the end groups of equal indices may occur, and I prove that there are actually groups of indices equal to

$$1, 2, 3, 4, \dots, \mu,$$

where μ again is a certain function of p. I also prove that if λ is fairly small in comparison with p, then

$$a_\lambda \log \lambda \sim \frac{\log p}{\log 2};$$

and that the later indices can be assigned with an error of at most unity.

I prove also that two successive highly composite numbers are asymptotically equivalent, i.e., that the ratio of two consecutive such numbers tends to unity. These are the most striking results. More precise ones will be found in the body of the paper. These results give us a fairly accurate idea of the structure of a highly composite number.

I then select from the general aggregate of highly composite numbers a special set which I call "superior highly composite numbers." I determine completely the general form of all such numbers, and I shew how a combination of the idea of a superior highly composite number with the assumption of the truth of the Riemann hypothesis concerning the roots of the ζ-function leads to even more precise results concerning the maximum order of $d(N)$. These results naturally differ from all which precede in that they depend on the truth of a hitherto unproved hypothesis.

II.

Elementary Results concerning the Order of $d(N)$.

2. Let $d(N)$ denote the number of divisors of N, and let

$$N = p_1^{a_1} p_2^{a_2} p_3^{a_3} \dots p_n^{a_n}, \qquad \text{.........................(1)}$$

where $p_1, p_2, p_3, \dots, p_n$ are a given set of n primes. Then

$$d(N) = (1+a_1)(1+a_2)(1+a_3)\dots(1+a_n). \qquad \text{............(2)}$$

From (1) we see that

$$\begin{aligned} &(1/n)\log(p_1 p_2 p_3 \dots p_n N) \\ &= (1/n)\{(1+a_1)\log p_1 + (1+a_2)\log p_2 + \dots + (1+a_n)\log p_n\} \\ &> \{(1+a_1)(1+a_2)(1+a_3)\dots(1+a_n)\log p_1 \log p_2 \dots \log p_n\}^{1/n}. \end{aligned}$$

Hence we have $$d(N) < \frac{\{(1/n)\log(p_1 p_2 p_3 \dots p_n N)\}^n}{\log p_1 \log p_2 \log p_3 \dots \log p_n}, \qquad \text{..................(3)}$$

for all values of N.

We shall now consider how near to this limit it is possible to make $d(N)$ by choice of the indices $a_1, a_2, a_3, \ldots, a_n$. Let us suppose that

$$1 + a_m = v\frac{\log p_n}{\log p_m} + \epsilon_m \quad (m = 1, 2, 3, \ldots, n), \qquad \ldots\ldots(4)$$

where v is a large integer and $-\frac{1}{2} < \epsilon_m < \frac{1}{2}$. Then, from (4), it is evident that

$$\epsilon_n = 0. \qquad \ldots\ldots(5)$$

Hence, by a well-known theorem due to Dirichlet*, it is possible to choose values of v as large as we please and such that

$$|\epsilon_1| < \epsilon, \quad |\epsilon_2| < \epsilon, \quad |\epsilon_3| < \epsilon, \ldots, \quad |\epsilon_{n-1}| < \epsilon, \qquad \ldots\ldots(6)$$

where $\epsilon \leqslant v^{-1/(n-1)}$. Now let

$$t = v \log p_n, \quad \delta_m = \epsilon_m \log p_m. \qquad \ldots\ldots(7)$$

Then from (1), (4) and (7) we have

$$\log(p_1 p_2 p_3 \ldots p_n N) = nt + \sum_1^n \delta_m. \qquad \ldots\ldots(8)$$

Similarly, from (2), (4) and (7), we see that

$$\begin{aligned} d(N) &= \frac{(t+\delta_1)(t+\delta_2)\ldots(t+\delta_n)}{\log p_1 \log p_2 \log p_3 \ldots \log p_n} \\ &= \frac{t^n \exp\left\{\frac{\Sigma\delta_m}{t} - \frac{\Sigma\delta_m^2}{2t^2} + \frac{\Sigma\delta_m^3}{3t^3} - \ldots\right\}}{\log p_1 \log p_2 \log p_3 \ldots \log p_n} \\ &= \left(t + \frac{\Sigma\delta_m}{n}\right)^n \frac{\exp\left\{-\frac{n\Sigma\delta_m^2 - (\Sigma\delta_m)^2}{2nt^2} + \frac{n^2\Sigma\delta_m^3 - (\Sigma\delta_m)^3}{3n^2t^3} - \ldots\right\}}{\log p_1 \log p_2 \log p_3 \ldots \log p_n} \\ &= \frac{\{(1/n)\log(p_1p_2p_3\ldots p_nN)\}^n}{\log p_1 \log p_2 \ldots \log p_n}[1 - \tfrac{1}{2}(\log N)^{-2}\{n^2\Sigma\delta_m^2 - n(\Sigma\delta_m)^2\} + \ldots], \end{aligned} \qquad \ldots\ldots(9)$$

in virtue of (8). From (6), (7) and (9) it follows that it is possible to choose the indices $a_1, a_2, \ldots, a_n$, so that

$$d(N) = \frac{\{(1/n)\log(p_1p_2p_3\ldots p_nN)\}^n}{\log p_1 \log p_2 \ldots \log p_n}\{1 - O(\log N)^{-2n/(n-1)}\}, \ldots(10)$$

where the symbol O has its ordinary meaning.

The following examples shew how close an approximation to $d(N)$ may be given by the right-hand side of (3). If

$$N = 2^{72}.7^{25},$$

then, according to (3), we have

$$d(N) < 1898{\cdot}00000685\ldots; \qquad \ldots\ldots(11)$$

and as a matter of fact $d(N) = 1898$. Similarly, taking

$$N = 2^{568}.3^{358},$$

* *Werke*, Vol. 1, p. 635.

we have, by (3), $d(N) < 204271{\cdot}000000372\ldots;$ ....(12)

while the actual value of $d(N)$ is 204271. In a similar manner, when

$$N = 2^{64} \,.\, 3^{40} \,.\, 5^{27},$$

we have, by (3), $d(N) < 74620{\cdot}00412\ldots;$ ....(13)

while actually $d(N) = 74620.$

3. Now let us suppose that, while the number n of different prime factors of N remains fixed, the primes p_ν, as well as the indices a_ν, are allowed to vary. It is evident that $d(N)$, considered as a function of N, is greatest when the primes p_ν are the first n primes, say 2, 3, 5, ..., p, where p is the nth prime. It therefore follows from (3) that

$$d(N) < \frac{\{(1/n)\log(2\,.\,3\,.\,5\,\ldots\,p\,.\,N)\}^n}{\log 2 \log 3 \log 5 \ldots \log p}, \qquad \text{..............(14)}$$

and from (10) that it is possible to choose the indices so that

$$d(N) = \frac{\{(1/n)\log(2\,.\,3\,.\,5\,\ldots\,p\,.\,N)\}^n}{\log 2 \log 3 \log 5 \ldots \log p}\{1 - O(\log N)^{-2n/(n-1)}\}. \quad \text{...(15)}$$

4. Before we proceed to consider the most general case, in which nothing is known about N, we must prove certain preliminary results. Let $\pi(x)$ denote the number of primes not exceeding x, and let

$$\vartheta(x) = \log 2 + \log 3 + \log 5 + \ldots + \log p,$$

and

$$\varpi(x) = \log 2 \,.\, \log 3 \,.\, \log 5 \ldots \log p,$$

where p is the largest prime not greater than x; also let $\phi(t)$ be a function of t such that $\phi'(t)$ is continuous between 2 and x. Then

$$\begin{aligned}\int_2^x \pi(t)\,\phi'(t)\,dt &= \int_2^3 \phi'(t)\,dt + 2\int_3^5 \phi'(t)\,dt + 3\int_5^7 \phi'(t)\,dt \\ &\quad + 4\int_7^{11} \phi'(t)\,dt + \ldots + \pi(x)\int_p^x \phi'(t)\,dt \\ &= \{\phi(3) - \phi(2)\} + 2\{\phi(5) - \phi(3)\} + 3\{\phi(7) - \phi(5)\} \\ &\quad + 4\{\phi(11) - \phi(7)\} + \ldots + \pi(x)\{\phi(x) - \phi(p)\} \\ &= \pi(x)\,\phi(x) - \{\phi(2) + \phi(3) + \phi(5) + \ldots + \phi(p)\}. \quad \text{...(16)}\end{aligned}$$

As an example let us suppose that $\phi(t) = \log t$. Then we have

$$\pi(x)\log x - \vartheta(x) = \int_2^x \frac{\pi(t)}{t}\,dt. \qquad \text{.................(17)}$$

Again let us suppose that $\phi(t) = \log\log t$. Then we see that

$$\pi(x)\log\log x - \log\varpi(x) = \int_2^x \frac{\pi(t)}{t\log t}\,dt. \qquad \text{..............(18)}$$

But $$\int_2^x \frac{\pi(t)}{t\log t}\,dt = \frac{1}{\log x}\int_2^x \frac{\pi(t)}{t}\,dt + \int_2^x \left(\frac{1}{u(\log u)^2}\int_2^u \frac{\pi(t)}{t}\,dt\right)du.$$

Hence we have

$$\pi(x)\log\left\{\frac{\vartheta(x)}{\pi(x)}\right\}-\log\varpi(x)$$

$$=\pi(x)\log\left\{\frac{\vartheta(x)}{\pi(x)\log x}\right\}+\frac{1}{\log x}\int_2^x\frac{\pi(t)}{t}dt+\int_2^x\left(\frac{1}{u(\log u)^2}\int_2^u\frac{\pi(t)}{t}dt\right)du. \quad \text{.........(19)}$$

But $$\pi(x)\log\left\{\frac{\vartheta(x)}{\pi(x)\log x}\right\}=\pi(x)\log\left\{1-\frac{\pi(x)\log x-\vartheta(x)}{\pi(x)\log x}\right\}$$

$$=\pi(x)\log\left\{1-\frac{1}{\pi(x)\log x}\int_2^x\frac{\pi(t)}{t}dt\right\}<-\frac{1}{\log x}\int_2^x\frac{\pi(t)}{t}dt;$$

and so $$\pi(x)\log\left\{\frac{\vartheta(x)}{\pi(x)\log x}\right\}+\frac{1}{\log x}\int_2^x\frac{\pi(t)}{t}dt<0. \quad \text{............(20)}$$

Again,

$$\pi(x)\log\left\{\frac{\vartheta(x)}{\pi(x)\log x}\right\}=-\pi(x)\log\left\{1+\frac{\pi(x)\log x-\vartheta(x)}{\vartheta(x)}\right\}$$

$$=-\pi(x)\log\left\{1+\frac{1}{\vartheta(x)}\int_2^x\frac{\pi(t)}{t}dt\right\}>-\frac{\pi(x)}{\vartheta(x)}\int_2^x\frac{\pi(t)}{t}dt;$$

and so

$$\pi(x)\log\left\{\frac{\vartheta(x)}{\pi(x)\log x}\right\}+\frac{1}{\log x}\int_2^x\frac{\pi(t)}{t}dt>-\frac{\pi(x)\log x-\vartheta(x)}{\vartheta(x)\log x}\int_2^x\frac{\pi(t)}{t}dt$$

$$=-\frac{1}{\vartheta(x)\log x}\left\{\int_2^x\frac{\pi(t)}{t}dt\right\}^2. \quad \text{............(21)}$$

It follows from (19), (20) and (21) that

$$\int_2^x\left(\frac{1}{u(\log u)^2}\int_2^u\frac{\pi(t)}{t}dt\right)du>\pi(x)\log\left\{\frac{\vartheta(x)}{\pi(x)}\right\}-\log\varpi(x)$$

$$>\int_2^x\left(\frac{1}{u(\log u)^2}\int_2^u\frac{\pi(t)}{t}dt\right)du-\frac{1}{\vartheta(x)\log x}\left\{\int_2^x\frac{\pi(t)}{t}dt\right\}^2.$$

Now it is easily proved by elementary methods* that

$$\pi(x)=O\left(\frac{x}{\log x}\right),\quad \frac{1}{\vartheta(x)}=O\left(\frac{1}{x}\right);$$

and so $$\int_2^x\frac{\pi(t)}{t}dt=O\left(\frac{x}{\log x}\right).$$

Hence $$\int_2^x\left(\frac{1}{u(\log u)^2}\int_2^u\frac{\pi(t)}{t}dt\right)du=\int_2^x O\left\{\frac{1}{(\log u)^3}\right\}du=O\left\{\frac{x}{(\log x)^3}\right\};$$

and $$\frac{1}{\vartheta(x)\log x}\left\{\int_2^x\frac{\pi(t)}{t}dt\right\}^2=\frac{1}{\vartheta(x)\log x}O\left\{\frac{x^2}{(\log x)^2}\right\}=O\left\{\frac{x}{(\log x)^3}\right\}.$$

Hence we see that $$\frac{\{\vartheta(x)/\pi(x)\}^{\pi(x)}}{\varpi(x)}=e^{O[x/(\log x)^3]}. \quad \text{....................(22)}$$

* See Landau, *Handbuch*, pp. 71 *et seq.*

5. We proceed to consider the case in which nothing is known about N. Let

$$N' = 2^{a_1} . 3^{a_2} . 5^{a_3} \ldots p^{a_n}.$$

Then it is evident that $d(N) = d(N')$, and that

$$\vartheta(p) \leqslant \log N' \leqslant \log N. \quad \ldots\ldots(23)$$

It follows from (3) that

$$d(N) = d(N') < \frac{1}{\varpi(p)} \left\{\frac{\vartheta(p) + \log N'}{\pi(p)}\right\}^{\pi(p)}$$

$$\leqslant \left\{1 + \frac{\log N}{\vartheta(p)}\right\}^{\pi(p)} \frac{\{\vartheta(p)/\pi(p)\}^{\pi(p)}}{\varpi(p)}$$

$$= \left\{1 + \frac{\log N}{\vartheta(p)}\right\}^{\pi(p)} e^{O[p/(\log p)^3]} = \left\{1 + \frac{\log N}{\vartheta(p)}\right\}^{\pi(p)+O[p/(\log p)^3]}, \quad \ldots(24)$$

in virtue of (22) and (23). But from (17) we know that

$$\pi(p) \log p - \vartheta(p) = O\left(\frac{p}{\log p}\right);$$

and so $\quad \vartheta(p) = \pi(p)\{\log p + O(1)\} = \pi(p)\{\log \vartheta(p) + O(1)\}.$

Hence $$\pi(p) = \vartheta(p)\left\{\frac{1}{\log \vartheta(p)} + O\frac{1}{\{\log \vartheta(p)\}^2}\right\}. \quad \ldots\ldots(25)$$

It follows from (24) and (25) that

$$d(N) \leqslant \left\{1 + \frac{\log N}{\vartheta(p)}\right\}^{\frac{\vartheta(p)}{\log \vartheta(p)} + O\frac{\vartheta(p)}{[\log \vartheta(p)]^2}}.$$

Writing t instead of $\vartheta(p)$, we have

$$d(N) \leqslant \left(1 + \frac{\log N}{t}\right)^{\frac{t}{\log t} + O\frac{t}{(\log t)^2}}; \quad \ldots\ldots(26)$$

and from (23) we have $$t \leqslant \log N. \quad \ldots\ldots(27)$$

Now, if N is a function of t, the order of the right-hand side of (26), considered as a function of N, is increased when N is decreased in comparison with t, and decreased when N is increased in comparison with t. Thus the most unfavourable hypothesis is that N, considered as a function of t, is as small as is compatible with the relation (27). We may therefore write $\log N$ for t in (26). Hence

$$d(N) < 2^{\frac{\log N}{\log \log N} + O\frac{\log N}{(\log \log N)^2}}, \quad \ldots\ldots(28)$$

for all values of N*.

* If we assume *nothing* about $\pi(x)$, we can shew that

$$d(N) < 2^{\frac{\log N}{\log \log N} + O\frac{\log N \log \log \log N}{(\log \log N)^2}}.$$

If we assume the prime number theorem, and nothing more, we can shew that

$$d(N) < 2^{\frac{\log N}{\log \log N} + [1 + o(1)]\frac{\log N}{(\log \log N)^2}}.$$

If we assume that $$\pi(x) = \frac{x}{\log x} + O\frac{x}{(\log x)^2},$$

we can shew that $$d(N) < 2^{\frac{\log N}{\log \log N} + \frac{\log N}{(\log \log N)^2} + O\frac{\log N}{(\log \log N)^3}}.$$

The inequality (28) has been proved by purely elementary reasoning. We have not assumed, for example, the prime number theorem, expressed by the relation

$$\pi(x) \sim \frac{x}{\log x}\,*.$$

We can also, without assuming this theorem, shew that the right-hand side of (28) is actually the order of $d(N)$ for an infinity of values of N. Let us suppose that

$$N = 2 \,.\, 3 \,.\, 5 \,.\, 7 \ldots p.$$

Then
$$d(N) = 2^{\pi(p)} = 2^{\frac{t}{\log t} + O\frac{t}{(\log t)^2}},$$

in virtue of (25). Since $\log N = \vartheta(p) = t$, we see that

$$d(N) = 2^{\frac{\log N}{\log\log N} + O\frac{\log N}{(\log\log N)^2}},$$

for an infinity of values of N. Hence the maximum order of $d(N)$ is

$$2^{\frac{\log N}{\log\log N} + O\frac{\log N}{(\log\log N)^2}}.$$

III.

The Structure of Highly Composite Numbers.

6. A number N may be said to be a highly composite number, if $d(N') < d(N)$ for all values of N' less than N. It is easy to see from the definition that, if N is highly composite and $d(N') > d(N)$, then there is at least one highly composite number M, such that

$$N < M \leqslant N'. \qquad \text{.............................}(29)$$

If N and N' are consecutive highly composite numbers, then $d(M) \leqslant d(N)$ for all values of M between N and N'. It is obvious that

$$d(N) < d(2N) \qquad \text{.............................}(30)$$

for all values of N. It follows from (29) and (30) that, if N is highly composite, then there is at least one highly composite number M such that $N < M \leqslant 2N$. That is to say, there is at least one highly composite number N, such that

$$x < N \leqslant 2x, \qquad \text{.............................}(31)$$

if
$$x \geqslant 1.$$

7. I do not know of any method for determining consecutive highly composite numbers except by trial. The following table gives the consecutive highly composite values of N, and the corresponding values of $d(N)$ and $dd(N)$, up to $d(N) = 10080$.

The numbers marked with the asterisk in the table are called superior highly composite numbers. Their definition and properties will be found in §§ 32, 33.

* $\phi(x) \sim \psi(x)$ means that $\phi(x)/\psi(x) \to 1$ as $x \to \infty$.

$dd(N)$	$d(N)$	N
2	$2=2$	*$2=2$
2	$3=3$	$4=2^2$
3	$4=2^2$	*$6=2.3$
4	$6=2.3$	*$12=2^2.3$
4	$8=2^3$	$24=2^3.3$
3	$9=3^2$	$36=2^2.3^2$
4	$10=2.5$	$48=2^4.3$
6	$12=2^2.3$	*$60=2^2.3.5$
5	$16=2^4$	*$120=2^3.3.5$
6	$18=2.3^2$	$180=2^2.3^2.5$
6	$20=2^2.5$	$240=2^4.3.5$
8	$24=2^3.3$	*$360=2^3.3^2.5$
8	$30=2.3.5$	$720=2^4.3^2.5$
6	$32=2^5$	$840=2^3.3.5.7$
9	$36=2^2.3^2$	$1260=2^2.3^2.5.7$
8	$40=2^3.5$	$1680=2^4.3.5.7$
10	$48=2^4.3$	*$2520=2^3.3^2.5.7$
12	$60=2^2.3.5$	*$5040=2^4.3^2.5.7$
7	$64=2^6$	$7560=2^3.3^3.5.7$
12	$72=2^3.3^2$	$10080=2^5.3^2.5.7$
10	$80=2^4.5$	$15120=2^4.3^3.5.7$
12	$84=2^2.3.7$	$20160=2^6.3^2.5.7$
12	$90=2.3^2.5$	$25200=2^4.3^2.5^2.7$
12	$96=2^5.3$	$27720=2^3.3^2.5.7.11$
9	$100=2^2.5^2$	$45360=2^4.3^4.5.7$
12	$108=2^2.3^3$	$50400=2^5.3^2.5^2.7$
16	$120=2^3.3.5$	*$55440=2^4.3^2.5.7.11$
8	$128=2^7$	$83160=2^3.3^3.5.7.11$
15	$144=2^4.3^2$	$110880=2^5.3^2.5.7.11$
12	$160=2^5.5$	$166320=2^4.3^3.5.7.11$
16	$168=2^3.3.7$	$221760=2^6.3^2.5.7.11$
18	$180=2^2.3^2.5$	$277200=2^4.3^2.5^2.7.11$
14	$192=2^6.3$	$332640=2^5.3^3.5.7.11$
12	$200=2^3.5^2$	$498960=2^4.3^4.5.7.11$
16	$216=2^3.3^3$	$554400=2^5.3^2.5^2.7.11$
12	$224=2^5.7$	$665280=2^6.3^3.5.7.11$
20	$240=2^4.3.5$	*$720720=2^4.3^2.5.7.11.13$
9	$256=2^8$	$1081080=2^3.3^3.5.7.11.13$
18	$288=2^5.3^2$	*$1441440=2^5.3^2.5.7.11.13$
14	$320=2^6.5$	$2162160=2^4.3^3.5.7.11.13$
20	$336=2^4.3.7$	$2882880=2^6.3^2.5.7.11.13$
24	$360=2^3.3^2.5$	$3603600=2^4.3^2.5^2.7.11.13$
16	$384=2^7.3$	*$4324320=2^5.3^3.5.7.11.13$
15	$400=2^4.5^2$	$6486480=2^4.3^4.5.7.11.13$
20	$432=2^4.3^3$	$7207200=2^5.3^2.5^2.7.11.13$
14	$448=2^6.7$	$8648640=2^6.3^3.5.7.11.13$
24	$480=2^5.3.5$	$10810800=2^4.3^3.5^2.7.11.13$
24	$504=2^3.3^2.7$	$14414400=2^6.3^2.5^2.7.11.13$
10	$512=2^9$	$17297280=2^7.3^3.5.7.11.13$
21	$576=2^6.3^2$	*$21621600=2^5.3^3.5^2.7.11.13$
24	$600=2^3.3.5^2$	$32432400=2^4.3^4.5^2.7.11.13$
16	$640=2^7.5$	$36756720=2^4.3^3.5.7.11.13.17$
24	$672=2^5.3.7$	$43243200=2^6.3^3.5^2.7.11.13$
30	$720=2^4.3^2.5$	$61261200=2^4.3^2.5^2.7.11.13.17$
18	$768=2^8.3$	$73513440=2^5.3^3.5.7.11.13.17$
18	$800=2^5.5^2$	$110270160=2^4.3^4.5.7.11.13.17$
24	$864=2^5.3^3$	$122522400=2^5.3^2.5^2.7.11.13.17$
16	$896=2^7.7$	$147026880=2^6.3^3.5.7.11.13.17$

$dd(N)$	$d(N)$	N
28	$960=2^6.3.5$	$183783600=2^4.3^3.5^2.7.11.13.17$
30	$1008=2^4.3^2.7$	$245044800=2^6.3^2.5^2.7.11.13.17$
11	$1024=2^{10}$	$294053760=2^7.3^3.5.7.11.13.17$
24	$1152=2^7.3^2$	$*367567200=2^5.3^3.5^2.7.11.13.17$
30	$1200=2^4.3.5^2$	$551350800=2^4.3^4.5^2.7.11.13.17$
18	$1280=2^8.5$	$698377680=2^4.3^3.5.7.11.13.17.19$
28	$1344=2^6.3.7$	$735134400=2^6.3^3.5^2.7.11.13.17$
36	$1440=2^5.3^2.5$	$1102701600=2^5.3^4.5^2.7.11.13.17$
20	$1536=2^9.3$	$1396755360=2^5.3^3.5.7.11.13.17.19$
21	$1600=2^6.5^2$	$2095133040=2^4.3^4.5.7.11.13.17.19$
40	$1680=2^4.3.5.7$	$2205403200=2^6.3^4.5^2.7.11.13.17$
28	$1728=2^6.3^3$	$2327925600=2^5.3^2.5^2.7.11.13.17.19$
18	$1792=2^8.7$	$2793510720=2^6.3^3.5.7.11.13.17.19$
32	$1920=2^7.3.5$	$3491888400=2^4.3^3.5^2.7.11.13.17.19$
36	$2016=2^5.3^2.7$	$4655851200=2^6.3^2.5^2.7.11.13.17.19$
12	$2048=2^{11}$	$5587021440=2^7.3^3.5.7.11.13.17.19$
27	$2304=2^8.3^2$	$*6983776800=2^5.3^3.5^2.7.11.13.17.19$
36	$2400=2^5.3.5^2$	$10475665200=2^4.3^4.5^2.7.11.13.17.19$
32	$2688=2^7.3.7$	$*13967553600=2^6.3^3.5^2.7.11.13.17.19$
42	$2880=2^6.3^2.5$	$20951330400=2^5.3^4.5^2.7.11.13.17.19$
22	$3072=2^{10}.3$	$27935107200=2^7.3^3.5^2.7.11.13.17.19$
48	$3360=2^5.3.5.7$	$41902660800=2^6.3^4.5^2.7.11.13.17.19$
32	$3456=2^7.3^3$	$48886437600=2^5.3^3.5^2.7^2.11.13.17.19$
20	$3584=2^9.7$	$64250746560=2^6.3^3.5.7.11.13.17.19.23$
45	$3600=2^4.3^2.5^2$	$73329656400=2^4.3^4.5^2.7^2.11.13.17.19$
36	$3840=2^8.3.5$	$80313433200=2^4.3^3.5^2.7.11.13.17.19.23$
42	$4032=2^6.3^2.7$	$97772875200=2^6.3^3.5^2.7^2.11.13.17.19$
13	$4096=2^{12}$	$128501493120=2^7.3^3.5.7.11.13.17.19.23$
48	$4320=2^5.3^3.5$	$146659312800=2^5.3^4.5^2.7^2.11.13.17.19$
30	$4608=2^9.3^2$	$160626866400=2^5.3^3.5^2.7.11.13.17.19.23$
42	$4800=2^6.3.5^2$	$240940299600=2^4.3^4.5^2.7.11.13.17.19.23$
60	$5040=7.5.3^2.2^4$	$\dagger 293318625600=2^6.3^4.5^2.7^2.11.13.17.19$
36	$5376=2^8.3.7$	$*321253732800=2^6.3^3.5^2.7.11.13.17.19.23$
48	$5760=2^7.3^2.5$	$481880599200=2^5.3^4.5^2.7.11.13.17.19.23$
24	$6144=2^{11}.3$	$642507465600=2^7.3^3.5^2.7.11.13.17.19.23$
56	$6720=2^6.3.5.7$	$963761198400=2^6.3^4.5^2.7.11.13.17.19.23$
36	$6912=2^8.3^3$	$1124388064800=2^5.3^3.5^2.7^2.11.13.17.19.23$
22	$7168=2^{10}.7$	$1606268664000=2^6.3^3.5^3.7.11.13.17.19.23$
54	$7200=2^5.3^2.5^2$	$1686582097200=2^4.3^4.5^2.7^2.11.13.17.19.23$
40	$7680=2^9.3.5$	$1927522396800=2^7.3^4.5^2.7.11.13.17.19.23$
48	$8064=2^7.3^2.7$	$*2248776129600=2^6.3^3.5^2.7^2.11.13.17.19.23$
14	$8192=2^{13}$	$3212537328000=2^7.3^3.5^3.7.11.13.17.19.23$
56	$8640=2^6.3^3.5$	$3373164194400=2^5.3^4.5^2.7^2.11.13.17.19.23$
33	$9216=2^{10}.3^2$	$4497552259200=2^7.3^3.5^2.7^2.11.13.17.19.23$
72	$10080=2^5.3^2.5.7$	$6746328388800=2^6.3^4.5^2.7^2.11.13.17.19.23$

8. Now let us consider what must be the nature of N in order that N should be a highly composite number. In the first place it must be of the form

$$2^{a_2}.3^{a_3}.5^{a_5}.7^{a_7}\dots p_1^{a_{p_1}},$$

where
$$a_2 \geqslant a_3 \geqslant a_5 \geqslant \dots \geqslant a_{p_1} \geqslant 1. \quad \dots\dots\dots\dots\dots(32)$$

This follows at once from the fact that

$$d(\varpi_2^{a_2}\varpi_3^{a_3}\varpi_5^{a_5}\dots\varpi_{p_1}^{a_{p_1}}) = d(2^{a_2}.3^{a_3}.5^{a_5}\dots p_1^{a_{p_1}}),$$

for all prime values of $\varpi_2, \varpi_3, \varpi_5, \dots, \varpi_{p_1}$.

† See Appendix, p. 339.

It follows from the definition that, if N is highly composite and $N' < N$, then $d(N')$ must be less than $d(N)$. For example, $\frac{5}{6}N < N$, and so $d(\frac{5}{6}N) < d(N)$. Hence

$$\left(1+\frac{1}{a_2}\right)\left(1+\frac{1}{a_3}\right) > \left(1+\frac{1}{1+a_5}\right),$$

provided that N is a multiple of 3.

It is convenient to write

$$a_\lambda = 0 \quad (\lambda > p_1). \qquad \text{.........................(33)}$$

Thus if N is not a multiple of 5 then a_5 should be considered as 0.

Again, a_{p_1} must be less than or equal to 2 for all values of p_1. For let P_1 be the prime next above p_1. Then it can be shewn that $P_1 < p_1^2$ for all values of p_1*. Now, if a_{p_1} is greater than 2, let

$$N' = \frac{NP_1}{p_1^2}.$$

Then N' is an integer less than N, and so $d(N') < d(N)$. Hence

$$(1+a_{p_1}) > 2(a_{p_1} - 1),$$

or

$$3 > a_{p_1},$$

which contradicts our hypothesis. Hence

$$a_{p_1} \leqslant 2, \qquad \text{...............................(34)}$$

for all values of p_1.

Now let p_1'', p_1', p_1, P_1, P_1' be consecutive primes in ascending order. Then, if $p_1 \geqslant 5$, $a_{p_1''}$ must be less than or equal to 4. For, if this were not so, we could suppose that

$$N' = \frac{NP_1}{(p_1'')^3}.$$

But it can easily be shewn that, if $p_1 \geqslant 5$, then

$$(p_1'')^3 > P_1;$$

and so $N' < N$ and $d(N') < d(N)$. Hence

$$(1+a_{p_1''}) > 2(a_{p_1''} - 2). \qquad \text{.......................(35)}$$

But since $a_{p_1''} \geqslant 5$, it is evident that

$$(1+a_{p_1''}) \leqslant 2(a_{p_1''} - 2),$$

* It can be proved by elementary methods that, if $x \geqslant 1$, there is at least one prime p such that $x < p \leqslant 2x$. This result is known as Bertrand's Postulate: for a proof, see Landau, *Handbuch*, p. 89. It follows at once that $P_1 < p_1^2$, if $p_1 > 2$; and the inequality is obviously true when $p_1 = 2$. Some similar results used later in this and the next section may be proved in the same kind of way. It is for some purposes sufficient to know that there is always a prime p such that $x < p < 3x$, and the proof of this is easier than that of Bertrand's Postulate. These inequalities are enough, for example, to shew that

$$\log P_1 = \log p_1 + O(1).$$

which contradicts (35); therefore, if $p_1 \geqslant 5$, then

$$a_{p_1''} \leqslant 4. \qquad \text{.................................(36)}$$

Now let $$N' = \frac{Np_1''P_1}{p_1'p_1}.$$

It is easy to verify that, if $5 \leqslant p_1 \leqslant 19$, then

$$p_1'p_1 > p_1''P_1;$$

and so $N' < N$ and $d(N') < d(N)$. Hence

$$(1 + a_{p_1})(1 + a_{p_1'})(1 + a_{p_1''}) > 2a_{p_1}a_{p_1'}(2 + a_{p_1''}),$$

or $$\left(1 + \frac{1}{a_{p_1}}\right)\left(1 + \frac{1}{a_{p_1'}}\right) > 2\left(1 + \frac{1}{1 + a_{p_1''}}\right).$$

But from (36) we know that $1 + a_{p_1''} \leqslant 5$. Hence

$$\left(1 + \frac{1}{a_{p_1}}\right)\left(1 + \frac{1}{a_{p_1'}}\right) > 2\tfrac{2}{5}. \qquad \text{.....................(37)}$$

From this it follows that $a_{p_1} = 1$. For, if $a_{p_1} \geqslant 2$, then

$$\left(1 + \frac{1}{a_{p_1}}\right)\left(1 + \frac{1}{a_{p_1'}}\right) \leqslant 2\tfrac{1}{4},$$

in virtue of (32). This contradicts (37). Hence, if $5 \leqslant p_1 \leqslant 19$, then

$$a_{p_1} = 1. \qquad \text{.................................(38)}$$

Next let $$N' = NP_1P_1'/(p_1p_1'p_1'').$$

It can easily be shewn that, if $p_1 \geqslant 11$, then

$$P_1P_1' < p_1p_1'p_1'';$$

and so $N' < N$ and $d(N') < d(N)$. Hence

$$(1 + a_{p_1})(1 + a_{p_1'})(1 + a_{p_1''}) > 4a_{p_1}a_{p_1'}a_{p_1''},$$

or $$\left(1 + \frac{1}{a_{p_1}}\right)\left(1 + \frac{1}{a_{p_1'}}\right)\left(1 + \frac{1}{a_{p_1''}}\right) > 4. \qquad \text{..................(39)}$$

From this we infer that a_{p_1} must be 1. For, if $a_{p_1} \geqslant 2$, it follows from (32) that

$$\left(1 + \frac{1}{a_{p_1}}\right)\left(1 + \frac{1}{a_{p_1'}}\right)\left(1 + \frac{1}{a_{p_1''}}\right) \leqslant 3\tfrac{3}{8},$$

which contradicts (39). Hence we see that, if $p_1 \geqslant 11$, then

$$a_{p_1} = 1. \qquad \text{.................................(40)}$$

It follows from (38) and (40) that, if $p_1 \geqslant 5$, then

$$a_{p_1} = 1. \qquad \text{.................................(41)}$$

But if $p_1 = 2$ or 3, then from (34) it is clear that

$$a_{p_1} = 1 \text{ or } 2. \qquad \text{.............................(42)}$$

It follows that $a_{p_1} = 1$ for all highly composite numbers, except for 2^2, and perhaps for certain numbers of the form $2^a . 3^2$. In the latter case $a \geqslant 2$.

It is easy to shew that, if $a \geqslant 3$, $2^a . 3^2$ cannot be highly composite. For if we suppose that

$$N' = 2^{a-1} . 3 . 5,$$

then it is evident that $N' < N$ and $d(N') < d(N)$, and so

$$3(1+a) > 4a,$$

or

$$a < 3.$$

Hence it is clear that a cannot have any other value except 2. Moreover we can see by actual trial that 2^2 and $2^2 . 3^2$ are highly composite. *Hence*

$$a_{p_1} = 1 \quad \text{.................................(43)}$$

for all highly composite values of N save 4 and 36, when

$$a_{p_1} = 2.$$

Hereafter when we use this result it is to be understood that 4 and 36 are exceptions.

9. It follows from (32) and (43) that N must be of the form

$$\begin{aligned} &2.3.5.7\ldots p_1 \\ \times &2.3.5.7\ldots p_2 \\ \times &2.3.5\ldots p_3 \\ \times &\ldots, \end{aligned} \quad \text{.................(44)}$$

where $p_1 > p_2 \geqslant p_3 \geqslant p_4 \geqslant \ldots$ and the number of rows is a_2.

Let P_r be the prime next above p_r, so that

$$\log P_r = \log p_r + O(1), \quad \text{........................(45)}$$

in virtue of Bertrand's Postulate. Then it is evident that

$$a_{p_r} \geqslant r, \quad a_{P_r} \leqslant r-1; \quad \text{..........................(46)}$$

and so

$$a_{P_r} \leqslant a_{p_r} - 1. \quad \text{..............................(47)}$$

It is to be understood that

$$a_{P_1} = 0, \quad \text{................................(48)}$$

in virtue of (33).

It is clear from the form of (44) that r can never exceed a_2, and that

$$p_{a_\lambda} = \lambda. \quad \text{................................(49)}$$

10. Now let

$$N' = \frac{N}{\nu} \lambda^{[\log \nu / \log \lambda]} *,$$

where $\nu \leqslant p_1$, so that N' is an integer. Then it is evident that $N' < N$ and $d(N') < d(N)$, and so

$$(1+a_\nu)(1+a_\lambda) > a_\nu \left(1 + a_\lambda + \left[\frac{\log \nu}{\log \lambda}\right]\right),$$

or

$$1 + a_\lambda > a_\nu \left[\frac{\log \nu}{\log \lambda}\right]. \quad \text{..........................(50)}$$

* $[x]$ denotes as usual the integral part of x.

Since the right-hand side vanishes when $\nu > p_1$, we see that (50) is true for all values of λ and ν*.

Again let $$N' = N\mu\lambda^{-1-[\log\mu/\log\lambda]},$$
where $[\log\mu/\log\lambda] < a_\lambda$, so that N' is an integer. Then it is evident that $N' < N$ and $d(N') < d(N)$, and so
$$(1+a_\mu)(1+a_\lambda) > (2+a_\mu)\left(a_\lambda - \left[\frac{\log\mu}{\log\lambda}\right]\right). \quad \text{.........(51)}$$
Since the right-hand side is less than or equal to 0 when
$$a_\lambda \leqslant [\log\mu/\log\lambda],$$
we see that (51) is true for all values of λ and μ. From (51) it evidently follows that
$$(1+a_\lambda) < (2+a_\mu)\left[\frac{\log(\lambda\mu)}{\log\lambda}\right]. \quad \text{.........(52)}$$
From (50) and (52) it is clear that
$$a_\nu\left[\frac{\log\nu}{\log\lambda}\right] \leqslant a_\lambda \leqslant a_\mu + (2+a_\mu)\left[\frac{\log\mu}{\log\lambda}\right], \quad \text{.........(53)}$$
for all values of λ, μ and ν.

Now let us suppose that $\nu = p_1$ and $\mu = P_1$, so that $a_\nu = 1$ and $a_\mu = 0$. Then we see that
$$\left[\frac{\log p_1}{\log\lambda}\right] \leqslant a_\lambda \leqslant 2\left[\frac{\log P_1}{\log\lambda}\right], \quad \text{.........(54)}$$
for all values of λ. Thus, for example, we have
$$p_1 = 3, \quad 1 \leqslant a_2 \leqslant 4;$$
$$p_1 = 5, \quad 2 \leqslant a_2 \leqslant 4;$$
$$p_1 = 7, \quad 2 \leqslant a_2 \leqslant 6;$$
$$p_1 = 11, \quad 3 \leqslant a_2 \leqslant 6;$$
and so on. It follows from (54) that, if $\lambda \leqslant p_1$, then
$$a_\lambda \log\lambda = O(\log p_1), \quad a_\lambda \log\lambda \neq o(\log p_1). \quad \text{.........(55)}$$

11. Again let
$$N' = N\lambda^{[\sqrt{\{(1+a_\lambda+a_\mu)\log\mu/\log\lambda\}}]}\mu^{-1-[\sqrt{\{(1+a_\lambda+a_\mu)\log\lambda/\log\mu\}}]},$$
and let us assume for the moment that
$$a_\mu > \sqrt{\{(1+a_\lambda+a_\mu)\log\lambda/\log\mu\}},$$
in order that N' may be an integer. Then $N' < N$ and $d(N') < d(N)$, and so
$$\begin{aligned}(1+a_\lambda)(1+a_\mu) > \; & \{1+a_\lambda+[\sqrt{\{(1+a_\lambda+a_\mu)\log\mu/\log\lambda\}}]\} \\ & \times\{a_\mu - [\sqrt{\{(1+a_\lambda+a_\mu)\log\lambda/\log\mu\}}]\} \\ > \; & \{a_\lambda + \sqrt{\{(1+a_\lambda+a_\mu)\log\mu/\log\lambda\}}\} \\ & \times\{a_\mu - \sqrt{\{(1+a_\lambda+a_\mu)\log\lambda/\log\mu\}}\}. \quad \text{......(56)}\end{aligned}$$

* That is to say all prime values of λ and ν, since λ in a_λ is by definition prime.

It is evident that the right-hand side of (56) becomes negative when

$$a_\mu < \sqrt{\{(1 + a_\lambda + a_\mu)\log\lambda/\log\mu\}},$$

while the left-hand side remains positive, and so the result is still true. Hence

$$a_\mu \log\mu - a_\lambda \log\lambda < 2\sqrt{\{(1 + a_\lambda + a_\mu)\log\lambda \log\mu\}}, \quad \text{.........(57)}$$

for all values of λ and μ. Interchanging λ and μ in (57), we obtain

$$a_\lambda \log\lambda - a_\mu \log\mu < 2\sqrt{\{(1 + a_\lambda + a_\mu)\log\lambda \log\mu\}}. \quad \text{.........(58)}$$

From (57) and (58) it evidently follows that

$$|a_\lambda \log\lambda - a_\mu \log\mu| < 2\sqrt{\{(1 + a_\lambda + a_\mu)\log\lambda \log\mu\}}, \quad \text{......(59)}$$

for all values of λ and μ. It follows from this and (55) that, if λ and μ are neither greater than p_1, then

$$a_\lambda \log\lambda - a_\mu \log\mu = O\sqrt{\{\log p_1 \log(\lambda\mu)\}}, \quad \text{...........(60)}$$

and so that, *if* $\log\lambda = o(\log p_1)$, *then*

$$a_2 \log 2 \sim a_3 \log 3 \sim a_5 \log 5 \sim \dots \sim a_\lambda \log\lambda. \quad \text{...........(61)}$$

12. It can easily be shewn by elementary algebra that, if x, y, m and n are not negative, and if

$$|x - y| < 2\sqrt{(mx + ny + mn)},$$

then

$$\left.\begin{aligned} |\sqrt{(x+n)} - \sqrt{(y+m)}| &< \sqrt{(m+n)}; \\ |\sqrt{(x+n)} - \sqrt{(m+n)}| &< \sqrt{(y+m)}. \end{aligned}\right\} \quad \text{..................(62)}$$

From (62) and (59) it follows that

$$|\sqrt{\{(1 + a_\lambda)\log\lambda\}} - \sqrt{\{(1 + a_\mu)\log\mu\}}| < \sqrt{\{\log(\lambda\mu)\}}, \quad \text{......(63)}$$

and

$$|\sqrt{\{(1 + a_\lambda)\log\lambda\}} - \sqrt{\{\log(\lambda\mu)\}}| < \sqrt{\{(1 + a_\mu)\log\mu\}}, \quad \text{......(64)}$$

for all values of λ and μ. If, in particular, we put $\mu = 2$ in (63), we obtain

$$\begin{aligned} \sqrt{\{(1 + a_2)\log 2\}} - \sqrt{\{\log(2\lambda)\}} &< \sqrt{\{(1 + a_\lambda)\log\lambda\}} \\ &< \sqrt{\{(1 + a_2)\log 2\}} + \sqrt{\{\log(2\lambda)\}}, \quad \text{...(65)} \end{aligned}$$

for all values of λ. Again, from (63), we have

$$(1 + a_\lambda)\log\lambda < (\sqrt{\{(1 + a_\nu)\log\nu\}} + \sqrt{\{\log(\lambda\nu)\}})^2,$$

or

$$a_\lambda \log\lambda < (1 + a_\nu)\log\nu + \log\nu + 2\sqrt{\{(1 + a_\nu)\log\nu \log(\lambda\nu)\}}. \quad \text{...(66)}$$

Now let us suppose that $\lambda \leqslant \mu$. Then, from (66), it follows that

$$\begin{aligned} a_\lambda \log\lambda + \log\mu &< (1 + a_\nu)\log\nu + \log(\mu\nu) + 2\sqrt{\{(1 + a_\nu)\log\nu \log(\lambda\nu)\}} \\ &\leqslant (1 + a_\nu)\log\nu + \log(\mu\nu) + 2\sqrt{\{(1 + a_\nu)\log\nu \log(\mu\nu)\}} \\ &= \{\sqrt{\{(1 + a_\nu)\log\nu\}} + \sqrt{\log(\mu\nu)}\}^2, \quad \text{.....................(67)} \end{aligned}$$

with the condition that $\lambda \leqslant \mu$. Similarly we can shew that

$$a_\lambda \log\lambda + \log\mu > \{\sqrt{\{(1 + a_\nu)\log\nu\}} - \sqrt{\log(\mu\nu)}\}^2, \quad \text{.........(67')}$$

with the condition that $\lambda \leqslant \mu$.

13. Now let

$$N' = \frac{N}{\lambda} 2^{[\log\lambda/\{\pi(\mu)\log 2\}]} 3^{[\log\lambda/\{\pi(\mu)\log 3\}]} \dots \mu^{[\log\lambda/\{\pi(\mu)\log\mu\}]},$$

where $\pi(\mu)\log\mu < \log\lambda \leqslant \log p_1$. Then it is evident that N' is an integer less than N, and so $d(N') < d(N)$. Hence

$$\left(1+\frac{1}{a_\lambda}\right)(1+a_2)(1+a_3)(1+a_5)\dots(1+a_\mu)$$

$$> \left\{a_2 + \frac{\log\lambda}{\pi(\mu)\log 2}\right\}\left\{a_3 + \frac{\log\lambda}{\pi(\mu)\log 3}\right\}\dots\left\{a_\mu + \frac{\log\lambda}{\mu(\mu)\log\mu}\right\};$$

that is

$$\left\{a_2\log 2 + \frac{\log\lambda}{\pi(\mu)}\right\}\left\{a_3\log 3 + \frac{\log\lambda}{\pi(\mu)}\right\}\dots\left\{a_\mu\log\mu + \frac{\log\lambda}{\pi(\mu)}\right\}$$

$$< \left(1+\frac{1}{a_\lambda}\right)(a_2\log 2 + \log 2)(a_3\log 3 + \log 3)\dots(a_\mu\log\mu + \log\mu)$$

$$\leqslant \left(1+\frac{1}{a_\lambda}\right)(a_2\log 2 + \log\mu)(a_3\log 3 + \log\mu)\dots(a_\mu\log\mu + \log\mu).$$

In other words,

$$\left(1+\frac{1}{a_\lambda}\right)$$

$$> \left\{1 + \frac{\frac{\log\lambda}{\pi(\mu)} - \log\mu}{a_2\log 2 + \log\mu}\right\}\left\{1 + \frac{\frac{\log\lambda}{\pi(\mu)} - \log\mu}{a_3\log 3 + \log\mu}\right\}\dots\left\{1 + \frac{\frac{\log\lambda}{\pi(\mu)} - \log\mu}{a_\mu\log\mu + \log\mu}\right\}$$

$$> \left\{1 + \frac{\frac{\log\lambda}{\pi(\mu)} - \log\mu}{[\sqrt{\{(1+a_\nu)\log\nu\}} + \sqrt{\log(\mu\nu)}]^2}\right\}^{\pi(\mu)}, \quad \dots\dots(68)$$

where ν is any prime, in virtue of (67). From (68) it follows that

$$\sqrt{\{(1+a_\nu)\log\nu\}} + \sqrt{\log(\mu\nu)} > \sqrt{\left\{\frac{\frac{\log\lambda}{\pi(\mu)} - \log\mu}{\left(1+\frac{1}{a_\lambda}\right)^{1/\pi(\mu)} - 1}\right\}}, \quad \dots\dots(69)$$

provided that $\pi(\mu)\log\mu < \log\lambda \leqslant \log p_1$.

14. Again let

$$N' = N\lambda 2^{-1-[\log\lambda/\{\pi(\mu)\log 2\}]} 3^{-1-[\log\lambda/\{\pi(\mu)\log 3\}]} \dots \mu^{-1-[\log\lambda/\{\pi(\mu)\log\mu\}]},$$

where $\mu \leqslant p_1$ and $\lambda > \mu$. Let us assume for the moment that

$$a_\kappa \log\kappa > \frac{\log\lambda}{\pi(\mu)},$$

for all values of κ less than or equal to μ, so that N' is an integer.

Then, by arguments similar to those of the previous section, we can shew that

$$\frac{1+a_\lambda}{2+a_\lambda} > \left\{1 - \frac{\dfrac{\log\lambda}{\pi(\mu)} + \log\mu}{\{\sqrt{\{(1+a_\nu)\log\nu\}} - \sqrt{\log(\mu\nu)}\}^2}\right\}^{\pi(\mu)} \quad \text{.........(70)}$$

From this it follows that

$$|\sqrt{\{(1+a_\nu)\log\nu\}} - \sqrt{\log(\mu\nu)}\,| < \sqrt{\left\{\frac{\dfrac{\log\lambda}{\pi(\mu)} + \log\mu}{1 - \left(\dfrac{1+a_\lambda}{2+a_\lambda}\right)^{1/\pi(\mu)}}\right\}}, \quad \text{...(71)}$$

provided that $\mu \leqslant p_1$ and $\mu < \lambda$. The condition that

$$a_\kappa \log\kappa > \{\log\lambda/\pi(\mu)\}$$

is unnecessary because we know from (67′) that

$$|\sqrt{\{(1+a_\nu)\log\nu\}} - \sqrt{\log(\mu\nu)}\,| < \sqrt{(a_\kappa\log\kappa + \log\mu)} \leqslant \sqrt{\left\{\frac{\log\lambda}{\pi(\mu)} + \log\mu\right\}}, \quad \text{.........(72)}$$

when $$a_\kappa \log\kappa \leqslant \{\log\lambda/\pi(\mu)\},$$

and the last term in (72) is evidently less than the right-hand side of (71).

15. We shall consider in this and the following sections some important deductions from the preceding formulæ. Putting $\nu = 2$ in (69) and (71), we obtain

$$\sqrt{\{(1+a_2)\log 2\}} > \sqrt{\left\{\frac{\dfrac{\log\lambda}{\pi(\mu)} - \log\mu}{\left(1 + \dfrac{1}{a_\lambda}\right)^{1/\pi(\mu)} - 1}\right\}} - \sqrt{\log(2\mu)}, \quad \text{...(73)}$$

provided that $\pi(\mu)\log\mu < \log\lambda \leqslant \log p_1$, and

$$\sqrt{\{(1+a_2)\log 2\}} < \sqrt{\left\{\frac{\dfrac{\log\lambda}{\pi(\mu)} + \log\mu}{1 - \left(\dfrac{1+a_\lambda}{2+a_\lambda}\right)^{1/\pi(\mu)}}\right\}} + \sqrt{\log(2\mu)}, \quad \text{...(74)}$$

provided that $\mu \leqslant p_1$, and $\mu < \lambda$. Now supposing that $\lambda = p_1$ in (73), and $\lambda = P_1$ in (74), we obtain

$$\sqrt{\{(1+a_2)\log 2\}} > \sqrt{\left\{\frac{\dfrac{\log p_1}{\pi(\mu)} - \log\mu}{2^{1/\pi(\mu)} - 1}\right\}} - \sqrt{\log(2\mu)}, \quad \text{.........(75)}$$

provided that $\pi(\mu)\log\mu < \log p_1$, and

$$\sqrt{\{(1+a_2)\log 2\}} < \sqrt{\left\{\frac{\dfrac{\log P_1}{\pi(\mu)} + \log\mu}{1 - 2^{-1/\pi(\mu)}}\right\}} + \sqrt{\log(2\mu)}, \quad \text{.........(76)}$$

provided that $\mu \leqslant p_1$. In (75) and (76) μ can be so chosen as to obtain the best possible inequality for a_2. If p_1 is too small, we may abandon this result in favour of

$$\left[\frac{\log p_1}{\log 2}\right] \leqslant a_2 \leqslant 2\left[\frac{\log P_1}{\log 2}\right], \quad \text{......................(77)}$$

which is obtained from (54) by putting $\lambda = 2$.

After having obtained in this way what information we can about a_2, we may use (73) and (74) to obtain information about a_λ. Here also we have to choose μ so as to obtain the best possible inequality for a_λ. But if λ is too small we may, instead of this, use

$$\sqrt{\{(1+a_2)\log 2\}} - \sqrt{\log(2\lambda)} < \sqrt{\{(1+a_\lambda)\log\lambda\}}$$
$$< \sqrt{\{(1+a_2)\log 2\}} + \sqrt{\log(2\lambda)}, \quad \text{......(78)}$$

which is obtained by putting $\mu = 2$ in (63).

16. Now let us consider the order of a_2. From (73) it is evident that, if $\pi(\mu)\log\mu < \log\lambda \leqslant \log p_1$, then

$$(1+a_2)\log 2 + \log(2\mu) + 2\sqrt{\{(1+a_2)\log 2\log(2\mu)\}} > \frac{\dfrac{\log\lambda}{\pi(\mu)} - \log\mu}{\left(1+\dfrac{1}{a_\lambda}\right)^{1/\pi(\mu)} - 1}.$$

But we know that for positive values of x, ...(79)

$$\frac{1}{e^x - 1} = \frac{1}{x} + O(1), \quad \frac{1}{e^x-1} = O\left(\frac{1}{x}\right).$$

Hence
$$\frac{\log\lambda}{\pi(\mu)}\frac{1}{\left(1+\dfrac{1}{a_\lambda}\right)^{1/\pi(\mu)} - 1} = \frac{\log\lambda}{\pi(\mu)}\left\{\frac{\pi(\mu)}{\log\left(1+\dfrac{1}{a_\lambda}\right)} + O(1)\right\}$$
$$= \frac{\log\lambda}{\log\left(1+\dfrac{1}{a_\lambda}\right)} + O\left\{\frac{\log\lambda}{\pi(\mu)}\right\};$$

and
$$\frac{\log\mu}{\left(1+\dfrac{1}{a_\lambda}\right)^{1/\pi(\mu)} - 1} = O\left\{\frac{\pi(\mu)\log\mu}{\log\left(1+\dfrac{1}{a_\lambda}\right)}\right\} = O(\mu a_\lambda).$$

Again from (55) we know that $a_2 = O(\log p_1)$. Hence (79) may be written as

$$a_2\log 2 + O\sqrt{(\log p_1\log\mu)} + O(\log\mu)$$
$$\geqslant \frac{\log\lambda}{\log\left(1+\dfrac{1}{a_\lambda}\right)} + O\left\{\frac{\log\lambda}{\pi(\mu)}\right\} + O(\mu a_\lambda). \quad \text{...(80)}$$

But
$$\log\mu = O(\mu a_\lambda),$$
$$\mu a_\lambda = \frac{\mu}{\log\lambda}\cdot a_\lambda\log\lambda = O\left(\frac{\mu\log p_1}{\log\lambda}\right),$$
$$\frac{\log\lambda}{\pi(\mu)} = O\left\{\frac{\log\lambda\log\mu}{\mu}\right\}.$$

Again
$$\frac{\log\lambda\log\mu}{\mu}+\frac{\mu\log p_1}{\log\lambda}>2\surd(\log p_1\log\mu);$$

and so
$$\surd(\log p_1\log\mu)=O\left(\frac{\log\lambda\log\mu}{\mu}+\frac{\mu\log p_1}{\log\lambda}\right).$$

Hence (80) may be replaced by

$$a_2\log 2\geqslant\frac{\log\lambda}{\log\left(1+\frac{1}{a_\lambda}\right)}+O\left(\frac{\log\lambda\log\mu}{\mu}+\frac{\mu\log p_1}{\log\lambda}\right),\quad\ldots\ldots(81)$$

provided that $\pi(\mu)\log\mu<\log\lambda\leqslant\log p_1$. Similarly, from (74), we can shew that

$$a_2\log 2\leqslant\frac{\log\lambda}{\log\left(1+\frac{1}{1+a_\lambda}\right)}+O\left(\frac{\log\lambda\log\mu}{\mu}+\frac{\mu\log p_1}{\log\lambda}\right),\ \ldots(82)$$

provided that $\mu\leqslant p_1$ and $\mu<\lambda$. Now supposing that $\lambda=p_1$ in (81), and $\lambda=P_1$ in (82), and also that

$$\mu=O\surd(\log p_1\log\log p_1),\quad \mu\neq o\surd(\log p_1\log\log p_1)^*,$$

we obtain
$$\left.\begin{aligned}a_2\log 2&\geqslant\frac{\log p_1}{\log 2}+O\surd(\log p_1\log\log p_1),\\ a_2\log 2&\leqslant\frac{\log p_1}{\log 2}+O\surd(\log p_1\log\log p_1).\end{aligned}\right\}\ \ldots\ldots\ldots\ldots\ldots(83)$$

From (83) it evidently follows that

$$a_2\log 2=\frac{\log p_1}{\log 2}+O\surd(\log p_1\log\log p_1).\quad\ldots\ldots\ldots\ldots\ldots(84)$$

And it follows from this and (60) that if $\lambda\leqslant p_1$ then

$$a_\lambda\log\lambda=\frac{\log p_1}{\log 2}+O\{\surd(\log p_1\log\lambda)+\surd(\log p_1\log\log p_1)\}.\ \ldots\ldots(85)$$

Hence, if $\log\lambda=o(\log p_1)$, we have

$$a_2\log 2\sim a_3\log 3\sim a_5\log 5\sim\ldots\sim a_\lambda\log\lambda\sim\frac{\log p_1}{\log 2}.\ \ldots\ldots\ldots(86)$$

17. The relations (86) give us information about the order of a_λ when λ is sufficiently small compared to p_1, in fact, when λ is of the form $p_1{}^\epsilon$, where $\epsilon\to 0$. Such values of λ constitute but a small part of its total range of variation, and it is clear that further formulæ must be proved before we can gain an adequate idea of the general behaviour of a_λ. From (81), (82)

* $f\neq o(\phi)$ is to be understood as meaning that $|f|>K\phi$, where K is a constant, and $f\neq O(\phi)$ as meaning that $|f|/\phi\to\infty$. They are *not* the mere negations of $f=o(\phi)$ and $f=O(\phi)$.

and (84) it follows that

$$\left.\begin{aligned}\frac{\log\lambda}{\log\left(1+\dfrac{1}{a_\lambda}\right)} &\leqslant \frac{\log p_1}{\log 2}+O\left\{\frac{\log\lambda\log\mu}{\mu}+\frac{\mu\log p_1}{\log\lambda}+\sqrt{(\log p_1\log\log p_1)}\right\},\\ \frac{\log\lambda}{\log\left(1+\dfrac{1}{1+a_\lambda}\right)} &\geqslant \frac{\log p_1}{\log 2}+O\left\{\frac{\log\lambda\log\mu}{\mu}+\frac{\mu\log p_1}{\log\lambda}+\sqrt{(\log p_1\log\log p_1)}\right\},\end{aligned}\right\} \quad \text{......(87)}$$

provided that $\pi(\mu)\log\mu<\log\lambda\leqslant\log p_1$. From this we can easily shew that if $\pi(\mu)\log\mu<\log\lambda\leqslant\log p_1$ then

$$\left.\begin{aligned}a_\lambda &\leqslant (2^{\log\lambda/\log p_1}-1)^{-1}+O\left\{\frac{\log\mu}{\mu}+\frac{\mu\log p_1}{(\log\lambda)^2}+\frac{\sqrt{(\log p_1\log\log p_1)}}{\log\lambda}\right\},\\ a_\lambda &\geqslant (2^{\log\lambda/\log p_1}-1)^{-1}-1+O\left\{\frac{\log\mu}{\mu}+\frac{\mu\log p_1}{(\log\lambda)^2}+\frac{\sqrt{(\log p_1\log\log p_1)}}{\log\lambda}\right\}.\end{aligned}\right\} \quad \text{...(88)}$$

Now let us suppose that

$$\log\lambda\neq o\sqrt{\left(\frac{\log p_1}{\log\log p_1}\right)}.$$

Then we can choose μ so that

$$\mu=O\left\{\log\lambda\sqrt{\left(\frac{\log\log p_1}{\log p_1}\right)}\right\},$$

$$\mu\neq o\left\{\log\lambda\sqrt{\left(\frac{\log\log p_1}{\log p_1}\right)}\right\}.$$

Now it is clear that $\log\mu=O(\log\log p_1)$, and so

$$\frac{\log\mu}{\mu}=O\left(\frac{\log\log p_1}{\mu}\right)=O\left\{\frac{\sqrt{(\log p_1\log\log p_1)}}{\log\lambda}\right\};$$

and

$$\frac{\mu\log p_1}{(\log\lambda)^2}=O\left\{\frac{\sqrt{(\log p_1\log\log p_1)}}{\log\lambda}\right\}.$$

From this and (88) it follows that, if

$$\log\lambda\neq o\sqrt{\left(\frac{\log p_1}{\log\log p_1}\right)},$$

then

$$\left.\begin{aligned}a_\lambda &\leqslant (2^{\log\lambda/\log p_1}-1)^{-1}+O\left\{\frac{\sqrt{(\log p_1\log\log p_1)}}{\log\lambda}\right\},\\ a_\lambda &\geqslant (2^{\log\lambda/\log p_1}-1)^{-1}-1+O\left\{\frac{\sqrt{(\log p_1\log\log p_1)}}{\log\lambda}\right\}.\end{aligned}\right\} \quad \text{.........(89)}$$

Now we shall divide the primes from 2 to p_1 into five ranges thus

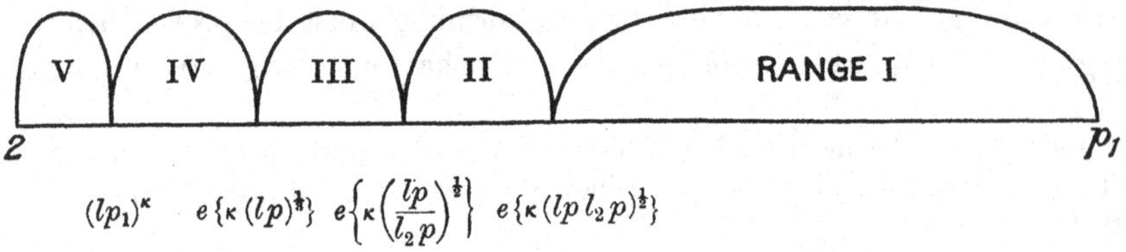

We shall use the inequalities (89) to specify the behaviour of a_λ in ranges I and II, and the formula (85) in ranges IV and V. Range III we shall deal with differently, by a different choice of μ in the inequalities (88). We can easily see that each result in the following sections gives the most information in its particular range.

18. Range I: $\log\lambda \neq O\sqrt{(\log p_1 \log\log p_1)}$*.

Let
$$\Lambda = [(2^{\log\lambda/\log p_1} - 1)^{-1}],$$
and let
$$(2^{\log\lambda/\log p_1} - 1)^{-1} + \epsilon_\lambda,$$
where $-\frac{1}{2} < \epsilon_\lambda < \frac{1}{2}$, be an integer, so that
$$(2^{\log\lambda/\log p_1} - 1)^{-1} = \Lambda + 1 - \epsilon_\lambda \qquad (90)$$
when $\epsilon_\lambda > 0$, and
$$(2^{\log\lambda/\log p_1} - 1)^{-1} = \Lambda - \epsilon_\lambda \qquad (91)$$
when $\epsilon_\lambda < 0$. By our supposition we have
$$\frac{\sqrt{(\log p_1 \log\log p_1)}}{\log\lambda} = o(1). \qquad (92)$$

First let us consider the case in which
$$\epsilon_\lambda \neq O\left\{\frac{\sqrt{(\log p_1 \log\log p_1)}}{\log\lambda}\right\},$$
so that
$$\frac{\sqrt{(\log p_1 \log\log p_1)}}{\log\lambda} = o(\epsilon_\lambda). \qquad (93)$$

It follows from (89), (90), and (93) that, if $\epsilon_\lambda > 0$, then
$$\left.\begin{aligned} a_\lambda &\leqslant \Lambda + 1 - \epsilon_\lambda + o(\epsilon_\lambda),\\ a_\lambda &\geqslant \Lambda - \epsilon_\lambda + o(\epsilon_\lambda).\end{aligned}\right\} \qquad (94)$$
Since $0 < \epsilon_\lambda < \frac{1}{2}$, and a_λ and Λ are integers, it follows from (94) that
$$a_\lambda \leqslant \Lambda, \quad a_\lambda > \Lambda - 1. \qquad (95)$$
Hence
$$a_\lambda = \Lambda. \qquad (96)$$
Similarly from (89), (91), and (93) we see that, if $\epsilon_\lambda < 0$, then
$$\left.\begin{aligned} a_\lambda &\leqslant \Lambda - \epsilon_\lambda + o(\epsilon_\lambda),\\ a_\lambda &\geqslant \Lambda - 1 - \epsilon_\lambda + o(\epsilon_\lambda).\end{aligned}\right\} \qquad (97)$$
Since $-\frac{1}{2} < \epsilon_\lambda < 0$, it follows from (97) that the inequalities (95), and therefore the equation (96), still hold. Hence (96) holds whenever
$$\epsilon_\lambda \neq O\left\{\frac{\sqrt{(\log p_1 \log\log p_1)}}{\log\lambda}\right\}. \qquad (98)$$

* We can with a little trouble replace all equations of the type $f = O(\phi)$ which occur by inequalities of the type $|f| < K\phi$, with definite numerical constants. This would enable us to extend all the different ranges a little. For example, an equation true for
$$\log\lambda \neq O\sqrt{(\log p_1)}$$
would be replaced by an inequality true for $\log\lambda > K\sqrt{(\log p_1)}$, where K is a definite constant, and similarly $\log\lambda = o\sqrt{(\log p_1)}$ would be replaced by $\log\lambda < k\sqrt{(\log p_1)}$.

In particular it holds whenever

$$\epsilon_\lambda \neq o(1), \qquad (99)$$

Now let us consider the case in which

$$\epsilon_\lambda = O\left\{\frac{\sqrt{(\log p_1 \log\log p_1)}}{\log \lambda}\right\}, \qquad (100)$$

so that $\epsilon_\lambda = o(1)$, in virtue of (92). It follows from this and (89) and (90) that, if $\epsilon_\lambda > 0$, then

$$\left.\begin{aligned} a_\lambda &\leqslant \Lambda + 1 + o(1), \\ a_\lambda &\geqslant \Lambda + o(1). \end{aligned}\right\} \qquad (101)$$

Hence $\qquad a_\lambda \leqslant \Lambda + 1, \quad a_\lambda \geqslant \Lambda;$

and so

$$a_\lambda = \Lambda \text{ or } \Lambda + 1. \qquad (102)$$

Similarly from (89), (91), and (100), we see that, if $\epsilon_\lambda < 0$, then

$$\left.\begin{aligned} a_\lambda &\leqslant \Lambda + o(1), \\ a_\lambda &\geqslant \Lambda - 1 + o(1). \end{aligned}\right\} \qquad (103)$$

Hence $\qquad a_\lambda \leqslant \Lambda, \quad a_\lambda \geqslant \Lambda - 1;$

and so

$$a_\lambda = \Lambda \text{ or } \Lambda - 1. \qquad (104)$$

For example, let us suppose that it is required to find a_λ when $\lambda \sim p_1^{\frac{1}{8}}$. We have

$$(2^{\log\lambda/\log p_1} - 1)^{-1} = (2^{1/8} - 1)^{-1} + o(1) = 11{\cdot}048 \ldots + o(1).$$

It is evident that $\Lambda = 11$ and $\epsilon_\lambda \neq o(1)$. Hence $a_\lambda = 11$.

19. The results in the previous section may be rewritten with slight modifications, in order that the transition of a_λ from one value to another may be more clearly expressed. Let

$$\lambda = p_1^{\frac{\log(1+1/x)}{\log 2}}, \qquad (105)$$

and let $x + \epsilon_x$, where $-\frac{1}{2} < \epsilon_x < \frac{1}{2}$, be an integer. Then the range of x which we are now considering is

$$x = o\sqrt{\left(\frac{\log p_1}{\log\log p_1}\right)}, \qquad (106)$$

and the results of the previous section may be stated as follows. If

$$\epsilon_x \neq O\left\{x\sqrt{\left(\frac{\log\log p_1}{\log p_1}\right)}\right\}, \qquad (107)$$

then

$$a_\lambda = [x]. \qquad (108)$$

As a particular case of this we have

$$a_\lambda = [x],$$

when $\epsilon_x \neq o(1)$. But if

$$\epsilon_x = O\left\{x\sqrt{\left(\frac{\log\log p_1}{\log p_1}\right)}\right\}, \qquad (109)$$

then when $\epsilon_x > 0$ $\qquad a_\lambda = [x] \text{ or } [x+1]; \qquad (110)$

and when $\epsilon_x < 0$ $\qquad a_\lambda = [x] \text{ or } [x-1]. \qquad (110')$

20. Range II:
$$\left.\begin{aligned} \log\lambda &= O\sqrt{(\log p_1 \log\log p_1)}, \\ \log\lambda &\neq o\sqrt{\left(\frac{\log p_1}{\log\log p_1}\right)}. \end{aligned}\right\}$$

From (89) it follows that

$$a_\lambda = (2^{\log\lambda/\log p_1} - 1)^{-1} + O\left\{\frac{\sqrt{(\log p_1 \log\log p_1)}}{\log\lambda}\right\}. \quad \text{.........(111)}$$

But
$$(2^{\log\lambda/\log p_1} - 1)^{-1} = \frac{\log p_1}{\log 2 \log\lambda} + O(1).$$

Hence
$$a_\lambda \log\lambda = \frac{\log p_1}{\log 2} + O\sqrt{(\log p_1 \log\log p_1)}. \quad \text{............(112)}$$

As an example we may suppose that

$$\lambda \sim e^{\sqrt{(\log p_1)}}.$$

Then from (112) it follows that

$$a_\lambda = \frac{\sqrt{(\log p_1)}}{\log 2} + O\sqrt{(\log\log p_1)}.$$

21. Range III:
$$\left.\begin{aligned} \log\lambda &= O\sqrt{\left(\frac{\log p_1}{\log\log p_1}\right)}, \\ \log\lambda &\neq o(\log p_1)^{\frac{1}{3}}. \end{aligned}\right\}$$

Let us suppose that $\mu = O(1)$ in (88). Then we see that

$$a_\lambda = \frac{\log p_1}{\log 2 \log\lambda} + O(1) + O\left\{\frac{\log\mu}{\mu} + \frac{\mu\log p_1}{(\log\lambda)^2} + \frac{\sqrt{(\log p_1 \log\log p_1)}}{\log\lambda}\right\}, \quad (113)$$

or
$$a_\lambda \log\lambda = \frac{\log p_1}{\log 2} + O\left\{\frac{\log\mu\log\lambda}{\mu} + \frac{\mu\log p_1}{\log\lambda} + \sqrt{(\log p_1 \log\log p_1)}\right\}. \quad \text{...(114)}$$

Now
$$\frac{\log\mu\log\lambda}{\mu} = O(\log\lambda) = o\left(\frac{\log p_1}{\log\lambda}\right),$$
$$\frac{\mu\log p_1}{\log\lambda} = O\left(\frac{\log p_1}{\log\lambda}\right),$$
$$\sqrt{(\log p_1 \log\log p_1)} = O\left(\frac{\log p_1}{\log\lambda}\right).$$

Hence
$$a_\lambda \log\lambda = \frac{\log p_1}{\log 2} + O\left(\frac{\log p_1}{\log\lambda}\right). \quad \text{.....................(115)}$$

For example, when
$$\lambda \sim e^{(\log p_1)^{\frac{3}{8}}},$$

we have
$$a_\lambda = \frac{(\log p_1)^{\frac{5}{8}}}{\log 2} + O(\log p_1)^{\frac{1}{4}}.$$

22. Range IV:
$$\left.\begin{aligned} \log\lambda &= O(\log p_1)^{\frac{1}{3}}, \\ \log\lambda &\neq o(\log\log p_1). \end{aligned}\right\}$$

In this case it follows from (85) that

$$a_\lambda \log\lambda = \frac{\log p_1}{\log 2} + O\sqrt{(\log p_1 \log\lambda)}. \quad \text{..............(116)}$$

As an example in this range, when we suppose that

$$\lambda \sim e^{(\log p_1)^{\frac{1}{4}}},$$

we obtain from (116)

$$a_\lambda = \frac{(\log p_1)^{\frac{3}{4}}}{\log 2} + O(\log p_1)^{\frac{3}{8}}.$$

23. Range V: $\log\lambda = O(\log\log p_1)$.

From (85) it follows that

$$a_\lambda \log\lambda = \frac{\log p_1}{\log 2} + O\sqrt{(\log p_1 \log\log p_1)}. \quad \text{............(117)}$$

For example, we may suppose that

$$\lambda \sim e^{\sqrt{(\log\log p_1)}}.$$

Then
$$a_\lambda = \frac{\log p_1}{\log 2\sqrt{(\log\log p_1)}} + O\sqrt{(\log p_1)}.$$

24. Let λ' be the prime next below λ, so that $\lambda' \leqslant \lambda - 1$. Then it follows from (63) that

$$\sqrt{\{(1+a_{\lambda'})\log\lambda'\}} - \sqrt{\{(1+a_\lambda)\log\lambda\}} > -\sqrt{\log(\lambda\lambda')}. \quad \text{......(118)}$$

Hence
$$\sqrt{\{(1+a_{\lambda'})\log(\lambda-1)\}} - \sqrt{\{(1+a_\lambda)\log\lambda\}} > -\sqrt{\{2\log\lambda\}}. \quad \text{...(119)}$$

But
$$\log(\lambda-1) < \log\lambda - \frac{1}{\lambda} < \log\lambda\left(1 - \frac{1}{2\lambda\log\lambda}\right)^2;$$

and so (119) may be replaced by

$$\sqrt{(1+a_{\lambda'})} - \sqrt{(1+a_\lambda)} > \frac{\sqrt{(1+a_{\lambda'})}}{2\lambda\log\lambda} - \sqrt{2}. \quad \text{............(120)}$$

But from (54) we know that

$$1 + a_{\lambda'} \geqslant 1 + \left[\frac{\log p_1}{\log\lambda'}\right] > \frac{\log p_1}{\log\lambda'} > \frac{\log p_1}{\log\lambda}.$$

From this and (120) it follows that

$$\sqrt{(1+a_{\lambda'})} - \sqrt{(1+a_\lambda)} > \frac{\sqrt{(\log p_1)}}{2\lambda(\log\lambda)^{\frac{3}{2}}} - \sqrt{2}. \quad \text{............(121)}$$

Now let us suppose that $\lambda^2(\log\lambda)^3 < \frac{1}{8}\log p_1$. Then, from (121), we have

$$\sqrt{(1+a_{\lambda'})} - \sqrt{(1+a_\lambda)} > 0,$$

or
$$a_{\lambda'} > a_\lambda. \quad \text{..................................(122)}$$

From (122) it follows that, if $\lambda^2(\log\lambda)^3 < \frac{1}{8}\log p_1$, then

$$a_2 > a_3 > a_5 > a_7 > \ldots > a_\lambda. \quad \text{....................(123)}$$

In other words, *in a large highly composite number*

$$2^{a_2} . 3^{a_3} . 5^{a_5} . 7^{a_7} \ldots p_1,$$

the indices comparatively near the beginning form a decreasing sequence in the strict sense which forbids equality. Later on groups of equal indices will in general occur.

To sum up, we have obtained fairly accurate information about a_λ for all possible values of λ. The range I is by far the most extensive, and throughout this range a_λ is known with an error never exceeding 1. The formulæ (86) hold throughout a range which includes all the remaining ranges II—V, and a considerable part of I as well, while we have obtained more precise formulæ for each individual range II—V.

25. Now let us consider the nature of p_r. It is evident that r cannot exceed a_2; i.e., r cannot exceed

$$\frac{\log p_1}{(\log 2)^2} + O\sqrt{(\log p_1 \log\log p_1)}. \quad\text{.................(124)}$$

From (55) it evidently follows that

$$\left.\begin{aligned} a_{p_r}\log p_r &= O(\log p_1), \\ a_{p_r}\log p_r &\neq o(\log p_1); \end{aligned}\right\} \quad\text{........................(125)}$$

$$\left.\begin{aligned} (1+a_{P_r})\log p_r &= O(\log p_1), \\ (1+a_{P_r})\log p_r &\neq o(\log p_1). \end{aligned}\right\} \quad\text{.................(126)}$$

But from (46) we know that

$$\left.\begin{aligned} a_{p_r}\log p_r &\geqslant r\log p_r, \\ (1+a_{P_r})\log p_r &\leqslant r\log p_r. \end{aligned}\right\} \quad\text{.....................(127)}$$

From (125)—(127) it follows that

$$\left.\begin{aligned} r\log p_r &= O(\log p_1), \\ r\log p_r &\neq o(\log p_1); \end{aligned}\right\} \quad\text{........................(128)}$$

and

$$\left.\begin{aligned} a_{p_r} &= O(r), \\ a_{p_r} &\neq o(r). \end{aligned}\right\} \quad\text{...........................(129)}$$

26. Supposing that $\lambda = p_r$ in (81) and $\lambda = P_r$ in (82), and remembering (128), we see that, if $r\mu = o(\log p_1)$, then

$$\log\left(1+\frac{1}{a_{p_r}}\right) \geqslant \frac{\log p_r}{a_2\log 2}\left\{1+O\left(\frac{\log\mu}{r\mu}+\frac{r\mu}{\log p_1}\right)\right\}, \quad\text{.........(130)}$$

and

$$\log\left(1+\frac{1}{1+a_{P_r}}\right) \leqslant \frac{\log P_r}{a_2\log 2}\left\{1+O\left(\frac{\log\mu}{r\mu}+\frac{r\mu}{\log p_1}\right)\right\}. \quad\text{......(131)}$$

But, from (47), we have

$$\log\left(1+\frac{1}{a_{p_r}}\right) \leqslant \log\left(1+\frac{1}{1+a_{P_r}}\right).$$

Also we know that

$$\log P_r = \log p_r + O(1) = \log p_r\left\{1+O\left(\frac{1}{\log p_r}\right)\right\} = \log p_r\left\{1+O\left(\frac{r}{\log p_1}\right)\right\}.$$

Hence (131) may be replaced by

$$\log\left(1+\frac{1}{a_{p_r}}\right) \leqslant \frac{\log p_r}{a_2 \log 2}\left\{1+O\left(\frac{\log\mu}{r\mu}+\frac{r\mu}{\log p_1}\right)\right\}. \quad \text{......(132)}$$

From (130) and (132) it is evident that

$$\log\left(1+\frac{1}{a_{p_r}}\right) = \frac{\log p_r}{a_2 \log 2}\left\{1+O\left(\frac{\log\mu}{r\mu}+\frac{r\mu}{\log p_1}\right)\right\}. \quad \text{......(133)}$$

In a similar manner

$$\log\left(1+\frac{1}{1+a_{P_r}}\right) = \frac{\log p_r}{a_2 \log 2}\left\{1+O\left(\frac{\log\mu}{r\mu}+\frac{r\mu}{\log p_1}\right)\right\}. \quad \text{......(134)}$$

Now supposing that

$$\left.\begin{array}{l} r\mu = o(\log p_1), \\ r\mu \neq O(\log\mu), \end{array}\right\} \quad \text{......(135)}$$

and dividing (134) by (133), we have

$$\frac{\log\left(1+\dfrac{1}{1+a_{P_r}}\right)}{\log\left(1+\dfrac{1}{a_{p_r}}\right)} = 1+O\left(\frac{\log\mu}{r\mu}+\frac{r\mu}{\log p_1}\right),$$

or

$$1+\frac{1}{1+a_{P_r}} = 1+\frac{1}{a_{p_r}}+O\left\{\left(\frac{\log\mu}{r\mu}+\frac{r\mu}{\log p_1}\right)\Big/ a_{p_r}\right\},$$

that is

$$\frac{1}{1+a_{P_r}} = \frac{1}{a_{p_r}}\left\{1+O\left(\frac{\log\mu}{r\mu}+\frac{r\mu}{\log p_1}\right)\right\}.$$

Hence

$$a_{p_r} = a_{P_r}+1+O\left(\frac{\log\mu}{\mu}+\frac{r^2\mu}{\log p_1}\right), \quad \text{......(136)}$$

in virtue of (129). But $a_{P_r} \leqslant r-1$, and so

$$a_{p_r} \leqslant r+O\left(\frac{\log\mu}{\mu}+\frac{r^2\mu}{\log p_1}\right). \quad \text{......(137)}$$

But we know that $a_{p_r} \geqslant r$. Hence it is clear that

$$a_{p_r} = r+O\left(\frac{\log\mu}{\mu}+\frac{r^2\mu}{\log p_1}\right). \quad \text{......(138)}$$

From this and (136) it follows that

$$a_{P_r} = r-1+O\left(\frac{\log\mu}{\mu}+\frac{r^2\mu}{\log p_1}\right), \quad \text{......(139)}$$

provided that the conditions (135) are satisfied.

Now let us suppose that $r = o\sqrt{(\log p_1)}$. Then we can choose μ such that $r^2\mu = o(\log p_1)$ and $\mu \neq O(1)$. Consequently we have

$$\frac{\log\mu}{\mu} = o(1), \quad \frac{r^2\mu}{\log p_1} = o(1);$$

and so it follows from (138) and (139) that

$$a_{p_r} = 1+a_{P_r} = r, \quad \text{......(140)}$$

provided that $r = o\sqrt{(\log p_1)}$. From this it is clear that, if $r = o\sqrt{(\log p_1)}$, then

$$p_1 > p_2 > p_3 > p_4 > \ldots > p_r. \qquad \ldots\ldots\ldots\ldots\ldots\ldots\ldots(141)$$

In other words, *in a large highly composite number*

$$2^{a_2} . 3^{a_3} . 5^{a_5} \ldots p_1,$$

the indices comparatively near the end form a sequence of the type

$$\ldots 5 \ldots 4 \ldots 3 \ldots 2 \ldots 1.$$

Near the beginning gaps in the indices will in general occur.

Again, let us suppose that $r = o(\log p_1)$, $r \neq o\sqrt{(\log p_1)}$, and $\mu = O(1)$ in (138) and (139). Then we see that

$$\left.\begin{aligned} a_{p_r} &= r + O\left(\frac{r^2}{\log p_1}\right), \\ a_{P_r} &= r + O\left(\frac{r^2}{\log p_1}\right); \end{aligned}\right\} \qquad \ldots\ldots\ldots\ldots\ldots\ldots\ldots(142)$$

provided that $r = o(\log p_1)$ and $r \neq o\sqrt{(\log p_1)}$. But when $r \neq o(\log p_1)$, we shall use the general result, viz.,

$$\left.\begin{aligned} a_{p_r} &= O(r), \quad a_{p_r} \neq o(r), \\ a_{P_r} &= O(r), \quad a_{P_r} \neq o(r), \end{aligned}\right\} \qquad \ldots\ldots\ldots\ldots\ldots\ldots\ldots(143)$$

which is true for all values of r except 1.

27. It follows from (87) and (128) that

$$\left.\begin{aligned} \frac{\log p_r}{\log\left(1 + \dfrac{1}{a_{p_r}}\right)} &\leqslant \frac{\log p_1}{\log 2} + O\left\{\frac{\log p_1 \log \mu}{r\mu} + r\mu + \sqrt{(\log p_1 \log\log p_1)}\right\}, \\ \frac{\log P_r}{\log\left(1 + \dfrac{1}{1 + a_{P_r}}\right)} &\geqslant \frac{\log p_1}{\log 2} + O\left\{\frac{\log p_1 \log \mu}{r\mu} + r\mu + \sqrt{(\log p_1 \log\log p_1)}\right\}, \end{aligned}\right\}$$

$$\ldots\ldots\ldots(144)$$

with the condition that $r\mu = o(\log p_1)$. From this it can easily be shewn, by arguments similar to those used in the beginning of the previous section, that

$$\frac{\log p_r}{\log(1 + 1/r)} = \frac{\log p_1}{\log 2} + O\left\{\frac{\log p_1 \log \mu}{r\mu} + r\mu + \sqrt{(\log p_1 \log\log p_1)}\right\}, \ldots(145)$$

provided that $r\mu = o(\log p_1)$.

Now let us suppose that $r = o(\log p_1)$; then we can choose μ such that

$$\mu = o\left(\frac{\log p_1}{r}\right), \quad \mu \neq O(1).$$

Consequently $r\mu = o(\log p_1)$ and $\log \mu = o(\mu)$, and so

$$\frac{\log p_1 \log \mu}{r\mu} = o(\log p_1).$$

From these relations and (145) it follows that, *if* $r = o(\log p_1)$, *then*

$$\frac{\log p_r}{\log(1+1/r)} \sim \frac{\log p_1}{\log 2}; \qquad \text{.........................(146)}$$

that is to say that, if $r = o(\log p_1)$, *then*

$$\frac{\log p_1}{\log 2} \sim \frac{\log p_2}{\log(1+\frac{1}{2})} \sim \frac{\log p_3}{\log(1+\frac{1}{3})} \sim \dots \sim \frac{\log p_r}{\log(1+1/r)}. \qquad \text{......(147)}$$

Again let us suppose that $r = O\sqrt{(\log p_1 \log\log p_1)}$ in (145). Then it is possible to choose μ such that

$$\left.\begin{aligned} r\mu &= O\sqrt{(\log p_1 \log\log p_1)}, \\ r\mu &\neq o\sqrt{(\log p_1 \log\log p_1)}. \end{aligned}\right\} \qquad \text{......................(148)}$$

It is evident that $\log\mu = O(\log\log p_1)$, and so

$$\frac{\log p_1 \log\mu}{r\mu} = O\left(\frac{\log p_1 \log\log p_1}{r\mu}\right) = O\sqrt{(\log p_1 \log\log p_1)},$$

in virtue of (148). Hence

$$\frac{\log p_r}{\log(1+1/r)} = \frac{\log p_1}{\log 2} + O\sqrt{(\log p_1 \log\log p_1)}, \qquad \text{.........(149)}$$

provided that $\qquad r = O\sqrt{(\log p_1 \log\log p_1)}$.

Now let us suppose that $r = o(\log p_1)$, $r \neq o\sqrt{(\log p_1 \log\log p_1)}$, and $\mu = O(1)$, in (145). Then it is evident that

$$\log p_1 = O(r^2), \quad \sqrt{(\log p_1 \log\log p_1)} = O(r),$$

and

$$\frac{\log p_1 \log\mu}{r\mu} = O\left(\frac{\log p_1}{r}\right) = O(r).$$

Hence we see that

$$\frac{\log p_r}{\log(1+1/r)} = \frac{\log p_1}{\log 2} + O(r), \qquad \text{.....................(150)}$$

if $\qquad r = o(\log p_1), \quad r \neq o\sqrt{(\log p_1 \log\log p_1)}$.

But, if $r \neq o(\log p_1)$, we see from (128) that

$$\left.\begin{aligned} \frac{\log p_r}{\log(1+1/r)} &= O(\log p_1), \\ \frac{\log p_r}{\log(1+1/r)} &\neq o(\log p_1). \end{aligned}\right\} \qquad \text{.....................(151)}$$

From (150) and (151) it follows that, if $r \neq o\sqrt{(\log p_1 \log\log p_1)}$, then

$$\frac{\log p_r}{\log(1+1/r)} = \frac{\log p_1}{\log 2} + O(r); \qquad \text{.....................(152)}$$

and from (149) and (152) that, if $r = o(\log p_1)$, then

$$\frac{\log p_r}{\log(1+1/r)} \sim \frac{\log p_1}{\log 2},$$

in agreement with (147). This result will, in general, fail for the largest possible values of r, which are of order $\log p_1$.

It must be remembered that all the results involving p_1 may be written in terms of N, since $p_1 = O(\log N)$ and $p_1 \neq o(\log N)$, and consequently

$$\log p_1 = \log\log N + O(1). \quad \text{.....................(153)}$$

28. We shall now prove that successive highly composite numbers are asymptotically equivalent. Let m and n be any two positive integers which are prime to each other, such that

$$\log mn = o(\log p_1) = o(\log\log N); \quad \text{..............(154)}$$

and let

$$\frac{m}{n} = 2^{\delta_2} . 3^{\delta_3} . 5^{\delta_5} \ldots \wp^{\delta_\wp}. \quad \text{........................(155)}$$

Then it is evident that

$$mn = 2^{|\delta_2|} . 3^{|\delta_3|} . 5^{|\delta_5|} \ldots \wp^{|\delta_\wp|}. \quad \text{..................(156)}$$

Hence

$$\delta_\lambda \log\lambda = O(\log mn) = o(\log p_1) = o(a_\lambda \log\lambda); \quad \text{.........(157)}$$

so that

$$\delta_\lambda = o(a_\lambda).$$

Now

$$d\left(\frac{m}{n} N\right) = d(N)\left(1 + \frac{\delta_2}{1+a_2}\right)\left(1 + \frac{\delta_3}{1+a_3}\right)\ldots\left(1 + \frac{\delta_\wp}{1+a_\wp}\right). \quad (158)$$

But, from (60), we know that

$$a_\lambda \log\lambda = a_2 \log 2 + O\sqrt{(\log p_1 \log\lambda)}.$$

Hence

$$\begin{aligned}
1 + \frac{\delta_\lambda}{1+a_\lambda} &= 1 + \frac{\delta_\lambda \log\lambda}{a_2 \log 2} + O\left\{|\delta_\lambda| \left(\frac{\log\lambda}{\log p_1}\right)^{\frac{3}{2}}\right\} \\
&= 1 + \frac{\delta_\lambda \log\lambda}{a_2 \log 2} + O\left\{|\delta_\lambda| \frac{\log\lambda}{\log p_1} \sqrt{\left(\frac{\log\wp}{\log p_1}\right)}\right\} \\
&= \exp\left\{\frac{\delta_\lambda \log\lambda}{a_2 \log 2} + O\frac{|\delta_\lambda| \log\lambda}{\log p_1} \sqrt{\left(\frac{\log\wp}{\log p_1}\right)} + O\left(\frac{\delta_\lambda \log\lambda}{\log p_1}\right)^2\right\} \\
&= \exp\left\{\frac{\delta_\lambda \log\lambda}{a_2 \log 2} + O\frac{|\delta_\lambda| \log\lambda}{\log p_1} \sqrt{\left(\frac{\log mn}{\log p_1}\right)}\right\}. \quad \text{......(159)}
\end{aligned}$$

It follows from (155), (156), (158), and (159) that

$$\begin{aligned}
d\left(\frac{m}{n} N\right) &= d(N) \exp\left\{\frac{\delta_2 \log 2 + \delta_3 \log 3 + \ldots + \delta_\wp \log\wp}{a_2 \log 2}\right. \\
&\qquad \left. + O\frac{|\delta_2| \log 2 + |\delta_3| \log 3 + \ldots + |\delta_\wp| \log\wp}{\log p_1} \sqrt{\left(\frac{\log mn}{\log p_1}\right)}\right\} \\
&= d(N)\, e^{\frac{\log(m/n)}{a_2 \log 2} + O\left(\frac{\log mn}{\log p_1}\right)^{\frac{3}{2}}} \\
&= d(N)\, e^{\frac{1}{a_2 \log 2}\left\{\log\frac{m}{n} + O\log mn \sqrt{\left(\frac{\log mn}{\log p_1}\right)}\right\}}. \quad \text{.....................(160)}
\end{aligned}$$

Putting $m = n + 1$, we see that, if

$$\log n = o(\log p_1) = o(\log\log N),$$

then $$d\left\{N\left(1+\frac{1}{n}\right)\right\}=d(N)\,e^{\frac{1}{a_2\log 2}\left\{\log\left(1+\frac{1}{n}\right)+O\left(\log n\sqrt{\frac{\log n}{\log p_1}}\right)\right\}}$$

$$=d(N)\left(1+\frac{1}{n}\right)^{\frac{1+O\left\{n\log n\sqrt{\left(\frac{\log n}{\log\log N}\right)}\right\}}{a_2\log 2}} \quad \text{.........(161)}$$

Now it is possible to choose n such that

$$n(\log n)^{\frac{3}{2}} \neq o\sqrt{(\log\log N)},$$

and $$1+O\left\{n\log n\sqrt{\left(\frac{\log n}{\log\log N}\right)}\right\}>0;$$

that is to say $$d\left\{N\left(1+\frac{1}{n}\right)\right\}>d(N). \quad \text{.......................(162)}$$

From this and (29) it follows that, *if N is a highly composite number, then the next highly composite number is of the form*

$$N+O\left\{\frac{N(\log\log\log N)^{\frac{3}{2}}}{\sqrt{(\log\log N)}}\right\}. \quad \text{.....................(163)}$$

Hence the ratio of two consecutive highly composite numbers tends to unity.

It follows from (163) that the number of highly composite numbers not exceeding x is not of the form

$$o\left\{\frac{\log x\sqrt{(\log\log x)}}{(\log\log\log x)^{\frac{3}{2}}}\right\}.$$

29. Now let us consider the nature of $d(N)$ for highly composite values of N. From (44) we see that

$$d(N)=2^{\pi(p_1)-\pi(p_2)}\cdot 3^{\pi(p_2)-\pi(p_3)}\cdot 4^{\pi(p_3)-\pi(p_4)}\ldots(1+a_2). \quad \text{...(164)}$$

From this it follows that

$$d(N)=2^{a_2}\cdot 3^{a_3}\cdot 5^{a_5}\ldots \varpi^{a_\varpi}, \quad \text{.....................(165)}$$

where ϖ is the largest prime not exceeding $1+a_2$; and

$$a_\lambda=\pi(p_{\lambda-1})+O(p_\lambda). \quad \text{.......................(166)}$$

It also follows that, if $\wp_1, \wp_2, \wp_3, \ldots, \wp_\lambda$ are a given set of primes, then a number $\bar{\mu}$ can be found such that the equation

$$d(N)=\wp_1^{\beta_1}\cdot\wp_2^{\beta_2}\cdot\wp_3^{\beta_3}\ldots,\wp_\mu^{\beta_\mu}\ldots\wp_\lambda^{\beta_\lambda}$$

is impossible if N is a highly composite number and $\beta_\mu>\bar{\mu}$. We may state this roughly by saying that as N (a highly composite number) tends to infinity, then, not merely in N itself, but also in $d(N)$, the number of prime factors, as well as the indices, must tend to infinity. In particular such an equation as

$$d(N)=k\cdot 2^m, \quad \text{.........................(167)}$$

where k is fixed, becomes impossible when m exceeds a certain limit depending on k.

It is easily seen from (153), (164), and (165) that

$$\left.\begin{aligned} \varpi = O(a_2) = O(\log p_1) = O(\log\log N) = O\{\log\log d(N)\}, \\ \varpi \neq o(a_2) = o(\log p_1) = o(\log\log N) = o\{\log\log d(N)\}. \end{aligned}\right\} \quad \text{...(168)*}$$

It follows from (147) that if $\lambda = o(\log p_1)$ then

$$\frac{\log \alpha_2}{\log(1-\frac{1}{2})} \sim \frac{\log \alpha_3}{\log(1-\frac{1}{3})} \sim \frac{\log \alpha_5}{\log(1-\frac{1}{5})} \sim \ldots \sim \frac{\log \alpha_\lambda}{\log(1-1/\lambda)}. \quad \text{...(169)}$$

Similarly, from (149), it follows that if $\lambda = O\sqrt{(\log p_1 \log\log p_1)}$ then

$$\frac{\log(1+\alpha_\lambda)}{\log(1-1/\lambda)} = -\frac{\log p_1}{\log 2} + O\sqrt{(\log p_1 \log\log p_1)}. \quad \text{.........(170)}$$

Again, from (152), we see that if $\lambda \neq o\sqrt{(\log p_1 \log\log p_1)}$ then

$$\frac{\log(1+\alpha_\lambda)}{\log(1-1/\lambda)} = -\frac{\log p_1}{\log 2} + O(\lambda). \quad \text{..................(171)}$$

In the left-hand side we cannot write α_λ instead of $1+\alpha_\lambda$, as α_λ may be zero for a few values of λ.

From (165) and (170) we can shew that

$$\log d(N) = \alpha_2 \log 2 + O(\alpha_3), \quad \log d(N) \neq \alpha_2 \log 2 + o(\alpha_3);$$

and so

$$\log d(N) = \alpha_2 \log 2 + e^{\frac{\log \frac{3}{2}}{\log 2}\log p_1 + O\sqrt{(\log p_1 \log\log p_1)}} \quad \text{......(172)}$$

But from (163) we see that

$$\log\log d(N) = \log p_1 + O(\log\log p_1).$$

From this and (172) it follows that

$$\alpha_2 \log 2 = \log d(N) - \{\log d(N)\}^{\frac{\log \frac{3}{2}}{\log 2} + o\sqrt{\left\{\frac{\log\log\log d(N)}{\log\log d(N)}\right\}}}. \quad \text{...(173)}$$

30. Now we shall consider the order of $dd(N)$ for highly composite values of N. It follows from (165) that

$$\log dd(N) = \log(1+\alpha_2) + \log(1+\alpha_3) + \ldots + \log(1+\alpha_\varpi). \quad \text{...(174)}$$

Now let $\lambda, \lambda', \lambda'', \ldots$ be consecutive primes in ascending order, and let

$$\lambda = O\sqrt{(\log p_1 \log\log p_1)},$$
$$\lambda \neq o\sqrt{(\log p_1 \log\log p_1)}.$$

Then, from (174), we have

$$\begin{aligned} \log dd(N) = {} & \log(1+\alpha_2) + \log(1+\alpha_3) + \ldots + \log(1+\alpha_\lambda) \\ & + \log(1+\alpha_{\lambda'}) + \log(1+\alpha_{\lambda''}) + \ldots + \log(1+\alpha_\varpi). \end{aligned} \quad \text{...(175)}$$

* More precisely $\varpi \sim a_2$. But this involves the assumption that two consecutive primes are asymptotically equivalent. This follows at once from the prime number theorem. It appears probable that such a result cannot really be as deep as the prime number theorem, but nobody has succeeded up to now in proving it by elementary reasoning.

But, from (170), we have

$$\log(1+\alpha_2)+\log(1+\alpha_3)+\ldots+\log(1+\alpha_\lambda)$$
$$=-\frac{\log p_1}{\log 2}\log\left\{(1-\tfrac{1}{2})(1-\tfrac{1}{3})(1-\tfrac{1}{5})\ldots\left(1-\frac{1}{\lambda}\right)\right\}$$
$$+O\sqrt{(\log p_1\log\log p_1)}\log\left\{(1-\tfrac{1}{2})(1-\tfrac{1}{3})\ldots\left(1-\frac{1}{\lambda}\right)\right\}. \qquad \ldots\ldots(176)$$

It can be shewn, without assuming the prime number theorem*, that

$$-\log\left\{(1-\tfrac{1}{2})(1-\tfrac{1}{3})(1-\tfrac{1}{5})\ldots\left(1-\frac{1}{p}\right)\right\}=\log\log p+\gamma+O\left(\frac{1}{\log p}\right), \qquad \ldots\ldots(177)$$

where γ is the Eulerian constant. Hence

$$\log\left\{(1-\tfrac{1}{2})(1-\tfrac{1}{3})(1-\tfrac{1}{5})\ldots\left(1-\frac{1}{p}\right)\right\}=O(\log\log p).$$

From this and (176) it follows that

$$\log(1+\alpha_2)+\log(1+\alpha_3)+\ldots+\log(1+\alpha_\lambda)$$
$$=-\frac{\log p_1}{\log 2}\log\left\{(1-\tfrac{1}{2})(1-\tfrac{1}{3})\ldots\left(1-\frac{1}{\lambda}\right)\right\}$$
$$+O\{\sqrt{(\log p_1\log\log p_1)}\log\log\lambda\}$$
$$=-\frac{\log p_1}{\log 2}\log\left\{(1-\tfrac{1}{2})(1-\tfrac{1}{3})\ldots\left(1-\frac{1}{\lambda}\right)\right\}$$
$$+O\{\sqrt{(\log p_1\log\log p_1)}\log\log\log p_1\}. \qquad \ldots\ldots\ldots\ldots(178)$$

Again, from (152), we see that

$$\log(1+\alpha_{\lambda'})+\log(1+\alpha_{\lambda''})+\ldots+\log(1+\alpha_{\varpi})$$
$$=-\frac{\log p_1}{\log 2}\log\left\{\left(1-\frac{1}{\lambda'}\right)\left(1-\frac{1}{\lambda''}\right)\ldots\left(1-\frac{1}{\varpi}\right)\right\}$$
$$+O\left\{\lambda'\log\left(1-\frac{1}{\lambda'}\right)+\lambda''\log\left(1-\frac{1}{\lambda''}\right)+\ldots+\varpi\log\left(1-\frac{1}{\varpi}\right)\right\}$$
$$=-\frac{\log p_1}{\log 2}\log\left\{\left(1-\frac{1}{\lambda'}\right)\left(1-\frac{1}{\lambda''}\right)\ldots\left(1-\frac{1}{\varpi}\right)\right\}+O\{\pi(\varpi)-\pi(\lambda)\}$$
$$=-\frac{\log p_1}{\log 2}\log\left\{\left(1-\frac{1}{\lambda'}\right)\left(1-\frac{1}{\lambda''}\right)\ldots\left(1-\frac{1}{\varpi}\right)\right\}+O\left(\frac{\log p_1}{\log\log p_1}\right). \qquad \ldots(179)$$

From (175), (178), and (179) it follows that

$$\log dd(N)=-\frac{\log p_1}{\log 2}\log\left\{(1-\tfrac{1}{2})(1-\tfrac{1}{3})\ldots\left(1-\frac{1}{\lambda}\right)\right\}$$
$$+O\{\sqrt{(\log p_1\log\log p_1)}\log\log\log p_1\}$$
$$-\frac{\log p_1}{\log 2}\log\left\{\left(1-\frac{1}{\lambda'}\right)\left(1-\frac{1}{\lambda''}\right)\ldots\left(1-\frac{1}{\varpi}\right)\right\}+O\left(\frac{\log p_1}{\log\log p_1}\right)$$

* See Landau, *Handbuch*, p. 139.

$$= -\frac{\log p_1}{\log 2}\log\left\{(1-\tfrac{1}{2})(1-\tfrac{1}{3})\ldots\left(1-\frac{1}{\varpi}\right)\right\} + O\left(\frac{\log p_1}{\log\log p_1}\right)$$

$$= \frac{\log p_1}{\log 2}\left\{\log\log\varpi + \gamma + O\left(\frac{1}{\log\varpi}\right)\right\} + O\left(\frac{\log p_1}{\log\log p_1}\right)$$

$$= \frac{\log p_1}{\log 2}\left\{\log\log\log p_1 + \gamma + O\left(\frac{1}{\log\log p_1}\right)\right\} + O\left(\frac{\log p_1}{\log\log p_1}\right)$$

$$= \frac{\log\log N}{\log 2}\left\{\log\log\log\log N + \gamma + O\left(\frac{1}{\log\log\log N}\right)\right\}, \quad \ldots\ldots(180)$$

in virtue of (177), (168), and (163). Hence, if N is a highly composite number, then

$$dd(N) = (\log N)^{\frac{1}{\log 2}\left\{\log\log\log\log N + \gamma + O\left(\frac{1}{\log\log\log N}\right)\right\}}. \quad \ldots(181)$$

31. It may be interesting to note that, as far as the table is constructed,

$$2,\ 2^2,\ 2^3,\ \ldots,\ 2^{13},\quad 3,\ 3.2,\ 3.2^2,\ \ldots,\ 3.2^{11},\quad 5.2,\ 5.2^2,\ \ldots,\ 5.2^8,$$
$$7.2^5,\ 7.2^6,\ \ldots,\ 7.2^{10},\quad 9,\ 9.2,\ 9.2^2,\ \ldots,\ 9.2^{10},$$

and so on, occur as values of $d(N)$. But we know from § 29 that $k.2^m$ cannot be the value of $d(N)$ for sufficiently large values of m; and so numbers of the form $k.2^m$ which occur as the value of $d(N)$ in the table must disappear sooner or later when the table is extended.

Thus numbers of the form 5.2^m have begun to disappear in the table itself. The powers of 2 disappear at any rate from 2^{18} onwards. The least number having 2^{18} divisors is

$$2^7.3^3.5^3.7.11.13\ldots 41.43,$$

while the smaller number, viz.,

$$2^8.3^4.5^3.7^2.11.13\ldots 41$$

has a larger number of divisors, viz. 135.2^{11}. The numbers of the form 7.2^m disappear at least from 7.2^{13} onwards. The least number having 7.2^{13} divisors is

$$2^6.3^3.5^3.7.11.13\ldots 31.37,$$

while the smaller number, viz.

$$2^9.3^4.5^2.7^2.11.13\ldots 31$$

has a larger number of divisors, viz. 225.2^8.

IV

Superior Highly Composite Numbers

32. A number N may be said to be a superior highly composite number if there is a positive number ϵ such that

$$\frac{d(N)}{N^\epsilon} \geqslant \frac{d(N')}{(N')^\epsilon}, \quad \ldots\ldots\ldots\ldots(182)$$

for all values of N' less than N, and

$$\frac{d(N)}{N^\epsilon} > \frac{d(N')}{(N')^\epsilon} \qquad \text{.........................(183)}$$

for all values of N' greater than N.

All superior highly composite numbers are also highly composite. For, if $N' < N$, it follows from (182) that

$$d(N) \geqslant d(N')\left(\frac{N}{N'}\right)^\epsilon > d(N');$$

and so N is highly composite.

33. Now let us consider what must be the nature of N in order that it should be a superior highly composite number. In the first place it must be of the form

$$2^{a_2}.3^{a_3}.5^{a_5}\ldots p_1^{a}, \qquad \text{.........................(184)}$$

or of the form

$$\begin{aligned} &2.3.5.7\ldots\ldots p_1 \\ \times\, &2.3.5.7\ldots p_2 \\ \times\, &2.3.5\ldots p_3 \\ \times\, &\ldots\ldots\ldots : \end{aligned}$$

i.e. it must satisfy the conditions for a highly composite number. Now let

$$N' = N/\lambda,$$

where $\lambda \leqslant p_1$. Then from (182) it follows that

$$\frac{1+a_\lambda}{\lambda^{\epsilon a_\lambda}} \geqslant \frac{a_\lambda}{\lambda^{\epsilon(a_\lambda - 1)}},$$

or

$$\lambda^\epsilon \leqslant \left(1 + \frac{1}{a_\lambda}\right). \qquad \text{.........................(185)}$$

Again let

$$N' = N\lambda.$$

Then, from (183), we see that

$$\frac{1+a_\lambda}{\lambda^{\epsilon a_\lambda}} > \frac{2+a_\lambda}{\lambda^{\epsilon(a_\lambda + 1)}},$$

or

$$\lambda^\epsilon > \left(1 + \frac{1}{1+a_\lambda}\right). \qquad \text{.........................(186)}$$

Now supposing that $\lambda = p_1$ in (185) and $\lambda = P_1$ in (186), we obtain

$$\frac{\log 2}{\log P_1} < \epsilon \leqslant \frac{\log 2}{\log p_1}. \qquad \text{.........................(187)}$$

Now let us suppose that $\epsilon = 1/x$. Then, from (187), we have

$$p_1 \leqslant 2^x < P_1. \qquad \text{.........................(188)}$$

That is, p_1 is the largest prime not exceeding 2^x. It follows from (185) that

$$a_\lambda \leqslant (\lambda^{1/x} - 1)^{-1}. \qquad \text{.........................(189)}$$

Similarly, from (186), $a_\lambda > (\lambda^{1/x}-1)^{-1}-1$. ....(190)

From (189) and (190) it is clear that

$$a_\lambda = [(\lambda^{1/x}-1)^{-1}]. \qquad \text{......................(191)}$$

Hence N is of the form

$$2^{[(2^{1/x}-1)^{-1}]} . 3^{[(3^{1/x}-1)^{-1}]} . 5^{[(5^{1/x}-1)^{-1}]} \ldots p_1, \qquad \text{..............(192)}$$

where p_1 is the largest prime not exceeding 2^x.

34. Now let us suppose that $\lambda = p_r$ in (189). Then

$$a_{p_r} \leqslant (p_r^{1/x}-1)^{-1}.$$

But we know that $r \leqslant a_{p_r}$. Hence

$$r \leqslant (p_r^{1/x}-1)^{-1},$$

or

$$p_r \leqslant \left(1+\frac{1}{r}\right)^x. \qquad \text{.........................(193)}$$

Similarly by supposing that $\lambda = P_r$ in (190), we see that

$$a_{P_r} > (P_r^{1/x}-1)^{-1}-1.$$

But we know that $r-1 \geqslant a_{P_r}$. Hence

$$r > (P_r^{1/x}-1)^{-1},$$

or

$$P_r > \left(1+\frac{1}{r}\right)^x. \qquad \text{.........................(194)}$$

From (193) and (194) it is clear that p_r is the largest prime not exceeding $(1+1/r)^x$. Hence N is of the form

$$\begin{aligned} &2.3.5.7\ldots\ldots p_1 \\ &\times 2.3.5.7\ldots p_2 \\ &\times 2.3.5\ldots p_3 \\ &\times \ldots\ldots\ldots, \end{aligned} \qquad \text{...................................(195)}$$

where p_1 is the largest prime not greater than 2^x, p_2 is the largest prime not greater than $(\frac{3}{2})^x$, and so on. In other words N is of the form

$$e^{\vartheta(2^x)+\vartheta(\frac{3}{2})^x+\vartheta(\frac{4}{3})^x+\ldots}; \qquad \text{........................(196)}$$

and $d(N)$ is of the form

$$2^{\pi(2^x)} . (\tfrac{3}{2})^{\pi(\frac{3}{2})^x} . (\tfrac{4}{3})^{\pi(\frac{4}{3})^x} \ldots \qquad \text{....................(197)}$$

Thus to every value of x not less than 1 corresponds one, and only one, value of N.

35. Since

$$\frac{d(N)}{N^{1/x}} \geqslant \frac{d(N')}{(N')^{1/x}},$$

for all values of N', it follows from (196) and (197) that

$$d(N) \leqslant N^{1/x} \frac{2^{\pi(2^x)}}{e^{(1/x)\vartheta(2^x)}} \frac{(\frac{3}{2})^{\pi(\frac{3}{2})^x}}{e^{(1/x)\vartheta(\frac{3}{2})^x}} \frac{(\frac{4}{3})^{\pi(\frac{4}{3})^x}}{e^{(1/x)\vartheta(\frac{4}{3})^x}} \ldots, \qquad \text{........(198)}$$

for all values of N and x; and $d(N)$ is equal to the right-hand side when

$$N = e^{\vartheta(2^x) + \vartheta(\frac{3}{2})^x + \vartheta(\frac{4}{3})^x + \dots}. \quad \text{......................(199)}$$

Thus, for example, putting $x = 2, 3, 4$ in (198), we obtain

$$\left.\begin{array}{l} d(N) \leqslant \sqrt{(3N)}, \\ d(N) \leqslant 8\,(3N/35)^{\frac{1}{3}}, \\ d(N) \leqslant 96\,(3N/50050)^{\frac{1}{4}}, \end{array}\right\} \quad \text{......................(200)}$$

for all values of N; and $d(N) = \sqrt{(3N)}$ when $N = 2^2 . 3$; $d(N) = 8\,(3N/35)^{\frac{1}{3}}$ when $N = 2^3 . 3^2 . 5 . 7$; $d(N) = 96\,(3N/50050)^{\frac{1}{4}}$ when

$$N = 2^5 . 3^3 . 5^2 . 7 . 11 . 13.$$

36. M and N are consecutive superior highly composite numbers if there are no superior highly composite numbers between M and N.

From (195) and (196) it is easily seen that, if M and N are any two superior highly composite numbers, and if $M > N$, then M is a multiple of N; and also that, if M and N are two consecutive superior highly composite numbers, and if $M > N$, then M/N is a prime number. From this it follows that consecutive superior highly composite numbers are of the form

$$\pi_1, \quad \pi_1\pi_2, \quad \pi_1\pi_2\pi_3, \quad \pi_1\pi_2\pi_3\pi_4, \quad \dots, \quad \text{............(201)}$$

where $\pi_1, \pi_2, \pi_3, \dots$ are primes. In order to determine $\pi_1, \pi_2, \dots$, we proceed as follows. Let x_1' be the smallest value of x such that $[2^x]$ is prime, x_2' the smallest value of x such that $[(\frac{3}{2})^x]$ is prime, and so on; and let $x_1, x_2, \dots$ be the numbers $x_1', x_2', \dots$ arranged in order of magnitude. Then π_n is the prime corresponding to x_n, and

$$N = \pi_1\pi_2\pi_3 \dots \pi_n, \quad \text{..........................(202)}$$

if $x_n \leqslant x < x_{n+1}$.

37. From the preceding results we see that the number of superior highly composite numbers not exceeding

$$e^{\vartheta(2^x) + \vartheta(\frac{3}{2})^x + \vartheta(\frac{4}{3})^x + \dots} \quad \text{........................(203)}$$

is

$$\pi(2^x) + \pi(\tfrac{3}{2})^x + \pi(\tfrac{4}{3})^x + \dots.$$

In other words if $x_n \leqslant x < x_{n+1}$ then

$$n = \pi(2^x) + \pi(\tfrac{3}{2})^x + \pi(\tfrac{4}{3})^x + \dots. \quad \text{..............(204)}$$

It follows from (192) and (202) that, of the primes $\pi_1, \pi_2, \pi_3, \dots, \pi_n$, the number of primes which are equal to a given prime ϖ is equal to

$$[(\varpi^{1/x} - 1)^{-1}]. \quad \text{..............................(205)}$$

Further, the greatest of the primes $\pi_1, \pi_2, \pi_3, \dots, \pi_n$ is the largest prime not greater than 2^x, and is asymptotically equivalent to the natural nth prime, in virtue of (204).

The following table gives the values of π_n and x_n for the first 50 values of n, that is till x_n reaches very nearly 7.

$$\pi_1 = 2 \qquad x_1 = \frac{\log 2}{\log 2} = 1$$

$$\pi_2 = 3 \qquad x_2 = \frac{\log 3}{\log 2} = 1{\cdot}5849\ldots$$

$$\pi_3 = 2 \qquad x_3 = \frac{\log 2}{\log(\frac{3}{2})} = 1{\cdot}7095\ldots$$

$$\pi_4 = 5 \qquad x_4 = \frac{\log 5}{\log 2} = 2{\cdot}3219\ldots$$

$$\pi_5 = 2 \qquad x_5 = \frac{\log 2}{\log(\frac{4}{3})} = 2{\cdot}4094\ldots$$

$$\pi_6 = 3 \qquad x_6 = \frac{\log 3}{\log(\frac{3}{2})} = 2{\cdot}7095\ldots$$

$$\pi_7 = 7 \qquad x_7 = \frac{\log 7}{\log 2} = 2{\cdot}8073\ldots$$

$$\pi_8 = 2 \qquad x_8 = \frac{\log 2}{\log(\frac{5}{4})} = 3{\cdot}1062\ldots$$

$$\pi_9 = 11 \qquad x_9 = \frac{\log 11}{\log 2} = 3{\cdot}4594\ldots$$

$$\pi_{10} = 13 \qquad x_{10} = \frac{\log 13}{\log 2} = 3{\cdot}7004\ldots$$

$$\pi_{11} = 2 \qquad x_{11} = \frac{\log 2}{\log(\frac{6}{5})} = 3{\cdot}8017\ldots$$

$$\pi_{12} = 3 \qquad x_{12} = \frac{\log 3}{\log(\frac{4}{3})} = 3{\cdot}8188\ldots$$

$$\pi_{13} = 5 \qquad x_{13} = \frac{\log 5}{\log(\frac{3}{2})} = 3{\cdot}9693\ldots$$

$$\pi_{14} = 17 \qquad x_{14} = \frac{\log 17}{\log 2} = 4{\cdot}0874\ldots$$

$$\pi_{15} = 19 \qquad x_{15} = \frac{\log 19}{\log 2} = 4{\cdot}2479\ldots$$

$$\pi_{16} = 2 \qquad x_{16} = \frac{\log 2}{\log(\frac{7}{6})} = 4{\cdot}4965\ldots$$

$$\pi_{17} = 23 \qquad x_{17} = \frac{\log 23}{\log 2} = 4{\cdot}5235\ldots$$

$$\pi_{18} = 7 \qquad x_{18} = \frac{\log 7}{\log(\frac{3}{2})} = 4{\cdot}7992\ldots$$

$$\pi_{19} = 29 \qquad x_{19} = \frac{\log 29}{\log 2} = 4{\cdot}8579\ldots$$

$$\pi_{20} = 3 \qquad x_{20} = \frac{\log 3}{\log(\frac{5}{4})} = 4{\cdot}9233\ldots$$

$$\pi_{21} = 31 \qquad x_{21} = \frac{\log 31}{\log 2} = 4{\cdot}9541\ldots$$

$$\pi_{22} = 2 \qquad x_{22} = \frac{\log 2}{\log(\frac{8}{7})} = 5{\cdot}1908\ldots$$

$$\pi_{23} = 37 \qquad x_{23} = \frac{\log 37}{\log 2} = 5{\cdot}2094\ldots$$

$$\pi_{24} = 41 \qquad x_{24} = \frac{\log 41}{\log 2} = 5{\cdot}3575\ldots$$

$$\pi_{25} = 43 \qquad x_{25} = \frac{\log 43}{\log 2} = 5{\cdot}4262\ldots$$

$$\pi_{26} = 47 \qquad x_{26} = \frac{\log 47}{\log 2} = 5{\cdot}5545\ldots$$

$$\pi_{27} = 5 \qquad x_{27} = \frac{\log 5}{\log(\frac{4}{3})} = 5{\cdot}5945\ldots$$

$$\pi_{28} = 53 \qquad x_{28} = \frac{\log 53}{\log 2} = 5{\cdot}7279\ldots$$

$$\pi_{29} = 59 \qquad x_{29} = \frac{\log 59}{\log 2} = 5{\cdot}8826\ldots$$

$$\pi_{30} = 2 \qquad x_{30} = \frac{\log 2}{\log(\frac{9}{8})} = 5{\cdot}8849\ldots$$

$$\pi_{31} = 11 \qquad x_{31} = \frac{\log 11}{\log(\frac{3}{2})} = 5{\cdot}9139\ldots$$

$$\pi_{32} = 61 \qquad x_{32} = \frac{\log 61}{\log 2} = 5{\cdot}9307\ldots$$

$$\pi_{33} = 3 \qquad x_{33} = \frac{\log 3}{\log(\frac{6}{5})} = 6{\cdot}0256\ldots$$

$$\pi_{34} = 67 \qquad x_{34} = \frac{\log 67}{\log 2} = 6{\cdot}0660\ldots$$

$$\pi_{35} = 71 \qquad x_{35} = \frac{\log 71}{\log 2} = 6{\cdot}1497\ldots$$

$$\pi_{36} = 73 \qquad x_{36} = \frac{\log 73}{\log 2} = 6{\cdot}1898\ldots$$

$$\pi_{37} = 79 \qquad x_{37} = \frac{\log 79}{\log 2} = 6{\cdot}3037\ldots$$

$$\pi_{38} = 13 \qquad x_{38} = \frac{\log 13}{\log(\frac{3}{2})} = 6{\cdot}3259\ldots$$

$$\pi_{39} = 83 \qquad x_{39} = \frac{\log 83}{\log 2} = 6{\cdot}3750\ldots$$

$$\pi_{40}=89 \qquad x_{40}=\frac{\log 89}{\log 2}=6{\cdot}4757\ldots$$

$$\pi_{41}=2 \qquad x_{41}=\frac{\log 2}{\log(\frac{10}{9})}=6{\cdot}5790\ldots$$

$$\pi_{42}=97 \qquad x_{42}=\frac{\log 97}{\log 2}=6{\cdot}5999\ldots$$

$$\pi_{43}=101 \qquad x_{43}=\frac{\log 101}{\log 2}=6{\cdot}6582\ldots$$

$$\pi_{44}=103 \qquad x_{44}=\frac{\log 103}{\log 2}=6{\cdot}6724\ldots$$

$$\pi_{45}=107 \qquad x_{45}=\frac{\log 107}{\log 2}=6{\cdot}7414\ldots$$

$$\pi_{46}=7 \qquad x_{46}=\frac{\log 7}{\log(\frac{4}{3})}=6{\cdot}7641\ldots$$

$$\pi_{47}=109 \qquad x_{47}=\frac{\log 109}{\log 2}=6{\cdot}7681\ldots$$

$$\pi_{48}=113 \qquad x_{48}=\frac{\log 113}{\log 2}=6{\cdot}8201\ldots$$

$$\pi_{49}=17 \qquad x_{49}=\frac{\log 17}{\log(\frac{3}{2})}=6{\cdot}9875\ldots$$

$$\pi_{50}=127 \qquad x_{50}=\frac{\log 127}{\log 2}=6{\cdot}9886\ldots$$

38. It follows from (17) and (198) that $\log d(N)\leqslant F(x)$, where

$$F(x)=\frac{1}{x}\log N+\frac{1}{x}\left\{\int_2^{2^x}\frac{\pi(t)}{t}\,dt+\int_2^{(\frac{3}{2})^x}\frac{\pi(t)}{t}\,dt+\int_2^{(\frac{4}{3})^x}\frac{\pi(t)}{t}\,dt+\ldots\right\};\qquad\ldots\ldots(206)$$

for all values of N and x. In order to obtain the best possible upper limit for $\log d(N)$, we must choose x so as to make the right-hand side a minimum.

The function $F(x)$ is obviously continuous unless $(1+1/r)^x=p$, where r is a positive integer and p a prime. It is easily seen to be continuous even then, and so continuous without exception. Also

$$\begin{aligned}F'(x)&=-\frac{1}{x^2}\log N-\frac{1}{x^2}\left\{\int_2^{2^x}\frac{\pi(t)}{t}\,dt+\int_2^{(\frac{3}{2})^x}\frac{\pi(t)}{t}\,dt+\ldots\right\}\\&\qquad+\frac{1}{x}\{\pi(2^x)\log 2+\pi(\tfrac{3}{2})^x\log\tfrac{3}{2}+\ldots\}\\&=\frac{1}{x^2}\{\vartheta(2^x)+\vartheta(\tfrac{3}{2})^x+\vartheta(\tfrac{4}{3})^x+\ldots-\log N\},\qquad\ldots\ldots\ldots\ldots\ldots\ldots(207)\end{aligned}$$

unless $(1+1/r)^x=p$, in virtue of (17).

Thus we see that $F(x)$ is continuous, and $F'(x)$ exists and is continuous except at certain isolated points. The sign of $F'(x)$, where it exists, is that of

$$\vartheta(2^x)+\vartheta(\tfrac{3}{2})^x+\vartheta(\tfrac{4}{3})^x+\ldots-\log N,$$

and

$$\vartheta(2^x)+\vartheta(\tfrac{3}{2})^x+\vartheta(\tfrac{4}{3})^x+\ldots,$$

is a monotonic function. Thus $F'(x)$ is first negative and then positive, changing sign once only, and so $F(x)$ has a unique minimum. Thus $F(x)$ is a minimum when x is a function of N defined by the inequalities

$$\vartheta(2^y)+\vartheta(\tfrac{3}{2})^y+\vartheta(\tfrac{4}{3})^y+\ldots \quad \begin{Bmatrix} < \log N\,(y<x) \\ > \log N\,(y>x) \end{Bmatrix}. \qquad \ldots\ldots(208)$$

Now let $D(N)$ be a function of N such that

$$D(N)=2^{\pi(2^x)}(\tfrac{3}{2})^{\pi(\frac{3}{2})^x}(\tfrac{4}{3})^{\pi(\frac{4}{3})^x}\ldots, \qquad \ldots\ldots\ldots\ldots\ldots\ldots(209)$$

where x is the function of N defined by the inequalities (208). Then, from (198), we see that

$$d(N)\leqslant D(N), \qquad \ldots\ldots\ldots\ldots\ldots\ldots\ldots\ldots\ldots(210)$$

for all values of N; and $d(N)=D(N)$ for all superior highly composite values of N. Hence $D(N)$ is the maximum order of $d(N)$. In other words, $d(N)$ will attain its maximum order when N is a superior highly composite number.

V.

Application to the Order of $d(N)$.

39. The most precise result known concerning the distribution of the prime numbers is that

$$\left.\begin{aligned} \pi(x) &= Li(x)+O(xe^{-a\sqrt{\log x}}), \\ \vartheta(x) &= \quad x \;\;+O(xe^{-a\sqrt{\log x}}), \end{aligned}\right\} \qquad \ldots\ldots\ldots\ldots\ldots\ldots(211)$$

where

$$Li(x)=\int^x \frac{dt}{\log t}$$

and a is a positive constant.

In order to find the maximum order of $d(N)$ we have merely to determine the order of $D(N)$ from the equations (208) and (209). Now, from (208), we have

$$\log N=\vartheta(2^x)+O(\tfrac{3}{2})^x=\vartheta(2^x)+o(2^{2x/3});$$

and so

$$\vartheta(2^x)=\log N+o(\log N)^{\frac{2}{3}}; \qquad \ldots\ldots\ldots\ldots\ldots\ldots\ldots(212)$$

and similarly from (209) we have

$$\pi(2^x)=\frac{\log D(N)}{\log 2}+o(\log N)^{\frac{2}{3}}. \qquad \ldots\ldots\ldots\ldots\ldots\ldots(213)$$

It follows from (211)—(213) that the maximum order of $d(N)$ is

$$2^{Li(\log N)+O[\log N e^{-a\sqrt{(\log\log N)}}]}. \qquad \ldots\ldots\ldots\ldots\ldots\ldots(214)$$

It does not seem to be possible to obtain an upper limit for $d(N)$ notably more precise than (214) without assuming results concerning the distribution of primes which depend on hitherto unproved properties of the Riemann ζ-function.

40. We shall now assume that the "Riemann hypothesis" concerning the ζ-function is true, i.e., that all the complex roots of $\zeta(s)$ have their real part equal to $\frac{1}{2}$. Then it is known that

$$\vartheta(x) = x - \sqrt{x} - \Sigma \frac{x^\rho}{\rho} + O(x^{\frac{1}{3}}), \quad \text{.................(215)}$$

where ρ is a complex root of $\zeta(s)$, and that

$$\begin{aligned}\pi(x) &= Li(x) - \tfrac{1}{2} Li(\sqrt{x}) - \Sigma\, Li(x^\rho) + O(x^{\frac{1}{3}}) \\ &= Li(x) - \frac{\sqrt{x}}{\log x} - \frac{2\sqrt{x}}{(\log x)^2} - \frac{1}{\log x} \Sigma \frac{x^\rho}{\rho} - \frac{1}{(\log x)^2} \Sigma \frac{x^\rho}{\rho^2} + O\left\{\frac{\sqrt{x}}{(\log x)^3}\right\},\end{aligned}$$

$$\text{......(216)}$$

since $\Sigma \frac{x^\rho}{\rho^k}$ is absolutely convergent when $k > 1$. Also it is known that

$$\Sigma \frac{x^\rho}{\rho} = O\{\sqrt{x}(\log x)^2\}; \quad \text{.......................(217)}$$

and so

$$\vartheta(x) - x = O\{\sqrt{x}(\log x)^2\}. \quad \text{.....................(218)}$$

From (215) and (216) it is clear that

$$\pi(x) = Li(x) + \frac{\vartheta(x) - x}{\log x} - R(x) + O\left\{\frac{\sqrt{x}}{(\log x)^3}\right\}, \quad \text{......(219)}$$

where

$$R(x) = \frac{2\sqrt{x} + \Sigma \frac{x^\rho}{\rho^2}}{(\log x)^2}. \quad \text{...........................(220)}$$

But it follows from Taylor's theorem and (218) that

$$Li\,\vartheta(x) - Li(x) = \frac{\vartheta(x) - x}{\log x} + O(\log x)^2, \quad \text{...........(221)}$$

and from (219) and (221) it follows that

$$\pi(x) = Li\,\vartheta(x) - R(x) + O\left\{\frac{\sqrt{x}}{(\log x)^3}\right\}. \quad \text{...........(222)}$$

41. It follows from the functional equation satisfied by $\zeta(s)$, viz.,

$$(2\pi)^{-s}\,\Gamma(s)\,\zeta(s)\cos\tfrac{1}{2}\pi s = \tfrac{1}{2}\zeta(1-s), \quad \text{..............(223)}$$

that

$$(1-s)\,\pi^{-\frac{1}{4}\sqrt{s}}\,\Gamma\left(\frac{1+\sqrt{s}}{4}\right)\zeta\left(\frac{1+\sqrt{s}}{2}\right)$$

is an integral function of s whose apparent order is less than 1, and hence is equal to

$$\Gamma(\tfrac{1}{4})\,\zeta(\tfrac{1}{2})\,\Pi\left\{1 - \frac{s}{(2\rho-1)^2}\right\},$$

[where ρ runs through the complex roots of $\zeta(s)$ whose imaginary parts are positive]. From this we can easily deduce that

$$s(1+s)\,\pi^{-\frac{1+s}{2}}\,\Gamma\left(\frac{1+s}{2}\right)\zeta(1+s)=\Pi\left(1+\frac{s}{\rho}\right), \quad \text{.........(224)}$$

[where ρ now runs through all the roots]. Subtracting 1 from both sides, dividing the result by s, and then making $s\to 0$, we obtain

$$\Sigma\frac{1}{\rho}=1+\tfrac{1}{2}(\gamma-\log 4\pi), \quad \text{.....................(225)}$$

where γ is the Eulerian constant. Hence we see that

$$\left|\Sigma\frac{x^{\rho}}{\rho^{2}}\right|\leqslant\Sigma\left|\frac{x^{\rho}}{\rho^{2}}\right|=\sqrt{x}\,\Sigma\frac{1}{\rho(1-\rho)}=\sqrt{x}\,\Sigma\left(\frac{1}{\rho}+\frac{1}{1-\rho}\right)$$

$$=2\sqrt{x}\,\Sigma\frac{1}{\rho}=\sqrt{x}\,(2+\gamma-\log 4\pi). \quad \text{.........(226)}$$

It follows from (220) and (226) that

$$(\log 4\pi-\gamma)\sqrt{x}\leqslant R(x)(\log x)^{2}\leqslant(4+\gamma-\log 4\pi)\sqrt{x}. \quad \text{...(227)}$$

It can easily be verified that

$$\left.\begin{aligned}\log 4\pi-\gamma&=1{\cdot}954,\\ 4+\gamma-\log 4\pi&=2{\cdot}046,\end{aligned}\right\} \quad \text{.....................(228)}$$

approximately.

42. Now
$$R(x)=\frac{2\sqrt{x}+S(x)}{(\log x)^{2}},$$
where
$$S(x)=\Sigma\frac{x^{\rho}}{\rho^{2}};$$
so that, considering $R(x)$ as a function of a continuous variable, we have

$$R'(x)=\frac{1}{\sqrt{x}(\log x)^{2}}-\frac{4\sqrt{x}+2S(x)}{x(\log x)^{3}}+\frac{S'(x)}{(\log x)^{2}}$$
$$=\frac{S'(x)}{(\log x)^{2}}+O\left\{\frac{1}{\sqrt{x}(\log x)^{2}}\right\},$$

for all values of x for which $S(x)$ possesses a differential coefficient.

Now the derived series of $S(x)$, viz.,

$$\bar{S}(x)=\frac{1}{x}\Sigma\frac{x^{\rho}}{\rho},$$

is uniformly convergent throughout any interval of positive values of x which does not include any value of x of the form $x=p^{m}$; and $S(x)$ is continuous for all values of x. It follows that

$$S(x_1)-S(x_2)=\int_{x_1}^{x_2}\bar{S}(x)\,dx,$$

for all positive values of x_1 and x_2, and that $S(x)$ possesses a derivative

$$S'(x)=\bar{S}(x),$$

whenever x is not of the form p^m. Also

$$\bar{S}(x) = O\left\{\frac{(\log x)^2}{\sqrt{x}}\right\}.$$

Hence

$$R(x+h) = R(x) + \int_x^{x+h} O\left(\frac{1}{\sqrt{t}}\right) dt = R(x) + O\left(\frac{h}{\sqrt{x}}\right). \quad \text{......(229)}$$

43. Now

$$\begin{aligned}\log N &= \vartheta(2^x) + \vartheta(\tfrac{3}{2})^x + O(\tfrac{4}{3})^x \\ &= \vartheta(2^x) + (\tfrac{3}{2})^x + O\{x^2(\tfrac{3}{2})^{\frac{1}{2}x}\} + O(\tfrac{4}{3})^x \\ &= \vartheta(2^x) + (\tfrac{3}{2})^x + O(2^{5x/12}).\end{aligned}$$

Similarly $\log D(N) = \log 2 \,.\, \pi(2^x) + \log(\tfrac{3}{2})\, Li(\tfrac{3}{2})^x + O(2^{5x/12})$.

Writing X for 2^x, we have

$$\left.\begin{aligned}\log N &= \vartheta(X) + X^{\log(\frac{3}{2})/\log 2} + O(X^{\frac{5}{12}}), \\ \log D(N) &= \log 2 \,.\, \pi(X) + \log(\tfrac{3}{2})\, Li\{X^{\log(\frac{3}{2})/\log 2}\} + O(X^{\frac{5}{12}}).\end{aligned}\right\} \quad \text{...(230)}$$

It follows that $\log N = X + O[X^{\log(\frac{3}{2})/\log 2}]$;

and so

$$X = \log N + O[(\log N)^{\log(\frac{3}{2})/\log 2}]. \quad \text{..................(231)}$$

Again, from (230) and (231), it follows that

$$\log N = \vartheta(X) + (\log N)^{\log(\frac{3}{2})/\log 2} + O\{(\log N)^{\frac{5}{12}}\}; \quad \text{......(232)}$$

and

$$\begin{aligned}\log D(N) &= \log 2 \,.\, \pi(X) + \log(\tfrac{3}{2})\, Li(\log N)^{\log(\frac{3}{2})/\log 2} + O\{(\log N)^{\frac{5}{12}}\} \\ &= \log 2\left\{Li\,\vartheta(X) - R(X) + O\left[\frac{\sqrt{X}}{(\log X)^3}\right]\right\} \\ &\quad + \log(\tfrac{3}{2})\, Li\{(\log N)^{\log(\frac{3}{2})/\log 2}\} + O\{(\log N)^{\frac{5}{12}}\}, \quad \text{.........(233)}\end{aligned}$$

in virtue of (222). From (231) and (233) it evidently follows that

$$\begin{aligned}\log D(N) &= \log 2 \,.\, Li\,\vartheta(X) - \log 2 \,.\, R(X) + \log(\tfrac{3}{2})\, Li\{(\log N)^{\log(\frac{3}{2})/\log 2}\} \\ &\qquad + O\left\{\frac{\sqrt{(\log N)}}{(\log\log N)^3}\right\} \\ &= \log 2 \,.\, Li\{\log N - (\log N)^{\log(\frac{3}{2})/\log 2} + O(\log N)^{\frac{5}{12}}\} \\ &\quad - \log 2 \,.\, R\{\log N + O(\log N)^{\log(\frac{3}{2})/\log 2}\} \\ &\quad + \log(\tfrac{3}{2})\, Li\{(\log N)^{\log(\frac{3}{2})/\log 2}\} + O\left\{\frac{\sqrt{(\log N)}}{(\log\log N)^3}\right\}, \quad \text{.........(234)}\end{aligned}$$

in virtue of (231) and (232). But

$$\begin{aligned}&Li\{\log N - (\log N)^{\log(\frac{3}{2})/\log 2} + O(\log N)^{\frac{5}{12}}\} \\ &= Li(\log N) - \frac{(\log N)^{\log(\frac{3}{2})/\log 2}}{\log\log N} + O\left\{\frac{(\log N)^{\frac{5}{12}}}{\log\log N}\right\} + O\left\{\frac{(\log N)^{\{2\log(\frac{3}{2})/\log 2\} - 1}}{(\log\log N)^2}\right\} \\ &= Li(\log N) - \frac{(\log N)^{\log(\frac{3}{2})/\log 2}}{\log\log N} + O(\log N)^{\frac{5}{12}};\end{aligned}$$

and

$$\begin{aligned}R\{\log N + O(\log N)^{\log(\frac{3}{2})/\log 2}\} &= R(\log N) + O\{(\log N)^{\{\log(\frac{3}{2})/\log 2\} - \frac{1}{2}}\} \\ &= R(\log N) + O(\log N)^{\frac{1}{10}},\end{aligned}$$

in virtue of (229). Hence (234) may be replaced by

$$\log D(N) = \log 2 \,.\, Li(\log N) + \log(\tfrac{3}{2})\, Li\{(\log N)^{\log(\frac{3}{2})/\log 2}\}$$
$$- \log 2 \frac{(\log N)^{\log(\frac{3}{2})/\log 2}}{\log\log N} - \log 2 \,.\, R(\log N) + O\left\{\frac{\sqrt{(\log N)}}{(\log\log N)^3}\right\}. \quad \text{...(235)}$$

That is to say the maximum order of $d(N)$ is

$$2^{Li(\log N) + \phi(N)}, \quad \text{.............................(236)}$$

where $$\phi(N) = \frac{\log(\frac{3}{2})}{\log 2} Li\{(\log N)^{\log(\frac{3}{2})/\log 2}\} - \frac{(\log N)^{\log(\frac{3}{2})/\log 2}}{\log\log N} - R(\log N)$$
$$+ O\left\{\frac{\sqrt{(\log N)}}{(\log\log N)^3}\right\}.$$

This order is actually attained for an infinity of values of N.

44. We can now find the order of the number of superior highly composite numbers not exceeding a given number N. Let N' be the smallest superior highly composite number greater than N, and let

$$N' = e^{\vartheta(2^x) + \vartheta(\frac{3}{2})^x + \vartheta(\frac{4}{3})^x + \dots}.$$

Then, from § 37, we know that

$$2N \leqslant N' \leqslant 2^x N, \quad \text{..........................(237)}$$

so that $N' = O(N \log N)$; and also that the number of superior highly composite numbers not exceeding N' is

$$n = \pi(2^x) + \pi(\tfrac{3}{2})^x + \pi(\tfrac{4}{3})^x + \dots.$$

By arguments similar to those of the previous section we can shew that

$$n = Li(\log N) + Li(\log N)^{\log(\frac{3}{2})/\log 2} - \frac{(\log N)^{\log(\frac{3}{2})/\log 2}}{\log\log N} - R(\log N)$$
$$+ O\left\{\frac{\sqrt{(\log N)}}{(\log\log N)^3}\right\}. \quad \text{...........(238)}$$

It is easy to see from § 37 that, if the largest superior highly composite number not exceeding N is

$$2^{a_2} \,.\, 3^{a_3} \,.\, 5^{a_5} \dots p^{a_p},$$

then the number of superior highly composite numbers not exceeding N is the sum of all the indices, viz.,

$$a_2 + a_3 + a_5 + \dots + a_p.$$

45. Proceeding as in § 28, we can shew that, if N is a superior highly composite number, and m and n are any two positive integers such that [n is a divisor of N, and]

$$\log mn = o(\log\log N),$$

then $$d\left(\frac{m}{n} N\right) = d(N)\, 2^{\frac{\log(m/n)}{\log\log N} + O\left(\frac{\log mn}{\log\log N}\right)^2}. \quad \text{...............(239)}$$

From this we can easily shew that the next highly composite number is of the form

$$N+O\left\{\frac{N(\log\log\log N)^2}{\log\log N}\right\}. \quad\text{.......................}(240)$$

Again, let S' and S be any two consecutive superior highly composite numbers, and let

$$S=e^{\vartheta(2^x)+\vartheta\left(\left(\frac{3}{2}\right)^x\right)+\vartheta\left(\left(\frac{4}{3}\right)^x\right)+\dots}.$$

Then it follows from § 35 that

$$d(N)<\left(\frac{N}{S}\right)^{1/x} d(S), \quad\text{........................}(241)$$

for all values of N except S and S'. Now, if S be the nth superior highly composite number, so that

$$x_n \leqslant x < x_{n+1},$$

where x_n is the same as in § 36, we see that

$$d(N)<\left(\frac{N}{S}\right)^{1/x_n} d(S), \quad\text{....................}(241')$$

for all values of N except S and S'. If N is S or S', then the inequality becomes an equality.

It follows from § 36 that $d(S)\leqslant 2d(S')$. Hence, if N be highly composite and $S'<N<S$, so that $d(S')<d(N)<d(S)$, then

$$\tfrac{1}{2}d(S)<d(N)<d(S), \quad d(S')<d(N)<2d(S').$$

From this it is easy to see that the order (236) is actually attained by $d(N)$, whenever N is a highly composite number. But it may also be attained when N is not a highly composite number. For example, if

$$N=(2.3.5\dots p_1)\times(2.3.5\dots p_2),$$

where p_1 is the largest prime not greater than 2^x, and p_2 the largest prime not greater than $\left(\frac{3}{2}\right)^x$, it is easily seen that $d(N)$ attains the order (236): and N is not highly composite.

VI.

Special Forms of N.

46. In §§ 33—38 we have indirectly solved the following problem: to find the relations which must hold between x_1, x_2, x_3, ... in order that

$$2^{\pi(x_1)}.\left(\tfrac{3}{2}\right)^{\pi(x_2)}.\left(\tfrac{4}{3}\right)^{\pi(x_3)}\dots$$

may be a maximum, when it is given that

$$\vartheta(x_1)+\vartheta(x_2)+\vartheta(x_3)+\dots$$

is a fixed number. The relations which we obtained are

$$\frac{\log 2}{\log x_1}=\frac{\log\left(\frac{3}{2}\right)}{\log x_2}=\frac{\log\left(\frac{4}{3}\right)}{\log x_3}=\dots.$$

This suggests the following more general problem. If N is an integer of the form

$$e^{c_1\vartheta(x_1)+c_2\vartheta(x_2)+c_3\vartheta(x_3)+\dots}, \qquad \text{.....................(242)}$$

where $c_1, c_2, c_3, \dots$ are any given positive integers, it is required to find the nature of N, that is to say the relations which hold between $x_1, x_2, x_3, \dots$, when $d(N)$ is of maximum order. From (242) we see that

$$d(N) = (1+c_1)^{\pi(x_1)} \left(\frac{1+c_1+c_2}{1+c_1}\right)^{\pi(x_2)} \left(\frac{1+c_1+c_2+c_3}{1+c_1+c_2}\right)^{\pi(x_3)} \dots \quad \dots(242')$$

If we define the "superior" numbers of the class (242) by the inequalities

$$\frac{d(N)}{N^{\epsilon}} \geqslant \frac{d(N')}{(N')^{\epsilon}},$$

for all values of N' less than N, and

$$\frac{d(N)}{N^{\epsilon}} > \frac{d(N')}{(N')^{\epsilon}},$$

for all values of N' greater than N, N and N' in the two inequalities being of the form (242), and proceed as in § 33, we can shew that

$$d(N) \leqslant N^{1/x} \frac{(1+c_1)^{\pi\{(1+c_1)^{x/c_1}\}}}{e^{(c_1/x)\vartheta\{(1+c_1)^{x/c_1}\}}} \frac{\left(\frac{1+c_1+c_2}{1+c_1}\right)^{\pi\left\{\left(\frac{1+c_1+c_2}{1+c_1}\right)^{x/c_2}\right\}}}{e^{(c_2/x)\vartheta\left\{\left(\frac{1+c_1+c_2}{1+c_1}\right)^{x/c_2}\right\}}} \dots, \quad \dots(243)$$

for all values of x, and for all values of N of the form (242). From this we can shew, by arguments similar to those of § 38, that N must be of the form

$$e^{c_1\vartheta\{(1+c_1)^{x/c_1}\}+c_2\vartheta\left\{\left(\frac{1+c_1+c_2}{1+c_1}\right)^{x/c_2}\right\}+c_3\vartheta\left\{\left(\frac{1+c_1+c_2+c_3}{1+c_1+c_2}\right)^{x/c_3}\right\}+\dots}, \quad \dots\dots(244)$$

and $d(N)$ of the form

$$(1+c_1)^{\pi\{(1+c_1)^{x/c_1}\}} \left(\frac{1+c_1+c_2}{1+c_1}\right)^{\pi\left\{\left(\frac{1+c_1+c_2}{1+c_1}\right)^{x/c_2}\right\}} \left(\frac{1+c_1+c_2+c_3}{1+c_1+c_2}\right)^{\pi\left\{\left(\frac{1+c_1+c_2+c_3}{1+c_1+c_2}\right)^{x/c_3}\right\}} \dots$$

$$\dots\dots(244')$$

From (244) and (244′) we can find the maximum order of $d(N)$, as in § 43.

47. We shall now consider the order of $d(N)$ for some special forms of N. The simplest case is that in which N is of the form

$$2 \,.\, 3 \,.\, 5 \,.\, 7 \dots p;$$

so that
$$\log N = \vartheta(p),$$

and
$$d(N) = 2^{\pi(p)}.$$

It is easy to shew that

$$d(N) = 2^{Li(\log N) - R(\log N) + O\left\{\frac{\sqrt{(\log N)}}{(\log\log N)^3}\right\}}. \quad \text{............(245)}$$

In this case $d(N)$ is exactly a power of 2, and this naturally suggests the question: what is the maximum order of $d(N)$ when $d(N)$ is exactly a power of 2?

It is evident that, if $d(N)$ is a power of 2, the indices of the prime divisors of N cannot be any other numbers except 1, 3, 7, 15, 31, ...; and so in order that $d(N)$ should be of maximum order, N must be of the form

$$e^{\vartheta(x_1)+2\vartheta(x_2)+4\vartheta(x_3)+8\vartheta(x_4)+\dots},$$

and $d(N)$ of the form $$2^{\pi(x_1)+\pi(x_2)+\pi(x_3)+\dots}.$$

It follows from §46 that, in order that $d(N)$ should be of maximum order, N must be of the form

$$e^{\vartheta(x)+2\vartheta(\sqrt{x})+4\vartheta(x^{\frac{1}{4}})+8\vartheta(x^{\frac{1}{8}})+\dots}, \quad \dots\dots\dots\dots\dots\dots\dots(246)$$

and $d(N)$ of the form $$2^{\pi(x)+\pi(\sqrt{x})+\pi(x^{\frac{1}{4}})+\pi(x^{\frac{1}{8}})+\dots}. \quad \dots\dots\dots\dots\dots\dots\dots(247)$$

Hence the maximum order of $d(N)$ can easily be shewn to be

$$2^{Li(\log N)+\frac{4\sqrt{(\log N)}}{(\log\log N)^2}-R(\log N)+O\left\{\frac{\sqrt{(\log N)}}{(\log\log N)^3}\right\}}. \quad \dots\dots\dots\dots(248)$$

It is easily seen from (246) that the least number having 2^n divisors is

$$2 \,.\, 3 \,.\, 4 \,.\, 5 \,.\, 7 \,.\, 9 \,.\, 11 \,.\, 13 \,.\, 16 \,.\, 17 \,.\, 19 \,.\, 23 \,.\, 25 \,.\, 29 \dots \text{ to } n \text{ factors}, \quad \dots(249)$$

where 2, 3, 4, 5, 7, ... are the natural primes, their squares, fourth powers, and so on, arranged according to order of magnitude.

48. We have seen that the last indices of the prime divisors of N must be 1, if $d(N)$ is of maximum order. Now we shall consider the maximum order of $d(N)$ when the indices of the prime divisors of N are never less than an integer n. In the first place, in order that $d(N)$ should be of maximum order, N must be of the form

$$e^{n\vartheta(x_1)+\vartheta(x_2)+\vartheta(x_3)+\dots},$$

and $d(N)$ of the form

$$(1+n)^{\pi(x_1)}\left(\frac{2+n}{1+n}\right)^{\pi(x_2)}\left(\frac{3+n}{2+n}\right)^{\pi(x_3)}\dots.$$

It follows from §46 that N must be of the form

$$e^{n\vartheta\{(1+n)^{x/n}\}+\vartheta\left\{\left(\frac{2+n}{1+n}\right)^x\right\}+\vartheta\left\{\left(\frac{3+n}{2+n}\right)^x\right\}+\dots}, \quad \dots\dots\dots\dots\dots(250)$$

and $d(N)$ of the form

$$(1+n)^{\pi\{(1+n)^{x/n}\}}\left(\frac{2+n}{1+n}\right)^{\pi\left\{\left(\frac{2+n}{1+n}\right)^x\right\}}\left(\frac{3+n}{2+n}\right)^{\pi\left\{\left(\frac{3+n}{2+n}\right)^x\right\}}\dots. \quad \dots(251)$$

Then, by arguments similar to those of §43, we can shew that the maximum order of $d(N)$ is

$$(n+1)^{Li\{(1/n)\log N\}+\phi(N)}, \quad \dots\dots\dots\dots\dots\dots\dots(252)$$

where $$\phi(N) = \left\{\frac{\log(n+2)}{\log(n+1)} - 1\right\} Li \left\{\left(\frac{1}{n}\log N\right)^{n\frac{\log(n+2)}{\log(n+1)} - n}\right\}$$

$$- \frac{\left(\frac{1}{n}\log N\right)^{n\frac{\log(n+2)}{\log(n+1)} - n}}{n \log\left(\frac{1}{n}\log N\right)} - R\left(\frac{1}{n}\log N\right) + O\left\{\frac{\sqrt{(\log N)}}{(\log\log N)^3}\right\}.$$

If $n \geqslant 3$, it is easy to verify that

$$n\frac{\log(n+2)}{\log(n+1)} - n < \tfrac{1}{2};$$

and so (252) reduces to

$$(n+1)^{Li\{(1/n)\log N\} - R\{(1/n)\log N\} + O\left\{\frac{\sqrt{(\log N)}}{(\log\log N)^3}\right\}}, \quad \text{.........(253)}$$

provided that $n \geqslant 3$.

49. Let us next consider the maximum order of $d(N)$ when N is a perfect nth power. In order that $d(N)$ should be of maximum order, N must be of the form

$$e^{n\vartheta(x_1) + n\vartheta(x_2) + n\vartheta(x_3) + \dots},$$

and $d(N)$ of the form

$$(1+n)^{\pi(x_1)} \left(\frac{1+2n}{1+n}\right)^{\pi(x_2)} \left(\frac{1+3n}{1+2n}\right)^{\pi(x_3)} \dots.$$

It follows from § 46 that N must be of the form

$$e^{n\vartheta\{(1+n)^x\} + n\vartheta\left\{\left(\frac{1+2n}{1+n}\right)^x\right\} + n\vartheta\left\{\left(\frac{1+3n}{1+2n}\right)^x\right\} + \dots}, \quad \text{..............(254)}$$

and $d(N)$ of the form

$$(1+n)^{\pi\{(1+n)^x\}} \left(\frac{1+2n}{1+n}\right)^{\pi\left\{\left(\frac{1+2n}{1+n}\right)^x\right\}} \left(\frac{1+3n}{1+2n}\right)^{\pi\left\{\left(\frac{1+3n}{1+2n}\right)^x\right\}} \dots \quad \text{...(255)}$$

Hence we can shew that the maximum order of $d(N)$ is

$$(n+1)^{Li\{(1/n)\log N\} - R\{(1/n)\log N\} + O\left\{\frac{\sqrt{(\log N)}}{(\log\log N)^3}\right\}}, \quad \text{.........(256)}$$

provided that $n > 1$.

50. Let $l(N)$ denote the least common multiple of the first N natural numbers. Then it can easily be shewn that

$$l(N) = 2^{[\log N/\log 2]} \cdot 3^{[\log N/\log 3]} \cdot 5^{[\log N/\log 5]} \dots p, \quad \text{.........(257)}$$

where p is the largest prime not greater than N. From this we can shew that

$$l(N) = e^{\vartheta(N) + \vartheta(\sqrt{N}) + \vartheta(N^{\frac{1}{3}}) + \vartheta(N^{\frac{1}{4}}) + \dots}; \quad \text{..............(258)}$$

and so $$d\{l(N)\} = 2^{\pi(N)} \left(\tfrac{3}{2}\right)^{\pi(\sqrt{N})} \left(\tfrac{4}{3}\right)^{\pi(N^{\frac{1}{3}})} \dots \quad \text{.................(259)}$$

From (258) and (259) we can shew that, if N is of the form $l(M)$, then

$$d(N) = 2^{Li(\log N) + \phi(N)}, \quad \text{...................(260)}$$

where

$$\phi(N)=\frac{\log(\frac{9}{8})}{\log 2}\frac{\sqrt{(\log N)}}{\log\log N}+\frac{4\log(\frac{3}{2})}{\log 2}\frac{\sqrt{(\log N)}}{(\log\log N)^2}-R(\log N)+O\left\{\frac{\sqrt{(\log N)}}{(\log\log N)^3}\right\}.$$

It follows from (258) that

$$l(N)=e^{N+O\{\sqrt{N}(\log N)^2\}}; \quad\text{.....................(261)}$$

and from (259) that $$d\{l(N)\}=2^{Li(N)+O(\sqrt{N}\log N)}. \quad\text{..................(262)}$$

51. Finally, we shall consider the number of divisors of $N!$. It is easily seen that

$$N!=2^{a_2}.3^{a_3}.5^{a_5}\dots p^{a_p}, \quad\text{.....................(263)}$$

where p is the largest prime not greater than N, and

$$a_\lambda=\left[\frac{N}{\lambda}\right]+\left[\frac{N}{\lambda^2}\right]+\left[\frac{N}{\lambda^3}\right]+\dots.$$

It is evident that the primes greater than $\frac{1}{2}N$ and not exceeding N appear once in $N!$, the primes greater than $\frac{1}{3}N$ and not exceeding $\frac{1}{2}N$ appear twice, and so on up to those greater than $N/[\sqrt{N}]$ and not exceeding $N/([\sqrt{N}]-1)$, appearing $[\sqrt{N}]-1$ times*. The indices of the smaller primes cannot be specified so simply. Hence it is clear that

$$N!=e^{\vartheta(N)+\vartheta(\frac{1}{2}N)+\vartheta(\frac{1}{3}N)+\dots+\vartheta\left(\frac{N}{[\sqrt{N}]-1}\right)}\times 2^{a_2}.3^{a_3}.5^{a_5}\dots\varpi^{a_\varpi}, \quad (264)$$

where ϖ is the largest prime not greater than $\sqrt{N}$, and

$$a_\lambda-1+[\sqrt{N}]=\left[\frac{N}{\lambda}\right]+\left[\frac{N}{\lambda^2}\right]+\left[\frac{N}{\lambda^3}\right]+\dots.$$

From (264) we see that

$$\begin{aligned}d(N!)&=2^{\pi(N)}(\tfrac{3}{2})^{\pi(\frac{1}{2}N)}(\tfrac{4}{3})^{\pi(\frac{1}{3}N)}\dots\text{ to }[\sqrt{N}]-1\text{ factors}\\&\quad\times e^{O\{\log(1+a_2)+\log(1+a_3)+\dots+\log(1+a_\varpi)\}}\\&=2^{\pi(N)}(\tfrac{3}{2})^{\pi(\frac{1}{2}N)}(\tfrac{4}{3})^{\pi(\frac{1}{3}N)}\dots\text{ to }[\sqrt{N}]-1\text{ factors}\\&\quad\times e^{O\{\varpi\log(1+a_2)\}}\\&=2^{Li(N)}(\tfrac{3}{2})^{Li(\frac{1}{2}N)}(\tfrac{4}{3})^{Li(\frac{1}{3}N)}\dots\text{ to }[\sqrt{N}]\text{ factors}\\&\quad\times e^{O(\sqrt{N}\log N)}.\end{aligned} \quad\text{..........(265)}$$

Since $$Li(N)=\frac{N}{\log N}+O\left\{\frac{N}{(\log N)^2}\right\},$$

we see that $$d(N!)=C^{\frac{N}{\log N}+O\left\{\frac{N}{(\log N)^2}\right\}}, \quad\text{.....................(266)}$$

where $$C=(1+1)^1(1+\tfrac{1}{2})^{\frac{1}{2}}(1+\tfrac{1}{3})^{\frac{1}{3}}(1+\tfrac{1}{4})^{\frac{1}{4}}\dots.$$

From this we can easily deduce that, if N is of the form $M!$, then

$$d(N)=C^{\frac{\log N}{(\log\log N)^2}+\frac{2\log N\log\log\log N}{(\log\log N)^3}+O\left\{\frac{\log N}{(\log\log N)^3}\right\}}, \quad\text{.........(267)}$$

where C is the same constant as in (266).

* Strictly speaking, this is true only when $N\geqslant 4$.

52. It is interesting in this connection to shew how, by considering numbers of certain special forms, we can obtain lower limits for the maximum orders of the iterated functions $dd(n)$ and $ddd(n)$. By supposing that

$$N = 2^{2-1} . 3^{3-1} \dots p^{p-1},$$

we can shew that
$$dd(n) > 4^{\frac{\sqrt{(2\log n)}}{\log\log n}} \qquad (268)$$
for an infinity of values of n. By supposing that

$$N = 2^{2^{a_2}-1} . 3^{3^{a_3}-1} \dots p^{p^{a_p}-1},$$

where
$$a_\lambda = \left[\frac{\log p}{\log \lambda}\right] - 1,$$

we can shew that
$$ddd(n) > (\log n)^{\log\log\log\log n} \qquad (269)$$
for an infinity of values of n.

16

ON CERTAIN INFINITE SERIES

(*Messenger of Mathematics*, XLV, 1916, 11—15)

1. This paper is merely a continuation of the paper on "Some definite integrals" published in this *Journal**. It deals with some series which resemble those definite integrals not merely in form but in many other respects. In each case there is a functional relation. In the case of the integrals there are special values of a parameter for which the integrals may be evaluated in finite terms. In the case of the series the corresponding results involve elliptic functions.

2. It can be shewn, by the theory of residues, that if α and β are real and $\alpha\beta = \frac{1}{4}\pi^2$, then

$$\frac{\alpha}{(\alpha+t)\cosh\alpha} - \frac{3\alpha}{(9\alpha+t)\cosh 3\alpha} + \frac{5\alpha}{(25\alpha+t)\cosh 5\alpha} - \dots$$
$$+ \frac{\beta}{(\beta-t)\cosh\beta} - \frac{3\beta}{(9\beta-t)\cosh 3\beta} + \frac{5\beta}{(25\beta-t)\cosh 5\beta} - \dots$$
$$= \frac{\pi}{4\cos\sqrt{(\alpha t)}\cosh\sqrt{(\beta t)}}. \qquad \dots\dots(1)$$

Now let

$$F(n) = \left\{\frac{\alpha e^{in\alpha}}{\cosh\alpha} - \frac{3\alpha e^{9in\alpha}}{\cosh 3\alpha} + \frac{5\alpha e^{25in\alpha}}{\cosh 5\alpha} - \dots\right\}$$
$$- \left\{\frac{\beta e^{-in\beta}}{\cosh\beta} - \frac{3\beta e^{-9in\beta}}{\cosh 3\beta} + \frac{5\beta e^{-25in\beta}}{\cosh 5\beta} - \dots\right\}. \dots(2)$$

Then we see that, if t is positive,

$$\int_0^\infty e^{-2tn} F(n)\,dn = \frac{\pi i}{4\cosh\{(1-i)\sqrt{(\alpha t)}\}\cosh\{(1+i)\sqrt{(\beta t)}\}} \dots\dots(3)$$

in virtue of (1). Again, let

$$f(n) = -\frac{1}{2n}\sqrt{\left(\frac{\pi}{2n}\right)}\Sigma\Sigma(-1)^{\frac{1}{2}(\mu+\nu)}\{\mu(1+i)\sqrt{\alpha} - \nu(1-i)\sqrt{\beta}\}$$
$$\times e^{-(\pi\mu\nu - i\mu^2\alpha + i\nu^2\beta)/4n} \qquad (\mu = 1, 3, 5, \dots;\ \nu = 1, 3, 5, \dots). \dots\dots(4)$$

Then it is easy to shew that

$$\int_0^\infty e^{-2tn} f(n)\,dn = \frac{\pi i}{4\cosh\{(1-i)\sqrt{(\alpha t)}\}\cosh\{(1+i)\sqrt{(\beta t)}\}}. \dots\dots(5)$$

Hence, by a theorem due to Lerch†, we obtain

$$F(n) = f(n) \qquad \dots\dots(6)$$

* [No. 11 of this volume (pp. 53–58); see also No. 12 (pp. 59—67).]

† See Mr Hardy's note at the end of my previous paper [*Messenger of Mathematics*, XLIV, pp. 18—21. See also Appendix, p. 338.]

for all positive values of n, provided that $\alpha\beta=\frac{1}{4}\pi^2$. In particular, when $\alpha=\beta=\frac{1}{2}\pi$, we have

$$\frac{\sin\frac{1}{2}\pi n}{\cosh\frac{1}{2}\pi}-\frac{3\sin\frac{9}{2}\pi n}{\cosh\frac{3}{2}\pi}+\frac{5\sin\frac{25}{2}\pi n}{\cosh\frac{5}{2}\pi}-\dots$$

$$=-\frac{1}{4n\sqrt{n}}\Sigma\Sigma(-1)^{\frac{1}{2}(\mu+\nu)}e^{-\pi\mu\nu/4n}\left[(\mu+\nu)\cos\frac{\pi(\mu^2-\nu^2)}{8n}+(\mu-\nu)\sin\frac{\pi(\mu^2-\nu^2)}{8n}\right]$$

$$(\mu=1,3,5,\dots;\ \nu=1,3,5,\dots)\quad\dots\dots(7)$$

for all positive values of n. As particular cases of (7), we have

$$\frac{\sin(\frac{1}{2}\pi/a)}{\cosh\frac{1}{2}\pi}-\frac{3\sin(\frac{9}{2}\pi/a)}{\cosh\frac{3}{2}\pi}+\frac{5\sin(\frac{25}{2}\pi/a)}{\cosh\frac{5}{2}\pi}-\dots$$

$$=\tfrac{1}{4}a\sqrt{a}\left(\frac{1}{\cosh\frac{1}{4}\pi a}-\frac{3}{\cosh\frac{3}{4}\pi a}+\frac{5}{\cosh\frac{5}{4}\pi a}-\dots\right)$$

$$=\tfrac{1}{2}a\sqrt{a}\,(e^{-\frac{1}{16}\pi a}-e^{-\frac{9}{16}\pi a}-e^{-\frac{25}{16}\pi a}+e^{-\frac{49}{16}\pi a}+\dots)^4,\dots(8)$$

if a is a positive even integer; and

$$\frac{\sin(\frac{1}{2}\pi/a)}{\cosh\frac{1}{2}\pi}-\frac{3\sin(\frac{9}{2}\pi/a)}{\cosh\frac{3}{2}\pi}+\frac{5\sin(\frac{25}{2}\pi/a)}{\cosh\frac{5}{2}\pi}-\dots$$

$$=\tfrac{1}{4}a\sqrt{a}\left(\frac{1}{\sinh\frac{1}{4}\pi a}+\frac{3}{\sinh\frac{3}{4}\pi a}+\frac{5}{\sinh\frac{5}{4}\pi a}+\dots\right)$$

$$=\tfrac{1}{2}a\sqrt{a}\,(e^{-\frac{1}{16}\pi a}+e^{-\frac{9}{16}\pi a}+e^{-\frac{25}{16}\pi a}+e^{-\frac{49}{16}\pi a}+\dots)^4,\dots(9)$$

if a is a positive odd integer; and so on*.

3. It is also easy to shew that if $\alpha\beta=\pi^2$, then

$$\left\{\frac{\alpha}{(\alpha+t)\sinh\alpha}-\frac{2\alpha}{(4\alpha+t)\sinh2\alpha}+\frac{3\alpha}{(9\alpha+t)\sinh3\alpha}-\dots\right\}$$

$$-\left\{\frac{\beta}{(\beta-t)\sinh\beta}-\frac{2\beta}{(4\beta-t)\sinh2\beta}+\frac{3\beta}{(9\beta-t)\sinh3\beta}-\dots\right\}$$

$$=\frac{1}{2t}-\frac{\pi}{2\sin\sqrt{(\alpha t)}\sinh\sqrt{(\beta t)}}.\quad\dots\dots\dots\dots\dots\dots(10)$$

From this we can deduce, as in the previous section, that if $\alpha\beta=\pi^2$, then

$$\frac{\alpha e^{in\alpha}}{\sinh\alpha}-\frac{2\alpha e^{4in\alpha}}{\sinh2\alpha}+\frac{3\alpha e^{9in\alpha}}{\sinh3\alpha}-\dots$$

$$+\frac{\beta e^{-in\beta}}{\sinh\beta}-\frac{2\beta e^{-4in\beta}}{\sinh2\beta}+\frac{3\beta e^{-9in\beta}}{\sinh3\beta}-\dots$$

$$=\frac{1}{2}-\frac{1}{n}\sqrt{\left(\frac{\pi}{2n}\right)}$$

$$\times\Sigma\Sigma\{\mu(1-i)\sqrt{\alpha}+\nu(1+i)\sqrt{\beta}\}e^{-(2\pi\mu\nu-i\mu^2\alpha+i\nu^2\beta)/4n}$$

$$(\mu=1,3,5,\dots;\ \nu=1,3,5,\dots)\quad\dots\dots(11)$$

* [Formulæ (9), (14), and (19) are incorrect; see Appendix, p. 339.]

for all positive values of n. If, in particular, we put $\alpha=\beta=\pi$, we obtain

$$\frac{1}{4\pi}-\frac{\cos\pi n}{\sinh\pi}+\frac{2\cos 4\pi n}{\sinh 2\pi}-\frac{3\cos 9\pi n}{\sinh 3\pi}+\dots$$

$$=\frac{1}{2n\sqrt{(2n)}}\Sigma\Sigma\, e^{-\pi\mu\nu/2n}\left\{(\mu+\nu)\cos\frac{\pi(\mu^2-\nu^2)}{4n}+(\mu-\nu)\sin\frac{\pi(\mu^2-\nu^2)}{4n}\right\}$$

$$(\mu=1,3,5,\dots;\ \nu=1,3,5,\dots)\quad\dots\dots(12)$$

for all positive values of n. Thus, for example, we have

$$\frac{1}{4\pi}-\frac{\cos(2\pi/a)}{\sinh\pi}+\frac{2\cos(8\pi/a)}{\sinh 2\pi}-\frac{3\cos(18\pi/a)}{\sinh 3\pi}+\dots$$

$$=\tfrac{1}{8}a\sqrt{a}\left(\frac{1}{\sinh\frac{1}{4}\pi a}+\frac{3}{\sinh\frac{3}{4}\pi a}+\frac{5}{\sinh\frac{5}{4}\pi a}+\dots\right)$$

$$=\tfrac{1}{4}a\sqrt{a}\,(e^{-\frac{1}{16}\pi a}+e^{-\frac{9}{16}\pi a}+e^{-\frac{25}{16}\pi a}+e^{-\frac{49}{16}\pi a}+\dots)^4,\quad\dots(13).$$

if a is a positive even integer; and

$$\frac{1}{4\pi}-\frac{\cos(2\pi/a)}{\sinh\pi}+\frac{2\cos(8\pi/a)}{\sinh 2\pi}-\frac{3\cos(18\pi/a)}{\sinh 3\pi}+\dots$$

$$=\tfrac{1}{8}a\sqrt{a}\left(\frac{1}{\cosh\frac{1}{4}\pi a}-\frac{3}{\cosh\frac{3}{4}\pi a}+\frac{5}{\cosh\frac{5}{4}\pi a}-\dots\right)$$

$$=\tfrac{1}{4}a\sqrt{a}\,(e^{-\frac{1}{16}\pi a}-e^{-\frac{9}{16}\pi a}-e^{-\frac{25}{16}\pi a}+e^{-\frac{49}{16}\pi a}+\dots)^4,\quad\dots(14)$$

if a is a positive odd integer.

4. In a similar manner we can shew that, if $\alpha\beta=\pi^2$, then

$$\frac{\alpha e^{in\alpha}}{e^{2\alpha}-1}+\frac{2\alpha e^{4in\alpha}}{e^{4\alpha}-1}+\frac{3\alpha e^{9in\alpha}}{e^{6\alpha}-1}+\dots$$

$$+\frac{\beta e^{-in\beta}}{e^{2\beta}-1}+\frac{2\beta e^{-4in\beta}}{e^{4\beta}-1}+\frac{3\beta e^{-9in\beta}}{e^{6\beta}-1}+\dots$$

$$=\alpha\int_0^\infty\frac{xe^{-in\alpha x^2}}{e^{2\pi x}-1}dx^*+\beta\int_0^\infty\frac{xe^{in\beta x^2}}{e^{2\pi x}-1}dx^*-\tfrac{1}{4}$$

$$+\frac{1}{n}\sqrt{\left(\frac{\pi}{2n}\right)}\sum_{\mu=1}^{\mu=\infty}\sum_{\nu=1}^{\nu=\infty}\{\mu(1-i)\sqrt{\alpha}+\nu(1+i)\sqrt{\beta}\}\,e^{-(2\pi\mu\nu-i\mu^2\alpha+i\nu^2\beta)/n}\quad\dots(15)$$

for all positive values of n. Putting $\alpha=\beta=\pi$ in (15) we see that, if $n>0$, then

$$\frac{1}{8\pi}+\frac{\cos\pi n}{e^{2\pi}-1}+\frac{2\cos 4\pi n}{e^{4\pi}-1}+\frac{3\cos 9\pi n}{e^{6\pi}-1}+\dots$$

$$=\int_0^\infty\frac{x\cos\pi nx^2}{e^{2\pi x}-1}dx+\frac{1}{2n\sqrt{(2n)}}\sum_{\mu=1}^{\mu=\infty}\sum_{\nu=1}^{\nu=\infty}e^{-2\pi\mu\nu/n}$$

$$\times\left[(\mu+\nu)\cos\left\{\frac{\pi(\mu^2-\nu^2)}{n}\right\}+(\mu-\nu)\sin\left\{\frac{\pi(\mu^2-\nu^2)}{n}\right\}\right].\quad\dots\dots(16)$$

* I shewed in my former paper [No. 12 of the present volume] that this integral can be calculated in finite terms whenever $n\alpha$ is a rational multiple of π.

As particular cases of (16) we have

$$\frac{1}{8\pi}+\frac{\cos(\pi/a)}{e^{2\pi}-1}+\frac{2\cos(4\pi/a)}{e^{4\pi}-1}+\frac{3\cos(9\pi/a)}{e^{6\pi}-1}+\ldots$$
$$=\int_0^\infty \frac{x\cos(\pi x^2/a)}{e^{2\pi x}-1}dx+a\sqrt{(\tfrac{1}{2}a)}\left(\frac{1}{e^{2\pi a}-1}+\frac{2}{e^{4\pi a}-1}+\frac{3}{e^{6\pi a}-1}+\ldots\right), \quad \ldots(17)$$

if a is a positive even integer;

$$\frac{1}{8\pi}+\frac{\cos(\pi/a)}{e^{2\pi}-1}+\frac{2\cos(4\pi/a)}{e^{4\pi}-1}+\frac{3\cos(9\pi/a)}{e^{6\pi}-1}+\ldots$$
$$=\int_0^\infty \frac{x\cos(\pi x^2/a)}{e^{2\pi x}-1}dx+a\sqrt{(\tfrac{1}{2}a)}\left(\frac{1}{e^{2\pi a}+1}-\frac{2}{e^{4\pi a}+1}+\frac{3}{e^{6\pi a}+1}-\ldots\right), \quad \ldots(18)$$

if a is a positive odd integer; and

$$\frac{1}{8\pi}+\frac{\cos(2\pi/a)}{e^{2\pi}-1}+\frac{2\cos(8\pi/a)}{e^{4\pi}-1}+\frac{3\cos(18\pi/a)}{e^{6\pi}-1}+\ldots$$
$$=\int_0^\infty \frac{x\cos(2\pi x^2/a)}{e^{2\pi x}-1}dx+\tfrac{1}{4}a\sqrt{a}\left(\frac{1}{e^{\pi a}+1}+\frac{3}{e^{3\pi a}+1}+\frac{5}{e^{5\pi a}+1}+\ldots\right), \quad \ldots(19)$$

if a is a positive odd integer.

It may be interesting to note that different functions dealt with in this paper have the same asymptotic expansion for small values of n. For example, the two different functions

$$\frac{1}{8\pi}+\frac{\cos n}{e^{2\pi}-1}+\frac{2\cos 4n}{e^{4\pi}-1}+\frac{3\cos 9n}{e^{6\pi}-1}+\ldots$$

and

$$\int_0^\infty \frac{x\cos nx^2}{e^{2\pi x}-1}dx$$

have the same asymptotic expansion, viz.

$$\frac{1}{24}-\frac{n^2}{1008}+\frac{n^4}{6336}-\frac{n^6}{17280}+\ldots^{*}. \quad \ldots\ldots\ldots\ldots\ldots\ldots(20)$$

* This series (in spite of the appearance of the first few terms) diverges for all values of n.

17

SOME FORMULÆ IN THE ANALYTIC THEORY OF NUMBERS*

(*Messenger of Mathematics*, XLV, 1916, 81—84)

I have found the following formulæ incidentally in the course of other investigations. None of them seem to be of particular importance, nor does their proof involve the use of any new ideas, but some of them are so curious that they seem to be worth printing. I denote by $d(x)$ the number of divisors of x, if x is an integer, and zero otherwise, and by $\zeta(s)$ the Riemann Zeta-function.

(A) $$\frac{\zeta^4(s)}{\zeta(2s)} = 1^{-s}d^2(1) + 2^{-s}d^2(2) + 3^{-s}d^2(3) + \dots, \qquad (1)$$

$$\frac{\eta^4(s)}{(1-2^{-2s})\zeta(2s)} = 1^{-s}d^2(1) - 3^{-s}d^2(3) + 5^{-s}d^2(5) - \dots, \qquad (2)$$

where $$\eta(s) = 1^{-s} - 3^{-s} + 5^{-s} - 7^{-s} + \dots.$$

(B) $$d^2(1) + d^2(2) + d^2(3) + \dots + d^2(n)$$
$$= An(\log n)^3 + Bn(\log n)^2 + Cn\log n + Dn + O(n^{\frac{3}{5}+\epsilon})\dagger, \qquad (3)$$

where $$A = \frac{1}{\pi^2}, \quad B = \frac{12\gamma - 3}{\pi^2} - \frac{36}{\pi^4}\zeta'(2),$$

γ is Euler's constant, C, D more complicated constants, and ϵ any positive number.

(C) $$d^3\left(\frac{n}{1}\right) + d^3\left(\frac{n}{2}\right) + d^3\left(\frac{n}{3}\right) + \dots = \left\{d\left(\frac{n}{1}\right) + d\left(\frac{n}{2}\right) + d\left(\frac{n}{3}\right) + \dots\right\}^2 \ddagger, \qquad (4)$$

$$\sum_1^\infty n^{-s}d^r(n) = \{\zeta(s)\}^{2^r}\phi(s), \qquad (5)$$

where $\phi(s)$ is absolutely convergent for $R(s) > \frac{1}{2}$, and in particular

$$\sum_1^\infty \frac{1}{n^s d(n)} = \prod_p \left\{p^s \log\left(\frac{1}{1-p^{-s}}\right)\right\} = \sqrt{\{\zeta(s)\}}\,\phi(s). \qquad (6)$$

(D) $$\frac{1}{d(1)} + \frac{1}{d(2)} + \frac{1}{d(3)} + \dots + \frac{1}{d(n)}$$
$$= n\left\{\frac{A_1}{(\log n)^{\frac{1}{2}}} + \frac{A_2}{(\log n)^{\frac{3}{2}}} + \dots + \frac{A_r}{(\log n)^{r-\frac{1}{2}}} + O\frac{1}{(\log n)^{r+\frac{1}{2}}}\right\}, \qquad (7)$$

where $$A_1 = \frac{1}{\sqrt{\pi}}\prod_p \left\{\sqrt{(p^2-p)}\log\left(\frac{p}{p-1}\right)\right\}$$

and $A_2, A_3, \dots A_r$ are more complicated constants.

* [See Appendix, p. 339.]

† If we assume the Riemann hypothesis, the error term here is of the form $O(n^{\frac{1}{2}+\epsilon})$.

‡ Mr Hardy has pointed out to me that this formula has been given already by Liouville, *Journal de Mathématiques*, Ser. 2, Vol. II (1857), p. 393.

More generally

$$d^s(1)+d^s(2)+d^s(3)+\dots+d^s(n)$$
$$=n\{A_1(\log n)^{2^s-1}+A_2(\log n)^{2^s-2}+\dots+A_{2^s}\}+O(n^{\frac{1}{2}+\epsilon})^*,\ \dots(8)$$

if 2^s is an integer, and

$$d^s(1)+d^s(2)+d^s(3)+\dots+d^s(n)$$
$$=n\left\{A_1(\log n)^{2^s-1}+A_2(\log n)^{2^s-2}+\dots+\frac{A_{r+2^s}}{(\log n)^r}+O\left[\frac{1}{(\log n)^{r+1}}\right]\right\},\ \dots(9)$$

if 2^s is not an integer, the A's being constants.

(E) $$d(1)\,d(2)\,d(3)\dots d(n)=2^{n\,(\log\log n+C)+\phi(n)},\ \dots\dots\dots\dots(10)$$

where $$C=\gamma+\sum_2^\infty\left\{\log_2\left(1+\frac{1}{\nu}\right)-\frac{1}{\nu}\right\}(2^{-\nu}+3^{-\nu}+5^{-\nu}+\dots).$$

Here 2, 3, 5, ... are the primes and

$$\frac{\phi(n)}{n}=\frac{\gamma-1}{\log n}+\frac{1!}{(\log n)^2}(\gamma+\gamma_1-1)+\frac{2!}{(\log n)^3}(\gamma+\gamma_1+\gamma_2-1)+\dots$$
$$+\frac{(r-1)!}{(\log n)^r}(\gamma+\gamma_1+\gamma_2+\dots+\gamma_{r-1}-1)+O\left\{\frac{1}{(\log n)^{r+1}}\right\},$$

where $$\zeta(1+s)=\frac{1}{s}+\gamma-\gamma_1 s+\gamma_2 s^2-\gamma_3 s^3+\dots$$

or $$r!\,\gamma_r=\lim_{\nu\to\infty}\left\{(\log 1)^r+\tfrac{1}{2}(\log 2)^r+\dots+\frac{1}{\nu}(\log\nu)^r-\frac{1}{r+1}(\log\nu)^{r+1}\right\}.$$

(F) $$d(uv)=\sum_1^\infty\mu(n)\,d\left(\frac{u}{n}\right)d\left(\frac{v}{n}\right)=\Sigma\mu(\delta)\,d\left(\frac{u}{\delta}\right)d\left(\frac{v}{\delta}\right),\ \dots\dots\dots(11)$$

where δ is a common factor of u and v, and

$$\frac{1}{\zeta(s)}=\sum_1^\infty\frac{\mu(n)}{n^s}.$$

(G) If $$D_v(n)=d(v)+d(2v)+\dots+d(nv),$$

we have $$D_v(n)=\Sigma\mu(\delta)\,d\left(\frac{v}{\delta}\right)D_1\left(\frac{n}{\delta}\right),\ \dots\dots\dots\dots\dots\dots\dots(12)$$

where δ is a divisor of v, and

$$D_v(n)=\alpha(v)\,n\,(\log n+2\gamma-1)+\beta(v)\,n+\Delta_v(n),\ \dots\dots\dots(13)$$

where $$\sum_1^\infty\frac{\alpha(\nu)}{\nu^s}=\frac{\zeta^2(s)}{\zeta(1+s)},\quad \sum_1^\infty\frac{\beta(\nu)}{\nu^s}=-\frac{\zeta^2(s)\,\zeta'(1+s)}{\zeta^2(1+s)},$$

and $$\Delta_v(n)=O(n^{\frac{1}{2}}\log n)\dagger.$$

(H) $$d(v+c)+d(2v+c)+d(3v+c)+\dots+d(nv+c)$$
$$=\alpha_c(v)\,n\,(\log n+2\gamma-1)+\beta_c(v)\,n\Delta_{v,c}(n),\ \dots(14)$$

where $$\sum_1^\infty\frac{\alpha_c(\nu)}{\nu^s}=\frac{\zeta(s)\,\sigma_{-s}(|c|)}{\zeta(1+s)},$$

$$\sum_1^\infty\frac{\beta_c(\nu)}{\nu^s}=-\frac{\zeta(s)\,\sigma_{-s}(|c|)}{\zeta(1+s)}\left\{\frac{\zeta'(s)}{\zeta(s)}+\frac{\zeta'(1+s)}{\zeta(1+s)}+\frac{\sigma_{-s}'(|c|)}{\sigma_{-s}(|c|)}\right\},$$

* Assuming the Riemann hypothesis.

† It seems not unlikely that $\Delta_v(n)$ is of the form $O(n^{\frac{1}{4}+\epsilon})$. Mr Hardy has recently shewn that $\Delta_1(n)$ is not of the form $o\{(n\log n)^{\frac{1}{4}}\log\log n\}$. The same is true in this case also.

$\sigma_s(n)$ being the sum of the sth powers of the divisors of n and $\sigma_s'(n)$ the derivative of $\sigma_s(n)$ with respect to s, and

$$\Delta_{v,c}(n) = O(n^{\frac{1}{2}} \log n)*.$$

(I) The formulæ (1) and (2) are special cases of

$$\frac{\zeta(s)\zeta(s-a)\zeta(s-b)\zeta(s-a-b)}{\zeta(2s-a-b)} = 1^{-s}\sigma_a(1)\sigma_b(1) + 2^{-s}\sigma_a(2)\sigma_b(2) + 3^{-s}\sigma_a(3)\sigma_b(3) + \ldots; \quad \ldots(15)$$

$$\frac{\eta(s)\eta(s-a)\eta(s-b)\eta(s-a-b)}{(1-2^{-2s+a+b})\zeta(2s-a-b)} = 1^{-s}\sigma_a(1)\sigma_b(1) - 3^{-s}\sigma_a(3)\sigma_b(3) + 5^{-s}\sigma_a(5)\sigma_b(5) - \ldots. \quad \ldots(16)$$

It is possible to find an approximate formula for the general sum

$$\sigma_a(1)\sigma_b(1) + \sigma_a(2)\sigma_b(2) + \ldots + \sigma_a(n)\sigma_b(n). \quad \ldots\ldots\ldots\ldots(17)$$

The general formula is complicated. The most interesting cases are $a=0$, $b=0$, when the formula is (3); $a=0$, $b=1$, when it is

$$\frac{\pi^4 n^2}{72\zeta(3)}(\log n + 2c) + nE(n), \quad \ldots\ldots\ldots\ldots(18)$$

where

$$c = \gamma - \tfrac{1}{4} + \frac{\zeta'(2)}{\zeta(2)} - \frac{\zeta'(3)}{\zeta(3)},$$

and the order of $E(n)$ is the same as that of $\Delta_1(n)$; and $a=1$, $b=1$, when it is

$$\tfrac{5}{6} n^3 \zeta(3) + E(n), \quad \ldots\ldots\ldots\ldots(19)$$

where

$$E(n) = O\{n^2(\log n)^2\}, \quad E(n) \neq o(n^2 \log n).$$

(J) If $s > 0$, then

$$\sigma_s(1)\sigma_s(2)\sigma_s(3)\sigma_s(4)\ldots\sigma_s(n) = \theta c^n (n!)^s, \quad \ldots\ldots\ldots\ldots(20)$$

where

$$1 > \theta > (1-2^{-s})(1-3^{-s})(1-5^{-s})\ldots(1-\varpi^{-s}),$$

ϖ is the greatest prime not exceeding n, and

$$c = \prod_p \left\{\left(\frac{p^{2s}-1}{p^{2s}-p^s}\right)^{1/p} \left(\frac{p^{3s}-1}{p^{3s}-p^s}\right)^{1/p^2} \left(\frac{p^{4s}-1}{p^{4s}-p^s}\right)^{1/p^3} \ldots\right\}.$$

(K) If

$$(\tfrac{1}{2} + q + q^4 + q^9 + q^{16} + \ldots)^2 = \tfrac{1}{4} + \sum_1^\infty r(n) q^n,$$

so that

$$\zeta(s)\eta(s) = \sum_1^\infty r(n) n^{-s},$$

then

$$\frac{\zeta^2(s)\eta^2(s)}{(1+2^{-s})\zeta(2s)} = 1^{-s} r^2(1) + 2^{-s} r^2(2) + 3^{-s} r^2(3) + \ldots. \quad \ldots\ldots(21)$$

$$r^2(1) + r^2(2) + r^2(3) + \ldots + r^2(n) = \frac{n}{4}(\log n + C) + O(n^{\frac{3}{5}+\epsilon}), \ldots(22)$$

where

$$C = 4\gamma - 1 + \tfrac{1}{3}\log 2 - \log \pi + 4 \log \Gamma(\tfrac{3}{4}) - \frac{12}{\pi^2}\zeta'(2).$$

These formulæ are analogous to (1) and (3).

* It is very likely that the order of $\Delta_{v,c}(n)$ is the same as that of $\Delta_1(n)$.

18

ON CERTAIN ARITHMETICAL FUNCTIONS

(*Transactions of the Cambridge Philosophical Society*, XXII, No. 9, 1916, 159—184)

1. Let $\sigma_s(n)$ denote the sum of the sth powers of the divisors of n (including 1 and n), and let

$$\sigma_s(0) = \tfrac{1}{2}\zeta(-s),$$

where $\zeta(s)$ is the Riemann Zeta-function. Further let

$$\Sigma_{r,s}(n) = \sigma_r(0)\,\sigma_s(n) + \sigma_r(1)\,\sigma_s(n-1) + \ldots + \sigma_r(n)\,\sigma_s(0). \quad \ldots(1)$$

In this paper I prove that

$$\Sigma_{r,s}(n) = \frac{\Gamma(r+1)\,\Gamma(s+1)}{\Gamma(r+s+2)}\,\frac{\zeta(r+1)\,\zeta(s+1)}{\zeta(r+s+2)}\,\sigma_{r+s+1}(n) + \frac{\zeta(1-r)+\zeta(1-s)}{r+s}\,n\sigma_{r+s-1}(n) + O\{n^{\frac{2}{3}(r+s+1)}\}, \quad (2)$$

whenever r and s are positive odd integers. I also prove that there is no error term on the right-hand side of (2) in the following nine cases: $r=1$, $s=1$; $r=1$, $s=3$; $r=1$, $s=5$; $r=1$, $s=7$; $r=1$, $s=11$; $r=3$, $s=3$; $r=3$, $s=5$; $r=3$, $s=9$; $r=5$, $s=7$. That is to say $\Sigma_{r,s}(n)$ has a finite expression in terms of $\sigma_{r+s+1}(n)$ and $\sigma_{r+s-1}(n)$ in these nine cases; but for other values of r and s it involves other arithmetical functions as well.

It appears probable, from the empirical results I obtain in §§ 18—23, that the error term on the right-hand side of (2) is of the form

$$O\{n^{\frac{1}{2}(r+s+1+\epsilon)}\}, \quad \ldots(3)$$

where ϵ is any positive number, and not of the form

$$o\{n^{\frac{1}{2}(r+s+1)}\}. \quad \ldots(4)$$

But all I can prove rigorously is (i) that the error is of the form

$$O\{n^{\frac{2}{3}(r+s+1)}\}$$

in all cases, (ii) that it is of the form

$$O\{n^{\frac{2}{3}(r+s+\frac{3}{4})}\} \quad \ldots(5)$$

if $r+s$ is of the form $6m$, (iii) that it is of the form

$$O\{n^{\frac{2}{3}(r+s+\frac{1}{2})}\} \quad \ldots(6)$$

if $r+s$ is of the form $6m+4$, and (iv) that it is not of the form

$$o\{n^{\frac{1}{2}(r+s)}\}. \quad \ldots(7)$$

It follows from (2) that, if r and s are positive odd integers, then

$$\Sigma_{r,s}(n) \sim \frac{\Gamma(r+1)\,\Gamma(s+1)}{\Gamma(r+s+2)}\,\frac{\zeta(r+1)\,\zeta(s+1)}{\zeta(r+s+2)}\,\sigma_{r+s+1}(n). \quad \text{......(8)}$$

It seems very likely that (8) is true for all positive values of r and s, but this I am at present unable to prove.

2. If $\Sigma_{r,s}(n)/\sigma_{r+s+1}(n)$ tends to a limit, then the limit must be

$$\frac{\Gamma(r+1)\,\Gamma(s+1)}{\Gamma(r+s+2)}\,\frac{\zeta(r+1)\,\zeta(s+1)}{\zeta(r+s+2)}.$$

For then

$$\begin{aligned}\operatorname*{Lim}_{n\to\infty}\frac{\Sigma_{r,s}(n)}{\sigma_{r+s+1}(n)} &= \operatorname*{Lim}_{n\to\infty}\frac{\Sigma_{r,s}(1)+\Sigma_{r,s}(2)+\ldots+\Sigma_{r,s}(n)}{\sigma_{r+s+1}(1)+\sigma_{r+s+1}(2)+\ldots+\sigma_{r+s+1}(n)}\\ &= \operatorname*{Lim}_{x\to 1}\frac{\Sigma_{r,s}(0)+\Sigma_{r,s}(1)\,x+\Sigma_{r,s}(2)\,x^2+\ldots}{\sigma_{r+s+1}(0)+\sigma_{r+s+1}(1)\,x+\sigma_{r+s+1}(2)\,x^2+\ldots}\\ &= \operatorname*{Lim}_{x\to 1}\frac{S_r S_s}{S_{r+s+1}},\end{aligned}$$

where

$$S_r = \tfrac{1}{2}\zeta(-r) + \frac{1^r x}{1-x} + \frac{2^r x^2}{1-x^2} + \frac{3^r x^3}{1-x^3} + \ldots \quad \text{............(9)}$$

Now it is known that, if $r > 0$, then

$$S_r \sim \frac{\Gamma(r+1)\,\zeta(r+1)}{(1-x)^{r+1}}, \quad \text{........................(10)}$$

as $x \to 1$*. Hence we obtain the result stated.

3. It is easy to see that

$$\begin{aligned}&\sigma_r(1)+\sigma_r(2)+\sigma_r(3)+\ldots+\sigma_r(n)\\ &= u_1+u_2+u_3+u_4+\ldots+u_n,\end{aligned}$$

where

$$u_t = 1^r + 2^r + 3^r + \ldots + \left[\frac{n}{t}\right]^r.$$

From this it is easy to deduce that

$$\sigma_r(1)+\sigma_r(2)+\ldots+\sigma_r(n) \sim \frac{n^{r+1}}{r+1}\zeta(r+1) \quad \text{...........(11)†}$$

and

$$\sigma_r(1)(n-1)^s+\sigma_r(2)(n-2)^s+\ldots+\sigma_r(n-1)\,1^s \sim \frac{\Gamma(r+1)\,\Gamma(s+1)}{\Gamma(r+s+2)}\zeta(r+1)\,n^{r+s+1},$$

provided $r>0$, $s \geqslant 0$. Now

$$\sigma_s(n) > n^s,$$

and

$$\sigma_s(n) < n^s(1^{-s}+2^{-s}+3^{-s}+\ldots) = n^s\zeta(s).$$

From these inequalities and (1) it follows that

$$\underline{\operatorname{Lim}}\,\frac{\Sigma_{r,s}(n)}{n^{r+s+1}} \geqslant \frac{\Gamma(r+1)\,\Gamma(s+1)}{\Gamma(r+s+2)}\zeta(r+1), \quad \text{...........(12)}$$

* Knopp, *Dissertation* (Berlin, 1907), p. 34.

† (10) follows from this as an immediate corollary.

if $r>0$ and $s\geqslant 0$; and

$$\overline{\operatorname{Lim}}\frac{\Sigma_{r,s}(n)}{n^{r+s+1}}\leqslant\frac{\Gamma(r+1)\Gamma(s+1)}{\Gamma(r+s+2)}\zeta(r+1)\zeta(s),\quad\ldots\ldots\ldots(13)$$

if $r>0$ and $s>1$. Thus $n^{-r-s-1}\Sigma_{r,s}(n)$ oscillates between limits included in the interval

$$\frac{\Gamma(r+1)\Gamma(s+1)}{\Gamma(r+s+2)}\zeta(r+1),\quad\frac{\Gamma(r+1)\Gamma(s+1)}{\Gamma(r+s+2)}\zeta(r+1)\zeta(s).$$

On the other hand $n^{-r-s-1}\sigma_{r+s+1}(n)$ oscillates between 1 and $\zeta(r+s+1)$, assuming values as near as we please to either of these limits. The formula (8) shews that the actual limits of indetermination of $n^{-r-s-1}\Sigma_{r,s}(n)$ are

$$\frac{\Gamma(r+1)\Gamma(s+1)}{\Gamma(r+s+2)}\frac{\zeta(r+1)\zeta(s+1)}{\zeta(r+s+2)},$$
$$\frac{\Gamma(r+1)\Gamma(s+1)}{\Gamma(r+s+2)}\frac{\zeta(r+1)\zeta(s+1)\zeta(r+s+1)}{\zeta(r+s+2)}.\quad(14)$$

Naturally

$$\zeta(r+1)<\frac{\zeta(r+1)\zeta(s+1)}{\zeta(r+s+2)}<\frac{\zeta(r+1)\zeta(s+1)\zeta(r+s+1)}{\zeta(r+s+2)}<\zeta(r+1)\zeta(s)^*.$$

What is remarkable about the formula (8) is that it shews the asymptotic equality of two functions neither of which itself increases in a regular manner.

4. It is easy to see that, if n is a positive integer, then

$$\cot\tfrac{1}{2}\theta\sin n\theta=1+2\cos\theta+2\cos 2\theta+\ldots+2\cos(n-1)\theta+\cos n\theta.$$

Suppose now that

$$\left(\frac{1}{4}\cot\frac{1}{2}\theta+\frac{x\sin\theta}{1-x}+\frac{x^2\sin 2\theta}{1-x^2}+\frac{x^3\sin 3\theta}{1-x^3}+\ldots\right)^2$$
$$=(\tfrac{1}{4}\cot\tfrac{1}{2}\theta)^2+C_0+C_1\cos\theta+C_2\cos 2\theta+C_3\cos 3\theta+\ldots,$$

where C_n is independent of θ. Then we have

$$C_0=\frac{1}{2}\left(\frac{x}{1-x}+\frac{x^2}{1-x^2}+\frac{x^3}{1-x^3}+\ldots\right)$$
$$+\frac{1}{2}\left\{\left(\frac{x}{1-x}\right)^2+\left(\frac{x^2}{1-x^2}\right)^2+\left(\frac{x^3}{1-x^3}\right)^2+\ldots\right\}$$
$$=\frac{1}{2}\left\{\frac{x}{(1-x)^2}+\frac{x^2}{(1-x^2)^2}+\frac{x^3}{(1-x^3)^2}+\ldots\right\}$$
$$=\frac{1}{2}\left\{\frac{x}{1-x}+\frac{2x^2}{1-x^2}+\frac{3x^3}{1-x^3}+\ldots\right\}.\quad\ldots\ldots\ldots\ldots\ldots\ldots(15)$$

* For example when $r=1$ and $s=9$ this inequality becomes

$$1{\cdot}64493\ldots<1{\cdot}64616\ldots<1{\cdot}64697\ldots<1{\cdot}64823\ldots.$$

Again

$$C_n = \frac{1}{2}\frac{x^n}{1-x^n} + \frac{x^{n+1}}{1-x^{n+1}} + \frac{x^{n+2}}{1-x^{n+2}} + \frac{x^{n+3}}{1-x^{n+3}} + \dots$$

$$+ \frac{x}{1-x}\cdot\frac{x^{n+1}}{1-x^{n+1}} + \frac{x^2}{1-x^2}\cdot\frac{x^{n+2}}{1-x^{n+2}} + \frac{x^3}{1-x^3}\cdot\frac{x^{n+3}}{1-x^{n+3}} + \dots$$

$$-\frac{1}{2}\left\{\frac{x}{1-x}\cdot\frac{x^{n-1}}{1-x^{n-1}} + \frac{x^2}{1-x^2}\cdot\frac{x^{n-2}}{1-x^{n-2}} + \dots + \frac{x^{n-1}}{1-x^{n-1}}\cdot\frac{x}{1-x}\right\}.$$

Hence

$$\frac{C_n}{x^n}(1-x^n) = \frac{1}{2} + \left(\frac{x}{1-x} - \frac{x^{n+1}}{1-x^{n+1}}\right) + \left(\frac{x^2}{1-x^2} - \frac{x^{n+2}}{1-x^{n+2}}\right) + \dots$$

$$-\frac{1}{2}\left\{\left(1 + \frac{x}{1-x} + \frac{x^{n-1}}{1-x^{n-1}}\right) + \left(1 + \frac{x^2}{1-x^2} + \frac{x^{n-2}}{1-x^{n-2}}\right) + \dots + \left(1 + \frac{x^{n-1}}{1-x^{n-1}} + \frac{x}{1-x}\right)\right\}$$

$$= \frac{1}{1-x^n} - \frac{n}{2}.$$

That is to say

$$C_n = \frac{x^n}{(1-x^n)^2} - \frac{nx^n}{2(1-x^n)}. \qquad \dots\dots(16)$$

It follows that

$$\left(\frac{1}{4}\cot\frac{1}{2}\theta + \frac{x\sin\theta}{1-x} + \frac{x^2\sin 2\theta}{1-x^2} + \frac{x^3\sin 3\theta}{1-x^3} + \dots\right)^2$$

$$= \left(\frac{1}{4}\cot\frac{1}{2}\theta\right)^2 + \frac{x\cos\theta}{(1-x)^2} + \frac{x^2\cos 2\theta}{(1-x^2)^2} + \frac{x^3\cos 3\theta}{(1-x^3)^2} + \dots$$

$$+\frac{1}{2}\left\{\frac{x}{1-x}(1-\cos\theta) + \frac{2x^2}{1-x^2}(1-\cos 2\theta) + \frac{3x^3}{1-x^3}(1-\cos 3\theta) + \dots\right\}. \qquad \dots\dots(17)$$

Similarly, using the equation

$$\cot^2\tfrac{1}{2}\theta\,(1-\cos n\theta) = (2n-1) + 4(n-1)\cos\theta + 4(n-2)\cos 2\theta + \dots + 4\cos(n-1)\theta + \cos n\theta,$$

we can shew that

$$\left\{\frac{1}{8}\cot^2\frac{1}{2}\theta + \frac{1}{12} + \frac{x}{1-x}(1-\cos\theta) + \frac{2x^2}{1-x^2}(1-\cos 2\theta) + \frac{3x^3}{1-x^3}(1-\cos 3\theta) + \dots\right\}^2$$

$$= \left(\frac{1}{8}\cot^2\frac{1}{2}\theta + \frac{1}{12}\right)^2 + \frac{1}{12}\left\{\frac{1^3x}{1-x}(5+\cos\theta) + \frac{2^3x^2}{1-x^2}(5+\cos 2\theta) + \frac{3^3x^3}{1-x^3}(5+\cos 3\theta) + \dots\right\}. \quad (18)$$

For example, putting $\theta = \frac{2}{3}\pi$ and $\theta = \frac{1}{2}\pi$ in (17), we obtain

$$\left(\frac{1}{6} + \frac{x}{1-x} - \frac{x^2}{1-x^2} + \frac{x^4}{1-x^4} - \frac{x^5}{1-x^5} + \dots\right)^2$$

$$= \frac{1}{36} + \frac{1}{3}\left(\frac{x}{1-x} + \frac{2x^2}{1-x^2} + \frac{4x^4}{1-x^4} + \frac{5x^5}{1-x^5} + \dots\right), \quad (19)$$

where 1, 2, 4, 5, ... are the natural numbers without the multiples of 3; and

$$\left(\frac{1}{4}+\frac{x}{1-x}-\frac{x^3}{1-x^3}+\frac{x^5}{1-x^5}-\frac{x^7}{1-x^7}+\dots\right)^2$$
$$=\frac{1}{16}+\frac{1}{2}\left(\frac{x}{1-x}+\frac{2x^2}{1-x^2}+\frac{3x^3}{1-x^3}+\frac{5x^5}{1-x^5}+\dots\right), \quad (20)$$

where 1, 2, 3, 5, ... are the natural numbers without the multiples of 4.

5. It follows from (18) that

$$\left(\frac{1}{2\theta^2}+\frac{\theta^2}{2!}S_3-\frac{\theta^4}{4!}S_5+\frac{\theta^6}{6!}S_7-\dots\right)^2$$
$$=\frac{1}{4\theta^4}+\frac{1}{2}S_3-\frac{1}{12}\left(\frac{\theta^2}{2!}S_5-\frac{\theta^4}{4!}S_7+\frac{\theta^6}{6!}S_9-\dots\right), \quad (21)$$

where S_r is the same as in (9). Equating the coefficients of θ^n in both sides in (21), we obtain

$$\frac{(n-2)(n+5)}{12(n+1)(n+2)}S_{n+3}={}^nC_2S_3S_{n-1}+{}^nC_4S_5S_{n-3}+{}^nC_6S_7S_{n-5}+\dots+{}^nC_{n-2}S_{n-1}S_3, \quad \dots\dots(22)$$

where

$${}^nC_r=\frac{n!}{r!\,(n-r)!},$$

if n is an even integer greater than 2.

Let us now suppose that

$$\Phi_{r,s}(x)=\sum_{m=1}^{m=\infty}\sum_{n=1}^{n=\infty} m^r n^s x^{mn}, \quad \dots\dots\dots\dots\dots\dots\dots(23)$$

so that

$$\Phi_{r,s}(x)=\Phi_{s,r}(x),$$

and

$$\left.\begin{aligned}\Phi_{0,s}(x)&=\frac{1^s x}{1-x}+\frac{2^s x^2}{1-x^2}+\frac{3^s x^3}{1-x^3}+\dots=S_s-\tfrac{1}{2}\zeta(-s),\\ \Phi_{1,s}(x)&=\frac{1^s x}{(1-x)^2}+\frac{2^s x^2}{(1-x^2)^2}+\frac{3^s x^3}{(1-x^3)^2}+\dots\end{aligned}\right\} \quad \dots\dots(24)$$

Further let

$$\left.\begin{aligned}P&=-\ 24S_1=1-\ 24\left(\frac{x}{1-x}+\frac{2x^2}{1-x^2}+\frac{3x^3}{1-x^3}+\dots\right)^*,\\ Q&=\ \ 240S_3=1+240\left(\frac{1^3x}{1-x}+\frac{2^3x^2}{1-x^2}+\frac{3^3x^3}{1-x^3}+\dots\right),\\ R&=-504S_5=1-504\left(\frac{1^5x}{1-x}+\frac{2^5x^2}{1-x^2}+\frac{3^5x^3}{1-x^3}+\dots\right)\end{aligned}\right\} \quad \dots(25)$$

* If $x=q^2$, then in the notation of elliptic functions

$$P=\frac{12\eta\omega}{\pi^2}=\left(\frac{2K}{\pi}\right)^2\left(\frac{3E}{K}+k^2-2\right),$$
$$Q=\frac{12g_2\omega^4}{\pi^4}=\left(\frac{2K}{\pi}\right)^4(1-k^2+k^4),$$
$$R=\frac{216g_3\omega^6}{\pi^6}=\left(\frac{2K}{\pi}\right)^6(1+k^2)(1-2k^2)(1-\tfrac{1}{2}k^2).$$

Then putting $n = 4, 6, 8, \ldots$ in (22) we obtain the results contained in the following table.

TABLE I.

1. $1 - 24\Phi_{0,1}(x) = P.$
2. $1 + 240\Phi_{0,3}(x) = Q.$
3. $1 - 504\Phi_{0,5}(x) = R.$
4. $1 + 480\Phi_{0,7}(x) = Q^2.$
5. $1 - 264\Phi_{0,9}(x) = QR.$
6. $691 + 65520\Phi_{0,11}(x) = 441Q^3 + 250R^2.$
7. $1 - 24\Phi_{0,13}(x) = Q^2R.$
8. $3617 + 16320\Phi_{0,15}(x) = 1617Q^4 + 2000QR^2.$
9. $43867 - 28728\Phi_{0,17}(x) = 38367Q^3R + 5500R^3.$
10. $174611 + 13200\Phi_{0,19}(x) = 53361Q^5 + 121250Q^2R^2.$
11. $77683 - 552\Phi_{0,21}(x) = 57183Q^4R + 20500QR^3.$
12. $236364091 + 131040\Phi_{0,23}(x) = 49679091Q^6 + 176400000Q^3R^2 + 10285000R^4.$
13. $657931 - 24\Phi_{0,25}(x) = 392931Q^5R + 265000Q^2R^3.$
14. $3392780147 + 6960\Phi_{0,27}(x) = 489693897Q^7 + 2507636250Q^4R^2 + 395450000QR^4.$
15. $1723168255201 - 171864\Phi_{0,29}(x) = 815806500201Q^6R + 881340705000Q^3R^3 + 26021050000R^5.$
16. $7709321041217 + 32640\Phi_{0,31}(x) = 764412173217Q^8 + 5323905468000Q^5R^2 + 1621003400000Q^2R^4.$

In general
$$\tfrac{1}{2}\zeta(-s) + \Phi_{0,s}(x) = \Sigma K_{m,n} Q^m R^n, \quad \ldots\ldots\ldots\ldots\ldots\ldots(26)$$
where $K_{m,n}$ is a constant and m and n are positive integers (including zero) satisfying the equation
$$4m + 6n = s + 1.$$
This is easily proved by induction, using (22).

6. Again from (17) we have
$$\left(\frac{1}{2\theta} + \frac{\theta}{1!}S_1 - \frac{\theta^3}{3!}S_3 + \frac{\theta^5}{5!}S_5 - \ldots\right)^2$$
$$= \frac{1}{4\theta^2} + S_1 - \frac{\theta^2}{2!}\Phi_{1,2}(x) + \frac{\theta^4}{4!}\Phi_{1,4}(x) - \frac{\theta^6}{6!}\Phi_{1,6}(x) + \ldots$$
$$+ \frac{1}{2}\left(\frac{\theta^2}{2!}S_3 - \frac{\theta^4}{4!}S_5 + \frac{\theta^6}{6!}S_7 - \ldots\right). \quad (27)$$

Equating the coefficients of θ^n in both sides in (27) we obtain
$$\frac{n+3}{2(n+1)}S_{n+1} - \Phi_{1,n}(x) = {}^nC_1S_1S_{n-1} + {}^nC_3S_3S_{n-3} + {}^nC_5S_5S_{n-5} + \ldots + {}^nC_{n-1}S_{n-1}S_1, \quad \ldots\ldots(28)$$
if n is a positive even integer. From this we deduce the results contained in Table II.

TABLE II.

1. $288\Phi_{1,2}(x) = Q - P^2.$
2. $720\Phi_{1,4}(x) = PQ - R.$
3. $1008\Phi_{1,6}(x) = Q^2 - PR.$
4. $720\Phi_{1,8}(x) = Q(PQ - R).$
5. $1584\Phi_{1,10}(x) = 3Q^3 + 2R^2 - 5PQR.$
6. $65520\Phi_{1,12}(x) = P(441Q^3 + 250R^2) - 691Q^2R.$
7. $144\Phi_{1,14}(x) = Q(3Q^3 + 4R^2 - 7PQR).$

In general $$\Phi_{1,s}(x) = \Sigma K_{l,m,n} P^l Q^m R^n, \quad \text{.....................(29)}$$
where $l \leqslant 2$ and $2l + 4m + 6n = s + 2$. This is easily proved by induction, using (28).

7. We have

$$\left.\begin{aligned} x\frac{dP}{dx} &= -\ \ 24\Phi_{1,2}(x) = \frac{P^2 - Q}{12}, \\ x\frac{dQ}{dx} &= \ \ 240\Phi_{1,4}(x) = \frac{PQ - R}{3}, \\ x\frac{dR}{dx} &= -\ 504\Phi_{1,6}(x) = \frac{PR - Q^2}{2} \end{aligned}\right\} \quad \text{..................(30)}$$

Suppose now that $r < s$ and that $r + s$ is even. Then

$$\Phi_{r,s}(x) = \left(x\frac{d}{dx}\right)^r \Phi_{0,s-r}(x), \quad \text{.....................(31)}$$

and $\Phi_{0,s-r}(x)$ is a polynomial in Q and R. Also

$$x\frac{dP}{dx}, \quad x\frac{dQ}{dx}, \quad x\frac{dR}{dx}$$

are polynomials in P, Q and R. Hence $\Phi_{r,s}(x)$ is a polynomial in P, Q and R. Thus we deduce the results contained in Table III.

TABLE III.

1. $1728\Phi_{2,3}(x) = 3PQ - 2R - P^3.$
2. $1728\Phi_{2,5}(x) = P^2Q - 2PR + Q^2.$
3. $1728\Phi_{2,7}(x) = 2PQ^2 - P^2R - QR.$
4. $8640\Phi_{2,9}(x) = 9P^2Q^2 - 18PQR + 5Q^3 + 4R^2.$
5. $1728\Phi_{2,11}(x) = 6PQ^3 - 5P^2QR + 4PR^2 - 5Q^2R.$
6. $6912\Phi_{3,4}(x) = 6P^2Q - 8PR + 3Q^2 - P^4.$
7. $3456\Phi_{3,6}(x) = P^3Q - 3P^2R + 3PQ^2 - QR.$
8. $5184\Phi_{3,8}(x) = 6P^2Q^2 - 2P^3R - 6PQR + Q^3 + R^2.$
9. $20736\Phi_{4,5}(x) = 15PQ^2 - 20P^2R + 10P^3Q - 4QR - P^5.$
10. $41472\Phi_{4,7}(x) = 7(P^4Q - 4P^3R + 6P^2Q^2 - 4PQR) + 3Q^3 + 4R^2.$

In general $$\Phi_{r,s}(x) = \Sigma K_{l,m,n} P^l Q^m R^n, \quad \text{.....................(32)}$$
where $l-1$ does not exceed the smaller of r and s and
$$2l + 4m + 6n = r + s + 1.$$

The results contained in these three tables are of course really results in the theory of elliptic functions. For example Q and R are substantially the invariants g_2 and g_3, and the formulæ of Table I are equivalent to the formulæ which express the coefficients in the series
$$\wp(u) = \frac{1}{u^2} + \frac{g_2 u^2}{20} + \frac{g_3 u^4}{28} + \frac{g_2{}^2 u^6}{1200} + \frac{3 g_2 g_3 u^8}{6160} + \dots$$
in terms of g_2 and g_3. The elementary proof of these formulæ given in the preceding sections seems to be of some interest in itself.

8. In what follows we shall require to know the form of $\Phi_{1,s}(x)$ more precisely than is shewn by the formula (29).

We have $$\tfrac{1}{2}\zeta(-s) + \Phi_{0,s}(x) = \Sigma K_{m,n} Q^m R^n, \quad \text{.................(33)}$$
where s is an odd integer greater than 1 and $4m + 6n = s + 1$. Also
$$x\frac{d}{dx}(Q^m R^n) = \left(\frac{m}{3} + \frac{n}{2}\right) P Q^m R^n - \left(\frac{m}{3} Q^{m-1} R^{n+1} + \frac{n}{3} Q^{m+2} R^{n-1}\right). \quad (34)$$
Differentiating (33) and using (34) we obtain
$$\Phi_{1,s+1}(x) = \tfrac{1}{12}(s+1) P \{\tfrac{1}{2}\zeta(-s) + \Phi_{0,s}(x)\} + \Sigma K_{m,n} Q^m R^n, \quad \text{...(35)}$$
where s is an odd integer greater than 1 and $4m + 6n = s + 3$. But when $s = 1$ we have
$$\Phi_{1,2}(x) = \frac{Q - P^2}{288}. \quad \text{..........................(36)}$$

9. Suppose now that
$$F_{r,s}(x) = \{\tfrac{1}{2}\zeta(-r) + \Phi_{0,r}(x)\}\{\tfrac{1}{2}\zeta(-s) + \Phi_{0,s}(x)\}$$
$$- \frac{\zeta(1-r) + \zeta(1-s)}{r+s} \Phi_{1,r+s}(x) - \frac{\Gamma(r+1)\Gamma(s+1)}{\Gamma(r+s+2)} \frac{\zeta(r+1)\zeta(s+1)}{\zeta(r+s+2)}$$
$$\times \{\tfrac{1}{2}\zeta(-r-s-1) + \Phi_{0,r+s+1}(x)\}. \quad (37)$$

Then it follows from (33), (35) and (36) that, if r and s are positive odd integers,
$$F_{r,s}(x) = \Sigma K_{m,n} Q^m R^n, \quad \text{........................(38)}$$
where $$4m + 6n = r + s + 2.$$
But it is easy to see, from the functional equation satisfied by $\zeta(s)$, viz.
$$(2\pi)^{-s}\Gamma(s)\zeta(s)\cos\tfrac{1}{2}\pi s = \tfrac{1}{2}\zeta(1-s), \quad \text{..............(39)}$$
that $$F_{r,s}(0) = 0. \quad \text{..............................(40)}$$

Hence $Q^3 - R^2$ is a factor of the right-hand side in (38), that is to say
$$F_{r,s}(x) = (Q^3 - R^2)\Sigma K_{m,n} Q^m R^n, \quad \text{.................(41)}$$
where $$4m + 6n = r + s - 10.$$

10. It is easy to deduce from (30) that

$$x\frac{d}{dx}\log(Q^3-R^2)=P. \qquad \text{(42)}$$

But is is obvious that

$$P=x\frac{d}{dx}\log[x\{(1-x)(1-x^2)(1-x^3)\ldots\}^{24}]; \qquad \text{(43)}$$

and the coefficient of x in Q^3-R^2 is 1728. Hence

$$Q^3-R^2=1728x\{(1-x)(1-x^2)(1-x^3)\ldots\}^{24}. \qquad \text{(44)}$$

But it is known that

$$\{(1-x)(1-x^2)(1-x^3)(1-x^4)\ldots\}^3=1-3x+5x^3-7x^6+9x^{10}-\ldots. \qquad \text{(45)}$$

Hence $$Q^3-R^2=1728x(1-3x+5x^3-7x^6+\ldots)^8. \qquad \text{(46)}$$

The coefficient of $x^{\nu-1}$ in $1-3x+5x^3-\ldots$ is numerically less than $\sqrt{(8\nu)}$, and the coefficient of x^ν in Q^3-R^2 is therefore numerically less than that of x^ν in

$$1728x\{\sqrt{(8\nu)}(1+x+x^3+x^6+\ldots)\}^8.$$

But $$x(1+x+x^3+x^6+\ldots)^8=\frac{1^3x}{1-x^2}+\frac{2^3x^2}{1-x^4}+\frac{3^3x^3}{1-x^6}+\ldots, \qquad \text{(47)}$$

and the coefficient of x^ν in the right-hand side is positive and less than

$$\nu^3\left(\frac{1}{1^3}+\frac{1}{3^3}+\frac{1}{5^3}+\ldots\right).$$

Hence the coefficient of x^ν in Q^3-R^2 is of the form

$$\nu^4 O(\nu^3)=O(\nu^7).$$

That is to say $$Q^3-R^2=\Sigma\, O(\nu^7)\,x^\nu. \qquad \text{(48)}$$

Differentiating (48) and using (42) we obtain

$$P(Q^3-R^2)=\Sigma\, O(\nu^8)\,x^\nu. \qquad \text{(49)}$$

Differentiating this again with respect to x we have

$$A(P^2-Q)(Q^3-R^2)+BQ(Q^3-R^2)=\Sigma\, O(\nu^9)\,x^\nu,$$

where A and B are constants. But

$$P^2-Q=-288\,\Phi_{1,2}(x)=-288\left\{\frac{1^2x}{(1-x)^2}+\frac{2^2x^2}{(1-x^2)^2}+\ldots\right\},$$

and the coefficient of x^ν in the right-hand side is a constant multiple of $\nu\sigma_1(\nu)$. Hence

$$(P^2-Q)(Q^3-R^2)=\Sigma\, O\nu\sigma_1(\nu)\,x^\nu\,\Sigma\, O(\nu^7)\,x^\nu$$
$$=\Sigma\, O(\nu^8)\{\sigma_1(1)+\sigma_1(2)+\ldots+\sigma_1(\nu)\}\,x^\nu=\Sigma\, O(\nu^{10})\,x^\nu,$$

and so $$Q(Q^3-R^2)=\Sigma\, O(\nu^{10})\,x^\nu. \qquad \text{(50)}$$

Differentiating this again with respect to x and using arguments similar to those used above, we deduce

$$R(Q^3-R^2)=\Sigma\, O(\nu^{12})\,x^\nu. \qquad \text{(51)}$$

Suppose now that m and n are any two positive integers including zero, and that $m+n$ is not zero. Then

$$\begin{aligned} Q^m R^n (Q^3 - R^2) &= Q (Q^3 - R^2) Q^{m-1} R^n \\ &= \Sigma O(\nu^{10}) x^\nu \{\Sigma O(\nu^3) x^\nu\}^{m-1} \{\Sigma O(\nu^5) x^\nu\}^n \\ &= \Sigma O(\nu^{10}) x^\nu \Sigma O(\nu^{4m-5}) x^\nu \Sigma O(\nu^{6n-1}) x^\nu \\ &= \Sigma O(\nu^{4m+6n+6}) x^\nu, \end{aligned}$$

if m is not zero. Similarly we can shew that

$$\begin{aligned} Q^m R^n (Q^3 - R^2) &= R (Q^3 - R^2) Q^m R^{n-1} \\ &= \Sigma O(\nu^{4m+6n+6}) x^\nu, \end{aligned}$$

if n is not zero. Therefore in any case

$$(Q^3 - R^2) Q^m R^n = \Sigma O(\nu^{4m+6n+6}) x^\nu. \qquad \text{.......(52)}$$

11. Now let r and s be any two positive odd integers including zero. Then, when $r+s$ is equal to 2, 4, 6, 8 or 12, there are no values of m and n satisfying the relation

$$4m + 6n = r + s - 10$$

in (41); consequently in these cases

$$F_{r,s}(x) = 0. \qquad \text{.......(53)}$$

When $r+s=10$, m and n must both be zero, and this result does not apply; but it follows from (41) and (48) that

$$F_{r,s}(x) = \Sigma O(\nu^7) x^\nu. \qquad \text{.......(54)}$$

And when $r+s \geqslant 14$ it follows from (52) that

$$F_{r,s}(x) = \Sigma O(\nu^{r+s-4}) x^\nu. \qquad \text{.......(55)}$$

Equating the coefficients of x^ν in both sides in (53), (54) and (55) we obtain

$$\begin{aligned} \Sigma_{r,s}(n) = {} & \frac{\Gamma(r+1)\Gamma(s+1)}{\Gamma(r+s+2)} \frac{\zeta(r+1)\zeta(s+1)}{\zeta(r+s+2)} \sigma_{r+s+1}(n) \\ & + \frac{\zeta(1-r)+\zeta(1-s)}{r+s} n\sigma_{r+s-1}(n) + E_{r,s}(n), \quad (56) \end{aligned}$$

where

$$\begin{aligned} E_{r,s}(n) &= 0, \quad r+s = 2,\ 4,\ 6,\ 8,\ 12; \\ E_{r,s}(n) &= O(n^7), \quad r+s = 10; \\ E_{r,s}(n) &= O(n^{r+s-4}), \quad r+s \geqslant 14. \end{aligned}$$

Since $\sigma_{r+s+1}(n)$ is of order n^{r+s+1}, it follows that in all cases

$$\Sigma_{r,s}(n) \sim \frac{\Gamma(r+1)\Gamma(s+1)}{\Gamma(r+s+2)} \frac{\zeta(r+1)\zeta(s+1)}{\zeta(r+s+2)} \sigma_{r+s+1}(n). \quad \text{...(57)}$$

The following table gives the values of $\Sigma_{r,s}(n)$ when $r+s=2, 4, 6, 8, 12$.

TABLE IV.

1. $\Sigma_{1,1}(n)=\dfrac{5\sigma_3(n)-6n\sigma_1(n)}{12}$.

2. $\Sigma_{1,3}(n)=\dfrac{7\sigma_5(n)-10n\sigma_3(n)}{80}$.

3. $\Sigma_{3,3}(n)=\dfrac{\sigma_7(n)}{120}$.

4. $\Sigma_{1,5}(n)=\dfrac{10\sigma_7(n)-21n\sigma_5(n)}{252}$.

5. $\Sigma_{3,5}(n)=\dfrac{11\sigma_9(n)}{5040}$.

6. $\Sigma_{1,7}(n)=\dfrac{11\sigma_9(n)-30n\sigma_7(n)}{480}$.

7. $\Sigma_{5,7}(n)=\dfrac{\sigma_{13}(n)}{10080}$.

8. $\Sigma_{3,9}(n)=\dfrac{\sigma_{13}(n)}{2640}$.

9. $\Sigma_{1,11}(n)=\dfrac{691\sigma_{13}(n)-2730n\sigma_{11}(n)}{65520}$.

12. In this connection it may be interesting to note that

$$\sigma_1(1)\sigma_3(n)+\sigma_1(3)\sigma_3(n-1)+\sigma_1(5)\sigma_3(n-2)+\ldots$$
$$+\sigma_1(2n+1)\sigma_3(0)=\tfrac{1}{240}\sigma_5(2n+1). \quad \ldots(58)$$

This formula may be deduced from the identity

$$\frac{1^5x}{1-x}+\frac{3^5x^2}{1-x^3}+\frac{5^5x^3}{1-x^5}+\ldots=Q\left(\frac{x}{1-x}+\frac{3x^2}{1-x^3}+\frac{5x^3}{1-x^5}+\ldots\right), \quad (59)$$

which can be proved by means of the theory of elliptic functions or by elementary methods.

13. More precise results concerning the order of $E_{r,s}(n)$ can be deduced from the theory of elliptic functions. Let

$$x=q^2.$$

Then we have

$$\left.\begin{aligned} Q&=\phi^8(q)\{1-(kk')^2\} \\ R&=\phi^{12}(q)(k'^2-k^2)\{1+\tfrac{1}{2}(kk')^2\} \\ &=\phi^{12}(q)\{1+\tfrac{1}{2}(kk')^2\}\sqrt{\{1-(2kk')^2\}} \end{aligned}\right\}, \quad \ldots\ldots\ldots\ldots(60)$$

where $\qquad \phi(q)=1+2q+2q^4+2q^9+\ldots.$

But, if $\qquad f(q)=q^{\frac{1}{24}}(1-q)(1-q^2)(1-q^3)\ldots,$

then we know that

$$\left.\begin{aligned} 2^{\frac{1}{6}}f(q) &= k^{\frac{1}{12}}k'^{-\frac{1}{3}}\phi(q) \\ 2^{\frac{1}{6}}f(-q) &= (kk')^{\frac{1}{12}}\phi(q) \\ 2^{\frac{1}{3}}f(q^2) &= (kk')^{\frac{1}{6}}\phi(q) \\ 2^{\frac{2}{3}}f(q^4) &= k^{\frac{1}{3}}k'^{\frac{1}{12}}\phi(q) \end{aligned}\right\}. \qquad (61)$$

It follows from (41), (60) and (61) that, if $r+s$ is of the form $4m+2$, but not equal to 2 or to 6, then

$$F_{r,s}(q^2) = \frac{f^{4(r+s-4)}(-q)}{f^{2(r+s-10)}(q^2)} \sum_{1}^{\frac{1}{4}(r+s-6)} K_n \frac{f^{24n}(q^2)}{f^{24n}(-q)}, \qquad (62)$$

and if $r+s$ is of the form $4m$, but not equal to 4, 8 or 12, then

$$F_{r,s}(q^2) = \frac{f^{4(r+s-6)}(-q)}{f^{2(r+s-10)}(q^2)} \{f^8(q) - 16f^8(q^4)\} \sum_{1}^{\frac{1}{4}(r+s-8)} K_n \frac{f^{24n}(q^2)}{f^{24n}(-q)}, \quad (63)$$

where K_n depends on r and s only. Hence it is easy to see that in all cases $F_{r,s}(q^2)$ can be expressed as

$$\Sigma K_{a,b,c,d,e,h,k} \{f^3(-q)\}^a \left\{\frac{f^5(-q)}{f^2(q^2)}\right\}^b \left\{\frac{f^5(q^2)}{f^2(-q)}\right\}^c \left\{\frac{f^5(q)}{f^2(q^2)}f^3(q)\right\}^d$$
$$\times \left\{\frac{f^5(q^4)}{f^2(q^2)}f^3(q^4)\right\}^e f^h(-q) f^k(q^2), \quad (64)$$

where a, b, c, d, e, h, k are zero or positive integers such that

$$a+b+c+2(d+e) = [\tfrac{2}{3}(r+s+2)],$$
$$h+k = 2(r+s+2) - 3[\tfrac{2}{3}(r+s+2)],$$

and $[x]$ denotes as usual the greatest integer in x. But

$$\left.\begin{aligned} f(q) &= q^{\frac{1^2}{24}} - q^{\frac{5^2}{24}} - q^{\frac{7^2}{24}} + q^{\frac{11^2}{24}} + \dots \\ f^3(q) &= q^{\frac{1^2}{8}} - 3q^{\frac{3^2}{8}} + 5q^{\frac{5^2}{8}} - 7q^{\frac{7^2}{8}} + \dots \\ \frac{f^5(q)}{f^2(q^2)} &= q^{\frac{1^2}{24}} - 5q^{\frac{5^2}{24}} + 7q^{\frac{7^2}{24}} - 11q^{\frac{11^2}{24}} + \dots \\ \frac{f^5(q^2)}{f^2(-q)} &= q^{\frac{1^2}{3}} - 2q^{\frac{2^2}{3}} + 4q^{\frac{4^2}{3}} - 5q^{\frac{5^2}{3}} + \dots \end{aligned}\right\}, \qquad (65)$$

where 1, 2, 4, 5, ... are the natural numbers without the multiples of 3, and 1, 5, 7, 11, ... are the natural odd numbers without the multiples of 3.

Hence it is easy to see that

$$n^{-\frac{1}{2}(a+b+c)-d-e} E_{r,s}(n)$$

is not of higher order than the coefficient of q^{2n} in

$$\phi^a(q^{\frac{1}{8}})\,\phi^b(q^{\frac{1}{24}})\,\phi^c(q^{\frac{1}{3}})\{\phi(q^{\frac{1}{24}})\,\phi(q^{\frac{1}{8}})\}^d\{\phi(q^{\frac{2}{3}})\,\phi(q^{\frac{1}{2}})\}^e\,\phi^h(q^{\frac{1}{24}})\,\phi^k(q^{\frac{1}{12}}),$$

or the coefficient of q^{48n} in

$$\phi^{a+d}(q^3)\,\phi^{b+d+h}(q)\,\phi^c(q^8)\,\phi^e(q^{16})\,\phi^e(q^{12})\,\phi^k(q^2).$$

But the coefficient of q^ν in $\phi^2(q^2)$ cannot exceed that of q^ν in $\phi^2(q)$, since

$$\phi^2(q)+\phi^2(-q)=2\phi^2(q^4); \qquad \text{.....................(66)}$$

and it is evident that the coefficient of q^ν in $\phi(q^{4\lambda})$ cannot exceed that of q^ν in $\phi(q^\lambda)$. Hence it follows that

$$n^{-\frac{1}{2}[\frac{2}{3}(r+s+2)]}\,E_{r,s}(n)$$

is not of higher order than the coefficient of q^{48n} in

$$\phi^A(q)\,\phi^B(q^3)\,\phi^C(q^2),$$

where A, B, C are zero or positive integers such that

$$A+B+C=2(r+s+2)-2[\tfrac{2}{3}(r+s+2)],$$

and C is 0 or 1.

Now, if $r+s \geqslant 14$, we have

$$A+B+C \geqslant 12,$$

and so

$$A+B \geqslant 11.$$

Therefore one at least of A and B is greater than 5. But

$$\phi^6(q)=\sum_0^\infty O(\nu^2)\,q^\nu\,*. \qquad \text{.........................(67)}$$

Hence it is easily deduced that

$$\phi^A(q)\,\phi^B(q^3)\,\phi^C(q^2)=\Sigma\, O\{\nu^{\frac{1}{2}(A+B+C)-1}\}\,q^\nu. \qquad \text{...........(68)}$$

It follows that

$$E_{r,s}(n)=O\{n^{r+s-\frac{1}{2}[\frac{2}{3}(r+s-1)]}\}, \qquad \text{.....................(69)}$$

if $r+s \geqslant 14$. We have already shewn in § 11 that, if $r+s=10$, then

$$E_{r,s}(n)=O(n^7). \qquad \text{.............................(70)}$$

This agrees with (69). Thus we see that in all cases

$$E_{r,s}(n)=O\{n^{\frac{2}{3}(r+s+1)}\}; \qquad \text{.......................(71)}$$

and that, if $r+s$ is of the form $6m$, then

$$E_{r,s}(n)=O\{n^{\frac{2}{3}(r+s+\frac{3}{4})}\}, \qquad \text{.......................(72)}$$

and if of the form $6m+4$, then

$$E_{r,s}(n)=O\{n^{\frac{2}{3}(r+s+\frac{1}{2})}\}. \qquad \text{.......................(73)}$$

14. I shall now prove that the order of $E_{r,s}(n)$ is not less than that of $n^{\frac{1}{2}(r+s)}$. In order to prove this result I shall follow the method used by Messrs Hardy and Littlewood in their paper "Some problems of Diophantine approximation" (II)†.

* See §§ 24—25.

† *Acta Mathematica*, Vol. XXXVII, pp. 193—238.

Let $$q = e^{\pi i\tau}, \quad q' = e^{\pi i T},$$
where $$T = \frac{c + d\tau}{a + b\tau},$$
and $$ad - bc = 1.$$

Also let $$V = \frac{v}{a + b\tau}.$$

Then we have $$\omega \sqrt{v e^{\pi i b v V}}\, \vartheta_1(v, \tau) = \sqrt{V}\, \vartheta_1(V, T), \qquad (74)$$
where ω is an eighth root of unity and
$$\vartheta_1(v, \tau) = 2 \sin \pi v \,.\, q^{\frac{1}{4}} \prod_1^\infty (1 - q^{2n})(1 - 2q^{2n} \cos 2\pi v + q^{4n}). \qquad (75)$$

From (75) we have
$$\log \vartheta_1(v, \tau) = \log(2 \sin \pi v) + \tfrac{1}{4} \log q - \sum_1^\infty \frac{q^{2n}(1 + 2\cos 2n\pi v)}{n(1 - q^{2n})}. \qquad (76)$$
It follows from (74) and (76) that
$$\log \sin \pi v + \tfrac{1}{2} \log v + \tfrac{1}{4} \log q + \log \omega - \sum_1^\infty \frac{q^{2n}(1 + 2\cos 2n\pi v)}{n(1 - q^{2n})}$$
$$= \log \sin \pi V + \tfrac{1}{2} \log V + \tfrac{1}{4} \log q' - \pi i b v V - \sum_1^\infty \frac{q'^{2n}(1 + 2\cos 2n\pi V)}{n(1 - q'^{2n})}. \qquad (77)$$

Equating the coefficients of v^{s+1} on the two sides of (77), we obtain
$$(a + b\tau)^{s+1} \left\{ \tfrac{1}{2}\zeta(-s) + \frac{1^s q^2}{1 - q^2} + \frac{2^s q^4}{1 - q^4} + \frac{3^s q^6}{1 - q^6} + \dots \right\}$$
$$= \tfrac{1}{2}\zeta(-s) + \frac{1^s q'^2}{1 - q'^2} + \frac{2^s q'^4}{1 - q'^4} + \frac{3^s q'^6}{1 - q'^6} + \dots, \qquad (78)$$

provided that s is an odd integer greater than 1. If, in particular, we put $s = 3$ and $s = 5$ in (78) we obtain
$$(a + b\tau)^4 \left\{ 1 + 240 \left(\frac{1^3 q^2}{1 - q^2} + \frac{2^3 q^4}{1 - q^4} + \frac{3^3 q^6}{1 - q^6} + \dots \right) \right\}$$
$$= \left\{ 1 + 240 \left(\frac{1^3 q'^2}{1 - q'^2} + \frac{2^3 q'^4}{1 - q'^4} + \frac{3^3 q'^6}{1 - q'^6} + \dots \right) \right\}, \qquad (79)$$
and
$$(a + b\tau)^6 \left\{ 1 - 504 \left(\frac{1^5 q^2}{1 - q^2} + \frac{2^5 q^4}{1 - q^4} + \frac{3^5 q^6}{1 - q^6} + \dots \right) \right\}$$
$$= \left\{ 1 - 504 \left(\frac{1^5 q'^2}{1 - q'^2} + \frac{2^5 q'^4}{1 - q'^4} + \frac{3^5 q'^6}{1 - q'^6} + \dots \right) \right\}. \qquad (80)$$

It follows from (38), (79) and (80) that
$$(a + b\tau)^{r+s+2} F_{r,s}(q^2) = F_{r,s}(q'^2). \qquad (81)$$
It can easily be seen from (56) and (37) that
$$F_{r,s}(x) = \sum_1^\infty E_{r,s}(n)\, x^n. \qquad (82)$$

Hence $$(a+b\tau)^{r+s+2}\sum_1^\infty E_{r,s}(n)\,q^{2n}=\sum_1^\infty E_{r,s}(n)\,q'^{2n}. \qquad (83)$$

It is important to observe that

$$E_{r,s}(1)=\frac{\zeta(-r)+\zeta(-s)}{2}-\frac{\zeta(1-r)+\zeta(1-s)}{r+s}$$
$$-\frac{\Gamma(r+1)\,\Gamma(s+1)}{\Gamma(r+s+2)}\frac{\zeta(r+1)\,\zeta(s+1)}{\zeta(r+s+2)}\neq 0, \qquad (84)$$

if $r+s$ is not equal to 2, 4, 6, 8 or 12. This is easily proved by the help of the equation (39).

15. Now let

$$\tau=u+iy,\quad t=e^{-\pi y}\quad (u>0,\ y>0,\ 0<t<1),$$

so that $$q=e^{\pi iu-\pi y}=te^{\pi iu};$$

and let us suppose that p_n/q_n is a convergent to

$$u=\frac{1}{a_1+}\,\frac{1}{a_2+}\,\frac{1}{a_3}+\ldots,$$

so that $$\eta_n=p_{n-1}q_n-p_nq_{n-1}=\pm 1.$$

Further, let us suppose that

$$a=p_n,\qquad b=-q_n,$$
$$c=\eta_n p_{n-1},\quad d=-\eta_n q_{n-1},$$

so that $$ad-bc=\eta_n^2=1.$$

Furthermore, let $$y=1/(q_n q'_{n+1}),$$

where $$q'_{n+1}=a'_{n+1}q_n+q_{n-1},$$

and a'_{n+1} is the complete quotient corresponding to a_{n+1}.

Then we have

$$|a+b\tau|=|p_n-q_n u-iq_n y|=\frac{|\pm 1-i|}{q'_{n+1}}=\frac{\sqrt{2}}{q'_{n+1}}, \qquad (85)$$

and $$|q'|=e^{-\pi\lambda},$$

where $$\lambda=\mathbf{I}(T)=\mathbf{I}\left(\frac{c+d\tau}{a+b\tau}\right)=\mathbf{I}\left\{\frac{d}{b}-\frac{1}{b(a+b\tau)}\right\}$$
$$=\frac{y}{(1/q'_{n+1})^2+q_n^2y^2}=\frac{q'_{n+1}}{2q_n}, \qquad (86)$$

and $\mathbf{I}(T)$ is the imaginary part of T. It follows from (83), (85) and (86) that

$$\left|\sum_1^\infty E_{r,s}(n)\,q^{2n}\right|=\left(\frac{q'_{n+1}}{\sqrt{2}}\right)^{r+s+2}\left|\sum_1^\infty E_{r,s}(n)\,q'^{2n}\right|$$
$$\geqslant\left(\frac{q'_{n+1}}{\sqrt{2}}\right)^{r+s+2}\{|E_{r,s}(1)|\,e^{-2\pi\lambda}-|E_{r,s}(2)|\,e^{-4\pi\lambda}-|E_{r,s}(3)|\,e^{-6\pi\lambda}-\ldots\}. \qquad (87)$$

We can choose a number λ_0, depending only on r and s, such that

$$|E_{r,s}(1)|\, e^{-2\pi\lambda} > 2\{|E_{r,s}(2)|\, e^{-4\pi\lambda} + |E_{r,s}(3)|\, e^{-6\pi\lambda} + \ldots\}$$

for $\lambda \geqslant \lambda_0$. Let us suppose $\lambda_0 > 10$. Let us also suppose that the continued fraction for u satisfies the condition

$$4\lambda_0 q_n > q'_{n+1} > 2\lambda_0 q_n \quad \text{.......................(88)}$$

for an infinity of values of n. Then

$$\left|\sum_1^\infty E_{r,s}(n)\, q^{2n}\right| \geqslant \tfrac{1}{2}\,|E_{r,s}(1)|\left(\frac{q'_{n+1}}{\sqrt{2}}\right)^{r+s+2} e^{-4\pi\lambda_0} > K(q'_{n+1})^{r+s+2}, \quad (89)$$

where K depends on r and s only. Also

$$q_n q'_{n+1} = 1/y,$$

$$q'_{n+1} > \frac{1}{\sqrt{y}} = \sqrt{\left\{\frac{\pi}{\log(1/t)}\right\}} > \frac{K}{\sqrt{(1-t)}}.$$

It follows that, if u is an irrational number such that the condition (88) is satisfied for an infinity of values of n, then

$$\left|\sum_1^\infty E_{r,s}(n)\, q^{2n}\right| > K(1-t)^{-\frac{1}{2}(r+s+2)} \quad \text{.................(90)}$$

for an infinity of values of t tending to unity.

But if we had $\qquad E_{r,s}(n) = o\{n^{\frac{1}{2}(r+s)}\}$,

then we should have

$$\left|\sum_1^\infty E_{r,s}(n)\, q^{2n}\right| = o\{(1-t)^{-\frac{1}{2}(r+s+2)}\},$$

which contradicts (90). It follows that the error term in $\Sigma_{r,s}(n)$ is not of the form

$$o\{n^{\frac{1}{2}(r+s)}\}. \quad \text{.............................(91)}$$

The arithmetical function $\tau(n)$.

16. We have seen that $\qquad E_{r,s}(n) = 0$,

if $r+s$ is equal to 2, 4, 6, 8 or 12. In these cases $\Sigma_{r,s}(n)$ has a finite expression in terms of $\sigma_{r+s+1}(n)$ and $\sigma_{r+s-1}(n)$. In other cases $\Sigma_{r,s}(n)$ involves other arithmetical functions as well. The simplest of these is the function $\tau(n)$ defined by

$$\sum_1^\infty \tau(n)\, x^n = x\{(1-x)(1-x^2)(1-x^3)\ldots\}^{24}. \quad \text{...........(92)}$$

These cases arise when $r+s$ has one of the values 10, 14, 16, 18, 20 or 24.

Suppose that $r+s$ has one of these values. Then

$$\frac{1728 \sum_1^\infty E_{r,s}(n)\, x^n}{(Q^3 - R^2)\, E_{r,s}(1)}$$

is, by (41) and (82), equal to the corresponding one of the functions

$$1,\ Q,\ R,\ Q^2,\ QR,\ Q^2R.$$

In other words

$$\sum_1^\infty E_{r,s}(n)\, x^n = E_{r,s}(1) \sum_1^\infty \tau(n)\, x^n \left\{1 + \frac{2}{\zeta(11-r-s)} \sum_1^\infty n^{r+s-11} \frac{x^n}{1-x^n}\right\}. \qquad \text{......(93)}$$

We thus deduce the formulæ

$$E_{r,s}(n) = E_{r,s}(1)\, \tau(n), \qquad \text{........................(94)}$$

if $r+s=10$; and

$$\sigma_{r+s-11}(0)\, E_{r,s}(n) = E_{r,s}(1) \{\sigma_{r+s-11}(0)\, \tau(n) + \sigma_{r+s-11}(1)\, \tau(n-1) + \dots + \sigma_{r+s-11}(n-1)\, \tau(1)\}, \qquad \text{...(95)}$$

if $r+s$ is equal to 14, 16, 18, 20 or 24. It follows from (94) and (95) that, if $r+s = r'+s'$, then

$$E_{r,s}(n)\, E_{r',s'}(1) = E_{r,s}(1)\, E_{r',s'}(n), \qquad \text{..................(96)}$$

and in general
$$E_{r,s}(m)\, E_{r',s'}(n) = E_{r,s}(n)\, E_{r',s'}(m), \qquad \text{..............(97)}$$

when $r+s$ has one of the values in question. The different cases in which $r+s$ has the same value are therefore not fundamentally distinct.

17. The values of $\tau(n)$ may be calculated as follows: differentiating (92) logarithmically with respect to x, we obtain

$$\sum_1^\infty n\, \tau(n)\, x^n = P \sum_1^\infty \tau(n)\, x^n. \qquad \text{.....................(98)}$$

Equating the coefficients of x^n in both sides in (98), we have

$$\tau(n) = \frac{24}{1-n} \{\sigma_1(1)\, \tau(n-1) + \sigma_1(2)\, \tau(n-2) + \dots + \sigma_1(n-1)\, \tau(1)\}. \qquad \text{......(99)}$$

If, instead of starting with (92), we start with

$$\sum_1^\infty \tau(n)\, x^n = x(1 - 3x + 5x^3 - 7x^6 + \dots)^8,$$

we can shew that

$$(n-1)\, \tau(n) - 3(n-10)\, \tau(n-1) + 5(n-28)\, \tau(n-3) - 7(n-55)\, \tau(n-6) + \dots \text{ to } [\tfrac{1}{2}\{1 + \sqrt{(8n-7)}\}] \text{ terms} = 0, \qquad \text{...........(100)}$$

where the rth term of the sequence 0, 1, 3, 6, ... is $\frac{1}{2}r(r-1)$, and the rth term of the sequence 1, 10, 28, 55, ... is $1 + \frac{9}{2}r(r-1)$. We thus obtain the values of $\tau(n)$ in the following table.

TABLE V.

n	$\tau(n)$	n	$\tau(n)$
1	+1	16	+987136
2	−24	17	−6905934
3	+252	18	+2727432
4	−1472	19	+10661420
5	+4830	20	−7109760
6	−6048	21	−4219488
7	−16744	22	−12830688
8	+84480	23	+18643272
9	−113643	24	+21288960
10	−115920	25	−25499225
11	+534612	26	+13865712
12	−370944	27	−73279080
13	−577738	28	+24647168
14	+401856	29	+128406630
15	+1217160	30	−29211840

18. Let us consider more particularly the case in which $r+s=10$. The order of $E_{r,s}(n)$ is then the same as that of $\tau(n)$. The determination of this order is a problem interesting in itself. We have proved that $E_{r,s}(n)$, and therefore $\tau(n)$, is of the form $O(n^7)$ and not of the form $o(n^6)$. There is reason for supposing that $\tau(n)$ is of the form $O(n^{\frac{11}{2}+\epsilon})$ and not of the form $o(n^{\frac{11}{2}})$. For it appears that

$$\sum_1^\infty \frac{\tau(n)}{n^t} = \prod_p \frac{1}{1-\tau(p)p^{-t}+p^{11-2t}}. \qquad \text{.............. (101)}$$

This assertion is equivalent to the assertion that, if

$$n = p_1^{a_1} p_2^{a_2} p_3^{a_3} \dots p_r^{a_r},$$

where $p_1, p_2, \dots p_r$ are the prime divisors of n, then

$$n^{-\frac{11}{2}}\tau(n) = \frac{\sin(1+a_1)\theta_{p_1}}{\sin\theta_{p_1}} \frac{\sin(1+a_2)\theta_{p_2}}{\sin\theta_{p_2}} \dots \frac{\sin(1+a_r)\theta_{p_r}}{\sin\theta_{p_r}}, \quad (102)$$

where $\cos\theta_p = \frac{1}{2}p^{-\frac{11}{2}}\tau(p)$.

It would follow that, if n and n' are prime to each other, we must have

$$\tau(nn') = \tau(n)\tau(n'). \qquad \text{........................ (103)}$$

Let us suppose that (102) is true, and also that (as appears to be highly probable)

$$\{2\tau(p)\}^2 \leqslant p^{11}, \qquad \text{......................... (104)}$$

so that θ_p is real. Then it follows from (102) that

$$n^{-\frac{11}{2}}|\tau(n)| \leqslant (1+a_1)(1+a_2)\dots(1+a_r),$$

that is to say
$$|\tau(n)| \leqslant n^{\frac{11}{2}} d(n), \qquad \text{......................... (105)}$$

where $d(n)$ denotes the number of divisors of n.

Now let us suppose that $n=p^a$, so that

$$n^{-\frac{11}{2}}\tau(n)=\frac{\sin(1+a)\theta_p}{\sin\theta_p}.$$

Then we can choose a as large as we please and such that

$$\left|\frac{\sin(1+a)\theta_p}{\sin\theta_p}\right|\geqslant 1.$$

Hence
$$|\tau(n)|\geqslant n^{\frac{11}{2}} \qquad\qquad (106)$$
for an infinity of values of n.

19. It should be observed that precisely similar questions arise with regard to the arithmetical function $\psi(n)$ defined by

$$\sum_0^\infty \psi(n)x^n=f^{a_1}(x^{c_1})f^{a_2}(x^{c_2})\dots f^{a_r}(x^{c_r}), \qquad\qquad (107)$$

where
$$f(x)=x^{\frac{1}{24}}(1-x)(1-x^2)(1-x^3)\dots,$$
the a's and c's are integers, the latter being positive,

$$\tfrac{1}{24}(a_1c_1+a_2c_2+\dots+a_rc_r)$$

is equal to 0 or 1, and
$$l\left(\frac{a_1}{c_1}+\frac{a_2}{c_2}+\dots+\frac{a_r}{c_r}\right),$$
where l is the least common multiple of $c_1, c_2, \dots c_r$, is equal to 0 or to a divisor of 24.

The arithmetical functions $\chi(n)$, $P(n)$, $\chi_4(n)$, $\Omega(n)$ and $\Theta(n)$, studied by Dr Glaisher in the *Quarterly Journal*, Vols. XXXVI—XXXVIII, are of this type. Thus

$$\sum_1^\infty \chi(n)x^n=f^6(x^4),$$

$$\sum_1^\infty P(n)x^n=f^4(x^2)f^4(x^4),$$

$$\sum_1^\infty \chi_4(n)x^n=f^4(x)f^2(x^2)f^4(x^4),$$

$$\sum_1^\infty \Omega(n)x^n=f^{12}(x^2),$$

$$\sum_1^\infty \Theta(n)x^n=f^8(x)f^8(x^2).$$

20. The results (101) and (104) may be written as

$$\sum_1^\infty \frac{E_{r,s}(n)}{n^t}=E_{r,s}(1)\prod_p \frac{1}{1-2c_p p^{-t}+p^{r+s+1-2t}}, \qquad\qquad (108)$$

where
$$c_p^2\leqslant p^{r+s+1},$$
and
$$2c_pE_{r,s}(1)=E_{r,s}(p).$$

It seems probable that the result (108) is true not only for $r+s=10$ but also when $r+s$ is equal to 14, 16, 18, 20 or 24, and that

$$\left|\frac{E_{r,s}(n)}{E_{r,s}(1)}\right| \leqslant n^{\frac{1}{2}(r+s+1)} d(n) \quad \text{.......................(109)}$$

for all values of n, and
$$\left|\frac{E_{r,s}(n)}{E_{r,s}(1)}\right| \geqslant n^{\frac{1}{2}(r+s+1)} \quad \text{..........................(110)}$$

for an infinity of values of n. If this be so, then

$$E_{r,s}(n) = O\{n^{\frac{1}{2}(r+s+1+\epsilon)}\}, \quad E_{r,s}(n) \neq o\{n^{\frac{1}{2}(r+s+1)}\}. \quad \text{........(111)}$$

And it seems very likely that these equations hold generally, whenever r and s are positive odd integers.

21. It is of some interest to see what confirmation of these conjectures can be found from a study of the coefficients in the expansion of

$$x\{(1-x^{24/a})(1-x^{48/a})(1-x^{72/a})\ldots\}^a = \sum_1^\infty \psi_a(n)\, x^n,$$

where a is a divisor of 24. When $a=1$ and $a=3$ we know the actual value of $\psi_a(n)$. For we have

$$\sum_1^\infty \psi_1(n)\, x^n = x^{1^2} - x^{5^2} - x^{7^2} + x^{11^2} + x^{13^2} - x^{17^2} - \ldots, \quad \text{........(112)}$$

where 1, 5, 7, 11, ... are the natural odd numbers without the multiples of 3; and

$$\sum_1^\infty \psi_3(n)\, x^n = x^{1^2} - 3x^{3^2} + 5x^{5^2} - 7x^{7^2} + \ldots. \quad \text{............(113)}$$

The corresponding Dirichlet's series are

$$\sum_1^\infty \frac{\psi_1(n)}{n^s} = \frac{1}{(1+5^{-2s})(1+7^{-2s})(1-11^{-2s})(1-13^{-2s})\ldots}, \quad \text{...(114)}$$

where 5, 7, 11, 13, ... are the primes greater than 3, those of the form $12n \pm 5$ having the plus sign and those of the form $12n \pm 1$ the minus sign; and

$$\sum_1^\infty \frac{\psi_3(n)}{n^s} = \frac{1}{(1+3^{1-2s})(1-5^{1-2s})(1+7^{1-2s})(1+11^{1-2s})\ldots}, \quad \text{...(115)}$$

where 3, 5, 7, 11, ... are the odd primes, those of the form $4n-1$ having the plus sign and those of the form $4n+1$ the minus sign.

It is easy to see that

$$|\psi_1(n)| \leqslant 1, \quad |\psi_3(n)| \leqslant \sqrt{n} \quad \text{.....................(116)}$$

for all values of n, and

$$|\psi_1(n)| = 1, \quad |\psi_3(n)| = \sqrt{n} \quad \text{.....................(117)}$$

for an infinity of values of n.

The next simplest case is that in which $\alpha = 2$. In this case it appears that

$$\sum_1^\infty \frac{\psi_2(n)}{n^s} = \Pi_1 \Pi_2, \quad \text{.....................(118)}$$

where
$$\Pi_1 = \frac{1}{(1+5^{-2s})(1-7^{-2s})(1-11^{-2s})(1+17^{-2s})\ldots},$$

5, 7, 11, ... being the primes of the forms $12n-1$ and $12n \pm 5$, those of the form $12n+5$ having the plus sign and the rest the minus sign; and

$$\Pi_2 = \frac{1}{(1+13^{-s})^2(1-37^{-s})^2(1-61^{-s})^2(1+73^{-s})^2\ldots},$$

13, 37, 61, ... being the primes of the form $12n+1$, those of the form $m^2+(6n-3)^2$ having the plus sign and those of the form $m^2+(6n)^2$ the minus sign.

This is equivalent to the assertion that if

$$n = (5^{a_5}.7^{a_7}.11^{a_{11}}.17^{a_{17}}\ldots)^2\, 13^{a_{13}}.37^{a_{37}}.61^{a_{61}}.73^{a_{73}}\ldots,$$

where a_p is zero or a positive integer, then

$$\psi_2(n) = (-1)^{a_5+a_{13}+a_{17}+a_{29}+a_{41}+\ldots}(1+a_{13})(1+a_{37})(1+a_{61})\ldots, \quad \text{...(119)}$$

where 5, 13, 17, 29, ... are the primes of the form $4n+1$, excluding those of the form $m^2+(6n)^2$; and that otherwise

$$\psi_2(n) = 0. \quad \text{..............................(120)}$$

It follows that
$$|\psi_2(n)| \leqslant d(n) \quad \text{............................(121)}$$
for all values of n, and
$$|\psi_2(n)| \geqslant 1 \quad \text{...............................(122)}$$
for an infinity of values of n. These results are easily proved to be actually true.

22. I have investigated also the cases in which α has one of the values 4, 6, 8 or 12. Thus for example, when $\alpha = 6$, I find

$$\sum_1^\infty \frac{\psi_6(n)}{n^s} = \Pi_1 \Pi_2{}^*, \quad \text{..........................(123)}$$

where
$$\Pi_1 = \frac{1}{(1-3^{2-2s})(1-7^{2-2s})(1-11^{2-2s})\ldots},$$

3, 7, 11, ... being the primes of the form $4n-1$; and

$$\Pi_2 = \frac{1}{(1-2c_5.5^{-s}+5^{2-2s})(1-2c_{13}.13^{-s}+13^{2-2s})\ldots},$$

5, 13, 17, ... being the primes of the form $4n+1$, and $c_p = u^2-(2v)^2$, where u and v are the unique pair of positive integers for which $p = u^2+(2v)^2$. This is equivalent to the assertion that if

$$n = (3^{a_3}.7^{a_7}.11^{a_{11}}\ldots)^2.5^{a_5}.13^{a_{13}}.17^{a_{17}}\ldots,$$

* $\psi_6(n)$ is Dr Glaisher's $\lambda(n)$.

then $$\frac{\psi_6(n)}{n} = \frac{\sin(1+a_5)\theta_5}{\sin\theta_5} \cdot \frac{\sin(1+a_{13})\theta_{13}}{\sin\theta_{13}} \cdot \frac{\sin(1+a_{17})\theta_{17}}{\sin\theta_{17}} \ldots, \quad \ldots(124)$$

where $$\tan\tfrac{1}{2}\theta_p = \frac{u}{2v} \quad (0 < \theta_p < \pi),$$

and that otherwise $\psi_6(n) = 0$. From these results it would follow that

$$|\psi_6(n)| \leqslant nd(n) \quad \ldots(125)$$

for all values of n, and $$|\psi_6(n)| \geqslant n \quad \ldots(126)$$

for an infinity of values of n. What can actually be proved to be true is that

$$|\psi_6(n)| < 2nd(n)$$

for all values of n, and $$|\psi_6(n)| \geqslant n$$

for an infinity of values of n.

23. In the case in which $\alpha = 4$ I find that, if

$$n = (5^{a_5}.11^{a_{11}}.17^{a_{17}}\ldots)^2.7^{a_7}.13^{a_{13}}.19^{a_{19}}\ldots,$$

where 5, 11, 17, ... are the primes of the form $6m-1$ and 7, 13, 19, ... are those of the form $6m+1$, then

$$\frac{\psi_4(n)}{\sqrt{n}} = (-1)^{a_5+a_{11}+a_{17}+\ldots} \frac{\sin(1+a_7)\theta_7}{\sin\theta_7} \cdot \frac{\sin(1+a_{13})\theta_{13}}{\sin\theta_{13}} \ldots, \quad \ldots(127)$$

where $$\tan\theta_p = \frac{u\sqrt{3}}{1 \pm 3v} \quad (0 < \theta_p < \pi),$$

and u and v are the unique pair of positive integers for which $p = 3u^2 + (1 \pm 3v)^2$; and that $\psi_4(n) = 0$ for other values.

In the case in which $\alpha = 8$ I find that, if

$$n = (2^{a_2}.5^{a_5}.11^{a_{11}}\ldots)^2.7^{a_7}.13^{a_{13}}.19^{a_{19}}\ldots,$$

where 2, 5, 11, ... are the primes of the form $3m-1$ and 7, 13, 19, ... are those of the form $6m+1$, then

$$\frac{\psi_8(n)}{n\sqrt{n}} = (-1)^{a_2+a_5+a_{11}+\ldots} \frac{\sin 3(1+a_7)\theta_7}{\sin 3\theta_7} \cdot \frac{\sin 3(1+a_{13})\theta_{13}}{\sin 3\theta_{13}} \ldots, \quad \ldots(128)$$

where θ_p is the same as in (127); and that $\psi_8(n) = 0$ for other values.

The case in which $\alpha = 12$ will be considered in § 28.

In short, such evidence as I have been able to find, while not conclusive, points to the truth of the results conjectured in § 18.

24. Analysis similar to that of the preceding sections may be applied to some interesting arithmetical functions of a different kind. Let

$$\phi^s(q) = 1 + 2\sum_1^\infty r_s(n) q^n, \quad \ldots(129)$$

where $$\phi(q) = 1 + 2q + 2q^4 + 2q^9 + \ldots,$$

so that $r_s(n)$ is the number of representations of n as the sum of s squares. Further let

$$\sum_1^\infty \delta_2(n)\, q^n = 2\left(\frac{q}{1-q} - \frac{q^3}{1-q^3} + \frac{q^5}{1-q^5} - \dots\right) = 2\left(\frac{q}{1+q^2} + \frac{q^2}{1+q^4} + \frac{q^3}{1+q^6} + \dots\right); \quad \dots\dots(130)$$

$$(2^s - 1) B_s \sum_1^\infty \delta_{2s}(n)\, q^n = s\left(\frac{1^{s-1} q}{1+q} + \frac{2^{s-1} q^2}{1-q^2} + \frac{3^{s-1} q^3}{1+q^3} + \dots\right), \quad \dots(131)$$

when s is a multiple of 4;

$$(2^s - 1) B_s \sum_1^\infty \delta_{2s}(n)\, q^n = s\left(\frac{1^{s-1} q}{1-q} + \frac{2^{s-1} q^2}{1+q^2} + \frac{3^{s-1} q^3}{1-q^3} + \dots\right), \quad \dots(132)$$

when $s+2$ is a multiple of 4;

$$E_s \sum_1^\infty \delta_{2s}(n)\, q^n = 2^s\left(\frac{1^{s-1} q}{1+q^2} + \frac{2^{s-1} q^2}{1+q^4} + \frac{3^{s-1} q^3}{1+q^6} + \dots\right) + 2\left(\frac{1^{s-1} q}{1-q} - \frac{3^{s-1} q^3}{1-q^3} + \frac{5^{s-1} q^5}{1-q^5} - \dots\right), \quad \dots(133)$$

when $s-1$ is a multiple of 4;

$$E_s \sum_1^\infty \delta_{2s}(n)\, q^n = 2^s\left(\frac{1^{s-1} q}{1+q^2} + \frac{2^{s-1} q^2}{1+q^4} + \frac{3^{s-1} q^3}{1+q^6} + \dots\right) - 2\left(\frac{1^{s-1} q}{1-q} - \frac{3^{s-1} q^3}{1-q^3} + \frac{5^{s-1} q^5}{1-q^5} - \dots\right), \quad \dots(134)$$

when $s+1$ is a multiple of 4. In these formulæ

$$B_2 = \tfrac{1}{6}, \quad B_4 = \tfrac{1}{30}, \quad B_6 = \tfrac{1}{42}, \quad B_8 = \tfrac{1}{30}, \quad B_{10} = \tfrac{5}{66}, \dots$$

are Bernoulli's numbers, and

$$E_1 = 1, \quad E_3 = 1, \quad E_5 = 5, \quad E_7 = 61, \quad E_9 = 1385, \dots$$

are Euler's numbers. Then $\delta_{2s}(n)$ is in all cases an arithmetical function depending on the real divisors of n; thus, for example, when $s+2$ is a multiple of 4, we have

$$(2^s - 1) B_s \delta_{2s}(n) = s\{\sigma_{s-1}(n) - 2^s \sigma_{s-1}(\tfrac{1}{4}n)\}, \quad \dots\dots\dots(135)$$

where $\sigma_s(x)$ should be considered as equal to zero if x is not an integer.

Now let
$$r_{2s}(n) = \delta_{2s}(n) + e_{2s}(n). \quad \dots\dots(136)$$

Then I can prove (see § 26) that

$$e_{2s}(n) = 0 \quad \dots\dots(137)$$

if $s = 1, 2, 3, 4$; and that

$$e_{2s}(n) = O(n^{s-1-\frac{1}{2}[\frac{2}{3}s]+\epsilon}) \quad \dots\dots(138)$$

for all positive integral values of s. But it is easy to see that, if $s \geqslant 3$, then

$$Hn^{s-1} < \delta_{2s}(n) < Kn^{s-1}, \quad \dots\dots(139)$$

where H and K are positive constants. It follows that

$$r_{2s}(n) \sim \delta_{2s}(n) \quad \dots\dots(140)$$

for all positive integral values of s.

It appears probable, from the empirical results I obtain at the end of this paper, that

$$e_{2s}(n) = O\{n^{\frac{1}{2}(s-1)+\epsilon}\} \quad \text{.......................... (141)}$$

for all positive integral values of s; and that

$$e_{2s}(n) \neq o\{n^{\frac{1}{2}(s-1)}\} \quad \text{.......................... (142)}$$

if $s \geqslant 5$. But all that I can actually prove is that

$$e_{2s}(n) = O(n^{s-1-\frac{1}{2}[\frac{2}{3}s]}) \quad \text{.......................... (143)}$$

if $s \geqslant 9$; and that

$$e_{2s}(n) \neq o(n^{\frac{1}{2}s-1}) \quad \text{.......................... (144)}$$

if $s \geqslant 5$.

25. Let

$$f_{2s}(q) = \sum_{1}^{\infty} e_{2s}(n)\, q^n = \sum_{1}^{\infty} \{r_{2s}(n) - \delta_{2s}(n)\}\, q^n. \quad \text{........... (145)}$$

Then it can be shewn by the theory of elliptic functions that

$$f_{2s}(q) = \phi^{2s}(q) \sum_{1 \leqslant n \leqslant \frac{1}{4}(s-1)} K_n (kk')^{2n}, \quad \text{.............. (146)}$$

that is to say that

$$f_{2s}(q) = \frac{f^{4s}(-q)}{f^{2s}(q^2)} \sum_{1 \leqslant n \leqslant \frac{1}{4}(s-1)} K_n \frac{f^{24n}(q^2)}{f^{24n}(-q)}, \quad \text{............. (147)}$$

where $\phi(q)$ and $f(q)$ are the same as in § 13. We thus obtain the results contained in the following table.

TABLE VI.

1. $f_2(q)=0,\quad f_4(q)=0,\quad f_6(q)=0,\quad f_8(q)=0.$

2. $5f_{10}(q)=16\dfrac{f^{14}(q^2)}{f^4(-q)},\quad f_{12}(q)=8f^{12}(q^2).$

3. $61f_{14}(q)=728f^4(-q)f^{10}(q^2),\quad 17f_{16}(q)=256f^8(-q)f^8(q^2).$

4. $1385f_{18}(q)=24416f^{12}(-q)f^6(q^2)-256\dfrac{f^{30}(q^2)}{f^{12}(-q)}.$

5. $31f_{20}(q)=616f^{16}(-q)f^4(q^2)-128\dfrac{f^{28}(q^2)}{f^8(-q)}.$

6. $50521f_{22}(q)=1103272f^{20}(-q)f^2(q^2)-821888\dfrac{f^{26}(q^2)}{f^4(-q)}.$

7. $691f_{24}(q)=16576f^{24}(-q)-32768f^{24}(q^2).$

It follows from the last formula of Table VI that

$$\tfrac{691}{64} e_{24}(n) = (-1)^{n-1} 259\tau(n) - 512\tau(\tfrac{1}{2}n), \quad \text{........... (148)}$$

where $\tau(n)$ is the same as in § 16, and $\tau(x)$ should be considered as equal to zero if x is not an integer.

Results equivalent to 1, 2, 3, 4 of Table VI were given by Dr Glaisher in the *Quarterly Journal*, Vol. XXXVIII. The arithmetical functions called by him

$$\chi_4(n),\quad \Omega(n),\quad W(n),\quad \Theta(n),\quad U(n)$$

are the coefficients of q^n in

$$\frac{f^{14}(q^2)}{f^4(-q)},\quad f^{12}(q^2),\quad f^4(-q)f^{10}(q^2),\quad f^8(q)f^8(q^2),\quad f^{12}(-q)f^6(q^2).$$

He gave reduction formulæ for these functions and observed how the functions which I call $e_{10}(n)$, $e_{12}(n)$ and $e_{16}(n)$ can be defined by means of the complex divisors of n. It is very likely that $\tau(n)$ is also capable of such a definition.

26. Now let us consider the order of $e_{28}(n)$. It is easy to see from (147) that $f_{28}(q)$ can be expressed in the form

$$\Sigma K_{a,b,c,h,k}\{f^3(-q)\}^a \left\{\frac{f^5(-q)}{f^2(q^2)}\right\}^b \left\{\frac{f^5(q^2)}{f^2(-q)}\right\}^c f^h(-q) f^k(q^2), \quad \ldots(149)$$

where a, b, c, h, k are zero or positive integers, such that

$$a+b+c=[\tfrac{2}{3}s], \quad h+k=2s-3[\tfrac{2}{3}s].$$

Proceeding as in § 13 we can easily shew that

$$n^{-\frac{1}{2}[\frac{2}{3}s]} e_{28}(n)$$

cannot be of higher order than the coefficient of q^{24n} in

$$\phi^A(q)\,\phi^B(q^3)\,\phi^C(q^2), \quad \ldots\ldots(150)$$

where C is 0 or 1 and $\quad A+B+C=2s-2[\frac{2}{3}s]$.

Now, if $s \geqslant 5$, $A+B+C \geqslant 4$; and so $A+B \geqslant 3$. Hence one at least of A and B is greater than 1. But we know that

$$\phi^2(q) = \Sigma O(\nu^\epsilon) q^\nu.$$

It follows that the coefficient of q^{24n} in (150) is of order not exceeding

$$n^{\frac{1}{2}(A+B+C)-1+\epsilon}.$$

Thus $\quad e_{28}(n) = O(n^{s-1-\frac{1}{2}[\frac{2}{3}s]+\epsilon}) \quad \ldots\ldots(151)$

for all positive integral values of s.

27. When $s \geqslant 9$ we can obtain a slightly more precise result.

If $s \geqslant 16$ we have $A+B+C \geqslant 12$; and so $A+B \geqslant 11$. Hence one at least of A and B is greater than 5. But

$$\phi^6(q) = \Sigma O(\nu^2) q^\nu.$$

It follows that the coefficient of q^{24n} in (150) is of order not exceeding

$$n^{\frac{1}{2}(A+B+C)-1},$$

or that $\quad e_{28}(n) = O(n^{s-1-\frac{1}{2}[\frac{2}{3}s]}), \quad \ldots\ldots(152)$

if $s \geqslant 16$. We can easily shew that (152) is true when $9 \leqslant s < 16$ considering all the cases separately, using the identities

$$f^{12}(-q) f^6(q^2) = \{f^3(-q)\}^4 \{f^3(q^2)\}^2,$$

$$\frac{f^{30}(q^2)}{f^{12}(-q)} = \left\{\frac{f^5(q^2)}{f^2(-q)}\right\}^6,$$

$$f^{16}(-q) f^4(q^2) = \left\{\frac{f^5(-q)}{f^2(q^2)}\right\}^4 \left\{\frac{f^5(q^2)}{f^2(-q)}\right\}^2 f^2(q^2),$$

$$\frac{f^{28}(q^2)}{f^8(-q)} = \left\{\frac{f^5(q^2)}{f^2(-q)}\right\}^4 \{f^3(q^2)\}^2 f^2(q^2), \ldots,$$

and proceeding as in the previous two sections.

The argument of §§ 14—15 may also be applied to the function $e_{2s}(n)$. We find that

$$e_{2s}(n) \neq o(n^{\frac{1}{2}s-1}). \quad \text{.........................(153)}$$

I leave the proof to the reader.

28. There is reason to suppose that

$$\left.\begin{array}{l} e_{2s}(n) = O\{n^{\frac{1}{2}(s-1+\epsilon)}\} \\ e_{2s}(n) \neq o\{n^{\frac{1}{2}(s-1)}\} \end{array}\right\}, \quad \text{.........................(154)}$$

if $s \geqslant 5$. I find, for example, that

$$\sum_{1}^{\infty} \frac{e_{10}(n)}{n^s} = \frac{e_{10}(1)}{1+2^{2-s}} \Pi_1 \Pi_2, \quad \text{....................(155)}$$

where $$\Pi_1 = \frac{1}{(1-3^{4-2s})(1-7^{4-2s})(1-11^{4-2s})\ldots},$$

3, 7, 11, ... being the primes of the form $4n-1$, and

$$\Pi_2 = \frac{1}{(1-2c_5 . 5^{-s} + 5^{4-2s})(1-2c_{13} . 13^{-s} + 13^{4-2s})\ldots},$$

5, 13, 17, ... being the primes of the form $4n+1$, and

$$c_p = u^2 - (4v)^2,$$

where u and v are the unique pair of positive integers satisfying the equation

$$u^2 + (4v)^2 = p^2.$$

The equation (155) is equivalent to the assertion that, if

$$n = (3^{a_3} . 7^{a_7} . 11^{a_{11}} \ldots)^2 . 2^{a_2} . 5^{a_5} . 13^{a_{13}} \ldots,$$

where a_p is zero or a positive integer, then

$$\frac{e_{10}(n)}{n^2 e_{10}(1)} = (-1)^{a_2} \frac{\sin 4(1+a_5)\theta_5}{\sin 4\theta_5} . \frac{\sin 4(1+a_{13})\theta_{13}}{\sin 4\theta_{13}} \ldots, \quad \text{...(156)}$$

where $$\tan\theta_p = \frac{u}{v} \qquad (0 < \theta_p < \tfrac{1}{2}\pi),$$

u and v being integers satisfying the equation $u^2 + v^2 = p$; and $e_{10}(n) = 0$ otherwise. If this is true then we should have

$$\left|\frac{e_{10}(n)}{e_{10}(1)}\right| \leqslant n^2 d(n) \quad \text{.........................(157)}$$

for all values of n, and $$\left|\frac{e_{10}(n)}{e_{10}(1)}\right| \geqslant n^2 \quad \text{..............................(158)}$$

for an infinity of values of n. In this case we can prove that, if n is the square of a prime of the form $4m-1$, then

$$\frac{e_{10}(n)}{e_{10}(1)} = n^2.$$

Similarly I find that

$$\sum_{1}^{\infty} \frac{e_{12}(n)}{n^s} = e_{12}(1) \prod_{p} \left(\frac{1}{1 + 2c_p \,.\, p^{-s} + p^{5-2s}}\right), \quad\text{.............(159)}$$

p being an odd prime and $c_p^2 \leqslant p^5$. From this it would follow that

$$\left|\frac{e_{12}(n)}{e_{12}(1)}\right| \leqslant n^{\frac{5}{2}} d(n) \quad\text{..............................(160)}$$

for all values of n, and

$$\left|\frac{e_{12}(n)}{e_{12}(1)}\right| \geqslant n^{\frac{5}{2}} \quad\text{......................................(161)}$$

for an infinity of values of n.

Finally I find that

$$\sum_{1}^{\infty} \frac{e_{16}(n)}{n^s} = \frac{e_{16}(1)}{1 + 2^{3-s}} \prod_{p} \left(\frac{1}{1 + 2c_p \,.\, p^{-s} + p^{7-2s}}\right), \quad\text{.........(162)}$$

p being an odd prime and $c_p^2 \leqslant p^7$. From this it would follow that

$$\left|\frac{e_{16}(n)}{e_{16}(1)}\right| \leqslant n^{\frac{7}{2}} d(n) \quad\text{..............................(163)}$$

for all values of n, and

$$\left|\frac{e_{16}(n)}{e_{16}(1)}\right| \geqslant n^{\frac{7}{2}} \quad\text{......................................(164)}$$

for an infinity of values of n.

In the case in which $2s = 24$ we have

$$\tfrac{691}{64} e_{24}(n) = (-1)^{n-1} 259 \tau(n) - 512 \tau(\tfrac{1}{2}n).$$

I have already stated the reasons for supposing that

$$|\tau(n)| \leqslant n^{\frac{11}{2}} d(n)$$

for all values of n, and

$$|\tau(n)| \geqslant n^{\frac{11}{2}}$$

for an infinity of values of n.

19

A SERIES FOR EULER'S CONSTANT γ

(*Messenger of Mathematics*, XLVI, 1917, 73—80)

1. In a paper recently published in this Journal (Vol. XLIV, pp. 1—10), Dr Glaisher proves a number of formulæ of the type

$$\gamma = 1 - 2\left(\frac{S_3}{3.4} + \frac{S_5}{5.6} + \frac{S_7}{7.8} + \ldots\right),$$

where $$S_n = 1^{-n} + 2^{-n} + 3^{-n} + 4^{-n} + \ldots,$$

and conjectures the existence of a general formula

$$\gamma = \lambda_r - (r+1)(r+2)\ldots(2r)$$
$$\times\left\{\frac{S_3}{3(r+3)(r+4)\ldots(2r+2)} + \frac{S_5}{5(r+5)(r+6)\ldots(2r+4)} + \ldots\right\},$$

where λ_r is a rational number. I propose now to prove the general formula of which Dr Glaisher's are particular cases: this formula is itself a particular case of still more general formulæ.

2. Let r and t be any two positive numbers. Then

$$\int_0^1 x^{r-1}(1-x)^{t-1}\log\Gamma(1-x)\,dx = \int_0^1 x^{t-1}(1-x)^{r-1}\log\Gamma(x)\,dx$$

$$= \int_0^1 x^{t-1}(1-x)^{r-1}\log\Gamma(1+x)\,dx - \int_0^1 x^{t-1}(1-x)^{r-1}\log x\,dx. \quad \ldots\ldots(1)$$

But $$\int_0^1 x^{r-1}(1-x)^{t-1}\log\Gamma(1-x)\,dx$$

$$= \int_0^1 x^{r-1}(1-x)^{t-1}\left\{\gamma x + S_2\frac{x^2}{2} + S_3\frac{x^3}{3} + \ldots\right\}dx$$

$$= \frac{\Gamma(1+r)\Gamma(t)}{\Gamma(1+r+t)}\gamma + \frac{\Gamma(2+r)\Gamma(t)}{\Gamma(2+r+t)}\frac{S_2}{2} + \frac{\Gamma(3+r)\Gamma(t)}{\Gamma(3+r+t)}\frac{S_3}{3} + \ldots\ldots\ldots(2)$$

Similarly $$\int_0^1 x^{t-1}(1-x)^{r-1}\log\Gamma(1+x)\,dx$$

$$= -\frac{\Gamma(1+t)\Gamma(r)}{\Gamma(1+r+t)}\gamma + \frac{\Gamma(2+t)\Gamma(r)}{\Gamma(2+r+t)}\frac{S_2}{2} - \frac{\Gamma(3+t)\Gamma(r)}{\Gamma(3+r+t)}\frac{S_3}{3} + \ldots\ldots(2')$$

And also

$$\int_0^1 x^{t-1}(1-x)^{r-1}\log x\,dx = \frac{d}{dt}\int_0^1 x^{t-1}(1-x)^{r-1}\,dx = \frac{d}{dt}\left\{\frac{\Gamma(t)\Gamma(r)}{\Gamma(r+t)}\right\}$$

$$= \frac{\Gamma(r)\Gamma(t)}{\Gamma(r+t)}\left\{\frac{\Gamma'(t)}{\Gamma(t)} - \frac{\Gamma'(r+t)}{\Gamma(r+t)}\right\} = -\frac{\Gamma(r)\Gamma(t)}{\Gamma(r+t)}\int_0^1 x^{t-1}\frac{1-x^r}{1-x}\,dx. \ldots(3)$$

It follows from (1)—(3) that, if r and t are positive, then

$$\frac{r}{1(r+t)}\gamma+\frac{r(r+1)}{2(r+t)(r+t+1)}S_2+\frac{r(r+1)(r+2)}{3(r+t)(r+t+1)(r+t+2)}S_3+\dots$$

$$+\frac{t}{1(r+t)}\gamma-\frac{t(t+1)}{2(r+t)(r+t+1)}S_2+\frac{t(t+1)(t+2)}{3(r+t)(r+t+1)(r+t+2)}S_3-\dots$$

$$=\int_0^1\frac{x^{t-1}(1-x^r)}{1-x}dx. \qquad \text{.............(4)}$$

Now, interchanging r and t in (4), and taking the sum and the difference of the two results, we see that, if r and t are positive, then

$$\frac{r+t}{1(r+t)}\gamma+\frac{r(r+1)(r+2)+t(t+1)(t+2)}{3(r+t)(r+t+1)(r+t+2)}S_3+\dots$$

$$=\tfrac{1}{2}\int_0^1\frac{x^{r-1}+x^{t-1}-2x^{r+t-1}}{1-x}dx; \qquad \text{......(5)}$$

and

$$\frac{r(r+1)-t(t+1)}{2(r+t)(r+t+1)}S_2$$

$$+\frac{r(r+1)(r+2)(r+3)-t(t+1)(t+2)(t+3)}{4(r+t)(r+t+1)(r+t+2)(r+t+3)}S_4+\dots=\tfrac{1}{2}\int_0^1\frac{x^{t-1}-x^{r-1}}{1-x}dx.$$

$$\text{......(6)}$$

The right-hand sides of (5) and (6) can be expressed in finite terms if r and t are rational. If, in particular, r and t are integers, then

$$\int_0^1\frac{x^{r-1}+x^{t-1}-2x^{r+t-1}}{1-x}dx=\frac{1}{r}+\frac{1}{r+1}+\frac{1}{r+2}+\dots+\frac{1}{r+t-1}$$

$$+\frac{1}{t}+\frac{1}{t+1}+\frac{1}{t+2}+\dots+\frac{1}{r+t-1};$$

and

$$\int_0^1\frac{x^{t-1}-x^{r-1}}{1-x}dx=\left(1+\tfrac{1}{2}+\tfrac{1}{3}+\tfrac{1}{4}+\dots+\frac{1}{r-1}\right)$$

$$-\left(1+\tfrac{1}{2}+\tfrac{1}{3}+\tfrac{1}{4}+\dots+\frac{1}{t-1}\right).$$

3. Let us now suppose that $t=r$ in (5). Then it is clear that

$$\gamma+\frac{(r+1)(r+2)}{3(2r+1)(2r+2)}S_3+\frac{(r+1)(r+2)(r+3)(r+4)}{5(2r+1)(2r+2)(2r+3)(2r+4)}S_5+\dots$$

$$=\int_0^1\frac{x^{r-1}(1-x^r)}{1-x}dx=\int_0^1\frac{1+x^{2r-1}}{1+x}dx, \qquad \text{.........(7)}$$

if $r>0$. If we suppose, in (7), that r is an integer, we obtain the formula conjectured by Dr Glaisher, the value of λ_r being

$$\int_0^1\frac{1+x^{2r-1}}{1+x}dx=1-\tfrac{1}{2}+\tfrac{1}{3}-\tfrac{1}{4}+\dots+\frac{1}{2r-1}.$$

Again, dividing both sides in (6) by $r-t$ and making $t \to r$, we see that, if $r>0$, then

$$\frac{r+1}{2(2r+1)}\left(\frac{1}{r}+\frac{1}{r+1}\right)S_2$$

$$+\frac{(r+1)(r+2)(r+3)}{4(2r+1)(2r+2)(2r+3)}\left(\frac{1}{r}+\frac{1}{r+1}+\frac{1}{r+2}+\frac{1}{r+3}\right)S_4+\dots$$

$$=-\int_0^1 \frac{x^{r-1}\log x}{1-x}dx=\frac{1}{r^2}+\frac{1}{(r+1)^2}+\frac{1}{(r+2)^2}+\frac{1}{(r+3)^2}+\dots \quad \dots\dots\dots(8)$$

Thus for example we have

$$\frac{\pi^2}{12}=(1+\tfrac{1}{2})\frac{S_2}{2.3}+(1+\tfrac{1}{2}+\tfrac{1}{3}+\tfrac{1}{4})\frac{S_4}{4.5}+(1+\tfrac{1}{2}+\tfrac{1}{3}+\tfrac{1}{4}+\tfrac{1}{5}+\tfrac{1}{6})\frac{S_6}{6.7}+\dots.$$

4. If we start with the integral

$$\int_0^1 x^{r-1}(1-x)^{t-1}\log\Gamma\left(1-\frac{x}{2}\right)dx,$$

and proceed as in § 2, we can shew that, if r and t are positive, then

$$\frac{r}{1(r+t)}S_1'+\frac{r(r+1)}{2(r+t)(r+t+1)}S_2'+\frac{r(r+1)(r+2)}{3(r+t)(r+t+1)(r+t+2)}S_3'+\dots$$

$$-\frac{t}{1(r+t)}S_1'+\frac{t(t+1)}{2(r+t)(r+t+1)}S_2'-\frac{t(t+1)(t+2)}{3(r+t)(r+t+1)(r+t+2)}S_3'+\dots$$

$$=\int_0^1 \frac{x^{t-1}(1-x^r)}{1-x}dx-\log\frac{\pi}{2}, \quad \dots\dots\dots(9)$$

where $\quad S_n'=1^{-n}-2^{-n}+3^{-n}-4^{-n}+\dots.$

From (9) we can easily deduce that, if r and t are positive, then

$$\frac{r(r+1)+t(t+1)}{2(r+t)(r+t+1)}S_2'+\frac{r(r+1)(r+2)(r+3)+t(t+1)(t+2)(t+3)}{4(r+t)(r+t+1)(r+t+2)(r+t+3)}S_4'+\dots$$

$$=\tfrac{1}{2}\int_0^1 \frac{x^{r-1}+x^{t-1}-2x^{r+t-1}}{1-x}dx-\log\frac{\pi}{2}; \quad \dots\dots(10)$$

and
$$\frac{r-t}{1(r+t)}S_1'+\frac{r(r+1)(r+2)-t(t+1)(t+2)}{3(r+t)(r+t+1)(r+t+2)}S_3'+\dots$$

$$=\tfrac{1}{2}\int_0^1 \frac{x^{t-1}-x^{r-1}}{1-x}dx. \quad \dots\dots\dots\dots\dots(11)$$

As particular cases of (10) and (11), we have

$$\log\frac{\pi}{2}+\frac{r+1}{2(2r+1)}S_2'+\frac{(r+1)(r+2)(r+3)}{4(2r+1)(2r+2)(2r+3)}S_4'+\dots=\int_0^1 \frac{1+x^{2r-1}}{1+x}dx;$$

$$\dots\dots(12)$$

and $$\frac{1}{r}S_1' + \frac{(r+1)(r+2)}{3(2r+1)(2r+2)}\left(\frac{1}{r}+\frac{1}{r+1}+\frac{1}{r+2}\right)S_3'$$
$$+\frac{(r+1)(r+2)(r+3)(r+4)}{5(2r+1)(2r+2)(2r+3)(2r+4)}$$
$$\times\left(\frac{1}{r}+\frac{1}{r+1}+\frac{1}{r+2}+\frac{1}{r+3}+\frac{1}{r+4}\right)S_5'+\dots$$
$$=\frac{1}{r^2}+\frac{1}{(r+1)^2}+\frac{1}{(r+2)^2}+\dots, \quad \dots\dots\dots\dots(13)$$

provided that $r > 0$. Thus for example we have

$$1 = \log\frac{\pi}{2} + 2\left(\frac{S_2'}{2.3}+\frac{S_4'}{4.5}+\frac{S_6'}{6.7}+\dots\right);$$

$$\frac{\pi^2}{12} = \frac{S_1'}{1.2} + \frac{S_3'}{3.4}(1+\tfrac{1}{2}+\tfrac{1}{3}) + \frac{S_5'}{5.6}(1+\tfrac{1}{2}+\tfrac{1}{3}+\tfrac{1}{4}+\tfrac{1}{5})+\dots.$$

5. The preceding results may be generalised as follows. Let $\zeta(s, x)$ denote the function represented by the series

$$x^{-s}+(x+1)^{-s}+(x+2)^{-s}+(x+3)^{-s}+\dots \quad (x>0)$$

and its analytical continuations, so that $\zeta(s,1)=\zeta(s)$ and $\zeta(s,\tfrac{1}{2})=(2^s-1)\zeta(s)$, $\zeta(s)$ being the Riemann ζ-function. Then

$$\int_0^1 x^{r-1}(1-x)^{t-1}\zeta(s,1-x)\,dx = \int_0^1 x^{t-1}(1-x)^{r-1}\zeta(s,x)\,dx$$
$$=\int_0^1 x^{t-1}(1-x)^{r-1}\zeta(s,1+x)\,dx + \int_0^1 x^{t-s-1}(1-x)^{r-1}\,dx, \quad \dots\dots\dots(14)$$

provided that r and t are positive. But we know that, if $|x|<1$, then

$$\zeta(s,1-x)=\zeta(s)+\frac{s}{1!}\zeta(s+1)x+\frac{s(s+1)}{2!}\zeta(s+2)x^2+\dots; \quad \dots(15)$$

and that $$\int_0^1 x^{t-s-1}(1-x)^{r-1}\,dx = \frac{\Gamma(t-s)\Gamma(r)}{\Gamma(r-s+t)}, \quad \dots\dots\dots\dots(16)$$

provided that $t > s$. It follows from (14)—(16) that, if r and t are positive and $t > s$, then

$$\left\{\zeta(s)+\frac{s}{1!}\frac{r}{r+t}\zeta(s+1)+\frac{s(s+1)}{2!}\frac{r(r+1)}{(r+t)(r+t+1)}\zeta(s+2)+\dots\right\}$$
$$-\left\{\zeta(s)-\frac{s}{1!}\frac{t}{r+t}\zeta(s+1)+\frac{s(s+1)}{2!}\frac{t(t+1)}{(r+t)(r+t+1)}\zeta(s+2)-\dots\right\}$$
$$=\frac{\Gamma(r+t)\Gamma(t-s)}{\Gamma(t)\Gamma(r-s+t)}. \quad \dots\dots\dots(17)$$

As particular cases of (17), we have

$$\frac{s}{1!}\frac{r+t}{r+t}\zeta(s+1)+\frac{s(s+1)(s+2)}{3!}\frac{r(r+1)(r+2)+t(t+1)(t+2)}{(r+t)(r+t+1)(r+t+2)}\zeta(s+3)+\dots$$

$$=\frac{1}{2}\frac{\Gamma(r+t)}{\Gamma(r-s+t)}\left\{\frac{\Gamma(t-s)}{\Gamma(t)}+\frac{\Gamma(r-s)}{\Gamma(r)}\right\}, \quad \dots\dots(18)$$

and

$$\frac{s(s+1)}{2!}\frac{r(r+1)-t(t+1)}{(r+t)(r+t+1)}\zeta(s+2)$$

$$+\frac{s(s+1)(s+2)(s+3)}{4!}\frac{r(r+1)(r+2)(r+3)-t(t+1)(t+2)(t+3)}{(r+t)(r+t+1)(r+t+2)(r+t+3)}\zeta(s+4)+\dots$$

$$=\frac{1}{2}\frac{\Gamma(r+t)}{\Gamma(r-s+t)}\left\{\frac{\Gamma(t-s)}{\Gamma(t)}-\frac{\Gamma(r-s)}{\Gamma(r)}\right\}, \quad \dots\dots\dots\dots(19)$$

provided that r and t are positive and greater than s. From (18) and (19) we deduce that, if r is positive and greater than s, then

$$\frac{s}{1!}\frac{r}{2r}\zeta(s+1)+\frac{s(s+1)(s+2)}{3!}\frac{r(r+1)(r+2)}{2r(2r+1)(2r+2)}\zeta(s+3)+\dots$$

$$=\frac{1}{2}\frac{\Gamma(2r)\,\Gamma(r-s)}{\Gamma(r)\,\Gamma(2r-s)}, \quad \dots\dots\dots\dots\dots\dots(20)$$

and

$$\frac{s(s+1)}{2!}\frac{r(r+1)}{2r(2r+1)}\left(\frac{1}{r}+\frac{1}{r+1}\right)\zeta(s+2)$$

$$+\frac{s(s+1)(s+2)(s+3)}{4!}\frac{r(r+1)(r+2)(r+3)}{2r(2r+1)(2r+2)(2r+3)}$$

$$\times\left(\frac{1}{r}+\frac{1}{r+1}+\frac{1}{r+2}+\frac{1}{r+3}\right)\zeta(s+4)+\dots$$

$$=\frac{1}{2}\frac{\Gamma(2r)\,\Gamma(r-s)}{\Gamma(r)\,\Gamma(2r-s)}\int_0^1\frac{x^{r-s-1}(1-x^s)}{1-x}\,dx. \quad \dots\dots\dots\dots(21)$$

6. If we start with the integral

$$\int_0^1 x^{r-1}(1-x)^{t-1}\zeta\left(s,\,1-\frac{x}{2}\right)dx,$$

and proceed as in § 5, we can shew that, if r and t are positive and $t>s$, then

$$\zeta_1(s)+\frac{s}{1!}\frac{r}{r+t}\zeta_1(s+1)+\frac{s(s+1)}{2!}\frac{r(r+1)}{(r+t)(r+t+1)}\zeta_1(s+2)+\dots$$

$$+\zeta_1(s)-\frac{s}{1!}\frac{t}{r+t}\zeta_1(s+1)+\frac{s(s+1)}{2!}\frac{t(t+1)}{(r+t)(r+t+1)}\zeta_1(s+2)-\dots$$

$$=\frac{\Gamma(r+t)\,\Gamma(t-s)}{\Gamma(t)\,\Gamma(r-s+t)}, \quad \dots\dots\dots\dots\dots(22)$$

where $\zeta_1(s)$ is the function represented by the series

$$1^{-s}-2^{-s}+3^{-s}-4^{-s}+\dots$$

and its analytical continuations. From (22) we deduce that, if r and t are positive and greater than s, then

$$(1+1)\zeta_1(s)+\frac{s(s+1)}{2!}\frac{r(r+1)+t(t+1)}{(r+t)(r+t+1)}\zeta_1(s+2)+\dots$$
$$=\frac{1}{2}\frac{\Gamma(r+t)}{\Gamma(r-s+t)}\left\{\frac{\Gamma(t-s)}{\Gamma(t)}+\frac{\Gamma(r-s)}{\Gamma(r)}\right\}; \quad \dots(23)$$

and
$$\frac{s}{1!}\frac{r-t}{r+t}\zeta_1(s+1)$$
$$+\frac{s(s+1)(s+2)}{3!}\frac{r(r+1)(r+2)-t(t+1)(t+2)}{(r+t)(r+t+1)(r+t+2)}\zeta_1(s+3)+\dots$$
$$=\frac{1}{2}\frac{\Gamma(r+t)}{\Gamma(r-s+t)}\left\{\frac{\Gamma(t-s)}{\Gamma(t)}-\frac{\Gamma(r-s)}{\Gamma(r)}\right\}. \quad \dots\dots\dots(24)$$

As particular cases of (23) and (24), we have

$$\zeta_1(s)+\frac{s(s+1)}{2!}\frac{r(r+1)}{2r(2r+1)}\zeta_1(s+2)+\dots=\frac{1}{2}\frac{\Gamma(2r)\Gamma(r-s)}{\Gamma(r)\Gamma(2r-s)}, \quad \dots(25)$$

and
$$\frac{s}{1!}\frac{r}{2r}\frac{1}{r}\zeta_1(s+1)$$
$$+\frac{s(s+1)(s+2)}{3!}\frac{r(r+1)(r+2)}{2r(2r+1)(2r+2)}\left(\frac{1}{r}+\frac{1}{r+1}+\frac{1}{r+2}\right)\zeta_1(s+3)+\dots$$
$$=\frac{1}{2}\frac{\Gamma(2r)\Gamma(r-s)}{\Gamma(r)\Gamma(2r-s)}\int_0^1\frac{x^{r-s-1}(1-x^s)}{1-x}\,dx, \quad \dots\dots\dots\dots(26)$$

provided that r is positive and greater than s.

20

ON THE EXPRESSION OF A NUMBER IN THE FORM $ax^2 + by^2 + cz^2 + du^2$

(*Proceedings of the Cambridge Philosophical Society*, XIX, 1917, 11—21)

1. It is well known that all positive integers can be expressed as the sum of four squares. This naturally suggests the question: *For what positive integral values of a, b, c, d can all positive integers be expressed in the form*

$$ax^2 + by^2 + cz^2 + du^2 ? \qquad (1 \cdot 1)$$

I prove in this paper that there are only 55 sets of values of a, b, c, d for which this is true.

The more general problem of finding all sets of values of a, b, c, d, for which all integers *with a finite number of exceptions* can be expressed in the form (1·1), is much more difficult and interesting. I have considered only very special cases of this problem, with two variables instead of four; namely, the cases in which (1·1) has one of the special forms

$$a(x^2 + y^2 + z^2) + bu^2, \qquad (1 \cdot 2)$$

and

$$a(x^2 + y^2) + b(z^2 + u^2). \qquad (1 \cdot 3)$$

These two cases are comparatively easy to discuss. In this paper I give the discussion of (1·2) only, reserving that of (1·3) for another paper.

2. Let us begin with the first problem. We can suppose, without loss of generality, that

$$a \leqslant b \leqslant c \leqslant d. \qquad (2 \cdot 1)$$

If $a > 1$, then 1 cannot be expressed in the form (1·1); and so

$$a = 1. \qquad (2 \cdot 2)$$

If $b > 2$, then 2 is an exception; and so

$$1 \leqslant b \leqslant 2. \qquad (2 \cdot 3)$$

We have therefore only to consider the two cases in which (1·1) has one or other of the forms

$$x^2 + y^2 + cz^2 + du^2, \quad x^2 + 2y^2 + cz^2 + du^2.$$

In the first case, if $c > 3$, then 3 is an exception; and so

$$1 \leqslant c \leqslant 3. \qquad (2 \cdot 31)$$

In the second case, if $c > 5$, then 5 is an exception; and so

$$2 \leqslant c \leqslant 5. \qquad (2 \cdot 32)$$

We can now distinguish 7 possible cases.

(2·41) $x^2 + y^2 + z^2 + du^2$.

If $d > 7$, 7 is an exception; and so

$$1 \leqslant d \leqslant 7. \qquad (2 \cdot 411)$$

(2·42) $x^2+y^2+2z^2+du^2$.

If $d>14$, 14 is an exception; and so

$$2 \leqslant d \leqslant 14. \qquad (2·421)$$

(2·43) $x^2+y^2+3z^2+du^2$.

If $d>6$, 6 is an exception; and so

$$3 \leqslant d \leqslant 6. \qquad (2·431)$$

(2·44) $x^2+2y^2+2z^2+du^2$.

If $d>7$, 7 is an exception; and so

$$2 \leqslant d \leqslant 7. \qquad (2·441)$$

(2·45) $x^2+2y^2+3z^2+du^2$.

If $d>10$, 10 is an exception; and so

$$3 \leqslant d \leqslant 10. \qquad (2·451)$$

(2·46) $x^2+2y^2+4z^2+du^2$.

If $d>14$, 14 is an exception; and so

$$4 \leqslant d \leqslant 14. \qquad (2·461)$$

(2·47) $x^2+2y^2+5z^2+du^2$.

If $d>10$, 10 is an exception; and so

$$5 \leqslant d \leqslant 10. \qquad (2·471)$$

We have thus eliminated all possible sets of values of a, b, c, d, except the following 55:

1, 1, 1, 1	1, 2, 3, 5	1, 2, 4, 8
1, 1, 1, 2	1, 2, 4, 5	1, 2, 5, 8
1, 1, 2, 2	1, 2, 5, 5	1, 1, 2, 9
1, 2, 2, 2	1, 1, 1, 6	1, 2, 3, 9
1, 1, 1, 3	1, 1, 2, 6	1, 2, 4, 9
1, 1, 2, 3	1, 2, 2, 6	1, 2, 5, 9
1, 2, 2, 3	1, 1, 3, 6	1, 1, 2, 10
1, 1, 3, 3	1, 2, 3, 6	1, 2, 3, 10
1, 2, 3, 3	1, 2, 4, 6	1, 2, 4, 10
1, 1, 1, 4	1, 2, 5, 6	1, 2, 5, 10
1, 1, 2, 4	1, 1, 1, 7	1, 1, 2, 11
1, 2, 2, 4	1, 1, 2, 7	1, 2, 4, 11
1, 1, 3, 4	1, 2, 2, 7	1, 1, 2, 12
1, 2, 3, 4	1, 2, 3, 7	1, 2, 4, 12
1, 2, 4, 4	1, 2, 4, 7	1, 1, 2, 13
1, 1, 1, 5	1, 2, 5, 7	1, 2, 4, 13
1, 1, 2, 5	1, 1, 2, 8	1, 1, 2, 14
1, 2, 2, 5	1, 2, 3 8	1, 2, 4, 14
1, 1, 3, 5		

Of these 55 forms, the 12 forms

1, 1, 1, 2	1, 1, 2, 4	1, 2, 4, 8
1, 1, 2, 2	1, 2, 2, 4	1, 1, 3, 3
1, 2, 2, 2	1, 2, 4, 4	1, 2, 3, 6
1, 1, 1, 4	1, 1, 2, 8	1, 2, 5, 10

have been already considered by Liouville and Pepin*.

3. I shall now prove that all integers can be expressed in each of the 55 forms. In order to prove this we shall consider the seven cases (2·41)—(2·47) of the previous section separately. We shall require the following results concerning ternary quadratic arithmetical forms.

The necessary and sufficient condition that a number *cannot* be expressed in the form

$$x^2 + y^2 + z^2 \quad \text{.............................(3·1)}$$

is that it should be of the form

$$4^\lambda(8\mu + 7), \qquad (\lambda = 0, 1, 2, \ldots, \quad \mu = 0, 1, 2, \ldots). \quad \text{......(3·11)}$$

Similarly the necessary and sufficient conditions that a number *cannot* be expressed in the forms

$$x^2 + y^2 + 2z^2, \quad \text{.............................(3·2)}$$
$$x^2 + y^2 + 3z^2, \quad \text{.............................(3·3)}$$
$$x^2 + 2y^2 + 2z^2, \quad \text{.............................(3·4)}$$
$$x^2 + 2y^2 + 3z^2, \quad \text{.............................(3·5)}$$
$$x^2 + 2y^2 + 4z^2, \quad \text{.............................(3·6)}$$
$$x^2 + 2y^2 + 5z^2, \quad \text{.............................(3·7)}$$

are that it should be of the forms

$$4^\lambda(16\mu + 14), \quad \text{..........................(3·21)}$$
$$9^\lambda(9\mu + 6), \quad \text{..........................(3·31)}$$
$$4^\lambda(8\mu + 7), \quad \text{..........................(3·41)}$$
$$4^\lambda(16\mu + 10), \quad \text{..........................(3·51)}$$
$$4^\lambda(16\mu + 14), \quad \text{..........................(3·61)}$$
$$25^\lambda(25\mu + 10) \text{ or } 25^\lambda(25\mu + 15)\dagger. \quad \text{..............(3·71)}$$

The result concerning $x^2 + y^2 + z^2$ is due to Cauchy: for a proof see Landau, *Handbuch der Lehre von der Verteilung der Primzahlen*, p. 550. The other results can be proved in an analogous manner. The form $x^2 + y^2 + 2z^2$ has been considered by Lebesgue, and the form $x^2 + y^2 + 3z^2$ by Dirichlet. For references see Bachmann, *Zahlentheorie*, Vol. IV, p. 149.

* There are a large number of short notes by Liouville in Vols. V—VIII of the second series of his Journal. See also Pepin, *ibid.*, Ser. 4, Vol. VI, pp. 1—67. The object of the work of Liouville and Pepin is rather different from mine, viz. to determine, in a number of special cases, explicit formulæ for the number of representations, in terms of other arithmetical functions.

† Results (3·11)—(3·71) may tempt us to suppose that there are similar simple results for the form $ax^2+by^2+cz^2$, whatever are the values of a, b, c. It appears, however, that

4. We proceed to consider the seven cases (2·41)—(2·47). In the first case we have to shew that any number N can be expressed in the form

$$N = x^2 + y^2 + z^2 + du^2, \qquad\qquad (4{\cdot}1)$$

d being any integer between 1 and 7 inclusive.

If N is not of the form $4^\lambda(8\mu + 7)$, we can satisfy (4·1) with $u = 0$. We may therefore suppose that $N = 4^\lambda(8\mu + 7)$.

First, suppose that d has one of the values 1, 2, 4, 5, 6. Take $u = 2^\lambda$. Then the number

$$N - du^2 = 4^\lambda(8\mu + 7 - d)$$

is plainly not of the form $4^\lambda(8\mu + 7)$, and is therefore expressible in the form $x^2 + y^2 + z^2$.

Next, let $d = 3$. If $\mu = 0$, take $u = 2^\lambda$. Then

$$N - du^2 = 4^{\lambda+1}.$$

If $\mu \geqslant 1$, take $u = 2^{\lambda+1}$. Then

$$N - du^2 = 4^\lambda(8\mu - 5).$$

In neither of these cases is $N - du^2$ of the form $4^\lambda(8\mu + 7)$, and therefore in either case it can be expressed in the form $x^2 + y^2 + z^2$.

in most cases there are no such simple results. For instance, the numbers which are not of the form $x^2+2y^2+10z^2$ are those belonging to one or other of the *four* classes

$$25^\lambda(8\mu+7), \quad 25^\lambda(25\mu+5), \quad 25^\lambda(25\mu+15), \quad 25^\lambda(25\mu+20).$$

Here some of the numbers of the first class belong also to one of the next three classes.

Again, the even numbers which are not of the form $x^2+y^2+10z^2$ are the numbers

$$4^\lambda(16\mu+6),$$

while the odd numbers that are not of that form, viz.

3, 7, 21, 31, 33, 43, 67, 79, 87, 133, 217, 219, 223, 253, 307, 391, ...

do not seem to obey any simple law.

I have succeeded in finding a law in the following six simple cases:

$$x^2 + y^2 + 4z^2,$$
$$x^2 + y^2 + 5z^2,$$
$$x^2 + y^2 + 6z^2,$$
$$x^2 + y^2 + 8z^2,$$
$$x^2 + 2y^2 + 6z^2,$$
$$x^2 + 2y^2 + 8z^2.$$

The numbers which are not of these forms are the numbers

$$4^\lambda(8\mu+7) \quad \textit{or} \quad 8\mu+3,$$
$$4^\lambda(8\mu+3),$$
$$9^\lambda(9\mu+3),$$
$$4^\lambda(16\mu+14), \quad 16\mu+6, \quad \textit{or} \quad 4\mu+3,$$
$$4^\lambda(8\mu+5),$$
$$4^\lambda(8\mu+7) \quad \textit{or} \quad 8\mu+5.$$

Finally, let $d=7$. If μ is equal to 0, 1, or 2, take $u=2^\lambda$. Then $N-du^2$ is equal to 0, $2.4^{\lambda+1}$, or $4^{\lambda+2}$. If $\mu \geqslant 3$, take $u=2^{\lambda+1}$. Then

$$N-du^2=4^\lambda(8\mu-21).$$

Therefore in either case $N-du^2$ can be expressed in the form $x^2+y^2+z^2$.

Thus in all cases N is expressible in the form (4·1). Similarly we can dispose of the remaining cases, with the help of the results stated in § 3. Thus in discussing (2·42) we use the theorem that every number not of the form (3·21) can be expressed in the form (3·2). The proofs differ only in detail, and it is not worth while to state them at length.

5. We have seen that all integers without any exception can be expressed in the form

$$m(x^2+y^2+z^2)+nu^2, \qquad (5\cdot 1)$$

when $m=1,\quad 1\leqslant n\leqslant 7,$

and $m=2,\quad n=1.$

We shall now consider the values of m and n for which all integers *with a finite number of exceptions* can be expressed in the form (5·1).

In the first place m must be 1 or 2. For, if $m>2$, we can choose an integer ν so that

$$nu^2 \not\equiv \nu \pmod m$$

for all values of u. Then

$$\frac{(m\mu+\nu)-nu^2}{m},$$

where μ is any positive integer, is not an integer; and so $m\mu+\nu$ can certainly not be expressed in the form (5·1).

We have therefore only to consider the two cases in which m is 1 or 2. First let us consider the form

$$x^2+y^2+z^2+nu^2. \qquad (5\cdot 2)$$

I shall shew that, when n has any of the values

$$1,\ 4,\ 9,\ 17,\ 25,\ 36,\ 68,\ 100, \qquad (5\cdot 21)$$

or is of any of the forms

$$4k+2,\quad 4k+3,\quad 8k+5,\quad 16k+12,\quad 32k+20, \qquad (5\cdot 22)$$

then all integers save a finite number, and in fact all integers from $4n$ onwards at any rate, can be expressed in the form (5·2); but that for the remaining values of n there is an infinity of integers which cannot be expressed in the form required.

In proving the first result we need obviously only consider numbers of the form $4^\lambda(8\mu+7)$ greater than n, since otherwise we may take $u=0$. The numbers of this form less than n are plainly among the exceptions.

6. I shall consider the various cases which may arise in order of simplicity.

(6·1) $n\equiv 0 \pmod 8$.

There are an infinity of exceptions. For suppose that

$$N=8\mu+7.$$

Then the number $$N - nu^2 \equiv 7 \pmod 8$$
cannot be expressed in the form $x^2 + y^2 + z^2$.

(6·2) $n \equiv 2 \pmod 4$.

There is only a finite number of exceptions. In proving this we may suppose that $N = 4^\lambda (8\mu + 7)$. Take $u = 1$. Then the number
$$N - nu^2 = 4^\lambda (8\mu + 7) - n$$
is congruent to 1, 2, 5, or 6 to modulus 8, and so can be expressed in the form $x^2 + y^2 + z^2$.

Hence the only numbers which cannot be expressed in the form (5·2) in this case are the numbers of the form $4^\lambda (8\mu + 7)$ not exceeding n.

(6·3) $n \equiv 5 \pmod 8$.

There is only a finite number of exceptions. We may suppose again that $N = 4^\lambda (8\mu + 7)$. First, let $\lambda \neq 1$. Take $u = 1$. Then
$$N - nu^2 = 4^\lambda (8\mu + 7) - n \equiv 2 \text{ or } 3 \pmod 8.$$
If $\lambda = 1$ we cannot take $u = 1$, since
$$N - n \equiv 7 \pmod 8;$$
so we take $u = 2$. Then
$$N - nu^2 = 4^\lambda (8\mu + 7) - 4n \equiv 8 \pmod{32}.$$
In either of these cases $N - nu^2$ is of the form $x^2 + y^2 + z^2$.

Hence the only numbers which cannot be expressed in the form (5·2) are those of the form $4^\lambda (8\mu + 7)$ not exceeding n, and those of the form $4(8\mu + 7)$ lying between n and $4n$.

(6·4) $n \equiv 3 \pmod 4$.

There is only a finite number of exceptions. Take
$$N = 4^\lambda (8\mu + 7).$$
If $\lambda \geqslant 1$, take $u = 1$. Then
$$N - nu^2 \equiv 1 \text{ or } 5 \pmod 8.$$
If $\lambda = 0$, take $u = 2$. Then
$$N - nu^2 \equiv 3 \pmod 8.$$
In either case the proof is completed as before.

In order to determine precisely which are the exceptional numbers, we must consider more particularly the numbers between n and $4n$ for which $\lambda = 0$. For these u must be 1, and
$$N - nu^2 \equiv 0 \pmod 4.$$
But the numbers which are multiples of 4 and which cannot be expressed in the form $x^2 + y^2 + z^2$ are the numbers
$$4^\kappa (8\nu + 7), \qquad (\kappa = 1, 2, 3, \ldots, \nu = 0, 1, 2, 3, \ldots).$$
The exceptions required are therefore those of the numbers
$$n + 4^\kappa (8\nu + 7) \quad \text{..........................(6·41)}$$
which lie between n and $4n$ and are of the form
$$8\mu + 7 \quad \text{............................(6·42)}.$$

Now in order that (6·41) may be of the form (6·42), κ must be 1 if n is of the form $8k+3$, and κ may have any of the values 2, 3, 4, ... if n is of the form $8k+7$. Thus the only numbers which cannot be expressed in the form (5·2), in this case, are those of the form $4^\lambda(8\mu+7)$ less than n and those of the form

$$n+4^\kappa(8\nu+7), \qquad (\nu=0,1,2,3,\dots),$$

lying between n and $4n$, where $\kappa=1$ if n is of the form $8k+3$, and $\kappa>1$ if n is of the form $8k+7$.

(6·5) $n\equiv 1 \pmod 8$.

In this case we have to prove that

(i) if $n\geqslant 33$, there is an infinity of integers which cannot be expressed in the form (5·2);

(ii) if n is 1, 9, 17, or 25, there is only a finite number of exceptions.

In order to prove (i) suppose that $N=7.4^\lambda$. Then obviously u cannot be zero. But if u is not zero u^2 is always of the form $4^\kappa(8\nu+1)$. Hence

$$N-nu^2=7.4^\lambda-n.4^\kappa(8\nu+1).$$

Since $n\geqslant 33$, λ must be greater than or equal to $\kappa+2$, to ensure that the right-hand side shall not be negative. Hence

$$N-nu^2=4^\kappa(8k+7),$$

where
$$k=14.4^{\lambda-\kappa-2}-n\nu-\tfrac{1}{8}(n+7)$$

is an integer; and so $N-nu^2$ is not of the form $x^2+y^2+z^2$.

In order to prove (ii) we may suppose, as usual, that

$$N=4^\lambda(8\mu+7).$$

If $\lambda=0$, take $u=1$. Then

$$N-nu^2=8\mu+7-n\equiv 6 \pmod 8.$$

If $\lambda\geqslant 1$, take $u=2^{\lambda-1}$. Then

$$N-nu^2=4^{\lambda-1}(8k+3),$$

where
$$k=4(\mu+1)-\tfrac{1}{8}(n+7).$$

In either case the proof may be completed as before. Thus the only numbers which cannot be expressed in the form (5·2), in this case, are those of the form $8\mu+7$ not exceeding n. In other words, there is no exception when $n=1$; 7 is the only exception when $n=9$; 7 and 15 are the only exceptions when $n=17$; 7, 15 and 23 are the only exceptions when $n=25$.

(6·6) $n\equiv 4 \pmod{32}$.

By arguments similar to those used in (6·5), we can shew that

(i) if $n\geqslant 132$, there is an infinity of integers which cannot be expressed in the form (5·2);

(ii) if n is equal to 4, 36, 68, or 100, there is only a finite number of exceptions, namely the numbers of the form $4^\lambda(8\mu+7)$ not exceeding n.

(6·7) $n \equiv 20 \pmod{32}$.

By arguments similar to those used in (6·3), we can shew that the only numbers which cannot be expressed in the form (5·2) are those of the form $4^\lambda(8\mu+7)$ not exceeding n, and those of the form $4^2(8\mu+7)$ lying between n and $4n$.

(6·8) $n \equiv 12 \pmod{16}$.

By arguments similar to those used in (6·4), we can shew that the only numbers which cannot be expressed in the form (5·2) are those of the form $4^\lambda(8\mu+7)$ less than n, and those of the form

$$n + 4^\kappa(8\nu+7), \qquad (\nu = 0, 1, 2, 3, \ldots),$$

lying between n and $4n$, where $\kappa = 2$ if n is of the form $4(8k+3)$ and $\kappa > 2$ if n is of the form $4(8k+7)$.

We have thus completed the discussion of the form (5·2), and determined the exceptional values of N precisely whenever they are finite in number.

7. We shall proceed to consider the form

$$2(x^2+y^2+z^2)+nu^2. \qquad \text{.........................(7·1)}$$

In the first place n must be odd; otherwise the odd numbers cannot be expressed in this form. Suppose then that n is odd. I shall shew that all integers save a finite number can be expressed in the form (7·1); and that the numbers which cannot be so expressed are

(i) the odd numbers less than n,

(ii) the numbers of the form $4^\lambda(16\mu+14)$ less than $4n$,

(iii) the numbers of the form $n+4^\lambda(16\mu+14)$ greater than n and less than $9n$,

(iv) the numbers of the form

$$cn + 4^\kappa(16\nu+14), \qquad (\nu = 0, 1, 2, 3, \ldots),$$

greater than $9n$ and less than $25n$, where $c = 1$ if $n \equiv 1 \pmod 4$, $c = 9$ if $n \equiv 3 \pmod 4$, $\kappa = 2$ if $n^2 \equiv 1 \pmod{16}$, and $\kappa > 2$ if $n^2 \equiv 9 \pmod{16}$.

First, let us suppose N even. Then, since n is odd and N is even, it is clear that u must be even. Suppose then that

$$u = 2v, \quad N = 2M.$$

We have to shew that M can be expressed in the form

$$x^2+y^2+z^2+2nv^2. \qquad \text{..........................(7·2)}$$

Since $2n \equiv 2 \pmod 4$, it follows from (6·2) that all integers except those which are less than $2n$ and of the form $4^\lambda(8\mu+7)$ can be expressed in the form (7·2). Hence the only *even* integers which cannot be expressed in the form (7·1) are those of the form $4^\lambda(16\mu+14)$ less than $4n$.

This completes the discussion of the case in which N is even. If N is odd the discussion is more difficult. In the first place, all odd numbers less than n are plainly among the exceptions. Secondly, since n and N are both odd, u must also be odd. We can therefore suppose that

$$N = n + 2M, \quad u^2 = 1 + 8\Delta,$$

where Δ is an integer of the form $\frac{1}{2}k(k+1)$, so that Δ may assume the values 0, 1, 3, 6, ... And we have to consider whether $n + 2M$ can be expressed in the form

$$2(x^2+y^2+z^2) + n(1+8\Delta),$$

or M in the form

$$x^2+y^2+z^2+4n\Delta. \qquad (7\cdot3)$$

If M is not of the form $4^\lambda(8\mu+7)$, we can take $\Delta = 0$. If it is of this form, and less than $4n$, it is plainly an exception. These numbers give rise to the exceptions specified in (iii) of section 7. We may therefore suppose that M is of the form $4^\lambda(8\mu+7)$ and greater than $4n$.

8. In order to complete the discussion, we must consider the three cases in which $n \equiv 1 \pmod 8$, $n \equiv 5 \pmod 8$, and $n \equiv 3 \pmod 4$ separately.

(8·1) $n \equiv 1 \pmod 8$.

If λ is equal to 0, 1, or 2, take $\Delta = 1$. Then

$$M - 4n\Delta = 4^\lambda(8\mu+7) - 4n$$

is of one of the forms

$$8\nu+3, \quad 4(8\nu+3), \quad 4(8\nu+6).$$

If $\lambda \geqslant 3$ we cannot take $\Delta = 1$, since $M - 4n\Delta$ assumes the form $4(8\nu+7)$; so we take $\Delta = 3$. Then

$$M - 4n\Delta = 4^\lambda(8\mu+7) - 12n$$

is of the form $4(8\nu+5)$. In either of these cases $M - 4n\Delta$ is of the form $x^2+y^2+z^2$. Hence the only values of M, other than those already specified, which cannot be expressed in the form (7·3), are those of the form

$$4^\kappa(8\nu+7), \qquad (\nu = 0, 1, 2, \ldots, \kappa > 2),$$

lying between $4n$ and $12n$. In other words, the only numbers greater than $9n$ which cannot be expressed in the form (7·1), in this case, are the numbers of the form

$$n + 4^\kappa(8\nu+7), \qquad (\nu = 0, 1, 2, \ldots, \kappa > 2),$$

lying between $9n$ and $25n$.

(8·2) $n \equiv 5 \pmod 8$.

If $\lambda \neq 2$, take $\Delta = 1$. Then

$$M - 4n\Delta = 4^\lambda(8\mu+7) - 4n$$

is of one of the forms

$$8\nu+3, \quad 4(8\nu+2), \quad 4(8\nu+3).$$

If $\lambda = 2$, we cannot take $\Delta = 1$, since $M - 4n\Delta$ assumes the form $4(8\nu + 7)$; so we take $\Delta = 3$. Then

$$M - 4n\Delta = 4^{\lambda}(8\mu + 7) - 12n$$

is of the form $4(8\nu + 5)$. In either of these cases $M - 4n\Delta$ is of the form $x^2 + y^2 + z^2$. Hence the only values of M, other than those already specified, which cannot be expressed in the form (7·3), are those of the form $16(8\mu + 7)$ lying between $4n$ and $12n$. In other words, the only numbers greater than $9n$ which cannot be expressed in the form (7·1), in this case, are the numbers of the form $n + 4^2(16\mu + 14)$ lying between $9n$ and $25n$.

(8·3) $n \equiv 3 \pmod 4$.

If $\lambda \neq 1$, take $\Delta = 1$. Then

$$M - 4n\Delta = 4^{\lambda}(8\mu + 7) - 4n$$

is of one of the forms $8\nu + 3, \quad 4(4\nu + 1)$.

If $\lambda = 1$, take $\Delta = 3$. Then

$$M - 4n\Delta = 4(8\mu + 7) - 12n$$

is of the form $4(4\nu + 2)$. In either of these cases $M - 4n\Delta$ is of the form $x^2 + y^2 + z^2$.

This completes the proof that there is only a finite number of exceptions. In order to determine what they are in this case, we have to consider the values of M, between $4n$ and $12n$, for which $\Delta = 1$ and

$$M - 4n\Delta = 4(8\mu + 7 - n) \equiv 0 \pmod{16}.$$

But the numbers which are multiples of 16 and which cannot be expressed in the form $x^2 + y^2 + z^2$ are the numbers

$$4^{\kappa}(8\nu + 7), \qquad (\kappa = 2, 3, 4, \ldots, \nu = 0, 1, 2, \ldots).$$

The exceptional values of M required are therefore those of the numbers

$$4n + 4^{\kappa}(8\nu + 7) \quad \ldots\ldots\ldots\ldots\ldots\ldots\ldots\ldots (8\cdot31)$$

which lie between $4n$ and $12n$ and are of the form

$$4(8\mu + 7). \quad \ldots\ldots\ldots\ldots\ldots\ldots\ldots\ldots (8\cdot32)$$

But in order that (8·31) may be of the form (8·32), κ must be 2 if n is of the form $8k + 3$, and κ may have any of the values 3, 4, 5, ... if n is of the form $8k + 7$. It follows that the only numbers greater than $9n$ which cannot be expressed in the form (7·1), in this case, are the numbers of the form

$$9n + 4^{\kappa}(16\nu + 14), \qquad (\nu = 0, 1, 2, \ldots),$$

lying between $9n$ and $25n$, where $\kappa = 2$ if n is of the form $8k + 3$, and $\kappa > 2$ if n is of the form $8k + 7$.

This completes the proof of the results stated in section 7.

21

ON CERTAIN TRIGONOMETRICAL SUMS AND THEIR APPLICATIONS IN THE THEORY OF NUMBERS

(*Transactions of the Cambridge Philosophical Society*, XXII, No. 13, 1918, 259—276)

1. The trigonometrical sums with which this paper is concerned are of the type

$$c_s(n) = \sum_{\lambda} \cos \frac{2\pi\lambda n}{s},$$

where λ is prime to s and not greater than s. It is plain that

$$c_s(n) = \Sigma\, \alpha^n,$$

where α is a primitive root of the equation

$$x^s - 1 = 0.$$

These sums are obviously of very great interest, and a few of their properties have been discussed already*. But, so far as I know, they have never been considered from the point of view which I adopt in this paper; and I believe that all the results which it contains are new.

My principal object is to obtain expressions for a variety of well-known arithmetical functions of n in the form of a series

$$\sum_{s} a_s c_s(n).$$

A typical formula is

$$\sigma(n) = \frac{\pi^2 n}{6}\left\{\frac{c_1(n)}{1^2} + \frac{c_2(n)}{2^2} + \frac{c_3(n)}{3^2} + \dots\right\},$$

where $\sigma(n)$ is the sum of the divisors of n. I give two distinct methods for the proof of this and a large variety of similar formulæ. The majority of my formulæ are "elementary" in the technical sense of the word—they can (that is to say) be proved by a combination of processes involving only finite algebra and simple general theorems concerning infinite series. There are however some which are of a "deeper" character, and can only be proved by means of theorems which seem to depend essentially on the theory of analytic functions. A typical formula of this class is

$$c_1(n) + \tfrac{1}{2}c_2(n) + \tfrac{1}{3}c_3(n) + \dots = 0,$$

a formula which depends upon, and is indeed substantially equivalent to, the "Prime Number Theorem" of Hadamard and de la Vallée Poussin.

* See, e.g., Dirichlet-Dedekind, *Vorlesungen über Zahlentheorie*, ed. 4, Supplement VII, pp. 360—370.

Many of my formulæ are intimately connected with those of my previous paper "On certain arithmetical functions", published in 1916 in these *Transactions**. They are also connected (in a manner pointed out in § 15) with a joint paper by Mr Hardy and myself, "Asymptotic Formulæ in Combinatory Analysis", in course of publication in the *Proceedings of the London Mathematical Society*†.

2. Let $F(u, v)$ be any function of u and v, and let

$$D(n) = \sum_{\delta} F(\delta, \delta'), \tag{2·1}$$

where δ is a divisor of n and $\delta\delta' = n$. For instance

$$D(1) = F(1, 1); \quad D(2) = F(1, 2) + F(2, 1);$$
$$D(3) = F(1, 3) + F(3, 1); \quad D(4) = F(1, 4) + F(2, 2) + F(4, 1);$$
$$D(5) = F(1, 5) + F(5, 1); \quad D(6) = F(1, 6) + F(2, 3) + F(3, 2) + F(6, 1); \ldots\ldots$$

It is clear that $D(n)$ may also be expressed in the form

$$D(n) = \sum_{\delta} F(\delta', \delta). \tag{2·2}$$

Suppose now that

$$\eta_s(n) = \sum_{0}^{s-1} \cos \frac{2\pi\nu n}{s}, \tag{2·3}$$

so that $\eta_s(n) = s$ if s is a divisor of n and $\eta_s(n) = 0$ otherwise. Then

$$D(n) = \sum_{1}^{t} \frac{1}{\nu} \eta_\nu(n) F\left(\nu, \frac{n}{\nu}\right)‡, \tag{2·4}$$

where t is any number not less than n. Now let

$$c_s(n) = \sum_{\lambda} \cos \frac{2\pi\lambda n}{s}, \tag{2·5}$$

where λ is prime to s and does not exceed s; e.g.

$$c_1(n) = 1; \quad c_2(n) = \cos n\pi; \quad c_3(n) = 2\cos \tfrac{2}{3}n\pi;$$
$$c_4(n) = 2\cos \tfrac{1}{2}n\pi; \quad c_5(n) = 2\cos \tfrac{2}{5}n\pi + 2\cos \tfrac{4}{5}n\pi;$$
$$c_6(n) = 2\cos \tfrac{1}{3}n\pi; \quad c_7(n) = 2\cos \tfrac{2}{7}n\pi + 2\cos \tfrac{4}{7}n\pi + 2\cos \tfrac{6}{7}n\pi;$$
$$c_8(n) = 2\cos \tfrac{1}{4}n\pi + 2\cos \tfrac{3}{4}n\pi; \quad c_9(n) = 2\cos \tfrac{2}{9}n\pi + 2\cos \tfrac{4}{9}n\pi + 2\cos \tfrac{8}{9}n\pi;$$
$$c_{10}(n) = 2\cos \tfrac{1}{5}n\pi + 2\cos \tfrac{3}{5}n\pi; \quad \ldots\ldots$$

It follows from (2·3) and (2·5) that

$$\eta_s(n) = \sum_{\delta} c_\delta(n), \tag{2·6}$$

where δ is a divisor of s; and hence§ that

$$c_s(n) = \sum_{\delta} \mu(\delta') \eta_\delta(n), \tag{2·7}$$

* [No. 18 of this volume]. † [No. 36 of this volume].

‡ $\sum_{1}^{t}$ is to be understood as meaning $\sum_{1}^{[t]}$, where $[t]$ denotes as usual the greatest integer in t.

§ See Landau, *Handbuch der Lehre von der Verteilung der Primzahlen*, p. 577.

where δ is a divisor of s, $\delta\delta' = s$, and

(2·8) $$\Sigma \frac{\mu(\nu)}{\nu^s} = \frac{1}{\zeta(s)},$$

$\zeta(s)$ being the Riemann Zeta-function. In particular

$$c_1(n) = \eta_1(n);\ c_2(n) = \eta_2(n) - \eta_1(n);\ c_3(n) = \eta_3(n) - \eta_1(n);$$
$$c_4(n) = \eta_4(n) - \eta_2(n);\ c_5(n) = \eta_5(n) - \eta_1(n);\ \ldots\ldots$$

But from (2·3) we know that $\eta_\delta(n) = 0$ if δ is not a divisor of n; and so we can suppose that, in (2·7), δ is a common divisor of n and s. It follows that

$$|c_s(n)| \leqslant \Sigma\delta,$$

where δ is a divisor of n; so that

(2·9) $$c_\nu(n) = O(1)$$

if n is fixed and $\nu \to \infty$. Since

$$\eta_s(n) = \eta_s(n+s);\ c_s(n) = c_s(n+s),$$

the values of $c_s(n)$ for $n = 1, 2, 3, \ldots$ can be shewn conveniently by writing

$$c_1(n) = \overline{1};\ c_2(n) = \overline{-1, 1};\ c_3(n) = \overline{-1, -1, 2};$$
$$c_4(n) = \overline{0, -2, 0, 2};\ c_5(n) = \overline{-1, -1, -1, -1, 4};$$
$$c_6(n) = \overline{1, -1, -2, -1, 1, 2};\ c_7(n) = \overline{-1, -1, -1, -1, -1, -1, 6};$$
$$c_8(n) = \overline{0, 0, 0, -4, 0, 0, 0, 4};\ c_9(n) = \overline{0, 0, -3, 0, 0, -3, 0, 0, 6};$$
$$c_{10}(n) = \overline{1, -1, 1, -1, -4, -1, 1, -1, 1, 4};\ \ldots\ldots$$

the meaning of the third formula, for example, being that $c_3(1) = -1$, $c_3(2) = -1$, $c_3(3) = 2$, and that these values are then repeated periodically.

It is plain that we have also

(2·91) $$c_\nu(n) = O(1),$$

when ν is fixed and $n \to \infty$.

3. Substituting (2·6) in (2·4), and collecting the coefficients of $c_1(n)$, $c_2(n)$, $c_3(n)$, ..., we find that

(3·1)
$$D(n) = c_1(n) \sum_1^t \frac{1}{\nu} F\left(\nu, \frac{n}{\nu}\right) + c_2(n) \sum_1^{\frac{1}{2}t} \frac{1}{2\nu} F\left(2\nu, \frac{n}{2\nu}\right) + c_3(n) \sum_1^{\frac{1}{3}t} \frac{1}{3\nu} F\left(3\nu, \frac{n}{3\nu}\right) + \ldots,$$

where t is any number not less than n. If we use (2·2) instead of (2·1) we obtain another expression, viz.

(3·2)
$$D(n) = c_1(n) \sum_1^t \frac{1}{\nu} F\left(\frac{n}{\nu}, \nu\right) + c_2(n) \sum_1^{\frac{1}{2}t} \frac{1}{2\nu} F\left(\frac{n}{2\nu}, 2\nu\right) + c_3(n) \sum_1^{\frac{1}{3}t} \frac{1}{3\nu} F\left(\frac{n}{3\nu}, 3\nu\right) + \ldots,$$

where t is any number not less than n.

Suppose now that

$$F_1(u,v)=F(u,v)\log u,\quad F_2(u,v)=F(u,v)\log v.$$

Then we have

$$\begin{aligned} D(n)\log n &= \sum_{\delta} F(\delta,\delta')\log n=\sum_{\delta} F(\delta,\delta')\log(\delta\delta') \\ &= \sum_{\delta} F_1(\delta,\delta')+\sum_{\delta} F_2(\delta,\delta'), \end{aligned}$$

where δ is a divisor of n and $\delta\delta'=n$.

Now for $\sum_{\delta} F_1(\delta,\delta')$ we shall write the expression corresponding to (3·1) and for $\sum_{\delta} F_2(\delta,\delta')$ the expression corresponding to (3·2). Then we have

$$\begin{aligned} \text{(3·3)}\qquad D(n)\log n &= c_1(n)\sum_{1}^{r}\frac{\log\nu}{\nu}F\left(\nu,\frac{n}{\nu}\right)+c_2(n)\sum_{1}^{\frac{1}{2}r}\frac{\log 2\nu}{2\nu}F\left(2\nu,\frac{n}{2\nu}\right) \\ &+c_3(n)\sum_{1}^{\frac{1}{3}r}\frac{\log 3\nu}{3\nu}F\left(3\nu,\frac{n}{3\nu}\right)+\ldots+c_1(n)\sum_{1}^{t}\frac{\log\nu}{\nu}F\left(\frac{n}{\nu},\nu\right) \\ &+c_2(n)\sum_{1}^{\frac{1}{2}t}\frac{\log 2\nu}{2\nu}F\left(\frac{n}{2\nu},2\nu\right)+c_3(n)\sum_{1}^{\frac{1}{3}t}\frac{\log 3\nu}{3\nu}F\left(\frac{n}{3\nu},3\nu\right)+\ldots, \end{aligned}$$

where r and t are any two numbers not less than n. If, in particular, $F(u,v)=F(v,u)$, then (3·3) reduces to

$$\begin{aligned} \text{(3·4)}\qquad \tfrac{1}{2}D(n)\log n &= c_1(n)\sum_{1}^{t}\frac{\log\nu}{\nu}F\left(\nu,\frac{n}{\nu}\right) \\ &+c_2(n)\sum_{1}^{\frac{1}{2}t}\frac{\log 2\nu}{2\nu}F\left(2\nu,\frac{n}{2\nu}\right)+c_3(n)\sum_{1}^{\frac{1}{3}t}\frac{\log 3\nu}{3\nu}F\left(3\nu,\frac{n}{3\nu}\right)+\ldots, \end{aligned}$$

where t is any number not less than n.

4. We may also write $D(n)$ in the form

$$\text{(4·1)}\qquad D(n)=\sum_{\delta=1}^{u} F(\delta,\delta')+\sum_{\delta=1}^{v} F(\delta',\delta),$$

where δ is a divisor of n, $\delta\delta'=n$, and u, v are any two positive numbers such that $uv=n$, it being understood that, if u and v are both integral, a term $F(u,v)$ is to be subtracted from the right-hand side. Hence (with the same conventions)

$$D(n)=\sum_{1}^{u}\frac{1}{\nu}\eta_\nu(n)F\left(\nu,\frac{n}{\nu}\right)+\sum_{1}^{v}\frac{1}{\nu}\eta_\nu(n)F\left(\frac{n}{\nu},\nu\right).$$

Applying to this formula transformations similar to those of § 3, we obtain

$$\begin{aligned} \text{(4·2)}\qquad D(n) &= c_1(n)\sum_{1}^{u}\frac{1}{\nu}F\left(\nu,\frac{n}{\nu}\right)+c_2(n)\sum_{1}^{\frac{1}{2}u}\frac{1}{2\nu}F\left(2\nu,\frac{n}{2\nu}\right)+\ldots \\ &+c_1(n)\sum_{1}^{v}\frac{1}{\nu}F\left(\frac{n}{\nu},\nu\right)+c_2(n)\sum_{1}^{\frac{1}{2}v}\frac{1}{2\nu}F\left(\frac{n}{2\nu},2\nu\right)+\ldots, \end{aligned}$$

where u and v are positive numbers such that $uv=n$. If u and v are integers then a term $F(u,v)$ should be subtracted from the right-hand side.

If we suppose that $0 < u \leqslant 1$ then (4·2) reduces to (3·2), and if $0 < v \leqslant 1$ it reduces to (3·1). Another particular case of interest is that in which $u = v$. Then

$$(4\cdot3)\qquad D(n) = c_1(n)\sum_1^{\sqrt{n}} \frac{1}{\nu}\left\{F\left(\nu, \frac{n}{\nu}\right) + F\left(\frac{n}{\nu}, \nu\right)\right\} + c_2(n)\sum_1^{\frac{1}{2}\sqrt{n}} \frac{1}{2\nu}\left\{F\left(2\nu, \frac{n}{2\nu}\right) + F\left(\frac{n}{2\nu}, 2\nu\right)\right\} + \dots.$$

If n is a perfect square then $F(\sqrt{n}, \sqrt{n})$ should be subtracted from the right-hand side.

5. We shall now consider some special forms of these general equations. Suppose that $F(u, v) = v^s$, so that $D(n)$ is the sum $\sigma_s(n)$ of the sth powers of the divisors of n. Then from (3·1) and (3·2) we have

$$(5\cdot1)\qquad \frac{\sigma_s(n)}{n^s} = c_1(n)\sum_1^{t}\frac{1}{\nu^{s+1}} + c_2(n)\sum_1^{\frac{1}{2}t}\frac{1}{(2\nu)^{s+1}} + c_3(n)\sum_1^{\frac{1}{3}t}\frac{1}{(3\nu)^{s+1}} + \dots,$$

$$(5\cdot2)\qquad \sigma_s(n) = c_1(n)\sum_1^{t}\nu^{s-1} + c_2(n)\sum_1^{\frac{1}{2}t}(2\nu)^{s-1} + c_3(n)\sum_1^{\frac{1}{3}t}(3\nu)^{s-1} + \dots,$$

where t is any number not less than n: from (3·3)

$$(5\cdot3)\qquad \sigma_s(n)\log n = c_1(n)\sum_1^{r}\nu^{s-1}\log\nu + c_2(n)\sum_1^{\frac{1}{2}r}(2\nu)^{s-1}\log 2\nu + \dots + n^s\left\{c_1(n)\sum_1^{t}\frac{\log\nu}{\nu^{s+1}} + c_2(n)\sum_1^{\frac{1}{2}t}\frac{\log 2\nu}{(2\nu)^{s+1}} + \dots\right\},$$

where r and t are any two numbers not less than n: and from (4·2)

$$(5\cdot4)\qquad \sigma_s(n) = c_1(n)\sum_1^{u}\nu^{s-1} + c_2(n)\sum_1^{\frac{1}{2}u}(2\nu)^{s-1} + c_3(n)\sum_1^{\frac{1}{3}u}(3\nu)^{s-1} + \dots + n^s\left\{c_1(n)\sum_1^{v}\frac{1}{\nu^{s+1}} + c_2(n)\sum_1^{\frac{1}{2}v}\frac{1}{(2\nu)^{s+1}} + c_3(n)\sum_1^{\frac{1}{3}v}\frac{1}{(3\nu)^{s+1}} + \dots\right\},$$

where $uv = n$. If u and v are integers then u^s should be subtracted from the right-hand side.

Let $d(n) = \sigma_0(n)$ denote the number of divisors of n and $\sigma(n) = \sigma_1(n)$ the sum of the divisors of n. Then from (5·1)—(5·4) we obtain

$$(5\cdot5)\qquad d(n) = c_1(n)\sum_1^{t}\frac{1}{\nu} + c_2(n)\sum_1^{\frac{1}{2}t}\frac{1}{2\nu} + c_3(n)\sum_1^{\frac{1}{3}t}\frac{1}{3\nu} + \dots,$$

$$(5\cdot6)\qquad \sigma(n) = c_1(n)[t] + c_2(n)[\tfrac{1}{2}t] + c_3(n)[\tfrac{1}{3}t] + \dots,$$

$$(5\cdot7)\qquad \tfrac{1}{2}d(n)\log n = c_1(n)\sum_1^{t}\frac{\log\nu}{\nu} + c_2(n)\sum_1^{\frac{1}{2}t}\frac{\log 2\nu}{2\nu} + c_3(n)\sum_1^{\frac{1}{3}t}\frac{\log 3\nu}{3\nu} + \dots,$$

(5·8)

$$d(n) = c_1(n)\left\{\sum_1^{u}\frac{1}{\nu} + \sum_1^{v}\frac{1}{\nu}\right\} + c_2(n)\left\{\sum_1^{\frac{1}{2}u}\frac{1}{2\nu} + \sum_1^{\frac{1}{2}v}\frac{1}{2\nu}\right\} + c_3(n)\left\{\sum_1^{\frac{1}{3}u}\frac{1}{3\nu} + \sum_1^{\frac{1}{3}v}\frac{1}{3\nu}\right\} + \dots,$$

where $t \geqslant n$ and $uv = n$. If u and v are integers then 1 should be subtracted from the right-hand side of (5·8). Putting $u = v = \sqrt{n}$ in (5·8) we obtain

$$(5{\cdot}9)\qquad \tfrac{1}{2}d(n) = c_1(n)\sum_1^{\sqrt{n}}\frac{1}{\nu} + c_2(n)\sum_1^{\frac{1}{2}\sqrt{n}}\frac{1}{2\nu} + c_3(n)\sum_1^{\frac{1}{3}\sqrt{n}}\frac{1}{3\nu} + \dots,$$

unless n is a perfect square, when $\frac{1}{2}$ should be subtracted from the right-hand side. It may be interesting to note that, if we replace the left-hand side in (5·9) by

$$[\tfrac{1}{2} + \tfrac{1}{2}d(n)],$$

then the formula is true without exception.

6. So far our work has been based on elementary formal transformations, and no questions of convergence have arisen. We shall now consider the equation (5·1) more carefully. Let us suppose that $s > 0$. Then

$$\sum_1^{t/k}\frac{1}{(k\nu)^{s+1}} = \sum_1^{\infty}\frac{1}{(k\nu)^{s+1}} + O\left(\frac{1}{kt^s}\right) = \frac{1}{k^{s+1}}\zeta(s+1) + O\left(\frac{1}{kt^s}\right).$$

The number of terms in the right-hand side of (5·1) is $[t]$. Also we know that $c_\nu(n) = O(1)$ as $n \to \infty$. Hence

$$\frac{\sigma_s(n)}{n^s} = \zeta(s+1)\sum_{\nu=1}^{t}\frac{c_\nu(n)}{\nu^{s+1}} + O\left\{\frac{1}{t^s}\sum_{\nu=1}^{t}\frac{1}{\nu}\right\} = \zeta(s+1)\sum_1^{\infty}\frac{c_\nu(n)}{\nu^{s+1}} + O\left(\frac{\log t}{t^s}\right).$$

Making $t \to \infty$, we obtain

$$(6{\cdot}1)\qquad \sigma_s(n) = n^s\zeta(s+1)\left\{\frac{c_1(n)}{1^{s+1}} + \frac{c_2(n)}{2^{s+1}} + \frac{c_3(n)}{3^{s+1}} + \dots\right\},$$

if $s > 0$. Similarly, if we make $t \to \infty$ in (5·3), we obtain

$$\sigma_s(n)\log n = c_1(n)\sum_1^{r}\nu^{s-1}\log\nu + c_2(n)\sum_1^{\frac{1}{2}r}(2\nu)^{s-1}\log 2\nu + \dots$$
$$+ n^s\left\{c_1(n)\sum_1^{\infty}\frac{\log\nu}{\nu^{s+1}} + c_2(n)\sum_1^{\infty}\frac{\log 2\nu}{(2\nu)^{s+1}} + \dots\right\}.$$

But
$$\sum_1^{\infty}\frac{\log k\nu}{(k\nu)^{s+1}} = \frac{\log k}{k^{s+1}}\zeta(s+1) - \frac{1}{k^{s+1}}\zeta'(s+1).$$

It follows from this and (6·1) that

(6·2)
$$\sigma_s(n)\left\{\frac{\zeta'(s+1)}{\zeta(s+1)} + \log n\right\} = c_1(n)\sum_1^{t}\nu^{s-1}\log\nu + c_2(n)\sum_1^{\frac{1}{2}t}(2\nu)^{s-1}\log 2\nu + \dots$$
$$+ n^s\zeta(s+1)\left\{\frac{c_1(n)\log 1}{1^{s+1}} + \frac{c_2(n)\log 2}{2^{s+1}} + \frac{c_3(n)\log 3}{3^{s+1}} + \dots\right\},$$

where $s > 0$ and $t \geqslant n$. Putting $s = 1$ in (6·1) and (6·2) we obtain

$$(6{\cdot}3)\qquad \sigma(n) = \frac{\pi^2}{6}n\left\{\frac{c_1(n)}{1^2} + \frac{c_2(n)}{2^2} + \frac{c_3(n)}{3^2} + \dots\right\},$$

(6·4) $$\sigma(n)\left\{\frac{\zeta'(2)}{\zeta(2)}+\log n\right\}=\frac{\pi^2}{6}n\left\{\frac{c_1(n)}{1^2}\log 1+\frac{c_2(n)}{2^2}\log 2+\dots\right\}$$
$$+c_1(n)[t]\log 1+c_2(n)[\tfrac{1}{2}t]\log 2+\dots$$
$$+c_1(n)\log[t]!+c_2(n)\log[\tfrac{1}{2}t]!+\dots,$$

where $t \geqslant n$.

7. Since

(7·1) $$\sigma_s(n)=n^s\sigma_{-s}(n),$$

we may write (6·1) in the form

(7·2) $$\frac{\sigma_{-s}(n)}{\zeta(s+1)}=\frac{c_1(n)}{1^{s+1}}+\frac{c_2(n)}{2^{s+1}}+\frac{c_3(n)}{3^{s+1}}+\dots,$$

where $s>0$. This result has been proved by purely elementary methods. But in order to know whether the right-hand side of (7·2) is convergent or not for values of s less than or equal to zero we require the help of theorems which have only been established by transcendental methods.

Now the right-hand side of (7·2) is an ordinary Dirichlet's series for

$$\sigma_{-s}(n)\times\frac{1}{\zeta(s+1)}.$$

The first factor is a finite Dirichlet's series and so an absolutely convergent Dirichlet's series. It follows that the right-hand side of (7·2) is convergent whenever the Dirichlet's series for $1/\zeta(s+1)$, viz.

(7·3) $$\Sigma\frac{\mu(n)}{n^{1+s}},$$

is convergent. But it is known* that the series (7·3) is convergent when $s=0$ and that its sum is 0.

It follows from this that

(7·4) $$c_1(n)+\tfrac{1}{2}c_2(n)+\tfrac{1}{3}c_3(n)+\dots=0.$$

Nothing is known about the convergence of (7·3) when $-\frac{1}{2}<s<0$. But with the assumption of the truth of the hitherto unproved Riemann hypothesis it has been proved† that (7·3) is convergent when $s>-\frac{1}{2}$. With this assumption we see that (7·2) is true when $s>-\frac{1}{2}$. In other words, if $-\frac{1}{2}<s<\frac{1}{2}$, then

(7·5) $$\sigma_s(n)=\zeta(1-s)\left\{\frac{c_1(n)}{1^{1-s}}+\frac{c_2(n)}{2^{1-s}}+\frac{c_3(n)}{3^{1-s}}+\dots\right\}$$
$$=n^s\zeta(1+s)\left\{\frac{c_1(n)}{1^{1+s}}+\frac{c_2(n)}{2^{1+s}}+\frac{c_3(n)}{3^{1+s}}+\dots\right\}.$$

8. It is known‡ that all the series obtained from (7·3) by term-by-term differentiation with respect to s are convergent when $s=0$; and it is obvious

* Landau, *Handbuch*, p. 591. † Littlewood, *Comptes Rendus*, 29 Jan. 1912.

‡ Landau, *Handbuch*, p. 594.

that the derivatives of $\sigma_{-s}(n)$ with respect to s are all finite Dirichlet's series and so absolutely convergent. It follows that all the derivatives of the right-hand side of (7·2) are convergent when $s=0$; and so we can equate the coefficients of like powers of s from the two sides of (7·2). Now

$$\frac{1}{\zeta(s+1)}=s-\gamma s^2+\dots, \tag{8·1}$$

where γ is Euler's constant. And

$$\sigma_{-s}(n)=\sum_{\delta}\delta^{-s}=\sum_{\delta}1-s\sum_{\delta}\log\delta+\dots,$$

where δ is a divisor of n. But

$$\sum_{\delta}\log\delta=\sum_{\delta}\log\delta'=\tfrac{1}{2}\sum_{\delta}\log(\delta\delta')=\tfrac{1}{2}\,d(n)\log n,$$

where $\delta\delta'=n$. Hence

$$\sigma_{-s}(n)=d(n)-\tfrac{1}{2}sd(n)\log n+\dots. \tag{8·2}$$

Now equating the coefficients of s and s^2 from the two sides of (7·2), and using (8·1) and (8·2), we obtain

$$c_1(n)\log 1+\tfrac{1}{2}c_2(n)\log 2+\tfrac{1}{3}c_3(n)\log 3+\dots=-d(n), \tag{8·3}$$

$$c_1(n)(\log 1)^2+\tfrac{1}{2}c_2(n)(\log 2)^2+\tfrac{1}{3}c_3(n)(\log 3)^2+\dots=-d(n)(2\gamma+\log n). \tag{8·4}$$

9. I shall now find an expression of the same kind for $\phi(n)$, the number of numbers prime to and not exceeding n. Let $p_1, p_2, p_3, \dots$ be the prime divisors of n, and let

$$\phi_s(n)=n^s(1-p_1^{-s})(1-p_2^{-s})(1-p_3^{-s})\dots, \tag{9·1}$$

so that $\phi_1(n)=\phi(n)$. Suppose that

$$F(u,v)=\mu(u)v^s.$$

Then it is easy to see that

$$D(n)=\phi_s(n).$$

Hence, from (3·1), we have

$$\frac{\phi_s(n)}{n^s}=c_1(n)\sum_{1}^{t}\frac{\mu(\nu)}{\nu^{s+1}}+c_2(n)\sum_{1}^{\frac{1}{2}t}\frac{\mu(2\nu)}{(2\nu)^{s+1}}+\dots, \tag{9·2}$$

where t is any number not less than n. If $s>0$ we can make $t\to\infty$, as in §6. Then we have

$$\frac{\phi_s(n)}{n^s}=c_1(n)\sum_{1}^{\infty}\frac{\mu(\nu)}{\nu^{s+1}}+c_2(n)\sum_{1}^{\infty}\frac{\mu(2\nu)}{(2\nu)^{s+1}}+\dots. \tag{9·3}$$

But it can easily be shewn that

$$\sum_{1}^{\infty}\frac{\mu(n\nu)}{\nu^s}=\frac{\mu(n)}{\zeta(s)(1-p_1^{-s})(1-p_2^{-s})(1-p_3^{-s})\dots}, \tag{9·4}$$

where $p_1, p_2, p_3, \dots$ are the prime divisors of n. In other words

$$\sum_{1}^{\infty}\frac{\mu(n\nu)}{\nu^s}=\frac{\mu(n)\,n^s}{\phi_s(n)\,\zeta(s)}. \tag{9·5}$$

It follows from (9·3) and (9·5) that

$$(9{\cdot}6)\qquad \frac{\phi_s(n)\,\zeta(s+1)}{n^s}=\frac{\mu(1)\,c_1(n)}{\phi_{s+1}(1)}+\frac{\mu(2)\,c_2(n)}{\phi_{s+1}(2)}+\frac{\mu(3)\,c_3(n)}{\phi_{s+1}(3)}+\dots.$$

In particular

$$(9{\cdot}7)\qquad \frac{\pi^2}{6n}\phi(n)=c_1(n)-\frac{c_2(n)}{2^2-1}-\frac{c_3(n)}{3^2-1}-\frac{c_5(n)}{5^2-1}$$
$$+\frac{c_6(n)}{(2^2-1)(3^2-1)}-\frac{c_7(n)}{7^2-1}+\frac{c_{10}(n)}{(2^2-1)(5^2-1)}-\dots.$$

10. I shall now consider an application of the main formulæ to the problem of the number of representations of a number as the sum of 2, 4, 6, 8, ... squares. We shall require the following preliminary results.

(1) Let

$$(10{\cdot}1)\qquad \Sigma\, D(n)\,x^n=X_1=\frac{1^{s-1}x}{1+x}+\frac{2^{s-1}x^2}{1-x^2}+\frac{3^{s-1}x^3}{1+x^3}+\dots.$$

We shall choose

$$F(u,v)=v^{s-1},\qquad u\equiv 1\ (\mathrm{mod}\ 2),$$
$$F(u,v)=-\,v^{s-1},\qquad u\equiv 2\ (\mathrm{mod}\ 4),$$
$$F(u,v)=(2^s-1)\,v^{s-1},\qquad u\equiv 0\ (\mathrm{mod}\ 4).$$

Then from (3·1) we can shew, by arguments similar to those used in § 6, that

$$(10{\cdot}11)\qquad D(n)=n^{s-1}(1^{-s}+3^{-s}+5^{-s}+\dots)\{1^{-s}c_1(n)+2^{-s}c_4(n)+3^{-s}c_3(n)$$
$$+4^{-s}c_8(n)+5^{-s}c_5(n)+6^{-s}c_{12}(n)+7^{-s}c_7(n)+8^{-s}c_{16}(n)+\dots\}$$

if $s>1$.

(2) Let

$$(10{\cdot}2)\qquad \Sigma\, D(n)\,x^n=X_2=\frac{1^{s-1}x}{1-x}+\frac{2^{s-1}x^2}{1+x^2}+\frac{3^{s-1}x^3}{1-x^3}+\dots.$$

We shall choose

$$F(u,v)=v^{s-1},\qquad u\equiv 1\ (\mathrm{mod}\ 2),$$
$$F(u,v)=v^{s-1},\qquad u\equiv 2\ (\mathrm{mod}\ 4),$$
$$F(u,v)=(1-2^s)\,v^{s-1},\qquad u\equiv 0\ (\mathrm{mod}\ 4).$$

Then we obtain as before

$$(10{\cdot}21)\qquad D(n)=n^{s-1}(1^{-s}+3^{-s}+5^{-s}+\dots)\{1^{-s}c_1(n)-2^{-s}c_4(n)+3^{-s}c_3(n)$$
$$-4^{-s}c_8(n)+5^{-s}c_5(n)-6^{-s}c_{12}(n)+7^{-s}c_7(n)-8^{-s}c_{16}(n)+\dots\}.$$

(3) Let

$$(10{\cdot}3)\qquad \Sigma\, D(n)\,x^n=X_3=\frac{1^{s-1}x}{1+x^2}+\frac{2^{s-1}x^2}{1+x^4}+\frac{3^{s-1}x^3}{1+x^6}+\dots.$$

We shall choose

$$F(u,v)=0, \qquad u\equiv 0 \pmod 2),$$
$$F(u,v)=v^{s-1}, \qquad u\equiv 1 \pmod 4),$$
$$F(u,v)=-v^{s-1}, \qquad u\equiv 3 \pmod 4).$$

Then we obtain as before

(10·31)

$$D(n)=n^{s-1}(1^{-s}-3^{-s}+5^{-s}-\dots)\{1^{-s}c_1(n)-3^{-s}c_3(n)+5^{-s}c_5(n)-\dots\}.$$

(4) We shall also require a similar formula for the function $D(n)$ defined by

(10·4) $$\Sigma D(n)x^n=X_4=\frac{1^{s-1}x}{1-x}-\frac{3^{s-1}x^3}{1-x^3}+\frac{5^{s-1}x^5}{1-x^5}-\dots.$$

The formula required is not a direct consequence of the preceding analysis, but if, instead of starting with the function

$$c_r(n)=\sum_\lambda \cos\frac{2\pi n\lambda}{r},$$

we start with the function

$$s_r(n)=\sum_\lambda (-1)^{\frac{1}{2}(\lambda-1)}\sin\frac{2\pi n\lambda}{r},$$

where λ is prime to r and does not exceed r, and proceed as in §§ 2—3, we can shew that

(10·41)

$$D(n)=\tfrac{1}{2}n^{s-1}(1^{-s}-3^{-s}+5^{-s}-\dots)\{1^{-s}s_4(n)+2^{-s}s_8(n)+3^{-s}s_{12}(n)+\dots\}.$$

It should be observed that there is a correspondence between $c_r(n)$ and the ordinary ζ-function on the one hand and $s_r(n)$ and the function

$$\eta(s)=1^{-s}-3^{-s}+5^{-s}-\dots$$

on the other. It is possible to define an infinity of systems of trigonometrical sums such as $c_r(n)$, $s_r(n)$, each corresponding to one of the general class of "*L*-functions*" of which $\zeta(s)$ and $\eta(s)$ are the simplest members.

We have shewn that (10·31) and (10·41) are true when $s>1$. But if we assume that the Dirichlet's series for $1/\eta(s)$ is convergent when $s=1$, a result which is precisely of the same depth as the prime number theorem and has only been established by transcendental methods, then we can shew by arguments similar to those of § 7 that (10·31) and (10·41) are true when $s=1$.

11. I have shewn elsewhere† that if s is a positive integer and

$$1+\Sigma r_s(n)x^n=(1+2x+2x^4+2x^9+\dots)^s,$$

then $$r_{28}(n)=\delta_{28}(n)+e_{28}(n),$$

* See Landau, *Handbuch*, pp. 414 *et seq.*

† *Transactions of the Cambridge Philosophical Society*, Vol. XXII, 1916, pp. 159—184. [No. 18 of this volume; see in particular §§ 24—28, pp. 157—162.]

where $e_{2s}(n)=0$ when $s=1, 2, 3$ or 4 and is of lower order* than $\delta_{2s}(n)$ in all cases; that if s is a multiple of 4 then

$$(1^{-s}+3^{-s}+5^{-s}+\ldots)\,\Sigma\,\delta_{2s}(n)\,x^n=\frac{\pi^s}{(s-1)!}X_1; \tag{11·1}$$

if s is of the form $4k+2$ then

$$(1^{-s}+3^{-s}+5^{-s}+\ldots)\,\Sigma\,\delta_{2s}(n)\,x^n=\frac{\pi^s}{(s-1)!}X_2; \tag{11·2}$$

if s is of the form $4k+1$ then

$$(1^{-s}-3^{-s}+5^{-s}-\ldots)\,\Sigma\,\delta_{2s}(n)\,x^n=\frac{\pi^s}{(s-1)!}(X_3+2^{1-s}X_4), \tag{11·3}$$

except when $s=1$; and if s is of the form $4k+3$ then

$$(1^{-s}-3^{-s}+5^{-s}-\ldots)\,\Sigma\,\delta_{2s}(n)\,x^n=\frac{\pi^s}{(s-1)!}(X_3-2^{1-s}X_4), \tag{11·4}$$

X_1, X_2, X_3, X_4 being the same as in § 10.

In the case in which $s=1$ it is well known that

$$\begin{aligned}\Sigma\,\delta_2(n)\,x^n&=4\left(\frac{x}{1-x}-\frac{x^3}{1-x^3}+\frac{x^5}{1-x^5}-\ldots\right)\\&=4\left(\frac{x}{1+x^2}+\frac{x^2}{1+x^4}+\frac{x^3}{1+x^6}+\ldots\right).\end{aligned} \tag{11·5}$$

It follows from § 10 that, if s is a multiple of 4 then

$$\begin{aligned}\delta_{2s}(n)=\frac{\pi^s n^{s-1}}{(s-1)!}\{&1^{-s}c_1(n)+2^{-s}c_4(n)+3^{-s}c_3(n)+4^{-s}c_8(n)\\&+5^{-s}c_5(n)+6^{-s}c_{12}(n)+7^{-s}c_7(n)+8^{-s}c_{16}(n)+\ldots\};\end{aligned} \tag{11·11}$$

if s is of the form $4k+2$ then

$$\begin{aligned}\delta_{2s}(n)=\frac{\pi^s n^{s-1}}{(s-1)!}\{&1^{-s}c_1(n)-2^{-s}c_4(n)+3^{-s}c_3(n)-4^{-s}c_8(n)\\&+5^{-s}c_5(n)-6^{-s}c_{12}(n)+7^{-s}c_7(n)-8^{-s}c_{16}(n)+\ldots\};\end{aligned} \tag{11·21}$$

if s is of the form $4k+1$ then

$$\begin{aligned}\delta_{2s}(n)=\frac{\pi^s n^{s-1}}{(s-1)!}\{&1^{-s}c_1(n)+2^{-s}s_4(n)-3^{-s}c_3(n)+4^{-s}s_8(n)\\&+5^{-s}c_5(n)+6^{-s}s_{12}(n)-7^{-s}c_7(n)+8^{-s}s_{16}(n)+\ldots\},\end{aligned} \tag{11·31}$$

except when $s=1$; and if s is of the form $4k+3$ then

$$\begin{aligned}\delta_{2s}(n)=\frac{\pi^s n^{s-1}}{(s-1)!}\{&1^{-s}c_1(n)-2^{-s}s_4(n)-3^{-s}c_3(n)-4^{-s}s_8(n)\\&+5^{-s}c_5(n)-6^{-s}s_{12}(n)-7^{-s}c_7(n)-8^{-s}s_{16}(n)+\ldots\}.\end{aligned} \tag{11·41}$$

* For a more precise result concerning the order of $e_{2s}(n)$ see § 15.

From (11·5) and the remarks at the end of the previous section, it follows that

(11·51) $$r_2(n) = \delta_2(n) = \pi\,\{c_1(n) - \tfrac{1}{3}c_3(n) + \tfrac{1}{5}c_5(n) - \ldots\}$$
$$= \pi\,\{\tfrac{1}{2}s_4(n) + \tfrac{1}{4}s_8(n) + \tfrac{1}{6}s_{12}(n) + \ldots\},$$

but this is of course not such an elementary result as the preceding ones.

We can combine all the formulæ (11·11)—(11·41) in one by writing

(11·6) $$\delta_{2s}(n) = \frac{\pi^s n^{s-1}}{(s-1)!}\{1^{-s}\mathbf{c}_1(n) + 2^{-s}\mathbf{c}_4(n) + 3^{-s}\mathbf{c}_3(n) + 4^{-s}\mathbf{c}_8(n)$$
$$+ 5^{-s}\mathbf{c}_5(n) + 6^{-s}\mathbf{c}_{12}(n) + 7^{-s}\mathbf{c}_7(n) + 8^{-s}\mathbf{c}_{16}(n) + \ldots\},$$

where s is an integer greater than 1 and

$$\mathbf{c}_r(n) = c_r(n)\cos\tfrac{1}{2}\pi s(r-1) - s_r(n)\sin\tfrac{1}{2}\pi s(r-1).$$

12. We can obtain analogous results concerning the number of representations of a number as the sum of 2, 4, 6, 8, ... triangular numbers. Equation (147) of my former paper* is equivalent to

(12·1) $$(1 - 2x + 2x^4 - 2x^9 + \ldots)^{2s}$$
$$= 1 + \sum_1^\infty \delta_{2s}(n)(-x)^n + \frac{f^{4s}(x)}{f^{2s}(x^2)} \sum_{1 \leqslant n \leqslant \frac{1}{4}(s-1)} K_n(-x)^n \frac{f^{24n}(x^2)}{f^{24n}(x)},$$

where K_n is a constant and

$$f(x) = (1-x)(1-x^2)(1-x^3)\ldots.$$

Suppose now that $$x = e^{-\pi\alpha}, \quad x' = e^{-2\pi/\alpha}.$$

Then we know that

(12·2) $$\sqrt{\alpha}\,(1 - 2x + 2x^4 - 2x^9 + \ldots) = 2x'^{\frac{1}{8}}(1 + x' + x'^3 + x'^6 + \ldots),$$

(12·3) $$\sqrt{(\tfrac{1}{2}\alpha)}\,x^{\frac{1}{24}} f(x) = x'^{\frac{1}{12}} f(x'^2), \quad \sqrt{\alpha}\,x^{\frac{1}{12}} f(x^2) = x'^{\frac{1}{24}} f(x').$$

Finally $1 + \sum_1^\infty \delta_{2s}(n)(-x)^n$ can be expressed in powers of x' by using the formulæ:

(12·4) $$\alpha^s\left\{\tfrac{1}{2}\zeta(1-2s) + \frac{1^{2s-1}}{e^{2\alpha}-1} + \frac{2^{2s-1}}{e^{4\alpha}-1} + \frac{3^{2s-1}}{e^{6\alpha}-1} + \ldots\right\}$$
$$= (-\beta)^s\left\{\tfrac{1}{2}\zeta(1-2s) + \frac{1^{2s-1}}{e^{2\beta}-1} + \frac{2^{2s-1}}{e^{4\beta}-1} + \frac{3^{2s-1}}{e^{6\beta}-1} + \ldots\right\},$$

where $\alpha\beta = \pi^2$ and s is an integer greater than 1; and

(12·5) $$(2\alpha)^{s+\frac{1}{2}}\left\{\frac{1^{2s}}{e^{\alpha}+e^{-\alpha}} + \frac{2^{2s}}{e^{2\alpha}+e^{-2\alpha}} + \frac{3^{2s}}{e^{3\alpha}+e^{-3\alpha}} + \ldots\right\}$$
$$= (-\beta)^s\sqrt{(2\beta)}\left\{\tfrac{1}{2}\eta(-2s) + \frac{1^{2s}}{e^{\beta}-1} - \frac{3^{2s}}{e^{3\beta}-1} + \frac{5^{2s}}{e^{5\beta}-1} - \ldots\right\},$$

where $\alpha\beta = \pi^2$, s is any positive integer, and $\eta(s)$ is the function represented by the series $1^{-s} - 3^{-s} + 5^{-s} - \ldots$ and its analytical continuations.

* *Loc. cit.*, p. 181 [p. 159 of this volume].

It follows from all these formulæ that, if s is a positive integer and

$$(12\cdot6)\quad (1+x+x^3+x^6+\dots)^{2s} = \Sigma r'_{2s}(n)\,x^n = \Sigma \delta'_{2s}(n)\,x^n + \Sigma e'_{2s}(n)\,x^n,$$

then
$$\Sigma e'_{2s}(n)\,x^n = \frac{f^{4s}(x^2)}{f^{2s}(x)} \sum_{1 \leqslant n \leqslant \frac{1}{4}(s-1)} K_n (-x)^{-n} \frac{f^{24n}(x)}{f^{24n}(x^2)},$$

where K_n is a constant, and $f(x)$ is the same as in (12·1);

(12·61)

$$(1^{-s}+3^{-s}+5^{-s}+\dots)\,\Sigma \delta'_{2s}(n)\,x^n = \frac{(\frac{1}{2}\pi)^s}{(s-1)!}\,x^{-\frac{1}{4}s}\left(\frac{1^{s-1}x}{1-x^2} + \frac{2^{s-1}x^2}{1-x^4} + \frac{3^{s-1}x^3}{1-x^6} + \dots\right)$$

if s is a multiple of 4;

(12·62)

$$(1^{-s}+3^{-s}+5^{-s}+\dots)\,\Sigma \delta'_{2s}(n)\,x^n = \frac{2(\frac{1}{4}\pi)^s}{(s-1)!}\,x^{-\frac{1}{4}s}\left(\frac{1^{s-1}x^{\frac{1}{2}}}{1-x} + \frac{3^{s-1}x^{\frac{3}{2}}}{1-x^3} + \frac{5^{s-1}x^{\frac{5}{2}}}{1-x^5} + \dots\right)$$

if s is of the form $4k+2$;

(12·63)

$$(1^{-s}-3^{-s}+5^{-s}-\dots)\,\Sigma \delta'_{2s}(n)\,x^n = \frac{2(\frac{1}{8}\pi)^s}{(s-1)!}\,x^{-\frac{1}{4}s}\left(\frac{1^{s-1}x^{\frac{1}{4}}}{1+x^{\frac{1}{2}}} + \frac{3^{s-1}x^{\frac{3}{4}}}{1+x^{\frac{3}{2}}} + \frac{5^{s-1}x^{\frac{5}{4}}}{1+x^{\frac{5}{2}}} + \dots\right.$$
$$\left. + \frac{1^{s-1}x^{\frac{1}{4}}}{1-x^{\frac{1}{2}}} - \frac{3^{s-1}x^{\frac{3}{4}}}{1-x^{\frac{3}{2}}} + \frac{5^{s-1}x^{\frac{5}{4}}}{1-x^{\frac{5}{2}}} - \dots\right)$$

if s is of the form $4k+1$ (except when $s=1$); and

(12·64)

$$(1^{-s}-3^{-s}+5^{-s}-\dots)\,\Sigma \delta'_{2s}(n)\,x^n = \frac{2(\frac{1}{8}\pi)^s}{(s-1)!}\,x^{-\frac{1}{4}s}\left(\frac{1^{s-1}x^{\frac{1}{4}}}{1+x^{\frac{1}{2}}} + \frac{3^{s-1}x^{\frac{3}{4}}}{1+x^{\frac{3}{2}}} + \frac{5^{s-1}x^{\frac{5}{4}}}{1+x^{\frac{5}{2}}} + \dots\right.$$
$$\left. - \frac{1^{s-1}x^{\frac{1}{4}}}{1-x^{\frac{1}{2}}} + \frac{3^{s-1}x^{\frac{3}{4}}}{1-x^{\frac{3}{2}}} - \frac{5^{s-1}x^{\frac{5}{4}}}{1-x^{\frac{5}{2}}} + \dots\right)$$

if s is of the form $4k+3$. In the case in which $s=1$ we have

$$(12\cdot65)\quad \Sigma \delta_2'(n)\,x^n = x^{-\frac{1}{4}}\left(\frac{x^{\frac{1}{4}}}{1+x^{\frac{1}{2}}} + \frac{x^{\frac{3}{4}}}{1+x^{\frac{3}{2}}} + \frac{x^{\frac{5}{4}}}{1+x^{\frac{5}{2}}} + \dots\right)$$
$$= x^{-\frac{1}{4}}\left(\frac{x^{\frac{1}{4}}}{1-x^{\frac{1}{2}}} - \frac{x^{\frac{3}{4}}}{1-x^{\frac{3}{2}}} + \frac{x^{\frac{5}{4}}}{1-x^{\frac{5}{2}}} - \dots\right).$$

It is easy to see that the principal results proved about $e_{2s}(n)$ in my former paper are also true of $e'_{2s}(n)$, and in particular that

$$e'_{2s}(n) = 0$$

when $s = 1$, 2, 3 or 4, and $\qquad r'_{2s}(n) \sim \delta'_{2s}(n)$

for all values of s.

13. It follows from (12·62) that, if s is of the form $4k+2$, then

$$(1^{-s}+3^{-s}+5^{-s}+\dots)\,\delta'_{2s}(n)$$

is the coefficient of x^n in

(13·1) $$\frac{2(\frac{1}{4}\pi)^s}{(s-1)!}x^{-\frac{1}{4}s}\left(\frac{1^{s-1}x^{\frac{1}{2}}}{1-x}+\frac{2^{s-1}x}{1-x^2}+\frac{3^{s-1}x^{\frac{3}{2}}}{1-x^3}+\dots\right).$$

Similarly from (12·63) and (12·64) it follows that, if s is an odd integer greater than 1, then $(1^{-s}-3^{-s}+5^{-s}-\dots)\,\delta'_{2s}(n)$ is the coefficient of x^n in

(13·2) $$\frac{4(\frac{1}{8}\pi)^s}{(s-1)!}x^{-\frac{1}{8}s}\left(\frac{1^{s-1}x^{\frac{1}{4}}}{1+x^{\frac{1}{2}}}+\frac{2^{s-1}x^{\frac{1}{2}}}{1+x}+\frac{3^{s-1}x^{\frac{3}{4}}}{1+x^{\frac{3}{2}}}+\dots\right).$$

Now by applying our main formulæ to (12·61), (13·1) and (13·2) we obtain:

(13·3)
$$\delta'_{2s}(n)=\frac{(\frac{1}{2}\pi)^s}{(s-1)!}(n+\tfrac{1}{4}s)^{s-1}\{1^{-s}c_1(n+\tfrac{1}{4}s)+3^{-s}c_3(n+\tfrac{1}{4}s)+5^{-s}c_5(n+\tfrac{1}{4}s)+\dots\}$$

if s is a multiple of 4;

(13·4) $$\delta'_{2s}(n)=\frac{(\frac{1}{2}\pi)^s}{(s-1)!}(n+\tfrac{1}{4}s)^{s-1}\{1^{-s}c_1(2n+\tfrac{1}{2}s)+3^{-s}c_3(2n+\tfrac{1}{2}s)+5^{-s}c_5(2n+\tfrac{1}{2}s)+\dots\}$$

if s is twice an odd number; and

(13·5) $$\delta'_{2s}(n)=\frac{(\frac{1}{2}\pi)^s}{(s-1)!}(n+\tfrac{1}{4}s)^{s-1}\{1^{-s}c_1(4n+s)-3^{-s}c_3(4n+s)+5^{-s}c_5(4n+s)-\dots\}$$

if s is an odd number greater than 1.

Since the coefficient of x^n in $(1+x+x^3+\dots)^2$ is that of x^{4n+1} in

$$(\tfrac{1}{2}+x+x^4+\dots)^2,$$

it follows from (11·51) that

(13·6) $$r_2'(n)=\delta_2'(n)=\frac{\pi}{4}\{c_1(4n+1)-\tfrac{1}{3}c_3(4n+1)+\tfrac{1}{5}c_5(4n+1)-\dots\}.$$

This result however depends on the fact that the Dirichlet's series for $1/\eta(s)$ is convergent when $s=1$.

14. The preceding formulæ for $\sigma_s(n)$, $\delta_{2s}(n)$, $\delta'_{2s}(n)$ may be arrived at by another method. We understand by

(14·1) $$\frac{\sin n\pi}{k\sin(n\pi/k)}$$

the limit of
$$\frac{\sin x\pi}{k\sin(x\pi/k)}$$

when $x\to n$. It is easy to see that, if n and k are positive integers, and k odd, then (14·1) is equal to 1 if k is a divisor of n and to 0 otherwise.

When k is even we have (with similar conventions)

(14·2) $$\frac{\sin n\pi}{k\tan(n\pi/k)}=1 \text{ or } 0$$

according as k is a divisor of n or not. It follows that

$$(14\cdot3)\qquad \sigma_{s-1}(n)=n^{s-1}\left\{1^{-s}\left(\frac{\sin n\pi}{\sin n\pi}\right)+2^{-s}\left(\frac{\sin n\pi}{\tan\frac{1}{2}n\pi}\right)\right.$$
$$\left.+3^{-s}\left(\frac{\sin n\pi}{\sin\frac{1}{3}n\pi}\right)+4^{-s}\left(\frac{\sin n\pi}{\tan\frac{1}{4}n\pi}\right)+\dots\right\}.$$

Similarly from the definitions of $\delta_{2s}(n)$ and $\delta'_{2s}(n)$ we find that

$$(14\cdot4)\qquad \{1^{-s}+(-3)^{-s}+5^{-s}+(-7)^{-s}+\dots\}\,\delta_{2s}(n)=\frac{\pi^s n^{s-1}}{(s-1)!}\left\{1^{-s}\left(\frac{\sin n\pi}{\sin n\pi}\right)\right.$$
$$\left.+2^{-s}\left(\frac{\sin n\pi}{\sin(\frac{1}{2}n\pi+\frac{1}{2}s\pi)}\right)+3^{-s}\left(\frac{\sin n\pi}{\sin(\frac{1}{3}n\pi+s\pi)}\right)+4^{-s}\left(\frac{\sin n\pi}{\sin(\frac{1}{4}n\pi+\frac{3}{2}s\pi)}\right)+\dots\right\}$$

if s is an integer greater than 1;

$$(14\cdot5)\qquad r_2(n)=\delta_2(n)=4\left\{\left(\frac{\sin n\pi}{\sin n\pi}\right)-\frac{1}{3}\left(\frac{\sin n\pi}{\sin\frac{1}{3}n\pi}\right)+\frac{1}{5}\left(\frac{\sin n\pi}{\sin\frac{1}{5}n\pi}\right)-\dots\right\}$$
$$=4\left\{\frac{1}{2}\left(\frac{\sin n\pi}{\cos\frac{1}{2}n\pi}\right)-\frac{1}{4}\left(\frac{\sin n\pi}{\cos\frac{1}{4}n\pi}\right)+\frac{1}{6}\left(\frac{\sin n\pi}{\cos\frac{1}{6}n\pi}\right)-\dots\right\};$$

(14·6)

$$(1^{-s}+3^{-s}+5^{-s}+\dots)\,\delta'_{2s}(n)=\frac{(\frac{1}{2}\pi)^s}{(s-1)!}(n+\tfrac{1}{4}s)^{s-1}\left\{1^{-s}\left(\frac{\sin(n+\frac{1}{4}s)\pi}{\sin(n+\frac{1}{4}s)\pi}\right)\right.$$
$$\left.+3^{-s}\left(\frac{\sin(n+\frac{1}{4}s)\pi}{\sin\frac{1}{3}(n+\frac{1}{4}s)\pi}\right)+5^{-s}\left(\frac{\sin(n+\frac{1}{4}s)\pi}{\sin\frac{1}{5}(n+\frac{1}{4}s)\pi}\right)+\dots\right\}$$

if s is a multiple of 4;

(14·7)

$$(1^{-s}+3^{-s}+5^{-s}+\dots)\,\delta'_{2s}(n)=\frac{(\frac{1}{2}\pi)^s}{(s-1)!}(n+\tfrac{1}{4}s)^{s-1}\left\{1^{-s}\left(\frac{\sin(2n+\frac{1}{2}s)\pi}{\sin(2n+\frac{1}{2}s)\pi}\right)\right.$$
$$\left.+3^{-s}\left(\frac{\sin(2n+\frac{1}{2}s)\pi}{\sin\frac{1}{3}(2n+\frac{1}{2}s)\pi}\right)+5^{-s}\left(\frac{\sin(2n+\frac{1}{2}s)\pi}{\sin\frac{1}{5}(2n+\frac{1}{2}s)\pi}\right)+\dots\right\}$$

if s is twice an odd number;

(14·8)

$$(1^{-s}-3^{-s}+5^{-s}-\dots)\,\delta'_{2s}(n)=\frac{(\frac{1}{2}\pi)^s}{(s-1)!}(n+\tfrac{1}{4}s)^{s-1}\left\{1^{-s}\left(\frac{\sin(4n+s)\pi}{\sin(4n+s)\pi}\right)\right.$$
$$\left.-3^{-s}\left(\frac{\sin(4n+s)\pi}{\sin\frac{1}{3}(4n+s)\pi}\right)+5^{-s}\left(\frac{\sin(4n+s)\pi}{\sin\frac{1}{5}(4n+s)\pi}\right)-\dots\right\}$$

if s is an odd number greater than 1; and

$$(14\cdot9)\qquad r_2'(n)=\delta_2'(n)=\left(\frac{\sin(4n+1)\pi}{\sin(4n+1)\pi}\right)-\frac{1}{3}\left(\frac{\sin(4n+1)\pi}{\sin\frac{1}{3}(4n+1)\pi}\right)$$
$$\left.+\frac{1}{5}\left(\frac{\sin(4n+1)\pi}{\sin\frac{1}{5}(4n+1)\pi}\right)-\dots\right\}.$$

In all these equations the series on the right hand are finite Dirichlet's series and therefore absolutely convergent.

But the series (14·3) is (as is easily shewn by actual multiplication) the product of the two series

$$1^{-s} c_1(n) + 2^{-s} c_2(n) + \ldots$$

and
$$n^{s-1}(1^{-s} + 2^{-s} + 3^{-s} + \ldots).$$

We thus obtain an alternative proof of the formulæ (7·5). Similarly taking the previous expression of $\delta_{2s}(n)$, viz. the right-hand side of (11·6), and multiplying it by the series

$$1^{-s} + (-3)^{-s} + 5^{-s} + (-7)^{-s} + \ldots$$

we can shew that the product is actually the right-hand side of (14·4). The formulæ for $\delta'_{2s}(n)$ can be disposed of similarly.

15. The formulæ which I have found are closely connected with a method used for another purpose by Mr Hardy and myself*. The function

$$(15·1) \qquad (1 + 2x + 2x^4 + 2x^9 + \ldots)^{2s} = \Sigma r_{2s}(n) x^n$$

has every point of the unit circle as a singular point. If x approaches a "rational point" $\exp(-2p\pi i/q)$ on the circle, the function behaves roughly like

$$(15·2) \qquad \frac{\pi^s (\omega_{p,q})^s}{\{-(2p\pi i/q) - \log x\}^s},$$

where $\omega_{p,q} = 1$, 0, or -1 according as q is of the form $4k+1$, $4k+2$ or $4k+3$, while if q is of the form $4k$ then $\omega_{p,q} = -2i$ or $2i$ according as p is of the form $4k+1$ or $4k+3$.

Following the argument of our paper referred to, we can construct simple functions of x which are regular except at one point of the circle of convergence, and there behave in a manner very similar to that of the function (15·1); for example at the point $\exp(-2p\pi i/q)$ such a function is

$$(15·3) \qquad \frac{\pi^s (\omega_{p,q})^s}{(s-1)!} \sum_1^\infty n^{s-1} e^{2np\pi i/q} x^n.$$

The method which we used, with particular reference to the function

$$(15·4) \qquad \frac{1}{(1-x)(1-x^2)(1-x^3)\ldots} = \Sigma p(n) x^n,$$

was to approximate to the coefficients by means of a sum of a large number of the coefficients of these auxiliary functions. This method leads, in the present problem, to formulæ of the type

$$r_{2s}(n) = \delta_{2s}(n) + O(n^{\frac{1}{2}s}),$$

the first term on the right-hand side presenting itself precisely in the form of the series (11·11) etc.

* "Asymptotic formulæ in Combinatory Analysis," *Proc. London Math. Soc.*, Ser. 2, Vol. XVII, 1918, pp. 75—115 [No. 36 of this volume].

It is a very interesting problem to determine in such cases whether the approximate formula gives an exact representation of such an arithmetical function. The results proved here shew that, in the case of $r_{2s}(n)$, this is in general not so. The formula represents not $r_{2s}(n)$ but (except when $s=1$) its dominant term $\delta_{2s}(n)$, which is equal to $r_{2s}(n)$ only when $s=1$, 2, 3, or 4. When $s=1$ the formula gives $2\delta_2(n)$*.

16. We shall now consider the sum

(16·1) $$\sigma_s(1)+\sigma_s(2)+\ldots+\sigma_s(n).$$

Suppose that

(16·2)
$$T_r(n)=\tfrac{1}{2}\sum_{\lambda}\left(\frac{\sin\{(2n+1)\pi\lambda/r\}}{\sin(\pi\lambda/r)}-1\right),\quad U_r(n)=\tfrac{1}{2}\sum_{\lambda}\frac{\sin\{(2n+1)\pi\lambda/r\}}{\sin(\pi\lambda/r)},$$

where λ is prime to r and does not exceed r, so that

$$T_r(n)=c_r(1)+c_r(2)+\ldots+c_r(n)$$

and
$$U_r(n)=T_r(n)+\tfrac{1}{2}\phi(r),$$

where $\phi(n)$ is the same as in § 9. Since $c_r(n)=O(1)$ as $r\to\infty$, it follows that

(16·21) $$T_r(n)=O(1),\qquad U_r(n)=O(r),$$

as $r\to\infty$. It follows from (7·5) that, if $s>0$, then

(16·3)
$$\sigma_{-s}(1)+\sigma_{-s}(2)+\ldots+\sigma_{-s}(n)=\zeta(s+1)\left\{n+\frac{T_2(n)}{2^{s+1}}+\frac{T_3(n)}{3^{s+1}}+\frac{T_4(n)}{4^{s+1}}+\ldots\right\}.$$

Since
$$\sum_{1}^{\infty}\frac{\phi(\nu)}{\nu^{s+1}}=\frac{\zeta(s)}{\zeta(s+1)}$$

if $s>1$, (16·3) can be written as

(16·31) $$\sigma_{-s}(1)+\sigma_{-s}(2)+\ldots+\sigma_{-s}(n)$$
$$=\zeta(s+1)\left\{n+\tfrac{1}{2}+\frac{U_2(n)}{2^{s+1}}+\frac{U_3(n)}{3^{s+1}}+\frac{U_4(n)}{4^{s+1}}+\ldots\right\}-\tfrac{1}{2}\zeta(s),$$

if $s>1$. Similarly from (8·3), (8·4) and (11·51) we obtain

(16·4) $$d(1)+d(2)+\ldots+d(n)$$
$$=-\tfrac{1}{2}T_2(n)\log 2-\tfrac{1}{3}T_3(n)\log 3-\tfrac{1}{4}T_4(n)\log 4-\ldots,$$

(16·5) $$d(1)\log 1+d(2)\log 2+\ldots+d(n)\log n$$
$$=\tfrac{1}{2}T_2(n)\{2\gamma\log 2-(\log 2)^2\}+\tfrac{1}{3}T_3(n)\{2\gamma\log 3-(\log 3)^2\}+\ldots,$$

(16·6) $$r_2(1)+r_2(2)+\ldots+r_2(n)=\pi\{n-\tfrac{1}{3}T_3(n)+\tfrac{1}{5}T_5(n)-\tfrac{1}{7}T_7(n)+\ldots\}.$$

* The method is also applicable to the problem of the representation of a number by the sum of an *odd* number of squares, and gives an exact result when the number of squares is 3, 5, or 7. See G. H. Hardy, "On the representation of a number as the sum of any number of squares, and in particular of five or seven," *Proc. London Math. Soc.* (*Records of proceedings at meetings*, March 1918). A fuller account of this paper will appear shortly in the *Proceedings of the National Academy of Sciences* (Washington, D.C.) [*loc. cit.*, Vol. IV, 1918, 189—193].

Suppose now that

$$T_{r,s}(n) = \sum_{\lambda} \left(1^s \cos \frac{2\pi\lambda}{r} + 2^s \cos \frac{4\pi\lambda}{r} + \dots + n^s \cos \frac{2n\pi\lambda}{r}\right),$$

where λ is prime to r and does not exceed r, so that

$$T_{r,s}(n) = 1^s c_r(1) + 2^s c_r(2) + \dots + n^s c_r(n).$$

Then it follows from (7·5) that

$$(16\cdot7) \quad \sigma_s(1) + \sigma_s(2) + \dots + \sigma_s(n)$$
$$= \zeta(s+1)\left\{(1^s + 2^s + \dots + n^s) + \frac{T_{2,s}(n)}{2^{s+1}} + \frac{T_{3,s}(n)}{3^{s+1}} + \frac{T_{4,s}(n)}{4^{s+1}} + \dots\right\}$$

if $s > 0$. Putting $s = 1$ in (16·3) and (16·7), we find that

$$(16\cdot8) \quad (n-1)\sigma_{-1}(1) + (n-2)\sigma_{-1}(2) + \dots + (n-n)\sigma_{-1}(n)$$
$$= \frac{\pi^2}{6}\left\{\frac{n(n-1)}{2} + \frac{\nu_2(n)}{2^2} + \frac{\nu_3(n)}{3^2} + \frac{\nu_4(n)}{4^2} + \dots\right\},$$

where
$$\nu_r(n) = \tfrac{1}{2}\sum_{\lambda}\left\{\frac{\sin^2(\pi n\lambda/r)}{\sin^2(\pi\lambda/r)} - n\right\},$$

λ being prime to r and not exceeding r.

It has been proved by Wigert*, by less elementary methods, that the left-hand side of (16·8) is equal to

$$(16\cdot9) \quad \frac{\pi^2}{12}n^2 - \tfrac{1}{2}n(\gamma - 1 + \log 2n\pi) - \tfrac{1}{24} + \frac{\sqrt{n}}{2\pi}\sum_{1}^{\infty}\frac{\sigma_{-1}(\nu)}{\sqrt{\nu}}J_1\{4\pi\sqrt{(\nu n)}\},$$

where J_1 is the ordinary Bessel's function.

17. We shall now find a relation between the functions (16·1) and (16·3) which enables us to determine the behaviour of the former for large values of n. It is easily shewn that this function is equal to

$$(17\cdot1) \quad \sum_{\nu=1}^{\sqrt{n}}\left(1^s + 2^s + 3^s + \dots + \left[\frac{n}{\nu}\right]^s\right) + \sum_{\nu=1}^{\sqrt{n}}\nu^s\left[\frac{n}{\nu}\right] - [\sqrt{n}]\sum_{\nu=1}^{\sqrt{n}}\nu^s.$$

Now
$$1^s + 2^s + \dots + k^s = \zeta(-s) + \frac{(k+\frac{1}{2})^{s+1}}{s+1} + O(k^{s-1})$$

for all values of s, it being understood that

$$\zeta(-s) + \frac{(k+\frac{1}{2})^{s+1}}{s+1}$$

denotes $\gamma + \log(k+\frac{1}{2})$ when $s = -1$. Let

$$\left[\frac{n}{\nu}\right] = \frac{n}{\nu} - \tfrac{1}{2} + \epsilon_\nu, \quad [\sqrt{n}] = t = \sqrt{n} - \tfrac{1}{2} + \epsilon.$$

* *Acta Mathematica*, Vol. XXXVII, 1914, pp. 113—140 (p. 140).

Then we have

$$1^s + 2^s + \dots + \left[\frac{n}{\nu}\right]^s = \zeta(-s) + \frac{1}{s+1}\left(\frac{n}{\nu}\right)^{s+1} + \epsilon_\nu\left(\frac{n}{\nu}\right)^s + O\left(\frac{n^{s-1}}{\nu^{s-1}}\right)$$

and

$$\nu^s\left[\frac{n}{\nu}\right] = n\nu^{s-1} - \tfrac{1}{2}\nu^s + \epsilon_\nu \nu^s.$$

It follows from these equations and (17·1) that

(17·2) $$\sigma_s(1) + \sigma_s(2) + \dots + \sigma_s(n) = \sum_{\nu=1}^{t}\left\{\zeta(-s) + \frac{1}{s+1}\left(\frac{n}{\nu}\right)^{s+1} + n\nu^{s-1}\right.$$
$$\left. + \epsilon_\nu\left(\frac{n}{\nu}\right)^s + \epsilon_\nu \nu^s - (\sqrt{n} + \epsilon)\nu^s + O\left(\frac{n^{s-1}}{\nu^{s-1}}\right)\right\}.$$

Changing s to $-s$ in (17·2) we have

(17·21)

$$n^s\{\sigma_{-s}(1) + \sigma_{-s}(2) + \dots + \sigma_{-s}(n)\} = \sum_{\nu=1}^{t}\left\{n^s\zeta(s) + \frac{n\nu^{s-1}}{1-s} + \left(\frac{n}{\nu}\right)^{s+1} + \epsilon_\nu \nu^s\right.$$
$$\left. + \epsilon_\nu\left(\frac{n}{\nu}\right)^s - (\sqrt{n} + \epsilon)\left(\frac{n}{\nu}\right)^s + O\left(\frac{\nu^{s+1}}{n}\right)\right\}.$$

It follows that

(17·3)

$$n^s\{\sigma_{-s}(1) + \sigma_{-s}(2) + \dots + \sigma_{-s}(n)\} - \{\sigma_s(1) + \sigma_s(2) + \dots + \sigma_s(n)\}$$
$$= \sum_{\nu=1}^{t}\left\{n^s\zeta(s) - \zeta(-s) + \frac{s}{1+s}\left(\frac{n}{\nu}\right)^{s+1} + \frac{s}{1-s}n\nu^{s-1} + (\sqrt{n} + \epsilon)\nu^s\right.$$
$$\left. - (\sqrt{n} + \epsilon)\left(\frac{n}{\nu}\right)^s + O\left(\frac{n^{s-1}}{\nu^{s-1}} + \frac{\nu^{s+1}}{n}\right)\right\}.$$

Suppose now that $s > 0$. Then, since ν varies from 1 to t, it is obvious that

$$\frac{\nu^{s+1}}{n} < \frac{n^{s-1}}{\nu^{s-1}}$$

and so

$$O\left(\frac{\nu^{s+1}}{n}\right) = O\left(\frac{n^{s-1}}{\nu^{s-1}}\right).$$

Also $\sum_{\nu=1}^{t}\{n^s\zeta(s) - \zeta(-s)\} = (\sqrt{n} - \tfrac{1}{2} + \epsilon)\{n^s\zeta(s) - \zeta(-s)\}$;

$$\sum_{\nu=1}^{t}\frac{s}{1+s}\left(\frac{n}{\nu}\right)^{s+1} = \frac{sn^{s+1}}{1+s}\zeta(1+s) - \frac{n^{s+1}}{s+1}(\sqrt{n} + \epsilon)^{-s} + O(n^{\frac{1}{2}s});$$

$$\sum_{\nu=1}^{t}\frac{s}{1-s}n\nu^{s-1} = \frac{ns}{1-s}\zeta(1-s) + \frac{n}{1-s}(\sqrt{n} + \epsilon)^s + O(n^{\frac{1}{2}s});$$

$$\sum_{\nu=1}^{t}(\sqrt{n} + \epsilon)\nu^s = (\sqrt{n} + \epsilon)\zeta(-s) + \frac{(\sqrt{n} + \epsilon)^{2+s}}{1+s} + O(n^{\frac{1}{2}s});$$

$$\sum_{\nu=1}^{t}(\sqrt{n} + \epsilon)\left(\frac{n}{\nu}\right)^s = n^s(\sqrt{n} + \epsilon)\zeta(s) + \frac{n^s}{1-s}(\sqrt{n} + \epsilon)^{2-s} + O(n^{\frac{1}{2}s});$$

and
$$\sum_{\nu=1}^{t} O\left(\frac{n^{s-1}}{\nu^{s-1}}\right) = O(m),$$
where

(17·4) $m = n^{\frac{1}{2}s}\,(s<2),\quad m = n\log n\,(s=2),\quad m = n^{s-1}\,(s>2).$

It follows that the right-hand side of (17·3) is equal to
$$\frac{sn^{1+s}}{1+s}\zeta(1+s) + \frac{sn}{1-s}\zeta(1-s) - \tfrac{1}{2}n^s\zeta(s) + \frac{(\sqrt{n}+\epsilon)^{2+s} - n^{s+1}(\sqrt{n}+\epsilon)^{-s}}{1+s}$$
$$+\frac{n(\sqrt{n}+\epsilon)^s - n^s(\sqrt{n}+\epsilon)^{2-s}}{1-s} + O(m).$$

But
$$\frac{(\sqrt{n}+\epsilon)^{2+s} - n^{s+1}(\sqrt{n}+\epsilon)^{-s}}{1+s} = 2\epsilon n^{\frac{1}{2}(1+s)} + O(n^{\frac{1}{2}s});$$
$$\frac{n(\sqrt{n}+\epsilon)^s - n^s(\sqrt{n}+\epsilon)^{2-s}}{1-s} = -2\epsilon n^{\frac{1}{2}(1+s)} + O(n^{\frac{1}{2}s}).$$

It follows that

(17·5)
$$\sigma_s(1)+\sigma_s(2)+\ldots+\sigma_s(n) = n^s\{\sigma_{-s}(1)+\sigma_{-s}(2)+\ldots+\sigma_{-s}(n)\}$$
$$-\frac{sn^{1+s}}{1+s}\zeta(1+s) + \tfrac{1}{2}n^s\zeta(s) - \frac{sn}{1-s}\zeta(1-s) + O(m)^*$$
if $s>0$, m being the same as in (17·4). If $s=1$, (17·5) reduces to

(17·6) $$(n-1)\sigma_{-1}(1) + (n-2)\sigma_{-1}(2) + \ldots + (n-n)\sigma_{-1}(n)$$
$$= \frac{\pi^2}{12}n^2 - \tfrac{1}{2}n(\gamma - 1 + \log 2n\pi) + O(\sqrt{n})\dagger.$$

From (16·2) and (17·5) it follows that

(17·7) $$\sigma_s(1)+\sigma_s(2)+\ldots+\sigma_s(n) = \frac{n^{1+s}}{1+s}\zeta(1+s) + \tfrac{1}{2}n^s\zeta(s)$$
$$+\frac{sn}{s-1}\zeta(1-s) + n^s\zeta(1+s)\left\{\frac{T_2(n)}{2^{s+1}} + \frac{T_3(n)}{3^{s+1}} + \frac{T_4(n)}{4^{s+1}} + \ldots\right\} + O(m),$$
for all positive values of s. If $s>1$, the right-hand side can be written as

(17·8)
$$\frac{ns}{s-1}\zeta(1-s) + n^s\zeta(1+s)\left\{\frac{n}{1+s} + \tfrac{1}{2} + \frac{U_2(n)}{2^{s+1}} + \frac{U_3(n)}{3^{s+1}} + \frac{U_4(n)}{4^{s+1}} + \ldots\right\} + O(m).$$

Putting $s=1$ in (17·7) we obtain

(17·9) $$\sigma_1(1)+\sigma_1(2)+\ldots+\sigma_1(n) = \frac{\pi^2}{12}n^2 + \tfrac{1}{2}n(\gamma - 1 + \log 2n\pi)$$
$$+\frac{\pi^2 n}{6}\left\{\frac{T_2(n)}{2^2} + \frac{T_3(n)}{3^2} + \frac{T_4(n)}{4^2} + \ldots\right\} + O(\sqrt{n}).$$

* [See Appendix, p. 343].

† This result has been proved by Landau. See his report on Wigert's memoir in the *Göttingische gelehrte Anzeigen*, 1915, pp. 377—414 (p. 402). Landau has also, by a more transcendental method, replaced $O(\sqrt{n})$ by $O(n^{\frac{2}{5}})$ (*loc. cit.* p. 414).

Additional note to § 7 (*May* 1, 1918).

From (7·2) it follows that

$$\frac{1}{\zeta(r)}\{1^{-s}\sigma_{1-r}(1)+2^{-s}\sigma_{1-r}(2)+\ldots\}=1^{-s}\sum_{1}^{\infty} m^{-r}c_m(1)+2^{-s}\sum_{1}^{\infty} m^{-r}c_m(2)+\ldots,$$

or

$$\frac{\zeta(s)\,\zeta(r+s-1)}{\zeta(r)}=\sum_{1}^{\infty}\sum_{1}^{\infty}\frac{c_m(n)}{m^r n^s},$$

from which we deduce

$$\zeta(s)\sum_{\delta}\mu(\delta)\,\delta'^{1-s}=\frac{c_m(1)}{1^s}+\frac{c_m(2)}{2^s}+\frac{c_m(3)}{3^s}+\ldots,$$

δ being a divisor of m and δ' its conjugate. The series on the right-hand side is convergent for $s>0$ (except when $m=1$, when it reduces to the ordinary series for $\zeta(s)$).

When $s=1$, $m>1$, we have to replace the left-hand side by its limit as $s\to 1$. We find that

(18) $$c_m(1)+\tfrac{1}{2}c_m(2)+\tfrac{1}{3}c_m(3)+\ldots=-\Lambda(m),$$

$\Lambda(m)$ being the well-known arithmetical function which is equal to $\log p$ if m is a power of a prime p and to zero otherwise.

22

SOME DEFINITE INTEGRALS

(*Proceedings of the London Mathematical Society*, 2, XVII, 1918, *Records for* 17 *Jan.* 1918)

Typical formulæ are:

$$(1)\quad \int_{-\infty}^{\infty} \frac{e^{nix}\,dx}{\Gamma(\alpha+x)\,\Gamma(\beta-x)} = \frac{(2\cos\frac{1}{2}n)^{\alpha+\beta-2}}{\Gamma(\alpha+\beta-1)}\, e^{\frac{1}{2}n(\beta-\alpha)i} \quad (or\ 0),$$

$$(2)\quad \int_{-\infty}^{\infty} \frac{\Gamma(\alpha+x)}{\Gamma(\beta+x)} e^{nix}\,dx = \pm\frac{2\pi i\,(2\sin\frac{1}{2}N)^{\beta-\alpha-1}}{\Gamma(\beta-\alpha)}\, e^{-nai+\frac{1}{2}(\pi-N)(\beta-\alpha-1)i} \quad (or\ 0),$$

$$(3)\quad \int_{-\infty}^{\infty} \Gamma(\alpha+x)\,\Gamma(\beta-x)\,e^{nix}\,dx$$

$$=\frac{2\pi i\,\Gamma(\alpha+\beta)}{(2\sin\frac{1}{2}N)^{\alpha+\beta}}\, e^{\frac{1}{2}n(\beta-\alpha)i}\left[\epsilon_n(\beta)\,e^{k\pi(\alpha+\beta)i} - \epsilon_n(-\alpha)\,e^{-k\pi(\alpha+\beta)i}\right].$$

Here n is real, $n = 2k\pi + N\ (0 \leqslant N < 2\pi)$ in (2), and $n = (2k-1)\pi + N$ $(0 \leqslant N < 2\pi)$ in (3). In (1) the zero value is to be taken if $|n| \geqslant \pi$, the non-zero value otherwise. In (2) α must be complex: the zero value is to be taken if n and $\mathfrak{I}(\alpha)$ have the same sign, the positive sign if $n \geqslant 0$ and $\mathfrak{I}(\alpha) < 0$, and the negative sign if $n \leqslant 0$ and $\mathfrak{I}(\alpha) > 0$. In (3) α and β must both be complex; and $\epsilon_n(\zeta)$ is 0, 1, or -1 according as (i) $\pi - n$ and $\mathfrak{I}(\zeta)$ have the same sign, (ii) $n \leqslant \pi$ and $\mathfrak{I}(\zeta) < 0$, (iii) $n \geqslant \pi$ and $\mathfrak{I}(\zeta) > 0$.

The convergence conditions are, in general, (1) $\mathfrak{R}(\alpha+\beta) > 1$, (2) $\mathfrak{R}(\alpha-\beta) < 0$, (3) $\mathfrak{R}(\alpha+\beta) < 1$. But there are certain special cases in which a more stringent condition is required.

A formula of a different character, deduced from (1), is

$$\int_{-\infty}^{\infty} \frac{J_{\alpha+x}(\lambda)}{\lambda^{\alpha+x}} \frac{J_{\beta-x}(\mu)}{\mu^{\beta-x}} e^{nix}\,dx = \left(\frac{2\cos\frac{1}{2}n}{\Omega}\right)^{\frac{1}{2}(\alpha+\beta)} e^{\frac{1}{2}n(\beta-\alpha)i} J_{\alpha+\beta}\{\sqrt{(2\Omega\cos\tfrac{1}{2}n)}\}\ (or\ 0).$$

Here

$$\Omega = \lambda^2 e^{\frac{1}{2}ni} + \mu^2 e^{-\frac{1}{2}ni};$$

the zero value is to be taken if $|n| \geqslant \pi$, the non-zero value otherwise; and the condition of convergence is, in general, that

$$\mathfrak{R}(\alpha+\beta) > -1.$$

The formulæ include a large number of interesting special cases, such as

$$\int_{-\infty}^{\infty} \frac{dx}{\Gamma(\alpha+x)\,\Gamma(\beta-x)} = \frac{2^{\alpha+\beta-2}}{\Gamma(\alpha+\beta-1)},$$

$$\int_0^{\infty} \frac{\sin\pi x\,dx}{x(x^2-1^2)(x^2-2^2)\dots(x^2-k^2)} = (-1)^k \frac{2^{2k-1}\pi}{(2k)!},$$

$$\int_{-\infty}^{\infty} J_{\alpha+x}(\lambda)\,J_{\beta-x}(\lambda)\,dx = J_{\alpha+\beta}(2\lambda).$$

The formula

$$\int_{-\infty}^{\infty} \frac{dx}{\Gamma(\alpha+x)\,\Gamma(\beta-x)\,\Gamma(\gamma+x)\,\Gamma(\delta-x)}$$

$$= \frac{\Gamma(\alpha+\beta+\gamma+\delta-3)}{\Gamma(\alpha+\beta-1)\,\Gamma(\beta+\gamma-1)\,\Gamma(\gamma+\delta-1)\,\Gamma(\delta+\alpha-1)},$$

may also be mentioned: it holds, in general, if

$$\mathfrak{R}(\alpha+\beta+\gamma+\delta) > 3.$$

A fuller account of these formulæ will be published in the *Quarterly Journal of Mathematics**.

* [See No. 27 of this volume.]

23

SOME DEFINITE INTEGRALS

(*Journal of the Indian Mathematical Society*, XI, 1919, 81—87)

I have shewn elsewhere* that the definite integrals

$$\phi_w(t) = \int_0^\infty \frac{\cos \pi t x}{\cosh \pi x} e^{-\pi w x^2} dx,$$

$$\psi_w(t) = \int_0^\infty \frac{\sin \pi t x}{\sinh \pi x} e^{-\pi w x^2} dx$$

can be evaluated in finite terms if w is any rational multiple of i.

In this paper I shall shew, by a much simpler method, that these integrals can be evaluated not only for these values but also for many other values of t and w.

Now we have

$$\phi_w(t) = 2\int_0^\infty \int_0^\infty \frac{\cos 2\pi x z}{\cosh \pi z} \cos \pi t x \, e^{-\pi w x^2} dx\, dz$$

$$= \frac{e^{-\frac{1}{4}\pi t^2 w'}}{\sqrt{w}} \int_0^\infty \frac{\cosh \pi t x w'}{\cosh \pi x} e^{-\pi x^2 w'} dx$$

where w' stands for $1/w$.

It follows that

$$\phi_w(t) = \frac{1}{\sqrt{w}} e^{-\frac{1}{4}\pi t^2 w'} \phi_{w'}(itw'). \qquad (1)$$

Again

$$\phi_w(t+w) = \frac{1}{\sqrt{w}} e^{-\frac{1}{4}\pi (t+w)^2 w'}$$

$$\times \int_0^\infty \frac{\cosh (\pi t x/w) \cosh \pi x + \sinh \pi t x/w \sinh \pi x}{\cosh \pi x} e^{-\pi x^2/w} dx$$

$$= \frac{1}{\sqrt{w}} e^{-\frac{1}{4}\pi (t+w)^2/w}$$

$$\times \left\{\tfrac{1}{2}\sqrt{w}\, e^{\frac{1}{4}\pi t^2/w} + 2\int_0^\infty \int_0^\infty \frac{\sin 2\pi x z}{\sinh \pi z} \sinh \frac{\pi t x}{w} e^{-\pi x^2/w} dx\, dz\right\}$$

$$= \frac{1}{\sqrt{w}} e^{-\frac{1}{4}\pi (t+w)^2/w}$$

$$\times \left\{\tfrac{1}{2}\sqrt{w}\, e^{\frac{1}{4}\pi t^2/w} + \sqrt{w}\, e^{\frac{1}{4}\pi t^2/w} \int_0^\infty \frac{\sin \pi t x}{\sinh \pi x} e^{-\pi w x^2} dx\right\}.$$

* *Messenger of Mathematics*, Vol. 44, 1915, pp. 75—85 [No. 12 of this volume].

In other words

$$e^{\frac{1}{4}\pi t^2/w}\{\tfrac{1}{2}+\psi_w(t)\}=e^{\frac{1}{4}\pi(t+w)^2/w}\phi_w(t+w). \qquad \text{.............(2)}$$

It is obvious that

$$\left.\begin{aligned}\phi_w(t)&=\phi_w(-t)\\ \psi_w(t)&=-\psi_w(-t)\end{aligned}\right\}. \qquad \text{.........................(3)}$$

From (1), (2) and (3) we easily find that

$$\tfrac{1}{2}+\psi_w(t+i)=\frac{i}{\sqrt{w}}e^{-\frac{1}{4}\pi t^2/w}\left\{\tfrac{1}{2}-\psi_{w'}\left(\frac{it}{w}+i\right)\right\}. \qquad \text{...........(4)}$$

It is easy to see that

$$\phi_w(i)=\frac{1}{2\sqrt{w}};\quad \psi_w(i)=\frac{i}{2\sqrt{w}};\quad \phi_w(w)=\tfrac{1}{2}e^{-\frac{1}{4}\pi w};$$

$$\tfrac{1}{2}-\psi_w(w)=e^{-\frac{1}{4}\pi w}\phi_w(0);\quad \phi_w(w\pm i)=\left(\frac{1}{2\sqrt{w}}\pm\frac{i}{2}\right)e^{-\frac{1}{4}\pi w};$$

$$\psi_w(w\pm i)=\tfrac{1}{2}\pm\frac{i}{2\sqrt{w}}e^{-\frac{1}{4}\pi w};\quad \phi_w(\tfrac{1}{2}w)+\psi_w(\tfrac{1}{2}w)=\tfrac{1}{2}.$$

Again we see that

$$\phi_w(t+i)+\phi_w(t-i)=\frac{1}{\sqrt{w}}e^{-\frac{1}{4}\pi t^2/w}; \qquad \text{..................(5)}$$

and

$$\psi_w(t+i)-\psi_w(t-i)=\frac{i}{\sqrt{w}}e^{-\frac{1}{4}\pi t^2/w}. \qquad \text{..................(6)}$$

From (1) and (5) we deduce that

$$e^{\frac{1}{4}\pi(t+w)^2/w}\phi_w(t+w)+e^{\frac{1}{4}\pi(t-w)^2/w}\phi_w(t-w)=e^{\frac{1}{4}\pi t^2/w}. \qquad \text{......(7)}$$

Similarly from (4) and (6) we obtain

$$e^{\frac{1}{4}\pi(t+w)^2/w}\{\tfrac{1}{2}-\psi_w(t+w)\}=e^{\frac{1}{4}\pi(t-w)^2/w}\{\tfrac{1}{2}+\psi_w(t-w)\}. \qquad \text{......(8)}$$

It is easy to deduce from (5) that if n is a positive integer, then

$$\phi_w(t)+(-1)^{n+1}\phi_w(t\pm 2ni)$$
$$=\frac{1}{\sqrt{w}}\left\{e^{-\frac{1}{4}\pi(t\pm i)^2/w}-e^{-\frac{1}{4}\pi(t\pm 3i)^2/w}+e^{-\frac{1}{4}\pi(t\pm 5i)^2/w}-\dots\text{ to } n \text{ terms}\right\}. \qquad \text{...(9)}$$

Similarly from (6) we have

$$\psi_w(t)-\psi_w(t\pm 2ni)$$
$$=\mp\frac{i}{\sqrt{w}}\left\{e^{-\frac{1}{4}\pi(t+i)^2/w}+e^{-\frac{1}{4}\pi(t+3i)^2/w}+e^{-\frac{1}{4}\pi(t+5i)^2/w}+\dots\text{ to } n \text{ terms}\right\}.$$
$$\text{......(10)}$$

Again from (7) we have

$$e^{\frac{1}{4}\pi t^2/w}\phi_w(t)+(-1)^{n+1}e^{\frac{1}{4}\pi(t+2nw)^2/w}\phi_w(t+2nw)$$
$$=e^{\frac{1}{4}\pi(t+w)^2/w}-e^{\frac{1}{4}\pi(t+3w)^2/w}+e^{\frac{1}{4}\pi(t+5w)^2/w}-\dots \text{ to } n \text{ terms}; \quad \dots(11)$$

and from (8)

$$e^{\frac{1}{4}\pi t^2/w}\{\tfrac{1}{2}+\psi_w(t)\}+(-1)^{n+1}e^{\frac{1}{4}\pi(t+2nw)^2/w}\{\tfrac{1}{2}+\psi_w(t+2nw)\}$$
$$=e^{\frac{1}{4}\pi(t+2w)^2/w}-e^{\frac{1}{4}\pi(t+4w)^2/w}+e^{\frac{1}{4}\pi(t+6w)^2/w}-\dots \text{ to } n \text{ terms}. \quad \dots(12)$$

Now, combining (9) and (11), we deduce that, if m and n are positive integers and $s=t+2mw\pm 2ni$, then

$$\phi_w(s)+(-1)^{(m+1)(n+1)}e^{-\frac{1}{2}\pi m(s+t)}\phi_w(t)$$
$$=e^{-\frac{1}{4}\pi s^2/w}\left\{e^{\frac{1}{4}\pi(s-w)^2/w}-e^{\frac{1}{4}\pi(s-3w)^2/w}+e^{\frac{1}{4}\pi(s-5w)^2/w}-\dots \text{ to } m \text{ terms}\right\}$$
$$+\frac{(-1)^{(m+1)(n+1)}}{\sqrt{w}}e^{-\frac{1}{2}\pi m(s+t)}$$
$$\times\left\{e^{-\frac{1}{4}\pi(t\pm i)^2/w}-e^{-\frac{1}{4}\pi(t\pm 3i)^2/w}+e^{-\frac{1}{4}\pi(t\pm 5i)^2/w}-\dots \text{ to } n \text{ terms}\right\}. \quad \dots\dots(13)$$

Similarly, combining (10) and (12), we obtain

$$\tfrac{1}{2}-\psi_w(s)+(-1)^{mn+m+1}e^{-\frac{1}{2}\pi m(s+t)}\{\tfrac{1}{2}-\psi_w(t)\}$$
$$=e^{-\frac{1}{4}\pi s^2/w}\left\{e^{\frac{1}{4}\pi(s-2w)^2/w}-e^{\frac{1}{4}\pi(s-4w)^2/w}+e^{\frac{1}{4}\pi(s-6w)^2/w}-\dots \text{ to } m \text{ terms}\right\}$$
$$\pm(-1)^{mn+m+1}\frac{i}{\sqrt{w}}e^{-\frac{1}{2}\pi m(s+t)}$$
$$\times\left\{e^{-\frac{1}{4}\pi(t\pm i)^2/w}+e^{-\frac{1}{4}\pi(t\pm 3i)^2/w}+e^{-\frac{1}{4}\pi(t\pm 5i)^2/w}+\dots \text{ to } n \text{ terms}\right\}, \quad \dots\dots(14)$$

where s and t have the same relation as in (13).

Suppose now that $s=t$ in (13) and (14). Then we see that, if $w=in/m$, then

$$\phi_w(t)\{1+(-1)^{(m+1)(n+1)}e^{-\pi mt}\}$$
$$=e^{-\frac{1}{4}\pi t^2/w}\left\{e^{\frac{1}{4}\pi(t-w)^2/w}-e^{\frac{1}{4}\pi(t-3w)^2/w}+e^{\frac{1}{4}\pi(t-5w)^2/w}-\dots \text{ to } m \text{ terms}\right\}$$
$$+\frac{(-1)^{(m+1)(n+1)}}{\sqrt{w}}e^{-\pi mt}\left\{e^{-\frac{1}{4}\pi(t-i)^2/w}-e^{-\frac{1}{4}\pi(t-3i)^2/w}+\dots \text{ to } n \text{ terms}\right\}; \quad \dots\dots(15)$$

$$\{\tfrac{1}{2} - \psi_w(t)\}\{1 + (-1)^{mn+m+1} e^{-\pi m t}\}$$

$$= e^{-\frac{1}{4}\pi t^2/w}\left\{e^{\frac{1}{4}\pi(t-2w)^2/w} - e^{\frac{1}{4}\pi(t-4w)^2/w} + \dots \text{ to } m \text{ terms}\right\}$$

$$+ (-1)^{mn+m}\frac{i}{\sqrt{w}}e^{-\pi m t}\left\{e^{-\frac{1}{4}\pi(t-i)^2/w} + e^{-\frac{1}{4}\pi(t-3i)^2/w} + \dots \text{ to } n \text{ terms}\right\}, \quad \dots\dots(16)$$

where $\sqrt{w}$ should be taken as

$$e^{\frac{1}{4}\pi i}\sqrt{\left(\frac{n}{m}\right)}.$$

In (15) and (16) there is no loss of generality in supposing that one of the two numbers m and n is odd.

Now equating the real and imaginary parts in (15), we deduce that, if m and n are positive integers of which one is odd, then

$$2\cosh nt\int_0^\infty \frac{\cos 2tx}{\cosh \pi x}\cos\left(\frac{\pi m x^2}{n}\right)dx$$

$$= [\cosh\{(1-n)t\}\cos(\pi m/4n) - \cosh\{(3-n)t\}\cos(9\pi m/4n) + \dots \text{ to } n \text{ terms}]$$

$$+\sqrt{\left(\frac{n}{m}\right)}\left[\cosh\left\{\left(1-\frac{1}{m}\right)nt\right\}\cos\left(\frac{\pi}{4} - \frac{nt^2}{\pi m} + \frac{\pi n}{4m}\right)\right.$$

$$\left. - \cosh\left\{\left(1-\frac{3}{m}\right)nt\right\}\cos\left(\frac{\pi}{4} - \frac{nt^2}{\pi m} + \frac{9\pi n}{4m}\right) + \dots \text{ to } m \text{ terms}\right]; \quad \dots\dots(17)$$

and

$$2\cosh nt\int_0^\infty \frac{\cos 2tx}{\cosh \pi x}\sin\left(\frac{\pi m x^2}{n}\right)dx$$

$$= -[\cosh\{(1-n)t\}\sin(\pi m/4n) - \cosh\{(3-n)t\}\sin(9\pi m/4n)$$

$$+ \cosh\{(5-n)t\}\sin(25\pi/4n) - \dots \text{ to } n \text{ terms}]$$

$$+\sqrt{\left(\frac{n}{m}\right)}\left[\cosh\left\{\left(1-\frac{1}{m}\right)nt\right\}\sin\left(\frac{\pi}{4} - \frac{nt^2}{\pi m} + \frac{\pi n}{4m}\right)\right.$$

$$\left. - \cosh\left\{\left(1-\frac{3}{m}\right)nt\right\}\sin\left(\frac{\pi}{4} - \frac{nt^2}{\pi m} + \frac{9\pi n}{\pi m}\right) + \dots \text{ to } n \text{ terms}\right]. \quad \dots\dots(18)$$

Equating the real and imaginary parts in (16), we can find similar expressions for the integrals

$$\int_0^\infty \frac{\sin tx}{\sinh \pi x}\sin\left(\frac{\pi m x^2}{n}\right)dx, \quad \int_0^\infty \frac{\sin tx}{\sinh \pi x}\cos\left(\frac{\pi m x^2}{n}\right)dx.$$

From these formulæ we can evaluate a number of definite integrals, such as

$$\int_0^\infty \frac{\cos 2\pi tx}{\cosh \pi x} \cos \pi x^2 dx = \frac{1+\sqrt{2}\sin \pi t^2}{2\sqrt{2}\cosh \pi t},$$

$$\int_0^\infty \frac{\cos 2\pi tx}{\cosh \pi x} \sin \pi x^2 dx = \frac{-1+\sqrt{2}\cos \pi t^2}{2\sqrt{2}\cosh \pi t},$$

$$\int_0^\infty \frac{\sin 2\pi tx}{\sinh \pi x} \cos \pi x^2 dx = \frac{\cosh \pi t - \cos \pi t^2}{2\sinh \pi t},$$

$$\int_0^\infty \frac{\sin 2\pi tx}{\sinh \pi x} \sin \pi x^2 dx = \frac{\sin \pi t^2}{2\sinh \pi t},$$

and so on.

Again supposing that $s = -t$ in (13), we deduce that if $t = mw \pm ni$, where m and n are positive integers of which one at least is odd, then

$$\phi_w(t) = \tfrac{1}{2} e^{-\frac{1}{4}\pi t^2/w} \left\{ e^{\frac{1}{4}\pi (t-w)^2/w} - e^{\frac{1}{4}\pi (t-3w)^2/w} + \dots \text{ to } m \text{ terms} \right\}$$
$$+ \frac{1}{2\sqrt{w}} \left\{ e^{-\frac{1}{4}\pi (t\mp i)^2/w} - e^{-\frac{1}{4}\pi (t\mp 3i)^2/w} + \dots \text{ to } n \text{ terms} \right\}. \quad \dots(19)$$

This formula is not true when both m and n are even.

If $t = mw \pm ni$, where m and n are both even, then

$$\phi_w(t) + (-1)^{(1+\frac{1}{2}m)(1+\frac{1}{2}n)} e^{-\frac{1}{2}\pi mt} \phi_w(0)$$
$$= e^{-\frac{1}{4}\pi t^2/w} \left\{ e^{\frac{1}{4}\pi (t-w)^2/w} - e^{\frac{1}{4}\pi (t-3w)^2/w} + \dots \text{ to } \tfrac{1}{2}m \text{ terms} \right\}$$
$$+ \frac{(-1)^{(1+\frac{1}{2}m)(1+\frac{1}{2}n)}}{\sqrt{w}} e^{-\frac{1}{2}\pi mt} \left\{ e^{\frac{1}{4}\pi/w} - e^{\frac{9}{4}\pi/w} + e^{\frac{25}{4}\pi/w} - \dots \text{ to } \tfrac{1}{2}n \text{ terms} \right\}. \quad \dots\dots(20)$$

This is easily obtained by putting $t = 0$ and then changing s to t in (13). Similarly from (14) we deduce that if $t = mw \pm ni$, where m and n are both even, or both odd, or m is even and n is odd, then

$$\psi_w(t) = -\tfrac{1}{2} e^{-\frac{1}{4}\pi t^2/w} \left\{ e^{\frac{1}{4}\pi (t-2w)^2/w} - e^{\frac{1}{4}\pi (t-4w)^2/w} + \dots \text{ to } m \text{ terms} \right\}$$
$$\pm \frac{i}{2\sqrt{w}} \left\{ e^{-\frac{1}{4}\pi (t\mp i)^2/w} + e^{-\frac{1}{4}\pi (t\mp 3i)^2/w} + \dots \text{ to } n \text{ terms} \right\}. \quad \dots(21)$$

If $t = mw \pm ni$, where m is odd and n is even, then

$$\tfrac{1}{2} - \psi_w(t) + \{(-1)^{1+\frac{1}{4}(m-1)(n+2)} e^{-\frac{1}{4}\pi\{(m-1)t+mw\}} \phi_w(0)$$

$$= e^{-\frac{1}{4}\pi t^2/w} \left\{ e^{\frac{1}{4}\pi(t-2w)^2/w} - e^{\frac{1}{4}\pi(t-4w)^2/w} \ldots + \text{to } \tfrac{1}{2}(m-1) \text{ terms} \right\}$$

$$\pm (-1)^{1+\frac{1}{4}(m-1)(n+2)} \frac{i}{\sqrt{w}} e^{-\frac{1}{4}\pi(m-1)(t+w)}$$

$$\times \left\{ e^{-\frac{1}{4}\pi(w\pm i)^2/w} - e^{-\frac{1}{4}\pi(w\pm 3i)^2/w} + \ldots \text{ to } \tfrac{1}{2}n \text{ terms} \right\}. \quad \ldots(22)$$

This is obtained by putting $t = w$ in (14). A number of definite integrals such as the following can be evaluated with the help of the above formulæ:

$$\int_0^\infty \frac{\cos \pi t x}{\cosh \pi x} e^{-\pi(t+i)x^2} dx = \frac{1+i}{2\sqrt{2}} e^{-\frac{1}{4}\pi t} \left\{ 1 - \frac{i}{\sqrt{(t+i)}} \right\},$$

$$\int_0^\infty \frac{\sin \pi t x}{\sinh \pi x} e^{-\pi(t+i)x^2} dx = \frac{1}{2} - \frac{1+i}{2\sqrt{2}} \cdot \frac{e^{-\frac{1}{4}\pi t}}{\sqrt{(t+i)}},$$

and so on.

24

A PROOF OF BERTRAND'S POSTULATE

(*Journal of the Indian Mathematical Society*, XI, 1919, 181—182)

1. Landau in his *Handbuch*, pp. 89—92, gives a proof of a theorem the truth of which was conjectured by Bertrand: namely that there is at least one prime p such that $x < p \leqslant 2x$, if $x \geqslant 1$. Landau's proof is substantially the same as that given by Tschebyschef. The following is a much simpler one.

Let $\nu(x)$ denote the sum of the logarithms of all the primes not exceeding x and let

$$\psi(x) = \nu(x) + \nu(x^{\frac{1}{2}}) + \nu(x^{\frac{1}{3}}) + \ldots, \qquad \text{..............(1)}$$

$$\log [x]! = \psi(x) + \psi(\tfrac{1}{2}x) + \psi(\tfrac{1}{3}x) + \ldots, \qquad \text{..............(2)}$$

where $[x]$ denotes as usual the greatest integer in x.

From (1) we have

$$\psi(x) - 2\psi(\sqrt{x}) = \nu(x) - \nu(x^{\frac{1}{2}}) + \nu(x^{\frac{1}{3}}) - \ldots, \qquad \text{...........(3)}$$

and from (2)

$$\log [x]! - 2 \log [\tfrac{1}{2}x]! = \psi(x) - \psi(\tfrac{1}{2}x) + \psi(\tfrac{1}{3}x) - \ldots. \qquad \text{.........(4)}$$

Now remembering that $\nu(x)$ and $\psi(x)$ are steadily increasing functions, we find from (3) and (4) that

$$\psi(x) - 2\psi(\sqrt{x}) \leqslant \nu(x) \leqslant \psi(x); \qquad \text{....................(5)}$$

and $$\psi(x) - \psi(\tfrac{1}{2}x) \leqslant \log [x]! - 2\log[\tfrac{1}{2}x]! \leqslant \psi(x) - \psi(\tfrac{1}{2}x) + \psi(\tfrac{1}{3}x). \qquad \text{....(6)}$$

But it is easy to see that

$$\log \Gamma(x) - 2 \log \Gamma(\tfrac{1}{2}x + \tfrac{1}{2}) \leqslant \log [x]! - 2 \log [\tfrac{1}{2}x]!$$
$$\leqslant \log \Gamma(x+1) - 2 \log \Gamma(\tfrac{1}{2}x + \tfrac{1}{2}). \qquad \text{...(7)}$$

Now using Stirling's approximation we deduce from (7) that

$$\log [x]! - 2 \log [\tfrac{1}{2}x]! < \tfrac{3}{4}x, \text{ if } x > 0; \qquad \text{..........................(8)}$$

and $$\log [x]! - 2 \log [\tfrac{1}{2}x]! > \tfrac{2}{3}x, \text{ if } x > 300. \qquad \text{.......................(9)}$$

It follows from (6), (8) and (9) that

$$\psi(x) - \psi(\tfrac{1}{2}x) < \tfrac{3}{4}x, \text{ if } x > 0; \qquad \text{.................(10)}$$

and $$\psi(x) - \psi(\tfrac{1}{2}x) + \psi(\tfrac{1}{3}x) > \tfrac{2}{3}x, \text{ if } x > 300. \qquad \text{...............(11)}$$

Now changing x to $\frac{1}{2}x$, $\frac{1}{4}x$, $\frac{1}{8}x$, ... in (10) and adding up all the results, we obtain

$$\psi(x) < \tfrac{3}{2}x, \text{ if } x > 0. \quad\text{.........................(12)}$$

Again we have

$$\psi(x) - \psi(\tfrac{1}{2}x) + \psi(\tfrac{1}{3}x) \leqslant \nu(x) + 2\psi(\sqrt{x}) - \nu(\tfrac{1}{2}x) + \psi(\tfrac{1}{3}x)$$
$$< \nu(x) - \nu(\tfrac{1}{2}x) + \tfrac{1}{2}x + 3\sqrt{x}, \text{ ...(13)}$$

in virtue of (5) and (12).

It follows from (11) and (13) that

$$\nu(x) - \nu(\tfrac{1}{2}x) > \tfrac{1}{6}x - 3\sqrt{x}, \text{ if } x > 300. \quad\text{..............(14)}$$

But it is obvious that $\quad \frac{1}{6}x - 3\sqrt{x} \geqslant 0$, if $x \geqslant 324$.

Hence $\quad \nu(2x) - \nu(x) > 0$, if $x \geqslant 162$. ....(15)

In other words there is at least one prime between x and $2x$ if $x \geqslant 162$. Thus Bertrand's Postulate is proved for all values of x not less than 162; and, by actual verification, we find that it is true for smaller values.

2. Let $\pi(x)$ denote the number of primes not exceeding x. Then, since $\pi(x) - \pi(\frac{1}{2}x)$ is the number of primes between x and $\frac{1}{2}x$, and $\nu(x) - \nu(\frac{1}{2}x)$ is the sum of logarithms of primes between x and $\frac{1}{2}x$, it is obvious that

$$\nu(x) - \nu(\tfrac{1}{2}x) \leqslant \{\pi(x) - \pi(\tfrac{1}{2}x)\} \log x, \quad\text{..............(16)}$$

for all values of x. It follows from (14) and (16) that

$$\pi(x) - \pi(\tfrac{1}{2}x) > \frac{1}{\log x}(\tfrac{1}{6}x - 3\sqrt{x}), \text{ if } x > 300. \quad\text{.........(17)}$$

From this we easily deduce that

$$\pi(x) - \pi(\tfrac{1}{2}x) \geqslant 1, 2, 3, 4, 5, \ldots, \text{ if } x \geqslant 2, 11, 17, 29, 41, \ldots, \text{ ...(18)}$$

respectively.

25

SOME PROPERTIES OF $p(n)$, THE NUMBER OF PARTITIONS OF n*

(*Proceedings of the Cambridge Philosophical Society*, XIX, 1919, 207—210)

1. A recent paper by Mr Hardy and myself† contains a table, calculated by Major MacMahon, of the values of $p(n)$, the number of unrestricted partitions of n, for all values of n from 1 to 200. On studying the numbers in this table I observed a number of curious congruence properties, apparently satisfied by $p(n)$. Thus

(1)	$p(4)$,	$p(9)$,	$p(14)$,	$p(19)$, ...	$\equiv 0 \pmod{5}$,
(2)	$p(5)$,	$p(12)$,	$p(19)$,	$p(26)$, ...	$\equiv 0 \pmod{7}$,
(3)	$p(6)$,	$p(17)$,	$p(28)$,	$p(39)$, ...	$\equiv 0 \pmod{11}$,
(4)	$p(24)$,	$p(49)$,	$p(74)$,	$p(99)$, ...	$\equiv 0 \pmod{25}$,
(5)	$p(19)$,	$p(54)$,	$p(89)$,	$p(124)$, ...	$\equiv 0 \pmod{35}$,
(6)	$p(47)$,	$p(96)$,	$p(145)$,	$p(194)$, ...	$\equiv 0 \pmod{49}$,
(7)	$p(39)$,	$p(94)$,	$p(149)$, ...		$\equiv 0 \pmod{55}$,
(8)	$p(61)$,	$p(138)$, ...			$\equiv 0 \pmod{77}$,
(9)	$p(116)$, ...				$\equiv 0 \pmod{121}$,
(10)	$p(99)$, ...				$\equiv 0 \pmod{125}$.

From these data I conjectured the truth of the following theorem:

If $\delta = 5^a 7^b 11^c$ *and* $24\lambda \equiv 1 \pmod{\delta}$, *then*

$$p(\lambda),\quad p(\lambda+\delta),\quad p(\lambda+2\delta),\ \ldots \equiv 0 \pmod{\delta}.$$

This theorem is supported by all the available evidence; but I have not yet been able to find a general proof.

I have, however, found quite simple proofs of the theorems expressed by (1) and (2), viz.

(1) $$p(5m+4) \equiv 0 \pmod{5}$$

and (2) $$p(7m+5) \equiv 0 \pmod{7}.$$

From these

(5) $$p(35m+19) \equiv 0 \pmod{35}$$

* [See also Ramanujan's posthumous paper "Congruence properties of partitions" in the *Math. Zeitschrift*, No. 30 of this volume.]

† G. H. Hardy and S. Ramanujan, "Asymptotic formulæ in Combinatory Analysis," *Proc. London Math. Soc.*, Ser. 2, Vol. XVII, 1918, pp. 75—115 (Table IV, pp. 114—115) [No. 36 of this volume].

follows at once as a corollary. These proofs I give in § 2 and § 3. I can also prove

(4) $$p(25n+24) \equiv 0 \pmod{25}$$

and (6) $$p(49n+47) \equiv 0 \pmod{49},$$

but only in a more recondite way, which I sketch in § 4.

2. *Proof of* (1). We have

(11) $$x\{(1-x)(1-x^2)(1-x^3)\ldots\}^4$$
$$= x(1-3x+5x^3-7x^6+\ldots)(1-x-x^2+x^5+\ldots)$$
$$= \Sigma(-1)^{\mu+\nu}(2\mu+1)\,x^{1+\frac{1}{2}\mu(\mu+1)+\frac{1}{2}\nu(3\nu+1)},$$

the summation extending from $\mu=0$ to $\mu=\infty$ and from $\nu=-\infty$ to $\nu=\infty$. Now if

$$1+\tfrac{1}{2}\mu(\mu+1)+\tfrac{1}{2}\nu(3\nu+1) \equiv 0 \pmod 5,$$

then $$8+4\mu(\mu+1)+4\nu(3\nu+1) \equiv 0 \pmod 5,$$

and therefore

(12) $$(2\mu+1)^2+2(\nu+1)^2 \equiv 0 \pmod 5.$$

But $(2\mu+1)^2$ is congruent to 0, 1, or 4, and $2(\nu+1)^2$ to 0, 2, or 3. Hence it follows from (12) that $2\mu+1$ and $\nu+1$ are both multiples of 5. That is to say, the coefficient of x^{5n} in (11) is a multiple of 5.

Again, all the coefficients in $(1-x)^{-5}$ are multiples of 5, except those of $1, x^5, x^{10}, \ldots$, which are congruent to 1: that is to say

$$\frac{1}{(1-x)^5} \equiv \frac{1}{1-x^5} \pmod 5,$$

or $$\frac{1-x^5}{(1-x)^5} \equiv 1 \pmod 5.$$

Thus all the coefficients in

$$\frac{(1-x^5)(1-x^{10})(1-x^{15})\ldots}{\{(1-x)(1-x^2)(1-x^3)\ldots\}^5}$$

(except the first) are multiples of 5. Hence the coefficient of x^{5n} in

$$\frac{x(1-x^5)(1-x^{10})\ldots}{(1-x)(1-x^2)(1-x^3)\ldots} = x\{(1-x)(1-x^2)\ldots\}^4 \frac{(1-x^5)(1-x^{10})\ldots}{\{(1-x)(1-x^2)\ldots\}^5}$$

is a multiple of 5. And hence, finally, the coefficient of x^{5n} in

$$\frac{x}{(1-x)(1-x^2)(1-x^3)\ldots}$$

is a multiple of 5; which proves (1).

3. *Proof of* (2). The proof of (2) is very similar. We have

$$(13)\quad x^2\{(1-x)(1-x^2)(1-x^3)\ldots\}^6$$
$$= x^2(1-3x+5x^3-7x^6+\ldots)^2$$
$$= \Sigma(-1)^{\mu+\nu}(2\mu+1)(2\nu+1)\,x^{2+\frac{1}{2}\mu(\mu+1)+\frac{1}{2}\nu(\nu+1)},$$

the summation now extending from 0 to ∞ for both μ and ν. If

$$2+\tfrac{1}{2}\mu(\mu+1)+\tfrac{1}{2}\nu(\nu+1)\equiv 0 \pmod 7,$$

then

$$16+4\mu(\mu+1)+4\nu(\nu+1)\equiv 0 \pmod 7,$$
$$(2\mu+1)^2+(2\nu+1)^2\equiv 0 \pmod 7,$$

and $2\mu+1$ and $2\nu+1$ are both divisible by 7. Thus the coefficient of x^{7n} in (13) is divisible by 49.

Again, all the coefficients in

$$\frac{(1-x^7)(1-x^{14})(1-x^{21})\ldots}{\{(1-x)(1-x^2)(1-x^3)\ldots\}^7}$$

(except the first) are multiples of 7. Hence (arguing as in § 2) we see that the coefficient of x^{7n} in

$$\frac{x^2}{(1-x)(1-x^2)(1-x^3)\ldots}$$

is a multiple of 7; which proves (2). As I have already pointed out, (5) is a corollary.

4. The proofs of (4) and (6) are more intricate, and in order to give them I have to consider a much more difficult problem, viz. that of expressing

$$p(\lambda)+p(\lambda+\delta)\,x+p(\lambda+2\delta)\,x^2+\ldots$$

in terms of Theta-functions, in such a manner as to exhibit explicitly the common factors of the coefficients, if such common factors exist. I shall content myself with sketching the method of proof, reserving any detailed discussion of it for another paper.

It can be shewn that

$$(14)\quad \frac{(1-x^5)(1-x^{10})(1-x^{15})\ldots}{(1-x^{\frac{1}{5}})(1-x^{\frac{2}{5}})(1-x^{\frac{3}{5}})\ldots}=\frac{1}{\xi^{-1}-x^{\frac{1}{5}}-\xi x^{\frac{2}{5}}}$$
$$=\frac{\xi^{-4}-3x\xi+x^{\frac{1}{5}}(\xi^{-3}+2x\xi^2)+x^{\frac{2}{5}}(2\xi^{-2}-x\xi^3)+x^{\frac{3}{5}}(3\xi^{-1}+x\xi^4)+5x^{\frac{4}{5}}}{\xi^{-5}-11x-x^2\xi^5},$$

where

$$\xi=\frac{(1-x)(1-x^4)(1-x^6)(1-x^9)\ldots}{(1-x^2)(1-x^3)(1-x^7)(1-x^8)\ldots},$$

the indices of the powers of x, in both numerator and denominator of ξ, forming two arithmetical progressions with common difference 5. It follows that

$$(15)\quad (1-x^5)(1-x^{10})(1-x^{15})\ldots\{p(4)+p(9)\,x+p(14)\,x^2+\ldots\}$$
$$=\frac{5}{\xi^{-5}-11x-x^2\xi^5}.$$

Again, if in (14) we substitute $\omega x^{\frac{1}{5}}$, $\omega^2 x^{\frac{1}{5}}$, $\omega^3 x^{\frac{1}{5}}$, and $\omega^4 x^{\frac{1}{5}}$, where $\omega^5 = 1$, for $x^{\frac{1}{5}}$, and multiply the resulting five equations, we obtain

$$(16) \qquad \left\{\frac{(1-x^5)(1-x^{10})(1-x^{15})\ldots}{(1-x)(1-x^2)(1-x^3)\ldots}\right\}^6 = \frac{1}{\xi^{-5} - 11x - x^2\xi^5}.$$

From (15) and (16) we deduce

$$(17) \quad p(4) + p(9)x + p(14)x^2 + \ldots = 5\frac{\{(1-x^5)(1-x^{10})(1-x^{15})\ldots\}^5}{\{(1-x)(1-x^2)(1-x^3)\ldots\}^6};$$

from which it appears directly that $p(5m+4)$ is divisible by 5.

The corresponding formula involving 7 is

$$(18) \quad p(5) + p(12)x + p(19)x^2 + \ldots = 7\frac{\{(1-x^7)(1-x^{14})(1-x^{21})\ldots\}^3}{\{(1-x)(1-x^2)(1-x^3)\ldots\}^4} + 49x\frac{\{(1-x^7)(1-x^{14})(1-x^{21})\ldots\}^7}{\{(1-x)(1-x^2)(1-x^3)\ldots\}^8},$$

which shews that $p(7m+5)$ is divisible by 7.

From (16) it follows that

$$\frac{p(4)x + p(9)x^2 + p(14)x^3 + \ldots}{5\{(1-x^5)(1-x^{10})(1-x^{15})\ldots\}^4} = \frac{x}{(1-x)(1-x^2)(1-x^3)\ldots}\,\frac{(1-x^5)(1-x^{10})(1-x^{15})\ldots}{\{(1-x)(1-x^2)(1-x^3)\ldots\}^5}.$$

As the coefficient of x^{5n} on the right-hand side is a multiple of 5, it follows that $p(25m+24)$ is divisible by 25.

Similarly

$$\frac{p(5)x + p(12)x^2 + p(19)x^3 + \ldots}{7\{(1-x^7)(1-x^{14})(1-x^{21})\ldots\}^2} = x(1 - 3x + 5x^3 - 7x^6 + \ldots)\frac{(1-x^7)(1-x^{14})\ldots}{\{(1-x)(1-x^2)\ldots\}^7} + 7x^2\frac{\{(1-x^7)(1-x^{14})\ldots\}^5}{\{(1-x)(1-x^2)\ldots\}^8};$$

from which it follows that $p(49m+47)$ is divisible by 49.

Another proof of (1) and (2) has been found by Mr H. B. C. Darling, to whom my conjecture had been communicated by Major MacMahon. This proof will also be published in these *Proceedings* [see Appendix, p. 343]. I have since found proofs of (3), (7) and (8).

26

PROOF OF CERTAIN IDENTITIES IN COMBINATORY ANALYSIS

(*Proceedings of the Cambridge Philosophical Society*, XIX, 1919, 214—216) *

Let $$G(x) = 1$$

$$+\sum_{1}^{\infty} (-1)^{\nu} x^{2\nu} q^{\frac{1}{2}\nu(5\nu-1)} (1 - xq^{2\nu}) \frac{(1-xq)(1-xq^2)\dots(1-xq^{\nu-1})}{(1-q)(1-q^2)(1-q^3)\dots(1-q^{\nu})}$$

$$= 1 - x^2q^2(1-xq^2)\frac{1}{1-q} + x^4q^9(1-xq^4)\frac{1-xq}{(1-q)(1-q^2)} - \dots \quad \dots(1)$$

If we write $$1 - xq^{2\nu} = 1 - q^{\nu} + q^{\nu}(1 - xq^{\nu}),$$

every term in (1) is split up into two parts. Associating the second part of each term with the first part of the succeeding term, we obtain

$$G(x) = (1 - x^2q^2) - x^2q^3(1 - x^2q^6)\frac{1-xq}{1-q}$$

$$+ x^4q^{11}(1 - x^2q^{10})\frac{(1-xq)(1-xq^2)}{(1-q)(1-q^2)} - \dots \quad \dots\dots(2)$$

Now consider $$H(x) = \frac{G(x)}{1-xq} - G(xq). \quad \dots\dots(3)$$

Substituting for the first term from (2) and for the second term from (1), we obtain

$$H(x) = xq - \frac{x^2q^3}{1-q}\{(1-q) + xq^4(1-xq^2)\}$$

$$+ \frac{x^4q^{11}(1-xq^2)}{(1-q)(1-q^2)}\{(1-q^2) + xq^7(1-xq^3)\}$$

$$- \frac{x^6q^{24}(1-xq^2)(1-xq^3)}{(1-q)(1-q^2)(1-q^3)}\{(1-q^3) + xq^{10}(1-xq^4)\} + \dots.$$

Associating, as before, the second part of each term with the first part of the succeeding term, we obtain

$$H(x) = xq(1-xq^2)\left\{1 - x^2q^6(1-xq^4)\frac{1}{1-q}\right.$$

$$+ x^4q^{17}(1-xq^6)\frac{1-xq^3}{(1-q)(1-q^2)}$$

$$\left. - x^6q^{33}(1-xq^8)\frac{(1-xq^3)(1-xq^4)}{(1-q)(1-q^2)(1-q^3)} + \dots\right\}$$

$$= xq(1-xq^2)\,G(xq^2). \quad \dots\dots(4)$$

* [See the Appendix, p. 344, for the history of this paper.]

If now we write
$$K(x) = \frac{G(x)}{(1-xq)\,G(xq)},$$
we obtain, from (3) and (4),
$$K(x) = 1 + \frac{xq}{K(xq)},$$
and so
$$K(x) = 1 + \frac{xq}{1+}\,\frac{xq^2}{1+}\,\frac{xq^3}{1+\ldots}. \qquad \ldots\ldots\ldots\ldots\ldots\ldots\ldots\ldots(5)$$
In particular we have
$$\frac{1}{1+}\,\frac{q}{1+}\,\frac{q^2}{1+\ldots} = \frac{1}{K(1)} = \frac{(1-q)\,G(q)}{G(1)}; \qquad \ldots\ldots\ldots\ldots\ldots(6)$$
or
$$\frac{1}{1+}\,\frac{q}{1+}\,\frac{q^2}{1+\ldots} = \frac{1-q-q^4+q^7+q^{13}-\ldots}{1-q^2-q^3+q^9+q^{11}-\ldots}. \qquad \ldots\ldots\ldots\ldots\ldots(7)$$
This equation may also be written in the form
$$\frac{1}{1+}\,\frac{q}{1+}\,\frac{q^2}{1+\ldots} = \frac{(1-q)(1-q^4)(1-q^6)(1-q^9)(1-q^{11})\ldots}{(1-q^2)(1-q^3)(1-q^7)(1-q^8)(1-q^{12})\ldots}. \qquad \ldots(8)$$

If we write
$$F(x) = \frac{G(x)}{(1-xq)(1-xq^2)(1-xq^3)\ldots},$$
then (4) becomes
$$F(x) = F(xq) + xqF(xq^2),$$
from which it readily follows that
$$F(x) = 1 + \frac{xq}{1-q} + \frac{x^2q^4}{(1-q)(1-q^2)} + \frac{x^3q^9}{(1-q)(1-q^2)(1-q^3)} + \ldots. \qquad (9)$$
In particular we have
$$1 + \frac{q}{1-q} + \frac{q^4}{(1-q)(1-q^2)} + \ldots = \frac{G(1)}{(1-q)(1-q^2)(1-q^3)\ldots}$$
$$= \frac{1-q^2-q^3+q^9+q^{11}-\ldots}{(1-q)(1-q^2)(1-q^3)\ldots}$$
$$= \frac{1}{(1-q)(1-q^4)(1-q^6)(1-q^9)(1-q^{11})\ldots}, \qquad \ldots\ldots(10)$$
and
$$1 + \frac{q^2}{1-q} + \frac{q^6}{(1-q)(1-q^2)} + \ldots = \frac{(1-q)\,G(q)}{(1-q)(1-q^2)(1-q^3)\ldots}$$
$$= \frac{1-q-q^4+q^7+q^{13}-\ldots}{(1-q)(1-q^2)(1-q^3)\ldots}$$
$$= \frac{1}{(1-q^2)(1-q^3)(1-q^7)(1-q^8)(1-q^{12})\ldots}. \qquad \ldots\ldots(11)$$

27

A CLASS OF DEFINITE INTEGRALS

(*Quarterly Journal of Mathematics*, XLVIII, 1920, 294—310)

1. It is well known that

$$(1\cdot1)\quad \int_{-\frac{1}{2}\pi}^{\frac{1}{2}\pi} (\cos x)^m e^{inx}\,dx = \frac{\pi}{2^m}\frac{\Gamma(1+m)}{\Gamma\{1+\frac{1}{2}(m+n)\}\Gamma\{1+\frac{1}{2}(m-n)\}}$$

if $R(m) > -1$. It follows from this and Fourier's Theorem that, if n is any real number except $\pm\pi$ and $R(\alpha+\beta)>1$, or if $n=\pm\pi$ and $R(\alpha+\beta)>2$, then

$$(1\cdot2)\quad \int_{-\infty}^{\infty} \frac{e^{inx}}{\Gamma(\alpha+x)\Gamma(\beta-x)}\,dx = \frac{(2\cos\frac{1}{2}n)^{\alpha+\beta-2}}{\Gamma(\alpha+\beta-1)}e^{\frac{1}{2}in(\beta-\alpha)} \text{ or } 0,$$

according as $|n|<\pi$ or $|n|\geqslant\pi$. In particular we have

$$(1\cdot21)\quad \int_{-\infty}^{\infty} \frac{dx}{\Gamma(\alpha+x)\Gamma(\beta-x)} = \frac{2^{\alpha+\beta-2}}{\Gamma(\alpha+\beta-1)}$$

if $R(\alpha+\beta)>1$; and

$$(1\cdot22)\quad \int_{0}^{\infty} \frac{dx}{\Gamma(\alpha+x)\Gamma(\alpha-x)} = \frac{2^{2\alpha-3}}{\Gamma(2\alpha-1)}$$

if $R(\alpha)>\frac{1}{2}$. If α is an integer $n+1$, (1·22) reduces to

$$(1\cdot23)\quad \int_{0}^{\infty} \frac{\sin\pi x}{x\{1-(x^2/1^2)\}\{1-(x^2/2^2)\}\ldots\{1-(x^2/n^2)\}}\,dx = \frac{\pi}{2}\frac{2^{2n}(n!)^2}{(2n)!}.$$

Again, if m is a positive integer, we have

$$\frac{\sin m\pi x}{\sin\pi x} = 1 + 2\cos 2\pi x + 2\cos 4\pi x + \ldots \text{ to } \tfrac{1}{2}(m+1) \text{ terms}$$

or

$$2\cos\pi x + 2\cos 3\pi x + \ldots \text{ to } \tfrac{1}{2}m \text{ terms},$$

according as m is odd or even. It follows from this and (1·2) that, if $R(\alpha)>1$,

$$\int_{0}^{\infty} \frac{\sin m\pi x}{\sin\pi x}\frac{dx}{\Gamma(\alpha+x)\Gamma(\alpha-x)} = \frac{2^{2\alpha-3}}{\Gamma(2\alpha-1)} \text{ or } 0,$$

according as m is odd or even. Hence, if m and n are positive integers, we have

(1·24)

$$\int_{0}^{\infty} \frac{\sin m\pi x}{x\{1-(x^2/1^2)\}\{1-(x^2/2^2)\}\ldots\{1-(x^2/n^2)\}}\,dx = \frac{\pi}{2}\frac{2^{2n}(n!)^2}{(2n)!} \text{ or } 0,$$

according as m is odd or even. From this we easily deduce that, if l, m, and n are positive integers,

(1·25)

$$\int_{0}^{\infty} \frac{(\sin m\pi x)^{2l+1}}{x\{1-(x^2/1^2)\}\{1-(x^2/2^2)\}\ldots\{1-(x^2/n^2)\}}\,dx = 2^{2(n-l)-1}\frac{(2l)!}{(2n)!}\left(\frac{n!}{l!}\right)^2\pi \text{ or } 0,$$

according as m is odd or even. It follows that

$$(1\cdot26)\quad \int_0^\infty \frac{(\sin m\pi x)^{2n+1}}{x\{1-(x^2/1^2)\}\{1-(x^2/2^2)\}\ldots\{1-(x^2/n^2)\}}\,dx = \frac{\pi}{2} \text{ or } 0,$$

according as m is odd or even. Similarly we can shew that

$$(1\cdot27)\quad \int_0^\infty \frac{(\sin m\pi x)^{2l}}{\{1-(x^2/1^2)\}\{1-(x^2/2^2)\}\ldots\{1-(x^2/n^2)\}}\,dx = 0$$

for all positive integral values of l, m, and n.

In this connection it is interesting to note the following results:

(i) If $R(\alpha+\beta)>1$ and $R(\gamma+\delta)>1$, then

$$(1\cdot3)\quad \Gamma(\alpha+\beta-1)\int_{-\frac{1}{2}}^{\frac{1}{2}} \frac{(2\cos\pi x)^{\gamma+\delta-2}\, e^{i\pi(\gamma-\delta)x}}{\Gamma(\alpha+x)\,\Gamma(\beta-x)}\,dx$$

$$= \Gamma(\gamma+\delta-1)\int_{-\frac{1}{2}}^{\frac{1}{2}} \frac{(2\cos\pi x)^{\alpha+\beta-2}\, e^{i\pi(\alpha-\beta)x}}{\Gamma(\gamma+x)\,\Gamma(\delta-x)}\,dx.$$

This is easily proved by writing

$$\int_{-\frac{1}{2}}^{\frac{1}{2}} (2\cos\pi z)^{\alpha+\beta-2}\, e^{i\pi z(\alpha-\beta+2x)}\,dz$$

instead of

$$\frac{\Gamma(\alpha+\beta-1)}{\Gamma(\alpha+x)\,\Gamma(\beta-x)}$$

in the left-hand side of (1·3).

(ii) If m and n are integers of which one is odd and the other even, and $m \geqslant 0$ and $R(\alpha+\beta)>2$, then

(1·4)

$$(\alpha+\beta-2)\int_\xi^\infty \frac{(\cos\pi x)^m\, e^{in\pi x}}{\Gamma(\alpha+x)\,\Gamma(\beta-x)}\,dx = \int_\xi^{\xi+1} \frac{(\cos\pi x)^m\, e^{in\pi x}}{\Gamma(\alpha-1+x)\,\Gamma(\beta-x)}\,dx$$

where ξ is any real number. This is proved as follows: Suppose that the left-hand side, minus the right-hand side, is $f(\xi)$. Then

$$f'(\xi) = -\frac{(\alpha+\beta-2)(\cos\pi\xi)^m}{\Gamma(\alpha+\xi)\,\Gamma(\beta-\xi)}\, e^{in\pi\xi}$$

$$-\frac{(-1)^{m+n}(\cos\pi\xi)^m\, e^{in\pi\xi}}{\Gamma(\alpha+\xi)\,\Gamma(\beta-\xi-1)} + \frac{(\cos\pi\xi)^m\, e^{in\pi\xi}}{\Gamma(\alpha-1+\xi)\,\Gamma(\beta-\xi)} = 0.$$

Hence $f(\xi)$ is a constant which is easily seen to be zero.

2. Before proceeding further I shall give a few general rules for generalising the results in the previous and the following sections. If

$$f_r(\zeta) = \int_\xi^\eta \frac{(x+\epsilon_1)(x+\epsilon_2)\ldots(x+\epsilon_r)}{\Gamma(\zeta\pm x)}\,F(x)\,dx,$$

where r is zero or a positive integer, and the ϵ's, ξ, η, ζ and $F(x)$ are all arbitrary, then it is easy to see that

$$(2\cdot1)\qquad f_{r+1}(\zeta+1)=\pm\{f_r(\zeta)-(\zeta\mp\epsilon_{r+1})f_r(\zeta+1)\},$$

provided the necessary convergence conditions are satisfied.

Similarly if

$$f_r(\zeta)=\int_\xi^\eta (x+\epsilon_1)(x+\epsilon_2)\dots(x+\epsilon_r)\,\Gamma(\zeta\pm x)\,F(x)\,dx,$$

then

$$(2\cdot2)\qquad f_{r+1}(\zeta)=\pm\{f_r(\zeta+1)-(\zeta\mp\epsilon_{r+1})f_r(\zeta)\}.$$

Thus we see that, if $f_0(\zeta)$ is known, $f_r(\zeta)$ can be easily determined.

Suppose now that $P(x)$ is a polynomial of the rth degree and N any integer greater than or equal to r. Let D, E and Δ denote the usual operators so that

$$E=1+\Delta=e^D.$$

Then, if

$$f(\zeta)=\int_\xi^\eta \frac{F(x)}{\Gamma(\zeta\pm x)}\,dx,$$

$$(2\cdot3)\qquad \int_\xi^\eta \frac{P(x)\,F(x)}{\Gamma(\zeta\pm x)}\,dx=\sum_0^N \frac{f(\zeta-\nu)}{\nu!}(\pm\Delta)^\nu P\{-\tfrac12\nu\pm(1-\zeta+\tfrac12\nu)\},$$

as is easily seen by replacing $P(x)$ by

$$(1\pm\Delta E^{-\frac12\pm\frac12})^{\zeta\pm x-1}P\{\pm(1-\zeta)\}.$$

Similarly, using the equation

$$P(x)=(1\mp\Delta E^{-\frac12\mp\frac12})^{-(\zeta\pm x)}P(\pm\zeta),$$

we find that, if

$$f(\zeta)=\int_\xi^\eta \Gamma(\zeta\pm x)\,F(x)\,dx,$$

then

$$(2\cdot4)\qquad \int_\xi^\eta \Gamma(\zeta\pm x)\,P(x)\,F(x)\,dx=\sum_0^N\frac{f(\zeta+\nu)}{\nu!}(\pm\Delta)^\nu P\{-\tfrac12\nu\mp(\zeta+\tfrac12\nu)\}.$$

As an illustration let us apply (2·3) to (1·2). We find that if n is any real number except $\pm\pi$, and $R(\alpha+\beta)>1+r$, or if $n=\pm\pi$ and $R(\alpha+\beta)>2+r$, and N is any integer greater than or equal to r, where r is the degree of the polynomial $P(x)$, then

$$(2\cdot5)\qquad \int_{-\infty}^\infty \frac{P(x)\,e^{inx}}{\Gamma(\alpha+x)\,\Gamma(\beta-x)}\,dx=0$$

$$or\ \sum_0^N\frac{k_\nu}{\nu!}\frac{(2\cos\frac12 n)^{\alpha+\beta-\nu-2}}{\Gamma(\alpha+\beta-\nu-1)}e^{\frac12 in(\beta-\alpha)},$$

according as $|n|\geqq\pi$ or $|n|<\pi$, k_ν being either $e^{\frac12 in\nu}\Delta^\nu P(1-\alpha)$ or

$$e^{-\frac12 in\nu}(-\Delta)^\nu P(\beta-\nu-1).$$

It is immaterial which value of k_ν we take.

If
$$P(x) = \frac{\Gamma(\zeta_1 + x)}{\Gamma(\zeta_2 + x)},$$
where $\zeta_1 - \zeta_2$ is a positive integer, then it is well known that

$$\Delta^\nu P(x) = \frac{(\zeta_1 - \zeta_2)!}{(\zeta_1 - \zeta_2 - \nu)!} \frac{\Gamma(\zeta_1 + x)}{\Gamma(\zeta_2 + x + \nu)}. \tag{2·6}$$

This affords a very good example for the previous formulæ.

It is easy to see that (1·2) can be restated as

$$\int_{-\infty}^{\infty} \frac{e^{inx}}{\Gamma(\alpha + x)\Gamma(\beta - x)} dx = 0 \tag{2·7}$$

or

$$\int_{-\infty}^{\infty} \frac{e^{inx}}{\Gamma(\alpha + x)\Gamma(\beta - x)} dx = \sum_1^\infty \frac{e^{in(\nu - \alpha)}}{\Gamma(\nu)\Gamma(\alpha + \beta - \nu)} = \sum_1^\infty \frac{e^{in(\beta - \nu)}}{\Gamma(\nu)\Gamma(\alpha + \beta - \nu)}, \tag{2·71}$$

according as $|n| \geqslant \pi$ or $|n| < \pi$ (n being of course real). But

$$\int_{-\infty}^{\infty} \frac{P(x)\, e^{inx}}{\Gamma(\alpha + x)\Gamma(\beta - x)} dx = \int_{-\infty}^{\infty} \frac{e^{(D+in)x} P(0)}{\Gamma(\alpha + x)\Gamma(\beta - x)} dx.$$

Hence, if the conditions stated for (2·5) are satisfied,

$$\int_{-\infty}^{\infty} \frac{P(x)\, e^{inx}}{\Gamma(\alpha + x)\Gamma(\beta - x)} dx = 0 \tag{2·8}$$

or

$$\int_{-\infty}^{\infty} \frac{P(x)\, e^{inx}}{\Gamma(\alpha + x)\Gamma(\beta - x)} dx = \sum_1^\infty \frac{e^{in(\nu - \alpha)} P(\nu - \alpha)}{\Gamma(\nu)\Gamma(\alpha + \beta - \nu)} = \sum_1^\infty \frac{e^{in(\beta - \nu)} P(\beta - \nu)}{\Gamma(\nu)\Gamma(\alpha + \beta - \nu)}, \tag{2·81}$$

according as $|n| \geqslant \pi$ or $|n| < \pi$.

3. We shall now consider an important extension of (1·2). Let $[x]$ denote the greatest integer not exceeding x, so that (e.g.) $[-5\tfrac{1}{2}] = -6$. Let us agree further that

$$\sum_\mu^\nu = 0,$$

if $\nu < \mu$. Then if n and s are real and $\phi(z)$ is a function that can be expanded in the form

$$\sum_{-\infty}^{\infty} C_\nu z^\nu$$

when $|z| = 1$, we have

$$\int_{-\infty}^{\infty} \frac{\phi(e^{isx})\, e^{inx}}{\Gamma(\alpha + x)\Gamma(\beta - x)} dx = \sum C_\nu \frac{\{2\cos\tfrac{1}{2}(n + \nu s)\}^{\alpha+\beta-2}}{\Gamma(\alpha + \beta - 1)} e^{\frac{1}{2}i(\beta-\alpha)(n+\nu s)}, \tag{3·1}$$

where the summation is bounded by

$$-\left[\frac{\pi}{|s|}+\frac{n}{s}\right] \leqslant \nu \leqslant \left[\frac{\pi}{|s|}-\frac{n}{s}\right],$$

provided that either

(i) $\pi+n$ and $\pi-n$ are not multiples of s, and $R(\alpha+\beta)>1$,

or (ii) $\pi+n$ or $\pi-n$ is a multiple of s, and $R(\alpha+\beta)>2$,

or (iii) $C_{(\pi-n)/s}=0$ and $C_{-(\pi+n)/s}=0$, whenever one or the other or both of the suffixes of C happen to be integral, and $R(\alpha+\beta)>1$.

or (iv) $\pi+n$ and $\pi-n$ are multiples of s, $\alpha+\beta$ is an integer greater than 1, and $C_{(\pi-n)/s}=e^{2i\pi\alpha}C_{-(\pi+n)/s}$.

The formula (3·1) is easily obtained by substituting the series for ϕ and integrating term-by-term, using (1·2).

It should be remembered that (3·1) is not true if n or s ceases to be real, though the integral may be convergent. In such cases, generally, the integral cannot be evaluated in finite terms.

The following integrals can be evaluated at once, in finite terms, with the help of (3·1):

(3·11) $$\int_{-\infty}^{\infty}\frac{dx}{(pe^{imx}+qe^{inx})\,\Gamma(\alpha+x)\,\Gamma(\beta-x)},$$

where m and n are real and $|p|\neq|q|$;

(3·12) $$\int_{-\infty}^{\infty}\frac{\begin{pmatrix}\cos\\ \sin\end{pmatrix}^{m} sx\; e^{inx}}{\Gamma(\alpha+x)\,\Gamma(\beta-x)}\,dx,$$

where n and s are real and m is a positive integer;

(3·13) $$\int_{-\infty}^{\infty}\frac{(1+\epsilon e^{isx})^{m}\,e^{inx}}{\Gamma(\alpha+x)\,\Gamma(\beta-x)}\,dx,$$

where m is real and $|\epsilon|<1$;

(3·14) $$\int_{-\infty}^{\infty}\frac{e^{p\cos sx+iq\sin sx+inx}}{\Gamma(\alpha+x)\,\Gamma(\beta-x)}\,dx.$$

For instance, the value of (3·14) is

$$\Sigma\left(\frac{q+p}{q-p}\right)^{\frac{1}{2}\nu} J_\nu\{\sqrt{(q^2-p^2)}\}\frac{\{2\cos\frac{1}{2}(n+\nu s)\}^{\alpha+\beta-2}}{\Gamma(\alpha+\beta-1)}e^{\frac{1}{2}i(\beta-\alpha)(n+\nu s)},$$

where $J_\nu(x)$ is the ordinary Bessel function of the νth order, and $\{(q+p)/(q-p)\}^{\frac{1}{2}\nu}$ should be interpreted so that the first term in the expansion of

$$\{(q+p)/(q-p)\}^{\frac{1}{2}\nu} J_\nu\{\sqrt{(q^2-p^2)}\}$$

is

$$\{\tfrac{1}{2}(p+q)\}^{\nu}/\nu!.$$

Putting $s=2\pi$ in (3·1) we obtain the following corollary. If ϕ is the same function as in (3·1), and n is any real number, then

$$(3\cdot2)\quad \int_{-\infty}^{\infty} \frac{\phi(e^{2i\pi x})}{\Gamma(\alpha+x)\,\Gamma(\beta-x)} e^{inx}\,dx$$

$$= C_\lambda \frac{\{2\cos(\tfrac{1}{2}n-\pi[(\pi+n)/2\pi])\}^{\alpha+\beta-2}}{\Gamma(\alpha+\beta-1)} e^{i(\beta-\alpha)\{\frac{1}{2}n-\pi[(\pi+n)/2\pi]\}},$$

where $$\lambda=-\left[\frac{\pi+n}{2\pi}\right],$$

provided that either

(i) n is not an odd multiple of π, and $R(\alpha+\beta)>1$,

or (ii) n is an odd multiple of π, and $R(\alpha+\beta)>2$,

or (iii) n is an odd multiple of π, $C_{(\pi-n)/2\pi}=0$, $C_{-(\pi+n)/2\pi}=0$, and $R(\alpha+\beta)>1$,

or (iv) n is an odd multiple of π, $\alpha+\beta$ is an integer greater than 1, and $C_{(\pi-n)/2\pi}=e^{2i\pi\alpha}C_{-(\pi+n)/2\pi}$.

Thus we see that the value of each of the integrals (3·12)—(3·14), when $s=2\pi$, reduces to a single term.

The next section will be devoted to the application of (3·2) in evaluating some special integrals.

4. Suppose that α is not real and

$$\phi(z)=1+e^{-2i\pi\alpha}z+e^{-4i\pi\alpha}z^2+\ldots\ (I(\alpha)<0),$$

and $$\phi(z)=-e^{2i\pi\alpha}z^{-1}-e^{4i\pi\alpha}z^{-2}-\ldots\ (I(\alpha)>0),$$

so that $\phi(z)$ is convergent when $|z|=1$. Then it is easy to see that

$$\int_{-\infty}^{\infty}\frac{\Gamma(\alpha+x)}{\Gamma(\beta+x)}e^{inx}\,dx=2i\pi e^{-i\pi\alpha}\int_{-\infty}^{\infty}\frac{\phi(e^{2i\pi x})}{\Gamma(1-\alpha+x)\,\Gamma(\beta-x)}e^{ix(\pi-n)}\,dx.$$

It follows from (3·2) that if α is not real, n real, and

(i) n is neither 0 nor any multiple of 2π, and $R(\alpha-\beta)<0$,

or (ii) n has the same sign as $I(\alpha)$ and $R(\alpha-\beta)<0$,

or (iii) n is 0, or a multiple of 2π, having the sign opposite to that of $I(\alpha)$, and $R(\alpha-\beta)<-1$,

or (iv) n is not 0 and $\alpha-\beta$ is a negative integer, then

$$(4\cdot1)\quad \int_{-\infty}^{\infty}\frac{\Gamma(\alpha+x)}{\Gamma(\beta+x)}e^{inx}\,dx=0$$

$$\text{or}\quad \pm\frac{2i\pi}{\Gamma(\beta-\alpha)}\left\{2\cos\left(\frac{\pi-n}{2}+\pi\left[\frac{n}{2\pi}\right]\right)\right\}^{\beta-\alpha-1}$$

$$\times\exp\left\{-in\alpha+i(\beta-\alpha-1)\left(\frac{\pi-n}{2}+\pi\left[\frac{n}{2\pi}\right]\right)\right\},$$

the zero value being taken when n and $I(\alpha)$ have the same sign, the plus sign when $n\geqslant0$ and $I(\alpha)<0$, and the minus sign when $n\leqslant0$ and $I(\alpha)>0$.

As particular cases of (4·1) we have

$$(4\cdot 11)\qquad \int_{-\infty}^{\infty} \frac{\Gamma(\alpha+x)}{\Gamma(\beta+x)}\,dx = 0,$$

if α is not real and $R(\alpha-\beta) < -1$. If α is not real, n real, and (i) r is a positive integer, *or* (ii) $r=0$ and $n \neq 0$, then

$$(4\cdot 12)\qquad \int_{-\infty}^{\infty} \frac{e^{inx}\,dx}{(x+\alpha)(x+\alpha+1)\ldots(x+\alpha+r)} = 0$$

$$or\ \pm\frac{2i\pi}{r!}(2\sin\tfrac{1}{2}n)^r e^{\frac{1}{2}ir(\pi-n)-in\alpha},$$

the different values being selected as in (4·1).

Similarly we can shew that if α and β are not real and n is real, and

(i) n is not an odd multiple of π and $R(\alpha+\beta) < 1$,

or (ii) n is an odd multiple of π which has either the sign of $I(\beta)$ or the sign opposite to that of $I(\alpha)$, and $R(\alpha+\beta) < 0$,

or (iii) n is an odd multiple of π which has neither the sign of $I(\beta)$ nor the sign opposite to that of $I(\alpha)$, and $R(\alpha+\beta) < 1$, then

(4·2)

$$\int_{-\infty}^{\infty} \Gamma(\alpha+x)\Gamma(\beta-x)e^{inx}\,dx = \frac{2i\pi\Gamma(\alpha+\beta)}{|2\cos\frac{1}{2}n|^{\alpha+\beta}}e^{\frac{1}{2}in(\beta-\alpha)}$$

$$\times\left(\eta_n(\beta)\exp\{i\pi(\alpha+\beta)[(\pi+n)/2\pi]\} - \eta_n(-\alpha)\exp\{-i\pi(\alpha+\beta)[(\pi+n)/2\pi]\}\right),$$

where $\eta_n(\zeta)$ is equal to 0 when $\pi-n$ and $I(\zeta)$ have the same sign, to 1 when $n \leqslant \pi$ and $I(\zeta) < 0$, and to -1 when $n \geqslant \pi$ and $I(\zeta) > 0$.

It should be remembered that for real values of n

$$|2\cos\tfrac{1}{2}n| = 2\cos(\tfrac{1}{2}n - \pi[(\pi+n)/2\pi]).$$

It follows, in particular, that if α and β are not real, and $R(\alpha+\beta) < 1$, then

$$(4\cdot 21)\qquad \int_{-\infty}^{\infty} \Gamma(\alpha+x)\Gamma(\beta-x)\,dx = 0\ \ or\ \ \pm 2^{1-\alpha-\beta}i\pi\Gamma(\alpha+\beta),$$

the zero value being chosen when $I(\alpha)$ and $I(\beta)$ have different signs, the plus sign when $I(\alpha)$ and $I(\beta)$ are both negative, and the minus sign when $I(\alpha)$ and $I(\beta)$ are both positive.

The following results can either be deduced from (1·5) or be proved independently in the same way as (1·4).

If n is zero or any multiple of 2π, ξ is real, α any number except the real numbers less than or equal to $-\xi$, and β is any number such that $R(\beta-\alpha) > 1$, then

$$(4\cdot 3)\qquad (\beta-\alpha-1)\int_{\xi}^{\infty} \frac{\Gamma(\alpha+x)}{\Gamma(\beta+x)}e^{inx}\,dx = \int_{\xi}^{\xi+1} \frac{\Gamma(\alpha+x)}{\Gamma(\beta-1+x)}e^{inx}\,dx.$$

If n is any odd multiple of π, α and ξ are the same as in (4·3), and β is not real and $R(\alpha+\beta)<0$, then

$$(4{\cdot}4)\qquad (\alpha+\beta)\int_{\xi}^{\infty}\Gamma(\alpha+x)\,\Gamma(\beta-x)\,e^{inx}\,dx$$
$$=\int_{\xi}^{\xi+1}\Gamma(\alpha+x)\,\Gamma(\beta+1-x)\,e^{inx}\,dx.$$

5. We now proceed to consider an application of (1·2) to some other functions. Suppose that $U_s(x)$, $V_s(x)$ and $W_s(x)$ are many-valued functions of x defined by

$$U_s(x)=\frac{u_0x^s}{\Gamma(1+s)}+\frac{u_1x^{s+\epsilon}}{\Gamma(1+s+\epsilon)}+\frac{u_2x^{s+2\epsilon}}{\Gamma(1+s+2\epsilon)}+\ldots,$$
$$V_s(x)=\frac{v_0x^s}{\Gamma(1+s)}+\frac{v_1x^{s+\epsilon}}{\Gamma(1+s+\epsilon)}+\frac{v_2x^{s+2\epsilon}}{\Gamma(1+s+2\epsilon)}+\ldots,$$
$$W_s(x)=\frac{w_0x^s}{\Gamma(1+s)}+\frac{w_1x^{s+\epsilon}}{\Gamma(1+s+\epsilon)}+\frac{w_2x^{s+2\epsilon}}{\Gamma(1+s+2\epsilon)}+\ldots,$$

where $R(\epsilon)\geqslant 0$, the u's, v's, and w's are any numbers connected by the relation to be found by equating the coefficients of the various powers of k in the equation

$$w_0+w_1k+w_2k^2+\ldots=(u_0+u_1k+u_2k^2+\ldots)(v_0+v_1k+v_2k^2+\ldots)^*,$$

and the series $U_s(x)$, $V_s(x)$ and $W_s(x)$ are convergent at least for the values of s and x that appear in the equation (5·2).

The functions U, V, and W are many valued. If $|x/y|=1$ and $|\arg(x/y)|<\pi$, then one value of $\arg(x+y)$ is given by the equation

$$(5{\cdot}1)\qquad \arg x+\arg y=2\arg(x+y).$$

If we choose $\arg x$ and $\arg y$ arbitrarily, and agree that

$$x^{s+\mu\epsilon}=\exp\{(s+\mu\epsilon)(\log|x|+i\arg x)\},$$

and that $y^{s+\mu\epsilon}$ and $(x+y)^{s+\mu\epsilon}$ are to be interpreted similarly, that value of $\arg(x+y)$ being chosen which is given by (5·1), then a definite branch of W is associated with any arbitrary pair of branches of U and V.

If α, β, x, y are any numbers such that $|x/y|=1$, and $R(\alpha+\beta)>0$ when $|\arg(x/y)|=\pi$ and $R(\alpha+\beta)>-1$ otherwise, then

$$(5{\cdot}2)\qquad \int_{-\infty}^{\infty}U_{\alpha+\xi}(x)\,V_{\beta-\xi}(y)\,(x/y)^{\xi}\,d\xi=0\ \textit{or}\ W_{\alpha+\beta}(x+y),$$

according as $|\arg(x/y)|\geqslant\pi$ or $<\pi$, whatever be the branches of $U(x)$ and $V(y)$, provided that the corresponding branch of $W(x+y)$ is fixed in accordance with the convention explained above. This is proved as follows. Suppose that

$$x=te^{\frac{1}{2}in},\quad y=te^{-\frac{1}{2}in},$$

* These series need not, of course, be convergent for any value of k.

where t is arbitrary and n is any real number. Then the integral becomes

$$\int_{-\infty}^{\infty} U_{\alpha+\xi}(te^{\frac{1}{2}in}) V_{\beta-\xi}(te^{-\frac{1}{2}in}) e^{in\xi} d\xi.$$

If we expand the integrand in powers of t, and integrate term by term with the help of (1·2), and then make use of the relations between the u's, v's, and w's, the result will be

$$W_{\alpha+\beta}(2t\cos\tfrac{1}{2}n) = W_{\alpha+\beta}(x+y),$$

or zero, according to the conditions stated with regard to (5·2).

In particular, if $R(\alpha+\beta) > -1$, we have

$$\int_{-\infty}^{\infty} U_{\alpha+\xi}(x) V_{\beta-\xi}(x) d\xi = W_{\alpha+\beta}(2x). \tag{5·21}$$

Suppose now that $G_s(p, x)$ is a many-valued function of x defined by

$$G_s(p, x) = \frac{x^s}{\Gamma(s+1)} - \frac{p}{1!}\frac{x^{s+1}}{\Gamma(s+2)} + \frac{p(p+1)}{2!}\frac{x^{s+2}}{\Gamma(s+3)} - \dots.$$

Then it follows from (5·2) that if α, β, x, y are any numbers such that $|x/y| = 1$, and $R(\alpha+\beta) > 0$ when $|\arg(x/y)| = \pi$ and $R(\alpha+\beta) > -1$ otherwise, we have

(5·3)

$$\int_{-\infty}^{\infty} G_{\alpha+\xi}(p, x) G_{\beta-\xi}(q, y)(x/y)^{\xi} d\xi = 0 \text{ or } G_{\alpha+\beta}(p+q, x+y),$$

according as $|\arg(x/y)| \geqslant \pi$ or $|\arg(x/y)| < \pi$, whatever be the branches of $G(p, x)$ and $G(q, y)$, provided the branch of $G(p+q, x+y)$ is chosen according to our former conventions. If, in particular, $R(\alpha+\beta) > -1$, then

$$\int_{-\infty}^{\infty} G_{\alpha+\xi}(p, x) G_{\beta-\xi}(q, x) d\xi = G_{\alpha+\beta}(p+q, 2x). \tag{5·31}$$

It may be interesting to note that the right-hand sides of (5·3) and (5·31) are of the form $G_{\alpha+\beta}(p+q, z)$, which reduces to

$$\frac{z^{\alpha+\beta}}{\Gamma(\alpha+\beta+1)}$$

when $p = -q$, becoming independent of p or q.

The ordinary Bessel's functions are particular cases of the function $G_s(p, x)$. Hence we have the following particular results. If n is real, and $R(\alpha+\beta) > 0$ when $n = \pm\pi$ and $R(\alpha+\beta) > -1$ otherwise, then

(5·4)

$$\int_{-\infty}^{\infty} \frac{J_{\alpha+\xi}(x)}{x^{\alpha+\xi}} \frac{J_{\beta-\xi}(y)}{y^{\beta-\xi}} e^{in\xi} d\xi = 0$$

$$\text{or } \left(\frac{2\cos\frac{1}{2}n}{x^2 e^{-\frac{1}{2}in} + y^2 e^{\frac{1}{2}in}}\right)^{\frac{1}{2}(\alpha+\beta)} e^{\frac{1}{2}in(\beta-\alpha)} J_{\alpha+\beta}\left[\sqrt{\{2\cos\tfrac{1}{2}n(x^2 e^{-\frac{1}{2}in} + y^2 e^{\frac{1}{2}in})\}}\right],$$

according as $|n| \geqslant \pi$ or $< \pi$. If n is real, and $R(\alpha+\beta)>0$ when $n=\pm\pi$ and $R(\alpha+\beta)>-1$ otherwise, then

$$(5{\cdot}41)\quad \int_{-\infty}^{\infty} J_{\alpha+\xi}(x)\, J_{\beta-\xi}(x)\, e^{in\xi}\, d\xi = 0 \;\; or \;\; e^{\frac{1}{2}in(\beta-\alpha)} J_{\alpha+\beta}(2x\cos\tfrac{1}{2}n),$$

according as $|n| \geqslant \pi$ or $< \pi$. If $R(\alpha+\beta)>-1$, then

$$(5{\cdot}42)\qquad \int_{-\infty}^{\infty} \frac{J_{\alpha+\xi}(x)}{x^{\alpha+\xi}} \frac{J_{\beta-\xi}(y)}{y^{\beta-\xi}}\, d\xi = \frac{J_{\alpha+\beta}\{\sqrt{(2x^2+2y^2)}\}}{(\frac{1}{2}x^2+\frac{1}{2}y^2)^{\frac{1}{2}(\alpha+\beta)}}$$

and

$$(5{\cdot}43)\qquad \int_{-\infty}^{\infty} J_{\alpha+\xi}(x)\, J_{\beta-\xi}(x)\, d\xi = J_{\alpha+\beta}(2x).$$

6. We shall now consider some special cases of the integral

$$(6{\cdot}1)\qquad \int_{-\infty}^{\infty} \frac{e^{inx}}{\Gamma(\alpha+x)\,\Gamma(\beta-x)\,\Gamma(\gamma+lx)\,\Gamma(\delta-lx)}\, dx,$$

l and n being real numbers.

Replacing $1/\{\Gamma(\gamma+lx)\,\Gamma(\delta-lx)\}$ by

$$\frac{1}{\pi\Gamma(\gamma+\delta-1)} \int_{-\frac{1}{2}\pi}^{\frac{1}{2}\pi} (2\cos z)^{\gamma+\delta-2}\, e^{-iz(\gamma-\delta+2lx)}\, dz,$$

it follows from (1·2) that (6·1) is equal to

$$(6{\cdot}11)\qquad \frac{1}{\pi\Gamma(\alpha+\beta-1)\,\Gamma(\gamma+\delta-1)} \int_u^v \{2\cos(\tfrac{1}{2}n-lz)\}^{\alpha+\beta-2}\,(2\cos z)^{\gamma+\delta-2} \times \exp\{i(\beta-\alpha)(\tfrac{1}{2}n-lz)+i(\delta-\gamma)z\}\, dz,$$

where u and v are the lower and upper extremities of the common part of the intervals

$$-\tfrac{1}{2}\pi < z < \tfrac{1}{2}\pi, \quad -\tfrac{1}{2}\pi < \tfrac{1}{2}n - lz < \tfrac{1}{2}\pi.$$

If the intervals do not overlap, the value of (6·1) is zero. It is easy to see that if

$$(6{\cdot}12)\qquad |n| \geqslant \pi(1+|l|),$$

the intervals do not overlap; and that, if they do overlap and $l>0$, then

$$u = \left|\frac{\pi}{4}+\frac{n-\pi}{4l}\right| - \left|\frac{\pi}{4}-\frac{n-\pi}{4l}\right|,$$

and

$$v = \left|\frac{\pi}{4}+\frac{n+\pi}{4l}\right| - \left|\frac{\pi}{4}-\frac{n+\pi}{4l}\right|.$$

It should also be observed that, though (6·11) may not be convergent for all the values of α, β, γ and δ for which (6·1) is convergent, yet we may evaluate (6·11) when it is convergent, and so obtain a formula for (6·1) which may be extended, by the theory of analytic continuation, to all values of the parameters for which the integral converges.

From (6·12) we see that, if l and n are any real numbers such that $|n| \geqslant \pi(1+|l|)$, then

$$(6\cdot2)\qquad \int_{-\infty}^{\infty} \frac{e^{inx}}{\Gamma(\alpha+x)\,\Gamma(\beta-x)\,\Gamma(\gamma+lx)\,\Gamma(\delta-lx)}\,dx = 0,$$

provided that (i) $R(\alpha+\beta+\gamma+\delta) > 2$

when $|n| > \pi(1+|l|)$,

and (ii) $R(\alpha+\beta+\gamma+\delta) > 3$

when $|n| = \pi(1+|l|)$.

7. Suppose now that $l=1$ and $n=0$; then (6·11) reduces to

$$\frac{1}{\pi\Gamma(\alpha+\beta-1)\,\Gamma(\gamma+\delta-1)}\int_{-\frac{1}{2}\pi}^{\frac{1}{2}\pi} (2\cos z)^{\alpha+\beta+\gamma+\delta-4}\, e^{iz(\alpha-\beta-\gamma+\delta)}\,dz,$$

which is easily evaluated by the help of (1·1). Hence

$$(7\cdot1)\quad \int_{-\infty}^{\infty} \frac{dx}{\Gamma(\alpha+x)\,\Gamma(\beta-x)\,\Gamma(\gamma+x)\,\Gamma(\delta-x)} = \frac{\Gamma(\alpha+\beta+\gamma+\delta-3)}{\Gamma(\alpha+\beta-1)\,\Gamma(\beta+\gamma-1)\,\Gamma(\gamma+\delta-1)\,\Gamma(\delta+\alpha-1)},$$

provided that (i) $R(\alpha+\beta+\gamma+\delta) > 3$, *or* (ii) $2(\alpha-\gamma)$ and $2(\beta-\delta)$ are odd integers and $R(\alpha+\beta+\gamma+\delta) > 2$.

It should be noted that the formula fails when $\alpha+\beta+\gamma+\delta=3$ and $2(\alpha-\gamma)$ and $2(\beta-\delta)$ are odd integers. The value of the integral in this case is sometimes $1/2\pi$ and sometimes $-1/2\pi$. The value to be selected may be fixed as follows. It is easy to see that, in this case, one and only one of the numbers $\alpha+\beta-1$, $\beta+\gamma-1$, $\gamma+\delta-1$, and $\delta+\alpha-1$ will be an integer less than or equal to zero. If $\alpha+\beta-1$ or $\beta+\gamma-1$ happens to be such a number, then the value of the integral is $\pm 1/2\pi$, according as $2(\beta-\delta) \equiv \mp 1 \pmod 4$. But if $\gamma+\delta-1$ or $\delta+\alpha-1$ happens to be such a number, the value of the integral is $\pm 1/2\pi$, according as $2(\beta-\delta) \equiv \pm 1$ (mod 4).

As particular cases of (7·1), we have

$$(7\cdot11)\qquad \int_{-\infty}^{\infty} \frac{dx}{\{\Gamma(\alpha+x)\,\Gamma(\beta-x)\}^2} = \frac{\Gamma(2\alpha+2\beta-3)}{\{\Gamma(\alpha+\beta-1)\}^4},$$

provided that $R(\alpha+\beta) > \frac{3}{2}$,

$$(7\cdot12)\quad \int_{0}^{\infty} \frac{dx}{\Gamma(\alpha+x)\,\Gamma(\alpha-x)\,\Gamma(\beta+x)\,\Gamma(\beta-x)} = \frac{\Gamma(2\alpha+2\beta-3)}{2\Gamma(2\alpha-1)\,\Gamma(2\beta-1)\{\Gamma(\alpha+\beta-1)\}^2},$$

provided that (i) $R(\alpha+\beta) > \frac{3}{2}$, *or* (ii) $2(\alpha-\beta)$ is an odd integer and $R(\alpha+\beta) > 1$. If $2(\alpha+\beta)=3$ and $2(\alpha-\beta)$ is an odd integer, then the value of the integral (7·12), when $\alpha \geqslant 1$, is $\pm 1/2\pi$, according as $2(\alpha-\beta) \equiv \pm 1$ (mod 4), and when $\alpha < 1$ it is $\pm 1/2\pi$, according as $2(\alpha-\beta) \equiv \mp 1$ (mod 4).

Putting $\alpha = \beta$ in (7·11) or in (7·12), we obtain

$$(7\cdot13)\qquad \int_0^\infty \frac{dx}{\{\Gamma(\alpha+x)\Gamma(\alpha-x)\}^2} = \frac{\Gamma(4\alpha-3)}{2\{\Gamma(2\alpha-1)\}^4},$$

if $R(\alpha) > \frac{3}{4}$. Suppose again that $l = 1$, $n = \pi$, and $\alpha + \delta = \beta + \gamma$. Then (6·11) reduces to

$$\frac{e^{\frac{1}{2}i\pi(\beta-\alpha)}}{\pi\Gamma(\alpha+\beta-1)\Gamma(\gamma+\delta-1)}\int_0^{\frac{1}{2}\pi}(2\sin z)^{\alpha+\beta-2}(2\cos z)^{\gamma+\delta-2}\,dz.$$

Hence we see that

$$(7\cdot2)\qquad \int_{-\infty}^\infty \frac{e^{\pm i\pi x}\,dx}{\Gamma(\alpha+x)\Gamma(\beta-x)\Gamma(\gamma+x)\Gamma(\delta-x)} = \frac{e^{\pm\frac{1}{2}i\pi(\beta-\alpha)}}{2\Gamma\{\frac{1}{2}(\alpha+\beta)\}\Gamma\{\frac{1}{2}(\gamma+\delta)\}\Gamma(\alpha+\delta-1)},$$

if $\alpha + \delta = \beta + \gamma$ and $R(\alpha+\beta+\gamma+\delta) > 2$. In particular

$$(7\cdot21)\qquad \int_{-\infty}^\infty \frac{e^{\pm i\pi x}}{\{\Gamma(\alpha+x)\Gamma(\beta-x)\}^2}\,dx = \frac{e^{\pm\frac{1}{2}i\pi(\beta-\alpha)}}{2\Gamma(\alpha+\beta-1)[\Gamma\{\frac{1}{2}(\alpha+\beta)\}]^2},$$

if $R(\alpha+\beta) > 1$, and

$$(7\cdot22)\qquad \int_0^\infty \frac{\cos\pi x}{\{\Gamma(\alpha+x)\Gamma(\alpha-x)\}^2}\,dx = \frac{1}{4\Gamma(2\alpha-1)\{\Gamma(\alpha)\}^2},$$

if $R(\alpha) > \frac{1}{2}$.

8. It follows from (6·2), (7·1), and (7·2) that, if

$$\phi(z) = \sum_{-\infty}^{\infty} c_{2\nu} z^{2\nu}, \quad \psi(z) = \sum_{-\infty}^{\infty} c_{2\nu+1} z^{2\nu+1},$$

the series being convergent when $|z| = 1$, then

$$(8\cdot1)\qquad \int_{-\infty}^\infty \frac{\phi(e^{i\pi x})}{\Gamma(\alpha+x)\Gamma(\beta-x)\Gamma(\gamma+x)\Gamma(\delta-x)}\,dx = \frac{c_0\Gamma(\alpha+\beta+\gamma+\delta-3)}{\Gamma(\alpha+\beta-1)\Gamma(\beta+\gamma-1)\Gamma(\gamma+\delta-1)\Gamma(\delta+\alpha-1)},$$

provided that (i) $R(\alpha+\beta+\gamma+\delta) > 3$ *or* (ii) $R(\alpha+\beta+\gamma+\delta) > 2$ and

$$c_2 e^{i\pi(\beta+\delta)} + c_{-2}e^{-i\pi(\beta+\delta)} = 2c_0\cos\pi(\beta-\delta),$$

$$c_2 e^{-i\pi(\alpha+\gamma)} + c_{-2}e^{i\pi(\alpha+\gamma)} = 2c_0\cos\pi(\alpha-\gamma);$$

and that

$$(8\cdot2)\qquad \int_{-\infty}^\infty \frac{\psi(e^{i\pi x})}{\Gamma(\alpha+x)\Gamma(\beta-x)\Gamma(\gamma+x)\Gamma(\delta-x)}\,dx = \frac{c_1 e^{\frac{1}{2}i\pi(\beta-\alpha)} + c_{-1}e^{-\frac{1}{2}i\pi(\beta-\alpha)}}{2\Gamma\{\frac{1}{2}(\alpha+\beta)\}\Gamma\{\frac{1}{2}(\gamma+\delta)\}\Gamma(\alpha+\delta-1)},$$

provided that $\alpha + \delta = \beta + \gamma$ and $R(\alpha+\beta+\gamma+\delta) > 2$.

If $\alpha + \delta - \beta - \gamma$ is an integer other than zero, it is possible to evaluate the integrals (7·2) and (8·2) in finite terms, but not as a single term.

The following integrals can be evaluated as a single term, with the help of (8·1) and (8·2), whenever they are convergent:

$$(8\cdot3)\qquad \int_{-\infty}^{\infty} \frac{\Gamma(\delta+x)\, e^{inx}}{\Gamma(\alpha+x)\,\Gamma(\beta-x)\,\Gamma(\gamma+x)}\, dx,$$

where (i) n is an odd multiple of π, *or* (ii) n is an even multiple of π and $\alpha+1=\beta+\gamma+\delta$;

$$(8\cdot4)\qquad \int_{-\infty}^{\infty} \frac{\Gamma(\gamma+x)\,\Gamma(\delta\pm x)}{\Gamma(\alpha+x)\,\Gamma(\beta\pm x)}\, e^{inx}\, dx,$$

where (i) n is an even multiple of π, *or* (ii) n is an odd multiple of π and $\alpha+\delta=\beta+\gamma$;

$$(8\cdot5)\qquad \int_{-\infty}^{\infty} \frac{\Gamma(\beta+x)\,\Gamma(\gamma-x)\,\Gamma(\delta+x)}{\Gamma(\alpha+x)}\, e^{inx}\, dx,$$

where (i) n is an odd multiple of π, *or* (ii) n is an even multiple of π and $\alpha+\beta+\gamma=1+\delta$;

$$(8\cdot6)\qquad \int_{-\infty}^{\infty} \Gamma(\alpha+x)\,\Gamma(\beta-x)\,\Gamma(\gamma+x)\,\Gamma(\delta-x)\, e^{inx}\, dx,$$

where (i) n is an even multiple of π, *or* (ii) n is an odd multiple of π and $\alpha+\delta=\beta+\gamma$. Thus, for instance, if δ is not real, $\alpha+1=\beta+\gamma+\delta$, and $R(\alpha+\beta+\gamma-\delta)>1$, then

$$(8\cdot31)\qquad \int_{-\infty}^{\infty} \frac{\Gamma(\delta+x)}{\Gamma(\alpha+x)\,\Gamma(\beta-x)\,\Gamma(\gamma+x)}\, dx = \frac{\pi e^{\pm\frac{1}{2}i\pi(\delta-\gamma)}}{\Gamma(\alpha-\delta)\,\Gamma\{\frac{1}{2}(\alpha+\beta)\}\,\Gamma\{\frac{1}{2}(\gamma-\delta+1)\}},$$

according as $I(\delta)$ is positive or negative; and if γ and δ are not real and $R(\alpha+\beta-\gamma-\delta)>1$, then

$$(8\cdot41)\qquad \int_{-\infty}^{\infty} \frac{\Gamma(\gamma+x)\,\Gamma(\delta+x)}{\Gamma(\alpha+x)\,\Gamma(\beta+x)}\, dx = 0$$

or

$$\pm\frac{2i\pi^2}{\sin\pi(\gamma-\delta)}\,\frac{\Gamma(\alpha+\beta-\gamma-\delta-1)}{\Gamma(\alpha-\gamma)\,\Gamma(\alpha-\delta)\,\Gamma(\beta-\gamma)\,\Gamma(\beta-\delta)},$$

the zero value being taken when $I(\gamma)$ and $I(\delta)$ have the same sign, the plus sign when $I(\gamma)>0$ and $I(\delta)<0$, and the minus sign when $I(\gamma)<0$ and $I(\delta)>0$.

9. The following results are easily obtained with the help of (6·11). If $2(\alpha-\beta)=\gamma-\delta$ and $R(\alpha+\beta+\gamma+\delta)>3$, then

$$(9\cdot1)\qquad \int_{-\infty}^{\infty} \frac{e^{\pm i\pi x}}{\Gamma(\alpha+x)\,\Gamma(\beta-x)\,\Gamma(\gamma+2x)\,\Gamma(\delta-2x)}\, dx = \frac{2^{\alpha+\beta+\gamma+\delta-5}\, e^{\pm\frac{1}{2}i\pi(\beta-\alpha)}\,\Gamma\{\frac{1}{2}(\alpha+\beta+\gamma+\delta-3)\}}{\sqrt{\pi}\,\Gamma\{\frac{1}{2}(\alpha+\beta)\}\,\Gamma(\gamma+\delta-1)\,\Gamma(2\alpha+\delta-2)}.$$

If $R(\alpha+\beta) > \frac{3}{2}$, then

(9·11) $$\int_0^\infty \frac{\cos \pi x}{\Gamma(\alpha+x)\Gamma(\alpha-x)\Gamma(\beta+2x)\Gamma(\beta-2x)}\,dx = \frac{2^{2\alpha+2\beta-6}\,\Gamma(\alpha+\beta-\frac{3}{2})}{\sqrt{\pi}\Gamma(\alpha)\Gamma(2\beta-1)\Gamma(2\alpha+\beta-2)}.$$

If $\alpha+\beta+\gamma+\delta=4$, then

(9·12) $$\int_{-\infty}^\infty \frac{\cos \pi(x+\beta+\gamma)}{\Gamma(\alpha+x)\Gamma(\beta-x)\Gamma(\gamma+2x)\Gamma(\delta-2x)}\,dx = \frac{1}{2\Gamma(\gamma+\delta-1)\Gamma(2\alpha+\delta-2)\Gamma(2\beta+\gamma-2)}.$$

If $2(\alpha-\beta)=\gamma-\delta+k$, where k is ± 1 or ± 2, then

(9·2) $$\int_{-\infty}^\infty \frac{\sin \pi(2x+\alpha-\beta)}{\Gamma(\alpha+x)\Gamma(\beta-x)\Gamma(\gamma+2x)\Gamma(\delta-2x)}\,dx = \pm\frac{2^{2\alpha-\gamma-3}}{\sqrt{\pi}\Gamma(\beta+\gamma-\alpha+\frac{1}{2})\Gamma(2\alpha+\delta-2)},$$

provided that $R(\alpha+\beta+\gamma+\delta) > 2$.

If $3(\alpha-\beta)=\gamma-\delta+k$, where k is ± 1 or ± 2, then

(9·3) $$\int_{-\infty}^\infty \frac{\sin \pi(2x+\alpha-\beta)}{\Gamma(\alpha+x)\Gamma(\beta-x)\Gamma(\gamma+3x)\Gamma(\delta-3x)}\,dx = \pm\frac{3^{3\alpha+\delta-4}\,\Gamma(2\alpha-\beta+\delta-2)}{4\pi\Gamma(\gamma+\delta-1)\Gamma(3\alpha+\delta-3)},$$

provided that (i) $R(\alpha+\beta+\gamma+\delta) > 3$, *or* (ii) $\beta+\gamma-2\alpha$ is integral and $R(\alpha+\beta+\gamma+\delta) > 2$.

In (9·2) and (9·3) the plus or minus sign on the right-hand side is to be taken according as k is positive or negative. If k is an integer other than ± 1 or ± 2, the integrals in (9·2) and (9·3) can still be evaluated in finite terms, but in a less simple form.

10. In this connection it may be interesting to note that, if n is an even multiple of π, and $\alpha+\beta+\gamma+\delta=4$, then

(10·1)
$$(\alpha+\beta-2)(\beta+\gamma-2)\int_\xi^\infty \frac{e^{inx}}{\Gamma(\alpha+x)\Gamma(\beta-x)\Gamma(\gamma+x)\Gamma(\delta-x)}\,dx = \int_\xi^{\xi+1} \frac{e^{inx}}{\Gamma(\alpha-1+x)\Gamma(\beta-x)\Gamma(\gamma-1+x)\Gamma(\delta-x)}\,dx$$

for all real values of ξ. The proof of this is the same as that of (1·4).

Finally I may mention the formula

(10·2) $$\int_{-\infty}^\infty J_{\alpha+\xi}(x)\,J_{\beta-\xi}(x)\,J_{\gamma+\xi}(x)\,J_{\delta-\xi}(x)\,d\xi = (\tfrac{1}{2}x)^{\alpha+\beta+\gamma+\delta}$$
$$\times\sum_{\nu=1}^{\nu=\infty} \frac{(-\frac{1}{4}x^2)^{\nu-1}\{\Gamma(\alpha+\beta+\gamma+\delta+2\nu-1)\}^2}{\Gamma(\nu)\Gamma(\alpha+\beta+\gamma+\delta+\nu)\Gamma(\alpha+\beta+\nu)\Gamma(\beta+\gamma+\nu)\Gamma(\gamma+\delta+\nu)\Gamma(\delta+\alpha+\nu)},$$

which holds if (i) $R(\alpha+\beta+\gamma+\delta) > -1$, *or* (ii) $2(\alpha-\gamma)$ and $2(\beta-\delta)$ are odd integers and $R(\alpha+\beta+\gamma+\delta) > -2$.

28

CONGRUENCE PROPERTIES OF PARTITIONS

(*Proceedings of the London Mathematical Society*, 2, XVIII, 1920, *Records for* 13 *March* 1919)

In a paper published recently in the *Proceedings of the Cambridge Philosophical Society**, I proved a number of arithmetical properties of $p(n)$, the number of unrestricted partitions of n, and in particular that

$$p(5n+4) \equiv 0 \pmod{5},$$

and

$$p(7n+5) \equiv 0 \pmod{7}.$$

Alternative proofs of these two theorems were found afterwards by Mr H. B. C. Darling†.

I have since found another method which enables me to prove all these properties and a variety of others, of which the most striking is

$$p(11n+6) \equiv 0 \pmod{11}.$$

There are also corresponding properties in which the moduli are powers of 5, 7, or 11; thus

$$p(25n+24) \equiv 0 \pmod{25},$$

$$p(49n+19),\quad p(49n+33),\quad p(49n+40),\quad p(49n+47) \equiv 0 \pmod{49},$$

$$p(121n+116) \equiv 0 \pmod{121}.$$

It appears that there are no equally simple properties for any moduli involving primes other than these three.

The function $\tau(n)$ defined by the equation

$$\sum_{1}^{\infty} \tau(n)\, x^n = x\{(1-x)(1-x^2)(1-x^3)\ldots\}^{24},$$

also possesses very remarkable arithmetical properties. Thus

$$\tau(5n) \equiv 0 \pmod{5},$$

$$\tau(7n),\quad \tau(7n+3),\quad \tau(7n+5),\quad \tau(7n+6) \equiv 0 \pmod{7},$$

while

$$\tau(23n+\nu) \equiv 0 \pmod{23},$$

if ν is any one of the numbers

5, 7, 10, 11, 14, 15, 17, 19, 20, 21, 22.

* Vol. XIX, 1919, pp. 207—210 [No. 25 of this volume: see also No. 30].

† *Ibid.*, pp. 217, 218.

29

ALGEBRAIC RELATIONS BETWEEN CERTAIN INFINITE PRODUCTS

(*Proceedings of the London Mathematical Society*, 2, XVIII, 1920, *Records for* 13 *March* 1919)

It was proved by Prof. L. J. Rogers* that

$$G(x) = 1 + \frac{1}{1-x} + \frac{x^4}{(1-x)(1-x^2)} + \frac{x^9}{(1-x)(1-x^2)(1-x^3)} + \dots$$

$$= \frac{1}{(1-x)(1-x^6)(1-x^{11})} \dots \times \frac{1}{(1-x^4)(1-x^9)(1-x^{14})\dots},$$

and $$H(x) = 1 + \frac{x^2}{1-x} + \frac{x^6}{(1-x)(1-x^2)} + \frac{x^{12}}{(1-x)(1-x^2)(1-x^3)} + \dots$$

$$= \frac{1}{(1-x^2)(1-x^7)(1-x^{12})\dots} \times \frac{1}{(1-x^3)(1-x^8)(1-x^{13})\dots}.$$

Simpler proofs were afterwards found by Prof. Rogers and myself†.

I have now found an algebraic relation between $G(x)$ and $H(x)$, viz.:

$$H(x)\{G(x)\}^{11} - x^2 G(x)\{H(x)\}^{11} = 1 + 11x\{G(x)H(x)\}^6.$$

Another noteworthy formula is

$$H(x)G(x^{11}) - x^2 G(x)H(x^{11}) = 1.$$

Each of these formulæ is the simplest of a large class.

* *Proc. London Math. Soc.*, Ser. 1, Vol. XXV, 1894, pp. 318—343.

† *Proc. Camb. Phil. Soc.*, Vol. XIX, 1919, pp. 211—216. A short account of the history of the theorems is given by Mr Hardy in a note attached to this paper. [For Ramanujan's proofs see No. 26 of this volume: those of Rogers, and the note by Hardy referred to, are reproduced in the notes on No. 26 in the Appendix.]

30

CONGRUENCE PROPERTIES OF PARTITIONS

(*Mathematische Zeitschrift*, IX, 1921, 147—153)

[Extracted from the manuscripts of the author by G. H. Hardy*]

1. Let

(1·11) $$P = 1 - 24\left(\frac{x}{1-x} + \frac{2x^2}{1-x^2} + \frac{3x^3}{1-x^3} + \dots\right),$$

(1·12) $$Q = 1 + 240\left(\frac{x}{1-x} + \frac{2^3x^2}{1-x^2} + \frac{3^3x^3}{1-x^3} + \dots\right),$$

(1·13) $$R = 1 - 504\left(\frac{x}{1-x} + \frac{2^5x^2}{1-x^2} + \frac{3^5x^3}{1-x^3} + \dots\right),$$

(1·2) $$f(x) = (1-x)(1-x^2)(1-x^3)\dots.$$

Then it is well known that

(1·3)
$$f(x) = 1 - x - x^2 + x^5 + x^7 - \dots = 1 + \sum_{n=1}^{\infty} (-1)^n \left(x^{\frac{1}{2}n(3n-1)} + x^{\frac{1}{2}n(3n+1)}\right),$$

(1·4) $$Q^3 - R^2 = 1728x\,(f(x))^{24}.$$

* Srinivasa Ramanujan, Fellow of Trinity College, Cambridge, and of the Royal Society of London, died in India on 26 April, 1920, aged 32. The manuscript from which this note is derived is a sequel to a short memoir "Some properties of $p(n)$, the number of partitions of n," *Proceedings of the Cambridge Philosophical Society*, Vol. XIX (1919), 207—210 [No. 25 of this volume]. In this memoir Ramanujan proves that

$$p(5n+4) \equiv 0 \pmod{5}$$

and

$$p(7n+5) \equiv 0 \pmod{7},$$

and states without proof a number of further congruences to moduli of the form $5^a\,7^b\,11^c$, of which the most striking is

$$p(11n+6) \equiv 0 \pmod{11}.$$

Here new proofs are given of the first two congruences, and the first published proof of the third.

The manuscript contains a large number of further results. It is very incomplete, and will require very careful editing before it can be published in full. I have taken from it the three simplest and most striking results, as a short but characteristic example of the work of a man who was beyond question one of the most remarkable mathematicians of his time.

I have adhered to Ramanujan's notation, and followed his manuscript as closely as I can. A few insertions of my own are marked by brackets. The most substantial of these is in § 5, where Ramanujan's manuscript omits the proof of (5·4). Whether I have reconstructed his argument correctly I cannot say.

The references given in the footnotes to "Ramanujan" are to his memoir "On certain arithmetical functions," *Transactions of the Cambridge Philosophical Society*, Vol. XXII, No. 9 (1916), 159—184 [No. 18 of this volume].

Further, let

(1·51) $$\Phi_{r,s}(x)=\sum_{m=1}^{\infty}\sum_{n=1}^{\infty} m^r n^s x^{mn}=\sum_{n=1}^{\infty} n^r\sigma_{s-r}(n)\,x^n,$$

where $\sigma_k(n)$ is the sum of the kth powers of the divisors of n; so that

(1·52) $$\Phi_{0,s}(x)=\frac{x}{1-x}+\frac{2^s x^2}{1-x^2}+\frac{3^s x^3}{1-x^3}+\ldots,$$

and in particular

(1·53) $$P=1-24\Phi_{0,1}(x),\quad Q=1+240\Phi_{0,3}(x),\quad R=1-504\Phi_{0,5}(x).$$

Then [it may be deduced from the theory of the elliptic modular functions, and has been shewn by the author in a direct and elementary manner*, that, when $r+s$ is odd, and $r<s$, $\Phi_{r,s}(x)$ is expressible as a polynomial in P, Q, and R, in the form

$$\Phi_{r,s}(x)=\Sigma k_{l,m,n}P^l Q^m R^n,$$

where $$l-1\leqq \operatorname{Min}(r,s),\quad 2l+4m+6n=r+s+1.$$

In particular†]

(1·61) $$Q^2=1+480\Phi_{0,7}(x)=1+480\left(\frac{x}{1-x}+\frac{2^7x^2}{1-x^2}+\ldots\right),$$

(1·62) $$QR=1-264\Phi_{0,9}(x)=1-264\left(\frac{x}{1-x}+\frac{2^9x^2}{1-x^2}+\ldots\right),$$

(1·63) $$441Q^3+250R^2=691+65520\Phi_{0,11}(x)$$
$$=691+65520\left(\frac{x}{1-x}+\frac{2^{11}x^2}{1-x^2}+\ldots\right),$$

(1·71) $$Q-P^2=288\Phi_{1,2}(x),$$

(1·72) $$PQ-R=720\Phi_{1,4}(x),$$

(1·73) $$Q^2-PR=1008\Phi_{1,6}(x),$$

(1·74) $$Q(PQ-R)=720\Phi_{1,8}(x),$$

(1·81) $$3PQ-2R-P^3=1728\Phi_{2,3}(x),$$

(1·82) $$P^2Q-2PR+Q^2=1728\Phi_{2,5}(x),$$

(1·83) $$2PQ^2-P^2R-QR=1728\Phi_{2,7}(x),$$

(1·91) $$6P^2Q-8PR+3Q^2-P^4=6912\Phi_{3,4}(x),$$

(1·92) $$P^3Q-3P^2R+3PQ^2-QR=3456\Phi_{3,6}(x),$$

(1·93) $$15PQ^2-20P^2R+10P^3Q-4QR-P^5=20736\Phi_{4,5}(x).$$

* Ramanujan, p. 165 [pp. 142–143].

† Ramanujan, pp. 163—165 [pp. 141—142] (Tables I to III). Ramanujan carried the calculation of formulæ of this kind to considerable lengths, the last formulæ of Table I being

$$7\,709\,321\,041\,217+32\,640\Phi_{0,31}(x)=764\,412\,173\,217Q^8$$
$$+5\,323\,905\,468\,000Q^5R^2+1\,621\,003\,400\,000Q^2R^4.$$

It is worth while to quote one such formula; for it is impossible to understand Ramanujan without realising his love of numbers for their own sake.

Modulus 5.

2. We denote generally by J an integral power-series in x whose coefficients are integers. It is obvious from (1·12) that

$$Q = 1 + 5J.$$

Also $n^5 - n \equiv 0 \pmod 5$, and so, from (1·11) and (1·13),

$$R = P + 5J.$$

Hence $$Q^3 - R^2 = Q(1+5J)^2 - (P+5J)^2 = Q - P^2 + 5J.$$

Using (1·4), (1·71), and (1·51), we obtain

$$(2·1) \qquad 1728x(f(x))^{24} = 288 \sum_{n=1}^{\infty} n\sigma_1(n)x^n + 5J.$$

Also
$$(1-x)^{25} = 1 - x^{25} + 5J,$$
$$(f(x))^{25} = f(x^{25}) + 5J,$$

and so

$$(2·2) \qquad (f(x))^{24} = \frac{f(x^{25})}{f(x)} + 5J.$$

But $$\frac{1}{f(x)} = 1 + p(1)x + p(2)x^2 + \ldots,$$

and therefore, by (2·1) and (2·2),

(2·3)

$$1728xf(x^{25})(1 + p(1)x + p(2)x^2 + \ldots)$$
$$= 1728x\frac{f(x^{25})}{f(x)} = 1728x(f(x))^{24} + 5J = 288 \sum_{n=1}^{\infty} n\sigma_1(n)x^n + 5J.$$

Multiplying by 2, rejecting multiples of 5, and replacing $f(x^{25})$ by its expansion given by (1·3), we obtain

$$(x - x^{26} - x^{51} + x^{126} + \ldots)(1 + p(1)x + p(2)x^2 + \ldots)$$
$$= \sum_{n=1}^{\infty} n\sigma_1(n)x^n + 5J.$$

Hence

$$(2·4) \quad p(n-1) - p(n-26) - p(n-51) + p(n-126) + p(n-176)$$
$$- p(n-301) - \ldots \equiv n\sigma_1(n) \pmod 5,$$

the numbers 1, 26, 51, ... being the numbers of the forms

$$\tfrac{25}{2}n(3n-1)+1, \quad \tfrac{25}{2}n(3n+1)+1,$$

or, what is the same thing, of the forms

$$\tfrac{1}{2}(5n-1)(15n-2), \quad \tfrac{1}{2}(5n+1)(15n+2).$$

In particular it follows from (2·3) that

$$(2·5) \qquad p(5m-1) \equiv 0 \pmod 5.$$

Modulus 7.

3. It is obvious from (1·13) that

$$R = 1 + 7J.$$

Also $n^7 - n \equiv 0 \pmod 7$, and so, from (1·11) and (1·61),

$$Q^2 = P + 7J.$$

Hence $$(Q^3 - R^2)^2 = (PQ - 1 + 7J)^2 = P^2Q^2 - 2PQ + 1 + 7J$$
$$= P^3 - 2PQ + R + 7J.$$

But, from (1·72) and (1·81),

$$P^3 - 2PQ + R = 144 \sum_{n=1}^{\infty} (5n\sigma_3(n) - 12n^2\sigma_1(n))\, x^n$$
$$= \sum_{n=1}^{\infty} (n^2\sigma_1(n) - n\sigma_3(n))\, x^n + 7J.$$

And therefore

$$(Q^3 - R^2)^2 = \sum_{n=1}^{\infty} (n^2\sigma_1(n) - n\sigma_3(n))\, x^n + 7J. \tag{3·1}$$

Again (by the same argument which led to (2·2)) we have

$$(f(x))^{48} = \frac{f(x^{49})}{f(x)} + 7J. \tag{3·2}$$

Combining (3·1) and (3·2), we obtain

$$x^2 \frac{f(x^{49})}{f(x)} = x^2 (f(x))^{48} + 7J = 1728^2 x^2 (f(x))^{48} + 7J \tag{3·3}$$
$$= (Q^3 - R^2)^2 + 7J$$
$$= \sum_{n=1}^{\infty} (n^2\sigma_1(n) - n\sigma_3(n))\, x^n + 7J.$$

From (3·3) it follows (just as (2·4) and (2·5) followed from (2·3)) that

$$p(n-2) - p(n-51) - p(n-100) + p(n-247) + p(n-345) - p(n-590) - \ldots \equiv n^2\sigma_1(n) - n\sigma_3(n) \pmod 7, \tag{3·4}$$

the numbers 2, 51, 100, ... being those of the forms

$$\tfrac{1}{2}(7n-1)(21n-4), \quad \tfrac{1}{2}(7n+1)(21n+4);$$

and that

$$p(7m-2) \equiv 0 \pmod 7. \tag{3·5}$$

Modulus 11.

4. It is obvious from (1·62) that

$$QR = 1 + 11J. \tag{4·1}$$

Also $n^{11}-n\equiv 0 \pmod{11}$, and so, from (1·11) and (1·63),

(4·2) $$Q^3-3R^2=441Q^3+250R^2+11J$$
$$=691+65520\left(\frac{x}{1-x}+\frac{2^{11}x^2}{1-x^2}+\dots\right)+11J$$
$$=-2+48\left(\frac{x}{1-x}+\frac{2x^2}{1-x^2}+\dots\right)+11J$$
$$=-2P+11J.$$

It is easily deduced that

(4·3)
$$(Q^3-R^2)^5=(Q^3-3R^2)^5-Q(Q^3-3R^2)^3-R(Q^3-3R^2)^2+6QR+11J$$
$$=P^5-3P^3Q-4P^2R+6QR+11J.$$

[For
$$(Q^3-3R^2)^5-Q(Q^3-3R^2)^3-R(Q^3-3R^2)^2+6QR$$
$$=(Q^3-3R^2)^5-Q^3R^2(Q^3-3R^2)^3-Q^3R^4(Q^3-3R^2)^2+6Q^6R^6+11J$$
$$=Q^{15}-16Q^{12}R^2+98Q^9R^4-285Q^6R^6+423Q^3R^8-243R^{10}+11J$$
$$=(Q^3-R^2)^5+11J$$
by (4·1), and (4·3) then follows from (4·2).]

Again, [if we multiply (1·74), (1·83), (1·92), and (1·93) by -1, 3, -4, and -1, and add, we obtain, on rejecting multiples of 11,]
$$P^5-3P^3Q-4P^2R+6QR=-5\Phi_{1,8}+3\Phi_{2,7}+3\Phi_{3,6}-\Phi_{4,5}+11J;$$
and from this and (4·3) follows

(4·4)
$$(Q^3-R^2)^5=-\sum_{n=1}^{\infty}(5n\sigma_7(n)-3n^2\sigma_5(n)-3n^3\sigma_3(n)+n^4\sigma_1(n))x^n+11J.$$

But (by the same argument which led to (2·2) and (3·2)) we have

(4·5) $$(f(x))^{120}=\frac{f(x^{121})}{f(x)}+11J.$$

From (4·4) and (4·5)
$$x^5\frac{f(x^{121})}{f(x)}=x^5(f(x))^{120}+11J=1728^5x^5(f(x))^{120}+11J$$
$$=(Q^3-R^2)^5+11J$$
$$=-\sum_{n=1}^{\infty}(5n\sigma_7(n)-3n^2\sigma_5(n)-3n^3\sigma_3(n)+n^4\sigma_1(n))x^n+11J.$$

It now follows as before that

(4·6)
$$p(n-5)-p(n-126)-p(n-247)+p(n-610)+p(n-852)$$
$$-p(n-1457)-\dots\equiv -n^4\sigma_1(n)+3n^3\sigma_3(n)+3n^2\sigma_5(n)$$
$$-5n\sigma_7(n)\pmod{11},$$

5, 126, 247, ... being the numbers of the forms

$$\tfrac{1}{2}(11n-2)(33n-5), \quad \tfrac{1}{2}(11n+2)(33n+5);$$

and in particular that

$$(4{\cdot}7) \qquad p(11m-5) \equiv 0 \pmod{11}.$$

5. If we are only concerned to prove (4·7), it is not necessary to assume quite so much.

Let us write ϑ for the operation $x\dfrac{d}{dx}$. Then* we have

$$(5{\cdot}11) \qquad \vartheta P = \tfrac{1}{12}(P^2 - Q),$$

$$(5{\cdot}12) \qquad \vartheta Q = \tfrac{1}{3}(PQ - R),$$

$$(5{\cdot}13) \qquad \vartheta R = \tfrac{1}{2}(PR - Q^2).$$

From these equations we deduce [by straightforward calculation

$$864\vartheta^4 P = P^5 - 10P^3Q - 15PQ^2 + 20P^2R + 4QR,$$

$$72\vartheta^3 Q = 5P^3Q + 15PQ^2 - 15P^2R - 5QR,$$

$$24\vartheta^2 R = -14PQ^2 + 7P^2R + 7QR.$$

The left-hand side of each of these equations is of the form

$$x\frac{dJ}{dx}.$$

Multiplying by 1, 8, and 2, adding, and rejecting multiples of 11, we find

$$(5{\cdot}2) \qquad P^5 - 3P^3Q + 2P^2R = x\frac{dJ}{dx} + 11J.$$

We have also, by (5·11),

$$6P^2R - 6QR = 72xR\frac{dP}{dx}.$$

But, differentiating (4·2), and using (4·1), we obtain

$$72xR\frac{dP}{dx} = 36xR\left(-3Q^2\frac{dQ}{dx} + 6R\frac{dR}{dx}\right) + 11J$$

$$= -108xQ\frac{dQ}{dx} + 216xR^2\frac{dR}{dx} + 11J$$

$$= x\frac{dJ}{dx} + 11J.$$

Hence

$$(5{\cdot}3) \qquad 6P^2R - 6QR = x\frac{dJ}{dx} + 11J.$$

* Ramanujan, p. 165 [p. 142].

From (5·2) and (5·3) we deduce

$$P^5 - 3P^3Q - 4P^2R + 6QR = x\frac{dJ}{dx} + 11J,$$

and from (4·3)]

$$(5{\cdot}4) \qquad (Q^3 - R^2)^5 = x\frac{dJ}{dx} + 11J.$$

Finally, from (4·5) and (5·4),

$$x^5 \frac{f(x^{121})}{f(x)} = x^5(f(x))^{120} + 11J = (Q^3 - R^2)^5 + 11J$$

$$= x\frac{dJ}{dx} + 11J.$$

As the coefficient of x^{11m} on the right-hand side is a multiple of 11, (4·7) follows immediately.

MEMOIRS WRITTEN IN COLLABORATION WITH G. H. HARDY (31—37)

31*

UNE FORMULE ASYMPTOTIQUE POUR LE NOMBRE DES PARTITIONS DE n

(*Comptes Rendus*, 2 Jan. 1917)

1. Les divers problèmes de la théorie de la partition des nombres ont été étudiés surtout par les mathématiciens anglais, Cayley, Sylvester et Macmahon†, qui les ont abordés d'un point de vue purement algébrique. Ces auteurs n'y ont fait aucune application des méthodes de la théorie des fonctions, de sorte qu'on ne trouve pas, dans la théorie en question, de formules asymptotiques, telles qu'on en rencontre, par exemple, dans la théorie des nombres premiers. Il nous semble donc que les résultats que nous allons faire connaître peuvent présenter quelque nouveauté.

2. Nous nous sommes occupés surtout de la fonction $p(n)$, nombre des partitions de n. On a

$$f(x) = \frac{1}{(1-x)(1-x^2)(1-x^3)\dots} = \sum_0^\infty p(n)\,x^n \qquad (|x|<1).$$

Nous avons pensé d'abord à faire usage de quelque théorème de caractère *Taubérien*: on désigne ainsi les théorèmes réciproques du théorème classique d'Abel et de ses généralisations. A cette catégorie appartient l'énoncé suivant:

Soit $g(x) = \Sigma a_n x^n$ une série de puissances à coefficients POSITIFS, *telle qu'on ait*

$$\log g(x) \sim \frac{A}{1-x},$$

quand x tend vers un par des valeurs positives. Alors on a

$$\log s_n = \log(a_0 + a_1 + \dots + a_n) \sim 2\sqrt{(An)},$$

quand n tend vers l'infini‡.

* [For proofs of the theorems enunciated in this note see No. 36 of this volume.]

† Voir le grand traité *Combinatory Analysis* de M. P. A. Macmahon (Cambridge, 1915-16).

‡ Nous avons donné des généralisations étendues de ce théorème dans un mémoire qui doit paraître dans un autre recueil. [The paper referred to is No. 34 of this volume; see, in particular, pp. 252—258.]

En posant $g(x) = (1-x) f(x)$, on a

$$A = \frac{\pi^2}{6};$$

et nous en tirons

$$p(n) = e^{\pi \sqrt{(\frac{2}{3}n)}\,(1+\epsilon)}, \quad \ldots\ldots\ldots\ldots\ldots\ldots\ldots\ldots\ldots(1)$$

où ϵ tend vers zéro avec $1/n$.

3. Pour pousser l'approximation plus loin, il faut recourir au théorème de Cauchy. Des formules

$$p(n) = \frac{1}{2\pi i}\int \frac{f(x)}{x^{n+1}}\,dx,$$

avec un chemin d'intégration convenable intérieur au cercle de rayon un, et

$$f(x) = \frac{x^{\frac{1}{24}}}{\sqrt{(2\pi)}}\sqrt{\left(\log\frac{1}{x}\right)}\exp\left(\frac{\pi^2}{6\log(1/x)}\right) f\left\{\exp\left(-\frac{4\pi^2}{\log(1/x)}\right)\right\} \ldots\ldots(2)$$

(fournie par la théorie de la transformation linéaire des fonctions elliptiques), nous avons tiré, en premier lieu, la formule vraiment asymptotique

$$p(n) \sim P(n) = \frac{1}{4n\sqrt{3}}\, e^{\pi\sqrt{(\frac{2}{3}n)}}. \quad \ldots\ldots\ldots\ldots\ldots\ldots\ldots(3)$$

On a

$p(10) = 42$, $p(20) = 627$, $p(50) = 204\,226$, $p(80) = 15\,796\,476$;

$P(10) = 48$, $P(20) = 692$, $P(50) = 217\,590$, $P(80) = 16\,606\,781$.

Les valeurs correspondantes de $P(n) : p(n)$ sont

1·145; 1·104; 1·065; 1·051:

la valeur approximative est toujours en excès.

4. Mais nous avons abouti plus tard à des résultats beaucoup plus satisfaisants. Nous considérons la fonction

$$F(x) = \frac{1}{\pi\sqrt{2}}\sum_{1}^{\infty}\frac{d}{dn}\left\{\frac{\cosh\left[\pi\sqrt{\{\frac{2}{3}(n-\frac{1}{24})\}}\right]-1}{\sqrt{(n-\frac{1}{24})}}\right\} x^n. \quad \ldots\ldots\ldots(4)$$

En faisant usage des formules sommatoires que démontre M. E. Lindelöf dans son beau livre *Le calcul des résidus*, on trouve aisément que $F(x)$ (on parle, il va sans dire, de la branche principale) a pour seul point singulier le point $x = 1$, et que la fonction

$$F(x) - \frac{x^{\frac{1}{24}}}{\sqrt{(2\pi)}}\sqrt{\left(\log\frac{1}{x}\right)}\left[\exp\left\{\frac{\pi^2}{6\log(1/x)}\right\}-1\right]$$

est régulière pour $x = 1$. On est conduit naturellement à appliquer le théorème de Cauchy à la fonction $f(x) - F(x)$, et l'on trouve

$$p(n) = \frac{1}{2\pi\sqrt{2}}\frac{d}{dn}\frac{e^{\pi\sqrt{\{\frac{2}{3}(n-\frac{1}{24})\}}}}{\sqrt{(n-\frac{1}{24})}} + O\left(e^{k\sqrt{n}}\right) = Q(n) + O\left(e^{k\sqrt{n}}\right), \ldots\ldots(5)$$

où k désigne un nombre quelconque supérieur à $\pi/\surd 6$. L'approximation, pour des valeurs assez grandes de n, est très bonne : on trouve, en effet,

$$p(61) = 1\,121\,505, \quad p(62) = 1\,300\,156, \quad p(63) = 1\,505\,499;$$
$$Q(61) = 1\,121\,539, \quad Q(62) = 1\,300\,121, \quad Q(63) = 1\,505\,536.$$

La valeur approximative est, pour les valeurs suffisamment grandes de n, alternativement en excès et en défaut.

5. On peut pousser ces calculs beaucoup plus loin. On forme des fonctions, analogues à $F(x)$, qui présentent, pour les valeurs

$$x = -1, \quad e^{\frac{2}{3}\pi i}, \quad e^{-\frac{2}{3}\pi i}, \quad i, \quad -i, \quad e^{\frac{2}{5}\pi i}, \quad \ldots,$$

des singularités d'un type très analogue à celles que présente $f(x)$. On soustrait alors de $f(x)$ une somme d'un nombre fini convenable de ces fonctions. On trouve ainsi, par exemple,

$$p(n) = \frac{1}{2\pi\surd 2}\frac{d}{dn}\frac{e^{\pi\surd\{\frac{2}{3}(n-\frac{1}{24})\}}}{\surd(n-\frac{1}{24})} + \frac{(-1)^n}{2\pi}\frac{d}{dn}\frac{e^{\frac{1}{2}\pi\surd\{\frac{2}{3}(n-\frac{1}{24})\}}}{\surd(n-\frac{1}{24})}$$
$$+ \frac{\surd 3}{\pi\surd 2}\cos\left(\frac{2n\pi}{3} - \frac{\pi}{18}\right)\frac{d}{dn}\frac{e^{\frac{1}{3}\pi\surd\{\frac{2}{3}(n-\frac{1}{24})\}}}{\surd(n-\frac{1}{24})} + O(e^{k\surd n}), \quad \ldots\ldots(6)$$

où k désigne un nombre quelconque plus grand que $\frac{1}{4}\pi\surd\frac{2}{3}$.

32

PROOF THAT ALMOST ALL NUMBERS n ARE COMPOSED OF ABOUT $\log\log n$ PRIME FACTORS

(*Proceedings of the London Mathematical Society*, 2, XVI, 1917, *Records for* 14 *Dec.* 1916)

A number n is described in popular language as a *round* number if it is composed of a considerable number of comparatively small factors: thus $1200 = 2^4 . 3 . 5^2$ would generally be said to be a round number. It is a matter of common experience that round numbers are exceedingly rare. The fact may be verified by anybody who will make a practice of factorising numbers, such as the numbers of taxi-cabs or railway carriages, which are presented to his attention in moments of leisure. The object of this paper* is to provide the mathematical explanation of this phenomenon.

Let $\pi_\nu(x)$ denote the number of numbers which do not exceed x and are formed of exactly ν prime factors. There is an ambiguity in this definition, for we may count multiple factors multiply or not. But the results are substantially the same on either interpretation.

It has been proved by Landau† that

$$\pi_\nu(x) \sim \frac{x}{\log x}\frac{(\log\log x)^{\nu-1}}{(\nu-1)!}, \quad \text{.......................(1)}$$

as $x \to \infty$, for every fixed value of ν. It is moreover obvious that

$$[x] = \pi_1(x) + \pi_2(x) + \pi_3(x) + \dots, \quad \text{..........................(2)}$$

and

$$x = \frac{x}{\log x}\left\{1 + \log\log x + \frac{(\log\log x)^2}{2!} + \dots\right\}. \quad \text{...........(3)}$$

Landau's result shews that there is a certain correspondence between the terms of the series (2) and (3). The correspondence is far from exact. The first series is finite, for it is obvious that $\pi_\nu(x) = 0$ if $\nu > (\log x)/(\log 2)$; and the second is infinite. But it is reasonable to anticipate a correspondence accurate enough to throw considerable light on the distribution of the numbers less than x in respect of the number of their prime factors.

The greatest term of the series (3) occurs when ν is about $\log\log x$. And if we consider the block of terms for which

$$\log\log x - \phi(x)\sqrt{(\log\log x)} < \nu < \log\log x + \phi(x)\sqrt{(\log\log x)}, \quad \text{...(4)}$$

* The paper has been published in the *Quarterly Journal of Mathematics*, Vol. XLVIII, pp. 76—92 [No. 35 of this volume].

† See *Handbuch der Lehre von der Verteilung der Primzahlen*, pp. 203—213.

where $\phi(x)$ is any function of x which tends to infinity with x, we find without difficulty that it is these terms which contribute almost all the sum of the series: the ratio of their sum to that of the remaining terms tends to infinity with x.

In this paper we shew that the same conclusion holds for the series (2). Let us consider all numbers n which do not exceed x, and denote by x_P the number of them which possess a property $P(n, x)$: this property may be a function of both n and x, or of one variable only. If then $x_P/x \to 1$ when $x \to \infty$, we say that *almost all numbers less than x possess the property P.* And if P is a function of n only, we say simply that *almost all numbers possess the property P.* This being so, we prove the following theorems.

1. *Almost all numbers n less than x are formed of more than*

$$\log\log x - \phi(x)\sqrt{(\log\log x)}$$

and less than

$$\log\log x + \phi(x)\sqrt{(\log\log x)}$$

prime factors.

2. *Almost all numbers n are formed of more than*

$$\log\log n - \phi(n)\sqrt{(\log\log n)}$$

and less than

$$\log\log n + \phi(n)\sqrt{(\log\log n)}$$

prime factors.

In these theorems ϕ is any function of x (or n) which tends to infinity with its argument: and either theorem is true in whichever manner the factors of n are counted. The only serious difficulty in the proofs lies in replacing Landau's asymptotic relations (1) by inequalities valid for all values of ν and x.

Since $\log\log n$ tends to infinity with extreme slowness, the theorems are fully sufficient to explain the observations which suggested them.

33

ASYMPTOTIC FORMULÆ IN COMBINATORY ANALYSIS

(*Proceedings of the London Mathematical Society*, 2, XVI, 1917, *Records for* 1 *March* 1917)

A preliminary account of some of the contents of this paper* appeared in the *Comptes Rendus* of January 2nd, 1917. The paper contains a full discussion and proof of the results there stated. The asymptotic formula for $p(n)$, the number of unrestricted partitions of n, of which only the first three terms were given, is completed; and it is shewn that, by taking a number of terms of order $\sqrt{n}$, the *exact* value of $p(n)$ can be obtained for all sufficiently large values of n. Some account is also given of actual or possible applications of the method used to other problems in Combinatory Analysis or the Analytic Theory of Numbers.

* [No. 36 of this volume.]

34

ASYMPTOTIC FORMULÆ FOR THE DISTRIBUTION OF INTEGERS OF VARIOUS TYPES*

(*Proceedings of the London Mathematical Society*, 2, XVI, 1917, 112—132)

1. *Statement of the problem.*

1·1. We denote by q a number of the form

(1·11) $$2^{a_2}3^{a_3}5^{a_5}\dots p^{a_p},$$

where 2, 3, 5, ..., p are primes and

(1·111) $$a_2 \geqslant a_3 \geqslant a_5 \geqslant \dots \geqslant a_p;$$

and by $Q(x)$ the number of such numbers which do not exceed x: and our problem is that of determining the order of $Q(x)$. We prove that

(1·12) $$Q(x) = \exp\left[\{1+o(1)\}\frac{2\pi}{\sqrt{3}}\sqrt{\left(\frac{\log x}{\log\log x}\right)}\right],$$

that is to say that to every positive ϵ corresponds an $x_0 = x_0(\epsilon)$, such that

(1·121)
$$\left(\frac{2\pi}{\sqrt{3}}-\epsilon\right)\sqrt{\left(\frac{\log x}{\log\log x}\right)} < \log Q(x) < \left(\frac{2\pi}{\sqrt{3}}+\epsilon\right)\sqrt{\left(\frac{\log x}{\log\log x}\right)},$$

for $x > x_0$. The function $Q(x)$ is thus of higher order than any power of $\log x$, but of lower order than any power of x.

The interest of the problem is threefold. In the first place the result itself, and the method by which it is obtained, are curious and interesting in themselves. Secondly, the method of proof is one which, as we shew at the end of the paper, may be applied to a whole class of problems in the analytic theory of numbers: it enables us, for example, to find asymptotic formulæ for the number of partitions of n into positive integers, or into different positive integers, or into primes. Finally, the class of numbers q includes as a sub-class the "highly composite" numbers recently studied by Mr Ramanujan in an elaborate memoir in these *Proceedings*†. The problem of determining, with any precision, the number $H(x)$ of highly composite numbers not exceeding x appears to be one of extreme difficulty. Mr Ramanujan has proved, by elementary methods, that the order of $H(x)$ is at any rate greater

* This paper was originally communicated under the title "A problem in the Analytic Theory of Numbers."

† Ramanujan, "Highly Composite Numbers," *Proc. London Math. Soc.*, Ser. 2, Vol. XIV 1915, pp. 347—409 [No. 15 of this volume].

than that of $\log x$*: but it is still uncertain whether or no the order of $H(x)$ is greater than that of any power of $\log x$. In order to apply transcendental methods to this problem, it would be necessary to study the properties of the function

$$\mathfrak{H}(s) = \Sigma \frac{1}{h^s},$$

where h is a highly composite number, and we have not been able to make any progress in this direction. It is therefore very desirable to study the distribution of wider classes of numbers which include the highly composite numbers and possess some at any rate of their characteristic properties. The simplest and most natural such class is that of the numbers q; and here progress is comparatively easy, since the function

$$\mathfrak{Q}(s) = \Sigma \frac{1}{q^s} \tag{1·13}$$

possesses a product expression analogous to Euler's product expression for $\zeta(s)$, viz.

$$\mathfrak{Q}(s) = \prod_1^\infty \left(\frac{1}{1 - l_n^{-s}}\right), \tag{1·14}$$

where $l_n = 2.3.5 \ldots p_n$ is the product of the first n primes.

We have not been able to apply to this problem the methods, depending on the theory of functions of a complex variable, by which the Prime Number Theorem was proved. The function $\mathfrak{Q}(s)$ has the line $\sigma = 0$† as a line of essential singularities, and we are not able to obtain sufficiently accurate information concerning the nature of these singularities. But it is easy enough to determine the behaviour of $\mathfrak{Q}(s)$ as a function of the *real* variable s; and it proves sufficient for our purpose to determine an asymptotic formula for $\mathfrak{Q}(s)$ when $s \to 0$, and then to apply a "Tauberian" theorem similar to those proved by Messrs Hardy and Littlewood in a series of papers published in these *Proceedings* and elsewhere‡.

This "Tauberian" theorem is in itself of considerable interest as being (so far as we are aware) the first such theorem which deals with functions or sequences tending to infinity more rapidly than any power of the variable.

* As great as that of $\dfrac{\log x \sqrt{(\log\log x)}}{(\log\log\log x)^{\frac{3}{2}}}$: see p. 385 of his memoir [p. 108 of this volume].

† We write as usual $s = \sigma + it$.

‡ See, in particular, Hardy and Littlewood, "Tauberian theorems concerning power series and Dirichlet's series whose coefficients are positive," *Proc. London Math. Soc.*, Ser. 2, Vol. XIII, 1914, pp. 174—191; and "Some theorems concerning Dirichlet's series," *Messenger of Mathematics*, Vol. XLIII, 1914, pp. 134—147.

2. *Elementary results.*

2·1. Let us consider, before proceeding further, what information concerning the order of $Q(x)$ can be obtained by purely elementary methods.

Let

$$l_n = 2.3.5 \ldots p_n = e^{\vartheta(p_n)}, \tag{2·11}$$

where $\vartheta(x)$ is Tschebyschef's function

$$\vartheta(x) = \sum_{p \leqslant x} \log p.$$

The class of numbers q is plainly identical with the class of numbers of the form

$$l_1^{b_1} l_2^{b_2} \ldots l_n^{b_n}, \tag{2·12}$$

where $\qquad b_1 \geqslant 0, \quad b_2 \geqslant 0, \quad \ldots, \quad b_n \geqslant 0.$

Now every b can be expressed in one and only one way in the form

$$b_i = c_{i,m} 2^m + c_{i,m-1} 2^{m-1} + \ldots + c_{i,0}, \tag{2·13}$$

where every c is equal to zero or to unity. We have therefore

$$q = \prod_{i=1}^{n} \left(l_i^{\sum\limits_{j=0}^{m} c_{i,j} 2^j} \right) = \prod_{j=0}^{m} \prod_{i=1}^{n} l_i^{c_{i,j} 2^j} = \prod_{j=0}^{m} r_j^{2^j}, \tag{2·14}$$

say, where

$$r_j = l_1^{c_{1,j}} l_2^{c_{2,j}} \ldots l_n^{c_{n,j}}. \tag{2·141}$$

Let r denote, generally, a number of the form

$$r = l_1^{c_1} l_2^{c_2} \ldots l_n^{c_n}, \tag{2·15}$$

where every c is zero or unity: and $R(x)$ the number of such numbers which do not exceed x. If $q \leqslant x$, we have

$$r_0 \leqslant x, \quad r_1^2 \leqslant x, \quad r_2^4 \leqslant x, \quad \ldots.$$

The number of possible values of r_0, in the formula (2·14), cannot therefore exceed $R(x)$; the number of possible values of r_1 cannot exceed $R(x^{\frac{1}{2}})$; and so on. The total number of values of q can therefore not exceed

$$S(x) = R(x) R(x^{\frac{1}{2}}) R(x^{\frac{1}{4}}) \ldots R(x^{2^{-\varpi}}), \tag{2·16}$$

where ϖ is the largest number such that

$$x^{2^{-\varpi}} \geqslant 2, \quad x \geqslant 2^{2^{\varpi}}. \tag{2·161}$$

Thus

$$Q(x) \leqslant S(x). \tag{2·17}$$

2·2. We denote by f and g the largest numbers such that

$$l_f \leqslant x, \tag{2·211}$$

$$l_1 l_2 \ldots l_g \leqslant x. \tag{2·212}$$

It is known* (and may be proved by elementary methods) that constants A and B exist, such that

$$\vartheta(x) \geqslant Ax \quad (x \geqslant 2), \tag{2·221}$$

and

$$p_n \geqslant Bn \log n \quad (n \geqslant 1). \tag{2·222}$$

We have therefore
$$e^{Ap_f} \leqslant x,$$
$$f \log f = O(\log x),$$
$$\log f = O(\log \log x); \tag{2·23}$$

and
$$\sum_1^g \vartheta(p_\nu) \leqslant \log x, \qquad \sum_1^g p_\nu = O(\log x),$$
$$\sum_1^g \nu \log \nu = O(\log x), \quad g^2 \log g = O(\log x),$$
$$g = O\sqrt{\left(\frac{\log x}{\log \log x}\right)}. \tag{2·24}$$

But it is easy to obtain an upper bound for $R(x)$ in terms of f and g. The number of numbers $l_1, l_2, \ldots$, not exceeding x, is not greater than f; the number of products, not exceeding x, of pairs of such numbers, is *a fortiori* not greater than $\frac{1}{2}f(f-1)$; and so on. Thus

$$R(x) \leqslant f + \frac{f(f-1)}{2!} + \frac{f(f-1)(f-2)}{3!} + \ldots,$$

where the summation need be extended to g terms only, since

$$l_1 l_2 \ldots l_g l_{g+1} > x.$$

A fortiori, we have

$$R(x) \leqslant 1 + f + \frac{f^2}{2!} + \ldots + \frac{f^g}{g!} < (1+f)^g = e^{g \log(1+f)}.$$

Thus

$$R(x) = e^{O(g \log f)} = e^{O\{\sqrt{(\log x \log \log x)}\}}, \tag{2·25}$$

by (2·23) and (2·24). Finally, since

$$\log \sqrt{x} \log \log \sqrt{x} < \tfrac{1}{2} \log x \log \log x,$$

it follows from (2·16) and (2·17) that

$$Q(x) = \exp\left[O\left\{\left(1 + \frac{1}{2} + \frac{1}{4} + \ldots + \frac{1}{2^\varpi}\right)\sqrt{(\log x \log \log x)}\right\}\right] \tag{2·26}$$
$$= e^{O\{\sqrt{(\log x \log \log x)}\}}.$$

2·3. A *lower* bound for $Q(x)$ may be found as follows. If g is defined as in 2·2, we have

$$l_1 l_2 \ldots l_g \leqslant x < l_1 l_2 \ldots l_g l_{g+1}.$$

* See Landau, *Handbuch*, pp. 79, 83, 214.

It follows from the analysis of 2·2 that

$$l_{g+1} = e^{\vartheta(p_{g+1})} = e^{O(g \log g)},$$

and

$$l_1 l_2 \ldots l_g = \exp\left\{\sum_1^g \vartheta(p_\nu)\right\} = e^{O(g^2 \log g)}.$$

Thus

$$x < e^{O(g^2 \log g)};$$

which is only possible if g is greater than a constant positive multiple of

$$\sqrt{\left(\frac{\log x}{\log\log x}\right)}.$$

Now the numbers $l_1, l_2, \ldots, l_g$ can be combined in 2^g different ways, and each such combination gives a number q not greater than x. Thus

$$(2{\cdot}31) \qquad Q(x) \geqslant 2^g > \exp\left\{K\sqrt{\left(\frac{\log x}{\log\log x}\right)}\right\},$$

where K is a positive constant. From (2·26) and (2·31) it follows that there are positive constants K and L such that

$$(2{\cdot}32) \qquad K\sqrt{\left(\frac{\log x}{\log\log x}\right)} < \log Q(x) < L\sqrt{(\log x \log\log x)}.$$

The inequalities (2·32) give a fairly accurate idea as to the order of magnitude of $Q(x)$. But they are much less precise than the inequalities (1·121). To obtain these requires the use of less elementary methods.

3. *The behaviour of* $\mathfrak{Q}(s)$ *when* $s \to 0$ *by positive values.*

3·1. From the fact, already used in 2·1, that the class of numbers q is identical with the class of numbers of the form (2·12), it follows at once that

$$(1{\cdot}14) \qquad \mathfrak{Q}(s) = \sum \frac{1}{q^s} = \prod_1^\infty \left(\frac{1}{1 - l_n^{-s}}\right).$$

Both series and product are absolutely convergent for $\sigma > 0$, and

$$(3{\cdot}11) \qquad \log \mathfrak{Q}(s) = \phi(s) + \tfrac{1}{2}\phi(2s) + \tfrac{1}{3}\phi(3s) + \ldots,$$

where

$$(3{\cdot}111) \qquad \phi(s) = \sum_1^\infty l_n^{-s}.$$

We have also

(3·12)

$$\phi(s) = \frac{1-2^{-s}}{2^s-1} + 2^{-s}\frac{1-3^{-s}}{3^s-1} + 2^{-s}3^{-s}\frac{1-5^{-s}}{5^s-1} + \ldots$$

$$= \frac{1}{2^s-1} + 2^{-s}\left(\frac{1}{3^s-1} - \frac{1}{2^s-1}\right) + 2^{-s}3^{-s}\left(\frac{1}{5^s-1} - \frac{1}{3^s-1}\right) + \ldots$$

$$= \frac{1}{2^s-1} + \sum_1^\infty e^{-s\vartheta(p_n)} \int_{p_n}^{p_{n+1}} \frac{d}{dx}\left(\frac{1}{x^s-1}\right) dx$$

$$= \frac{1}{2^s-1} - s\int_2^\infty \frac{x^{s-1}}{(x^s-1)^2} e^{-s\vartheta(x)}\, dx.$$

3·2. LEMMA.—*If* $x > 1$, $s > 0$, *then*

$$\frac{1}{(s\log x)^2} - \frac{1}{12} < \frac{x^s}{(x^s-1)^2} < \frac{1}{(s\log x)^2}. \tag{3·21}$$

Write $x^s = e^u$: then we have to prove that

$$\frac{1}{u^2} - \frac{1}{12} < \frac{e^u}{(e^u-1)^2} < \frac{1}{u^2} \tag{3·22}$$

for all positive values of u; or (writing w for $\frac{1}{2}u$) that

$$\frac{1}{w^2} - \frac{1}{3} < \frac{1}{\sinh^2 w} < \frac{1}{w^2} \tag{3·23}$$

for all positive values of w. But it is easy to prove that the function

$$f(w) = \frac{1}{w^2} - \frac{1}{\sinh^2 w}$$

is a steadily decreasing function of w, and that its limit when $w \to 0$ is $\frac{1}{3}$; and this establishes the truth of the lemma.

3·3. We have therefore

$$\phi(s) = \frac{1}{2^s-1} - \phi_1(s) = \frac{1}{s\log 2} - \phi_1(s) + O(1), \tag{3·31}$$

where

$$\frac{1}{s}\int_2^\infty \left\{\frac{1}{(\log x)^2} - \frac{s^2}{12}\right\} e^{-s\vartheta(x)} \frac{dx}{x} < \phi_1(s) < \frac{1}{s}\int_2^\infty \frac{e^{-s\vartheta(x)}}{(\log x)^2}\frac{dx}{x}. \tag{3·311}$$

From the second of these inequalities, and (2·221), it follows that

$$\phi_1(s) < \frac{1}{s}\int_2^\infty \frac{e^{-Asx}}{(\log x)^2}\frac{dx}{x} = \frac{e^{-2As}}{s\log 2} - A\int_2^\infty \frac{e^{-Asx}}{\log x}dx$$
$$= \frac{1}{s\log 2} - A\int_2^\infty \frac{e^{-Asx}}{\log x}dx + O(1);$$

and so that

$$\phi(s) > A\int_2^\infty \frac{e^{-Asx}}{\log x}dx + O(1). \tag{3·32}$$

On the other hand there is a positive constant B, such that

$$\vartheta(x) < Bx \quad (x \geqslant 2)^*.$$

Thus

$$\phi_1(s) > \frac{1}{s}\int_2^\infty \left\{\frac{1}{(\log x)^2} - \frac{s^2}{12}\right\} e^{-Bsx}\frac{dx}{x} = \frac{1}{s}\int_2^\infty \frac{e^{-Bsx}}{(\log x)^2}\frac{dx}{x} + O(1)$$
$$= \frac{1}{s\log 2} - B\int_2^\infty \frac{e^{-Bsx}}{\log x}dx + O(1);$$

and so

$$\phi(s) < B\int_2^\infty \frac{e^{-Bsx}}{\log x}dx + O(1). \tag{3·33}$$

* Landau, *Handbuch*, *loc. cit.*

3·4. LEMMA.—*If H is any positive number, then*

$$J(s) = H\int_2^\infty \frac{e^{-Hsx}}{\log x}\,dx \sim \frac{1}{s\log(1/s)},$$

when $s \to 0$.

Given any positive number ϵ, we can choose ξ and X, so that

$$\int_0^\xi He^{-Hx}\,dx < \epsilon, \quad \int_X^\infty He^{-Hx}\,dx < \epsilon.$$

Now $$s\log\left(\frac{1}{s}\right)J(s) = \int_{2s}^\infty \frac{He^{-Hu}\log(1/s)}{\log u + \log(1/s)}\,du = \int_{2s}^{\sqrt{s}} + \int_{\sqrt{s}}^{\xi} + \int_\xi^X + \int_X^\infty$$
$$= j_1(s) + j_2(s) + j_3(s) + j_4(s),$$

say. And we have

$$0 < j_1(s) < \frac{\log(1/s)}{\log 2}\int_{2s}^{\sqrt{s}} He^{-Hu}\,du = O\{\sqrt{s}\log(1/s)\} = o(1),$$

$$0 < j_2(s) < 2\int_0^\xi He^{-Hu}\,du < 2\epsilon,$$

$$j_3(s) = \int_\xi^X He^{-Hu}\,du + o(1),$$

$$0 < j_4(s) < \int_X^\infty He^{-Hu}\,du < \epsilon;$$

and so

$$\left|1 - s\log\left(\frac{1}{s}\right)J(s)\right| = \left|\int_0^\infty He^{-Hu}\,du - j_1(s) - j_2(s) - j_3(s) - j_4(s)\right|$$
$$< 5\epsilon + o(1) < 6\epsilon,$$

for all sufficiently small values of s.

3·5. From (3·32), (3·33), and the lemma just proved, it follows that

(3·51) $$\phi(s) = \Sigma l_n^{-s} \sim \frac{1}{s\log(1/s)}.$$

From this formula we can deduce an asymptotic formula for $\log \mathfrak{Q}(s)$. We choose N so that

(3·52) $$\sum_{N<n} \frac{1}{n^2} < \epsilon,$$

and we write

(3·53) $$\log \mathfrak{Q}(s) = \Sigma \frac{1}{n}\phi(ns) = \sum_{1\leqq n\leqq N} + \sum_{N<n<1/\sqrt{s}} + \sum_{1/\sqrt{s}\leqq n\leqq 1/s} + \sum_{1/s<n}$$
$$= \Phi_1(s) + \Phi_2(s) + \Phi_3(s) + \Phi_4(s),$$

say.

In the first place

(3·541) $$\Phi_1(s) = \frac{1+o(1)}{s\log(1/s)}\sum_1^N \frac{1}{n^2}.$$

In the second place

$$\phi(ns) = O\left\{\frac{1}{ns \log(1/ns)}\right\},$$

and $$\log(1/ns) > \tfrac{1}{2}\log(1/s),$$

if $N < n < 1/\sqrt{s}$. It follows that a constant K exists such that

$$(3{\cdot}542) \qquad \Phi_2(s) < \frac{K}{s\log(1/s)} \sum_{N<n} \frac{1}{n^2} < \frac{K\epsilon}{s\log(1/s)}.$$

Thirdly, $\sqrt{s} \leqslant ns \leqslant 1$ in $\Phi_3(s)$, and a constant L exists such that

$$\phi(ns) < \frac{L}{\sqrt{s}\log(1/s)}.$$

Thus

$$(3{\cdot}543) \qquad \Phi_3(s) < \frac{L}{\sqrt{s}\log(1/s)} \sum_{1}^{1/s} \frac{1}{n} < \frac{2L}{\sqrt{s}},$$

for all sufficiently small values of s.

Finally, in $\Phi_4(s)$ we have $ns > 1$, and a constant M exists such that

$$\phi(ns) < M2^{-ns}.$$

Thus

$$(3{\cdot}544) \quad \Phi_4(s) < M \sum_{1/s<n} \frac{2^{-ns}}{n} < sM \sum_{1/s<n} 2^{-ns} < \frac{sM}{1-2^{-s}} = O(1).$$

From (3·53), (3·541)—(3 544), and (3·52) it follows that

$$(3{\cdot}55) \qquad \log \mathfrak{Q}(s) = \frac{1}{s\log(1/s)}\left[\{1+o(1)\}\left(\frac{\pi^2}{6}+\rho\right)+\rho'\right] + O\left\{\frac{1}{\sqrt{s}\log(1/s)}\right\} + O(1),$$

where $$|\rho| < \epsilon, \quad |\rho'| < K\epsilon.$$

Thus

$$(3{\cdot}56) \qquad \log \mathfrak{Q}(s) \sim \frac{\pi^2}{6s\log(1/s)},$$

or

$$(3{\cdot}57) \qquad \mathfrak{Q}(s) = \exp\left[\{1+o(1)\}\frac{\pi^2}{6s\log(1/s)}\right].$$

4. *A Tauberian theorem.*

4·1. The passage from (3·57) to (1·12) depends upon a theorem of the "Tauberian" type.

THEOREM A. *Suppose that*

(1) $\lambda_1 \geqslant 0, \quad \lambda_n > \lambda_{n-1}, \quad \lambda_n \to \infty;$

(2) $\lambda_n/\lambda_{n-1} \to 1;$

(3) $a_n \geqslant 0;$

(4) $A > 0, \quad \alpha > 0$;

(5) $\Sigma a_n e^{-\lambda_n s}$ *is convergent for* $s > 0$;

(6) $f(s) = \Sigma a_n e^{-\lambda_n s} = \exp\left[\{1 + o(1)\} A s^{-\alpha} \left\{\log\left(\frac{1}{s}\right)\right\}^{-\beta}\right]$,

when $s \to 0$. *Then*

$$A_n = a_1 + a_2 + \ldots + a_n = \exp\left[\{1 + o(1)\} B \lambda_n^{\alpha/(1+\alpha)} (\log \lambda_n)^{-\beta/(1+\alpha)}\right],$$

where $$B = A^{1/(1+\alpha)} \alpha^{-\alpha/(1+\alpha)} (1 + \alpha)^{1+[\beta/(1+\alpha)]},$$

when $n \to \infty$.

We are given that

(4·11) $$(1 - \delta) A s^{-\alpha} \left(\log \frac{1}{s}\right)^{-\beta} < \log f(s) < (1 + \delta) A s^{-\alpha} \left(\log \frac{1}{s}\right)^{-\beta},$$

for every positive δ and all sufficiently small values of s; and we have to shew that

(4·12)
$$(1 - \epsilon) B \lambda_n^{\alpha/(1+\alpha)} (\log \lambda_n)^{-\beta/(1+\alpha)} < \log A_n < (1 + \epsilon) B \lambda_n^{\alpha/(1+\alpha)} (\log \lambda_n)^{-\beta/(1+\alpha)},$$

for every positive ϵ and all sufficiently large values of n.

In the argument which follows we shall be dealing with three variables, δ, s, and n (or m), the two latter variables being connected by an equation or by inequalities, and with an auxiliary parameter ζ. We shall use the letter η, without a suffix, to denote generally a function of δ, s, and n (or m)*, which is not the same in different formulæ, but in all cases tends to zero when δ and s tend to zero and n (or m) to infinity; so that, given any positive ϵ, we have

$$0 < |\eta| < \epsilon,$$

for $$0 < \delta < \delta_0, \quad 0 < s < s_0, \quad n > n_0.$$

We shall use the symbol η_ζ to denote a function *of* ζ *only* which tends to zero with ζ, so that

$$0 < |\eta_\zeta| < \epsilon,$$

if ζ is small enough. It is to be understood that the choice of a ζ to satisfy certain conditions is in all cases prior to that of δ, s, and n (or m). Finally, we use the letters H, K, ... to denote positive numbers independent of these variables *and of* ζ.

The *second* of the inequalities (4·12) is very easily proved. For

(4·131) $$A_n e^{-\lambda_n s} < a_1 e^{-\lambda_1 s} + a_2 e^{-\lambda_2 s} + \ldots + a_n e^{-\lambda_n s} < f(s) < \exp\left\{(1 + \delta) A s^{-\alpha} \left(\log \frac{1}{s}\right)^{-\beta}\right\},$$

(4·132) $$A_n < \exp \chi,$$

where

(4·1321) $$\chi = (1 + \delta) A s^{-\alpha} \left(\log \frac{1}{s}\right)^{-\beta} + \lambda_n s.$$

* η may, of course, in some cases be a function of some of these variables only.

We can choose a value of s, corresponding to every large value of n, such that

$$(4{\cdot}14)\qquad (1-\delta)\,A\alpha s^{-1-\alpha}\left(\log\frac{1}{s}\right)^{-\beta} < \lambda_n < (1+\delta)\,A\alpha s^{-1-\alpha}\left(\log\frac{1}{s}\right)^{-\beta}.$$

From these inequalities we deduce, by an elementary process of approximation,

$$(4{\cdot}151)\qquad (1-\eta)\,(A\alpha)^{-1/(1+\alpha)}\lambda_n^{1/(1+\alpha)}\left(\log\frac{1}{s}\right)^{\beta/(1+\alpha)} < \frac{1}{s} < (1+\eta)\,(A\alpha)^{-1/(1+\alpha)}\lambda_n^{1/(1+\alpha)}\left(\log\frac{1}{s}\right)^{\beta/(1+\alpha)},$$

$$(4{\cdot}152)\qquad \frac{1-\eta}{1+\alpha}\log\lambda_n < \log\frac{1}{s} < \frac{1+\eta}{1+\alpha}\log\lambda_n,$$

$$(4{\cdot}153)\qquad (1-\eta)\frac{\alpha B}{1+\alpha}\lambda_n^{-1/(1+\alpha)}(\log\lambda_n)^{-\beta/(1+\alpha)} < s < (1+\eta)\frac{\alpha B}{1+\alpha}\lambda_n^{-1/(1+\alpha)}(\log\lambda_n)^{-\beta/(1+\alpha)},$$

$$(4{\cdot}154)\qquad \chi < (1+\eta)\,B\lambda_n^{\alpha/(1+\alpha)}(\log\lambda_n)^{-\beta/(1+\alpha)}.$$

We have therefore

$$(4{\cdot}16)\qquad \log A_n < (1+\epsilon)\,B\lambda_n^{\alpha/(1+\alpha)}(\log\lambda_n)^{-\beta/(1+\alpha)},$$

for every positive ϵ and all sufficiently large values of n*.

4·2. We have

$$(4{\cdot}21)\qquad f(s) = \Sigma a_n e^{-\lambda_n s} = \Sigma A_n\left(e^{-\lambda_n s} - e^{-\lambda_{n+1}s}\right) = s\sum_1^\infty A_n\int_{\lambda_n}^{\lambda_{n+1}} e^{-sx}dx = s\int_0^\infty \mathcal{A}(x)\,e^{-sx}dx,$$

where $\mathcal{A}(x)$ is the discontinuous function defined by

$$\mathcal{A}(x) = A_n \quad (\lambda_n \leqslant x < \lambda_{n+1})\dagger,$$

so that, by (4·16),

$$(4{\cdot}22)\qquad \log\mathcal{A}(x) < (1+\epsilon)\,Bx^{\alpha/(1+\alpha)}(\log x)^{-\beta/(1+\alpha)}$$

for every positive ϵ and all sufficiently large values of x. We have therefore

$$(4{\cdot}23)\qquad \exp\left\{(1-\delta)\,As^{-\alpha}\left(\log\frac{1}{s}\right)^{-\beta}\right\} < s\int_0^\infty \mathcal{A}(x)\,e^{-sx}dx < \exp\left\{(1+\delta)\,As^{-\alpha}\left(\log\frac{1}{s}\right)^{-\beta}\right\}$$

for every positive δ and all sufficiently small values of s.

* We use the second inequality (4·12) in the proof of the first. It would be sufficient for our purpose to begin by proving a result cruder than (4·16), with any constant K on the right-hand side instead of $(1+\epsilon)\,B$. But it is equally easy to obtain the more precise inequality. Compare the argument in the second of the two papers by Hardy and Littlewood quoted on p. 114 [p. 246 of this volume] (pp. 143 *et seq.*).

† Compare Hardy and Riesz, "The General Theory of Dirichlet's Series," *Cambridge Tracts in Mathematics*, No. 18, 1915, p. 24.

We define λ_x, a steadily increasing and continuous function of the continuous variable x, by the equation

$$\lambda_x = \lambda_n + (x-n)(\lambda_{n+1} - \lambda_n) \quad (n \leqslant x \leqslant n+1).$$

We can then choose m so that

$$\frac{1}{s} = \frac{1+\alpha}{\alpha B} \lambda_m^{1/(1+\alpha)} (\log \lambda_m)^{\beta/(1+\alpha)}. \tag{4·24}$$

We shall now shew that the limits of the integral in (4·23) may be replaced by $(1-\zeta)\lambda_m$ and $(1+\zeta)\lambda_m$, where ζ is an arbitrary positive number less than unity.

We write

(4·25)

$$J(s) = s\int_0^\infty \mathscr{A}(x) e^{-sx} dx = s\left\{\int_0^{\lambda_m/H} + \int_{\lambda_m/H}^{(1-\zeta)\lambda_m} + \int_{(1-\zeta)\lambda_m}^{(1+\zeta)\lambda_m} + \int_{(1+\zeta)\lambda_m}^{H\lambda_m} + \int_{H\lambda_m}^{\infty}\right\}$$
$$= J_1 + J_2 + J_3 + J_4 + J_5,$$

where H is a constant, in any case greater than 1, and large enough to satisfy certain further conditions which will appear in a moment; and we proceed to shew that J_1, J_2, J_4, and J_5 are negligible in comparison with the exponentials which occur in (4·23), and so in comparison with J_3.

4·3. The integrals J_1 and J_5 are easily disposed of. In the first place we have

$$J_1 = s\int_0^{\lambda_m/H} \mathscr{A}(x) e^{-sx} dx < \mathscr{A}\left(\frac{\lambda_m}{H}\right) \tag{4·31}$$
$$< \exp\left\{(1+\delta) B \left(\frac{\lambda_m}{H}\right)^{\alpha/(1+\alpha)} \left(\log \frac{\lambda_m}{H}\right)^{-\beta/(1+\alpha)}\right\},$$

by (4·22)*. It will be found, by a straightforward calculation, that this expression is less than

$$\exp\left\{(1+\eta) A (1+\alpha) H^{-\alpha/(1+\alpha)} s^{-\alpha} \left(\log \frac{1}{s}\right)^{-\beta}\right\}, \tag{4·32}$$

and is therefore certainly negligible if H is sufficiently large.

Thus J_1 is negligible. To prove that J_5 is negligible we prove first that

$$sx > 4Bx^{\alpha/(1+\alpha)} (\log x)^{-\beta/(1+\alpha)},$$

if $x > H\lambda_m$ and H is large enough†. It follows that

$$J_5 = s\int_{H\lambda_m}^\infty \mathscr{A}(x) e^{-sx} dx < s\int_{H\lambda_m}^\infty \exp\{(1+\delta) Bx^{\alpha/(1+\alpha)} (\log x)^{-\beta/(1+\alpha)} - sx\}\, dx$$
$$< s\int_0^\infty e^{-\frac{1}{2}sx} dx = \tfrac{1}{2},$$

and is therefore negligible.

* With δ in the place of ϵ.

† We suppress the details of the calculation, which is quite straightforward.

4·4. The integrals J_2 and J_4 may be discussed in practically the same way, and we may confine ourselves to the latter.

We have

(4·41) $$J_4(s) = s\int_{(1+\zeta)\lambda_m}^{H\lambda_m} \mathscr{A}(x)\, e^{-sx}\, dx < s\int_{(1+\zeta)\lambda_m}^{H\lambda_m} e^{\psi}\, dx,$$

where

(4·411) $$\psi = (1+\delta)\, B x^{\alpha/(1+\alpha)} (\log x)^{-\beta/(1+\alpha)} - sx.$$

The maximum of the function ψ occurs for $x = x_0$, where

(4·42) $$\frac{1}{s} = (1+\eta)\frac{1+\alpha}{\alpha B} x_0^{1/(1+\alpha)} (\log x_0)^{\beta/(1+\alpha)}.$$

From this equation, and (4·24), it plainly results that

(4·43) $$(1-\eta)\lambda_m < x_0 < (1+\eta)\lambda_m,$$

and that x_0 falls (when δ and s are small enough) between $(1-\zeta)\lambda_m$ and $(1+\zeta)\lambda_m$.

Let us write $$x = x_0 + \xi$$ in J_4. Then

$$\psi(x) = \psi(x_0) + \tfrac{1}{2}(1+\delta)\, B\xi^2 \frac{d^2}{dx_1^2}\{x_1^{\alpha/(1+\alpha)} (\log x_1)^{-\beta/(1+\alpha)}\},$$

where $x_0 < x_1 < x$ and *a fortiori*

$$(1-\zeta)\lambda_m < x_1 < H\lambda_m.$$

It follows that

(4·44) $$\frac{d^2}{dx_1^2}\{x_1^{\alpha/(1+\alpha)} (\log x_1)^{-\beta/(1+\alpha)}\} < -K\lambda_m^{\alpha/(1+\alpha)-2} (\log\lambda_m)^{-\beta/(1+\alpha)}.$$

On the other hand, an easy calculation shews that

(4·45) $$(1-\eta)\, A s^{-\alpha}\left(\log\frac{1}{s}\right)^{-\beta} < \psi(x_0) < (1+\eta)\, A s^{-\alpha}\left(\log\frac{1}{s}\right)^{-\beta}.$$

Thus

(4·46) $$J_4 < \exp\left\{(1+\eta)\, A s^{-\alpha}\left(\log\frac{1}{s}\right)^{-\beta}\right\}$$
$$\times \int_{(\zeta-\eta)\lambda_m}^{\infty} \exp\{-L\xi^2\lambda_m^{\alpha/(1+\alpha)-2} (\log\lambda_m)^{-\beta/(1+\alpha)}\}\, d\xi$$
$$< \exp\left\{(1+\eta)\, A s^{-\alpha}\left(\log\frac{1}{s}\right)^{-\beta} - M\zeta^2\lambda_m^{\alpha/(1+\alpha)} (\log\lambda_m)^{-\beta/(1+\alpha)}\right\}$$
$$< \exp\left\{(1+\eta-N\zeta^2)\, A s^{-\alpha}\left(\log\frac{1}{s}\right)^{-\beta}\right\}.$$

Since ζ is independent of δ and s, this inequality shews that J_4 is negligible; and a similar argument may be applied to J_2.

4·5. We may therefore replace the inequalities (4·23) by

(4·51)
$$\exp\left\{(1-\delta)\,A s^{-\alpha}\left(\log\frac{1}{s}\right)^{-\beta}\right\} < s\int_{(1-\zeta)\lambda_m}^{(1+\zeta)\lambda_m} \mathscr{A}(x)\,e^{-sx}\,dx < \exp\left\{(1+\delta)\,A s^{-\alpha}\left(\log\frac{1}{s}\right)^{-\beta}\right\}.$$

Since $\mathscr{A}(x)$ is a steadily increasing function of x, it follows that

(4·521)
$$\exp\left\{(1-\delta)\,A s^{-\alpha}\left(\log\frac{1}{s}\right)^{-\beta}\right\} < s\mathscr{A}\{(1+\zeta)\lambda_m\}\int_{(1-\zeta)\lambda_m}^{(1+\zeta)\lambda_m} e^{-sx}\,dx,$$

(4·522)
$$\exp\left\{(1+\delta)\,A s^{-\alpha}\left(\log\frac{1}{s}\right)^{-\beta}\right\} > s\mathscr{A}\{(1-\zeta)\lambda_m\}\int_{(1-\zeta)\lambda_m}^{(1+\zeta)\lambda_m} e^{-sx}\,dx;$$

or

(4·531)
$$(e^{\zeta s\lambda_m}-e^{-\zeta s\lambda_m})\,\mathscr{A}\{(1-\zeta)\lambda_m\} < \exp\left\{(1+\delta)\,A s^{-\alpha}\left(\log\frac{1}{s}\right)^{-\beta}+\lambda_m s\right\},$$

(4·532)
$$(e^{\zeta s\lambda_m}-e^{-\zeta s\lambda_m})\,\mathscr{A}\{(1+\zeta)\lambda_m\} > \exp\left\{(1-\delta)\,A s^{-\alpha}\left(\log\frac{1}{s}\right)^{-\beta}+\lambda_m s\right\}.$$

If we substitute for s, in terms of λ_m, in the right-hand sides of (4·531) and (4·532), we obtain expressions of the form

$$\exp\{(1+\eta)\,B\lambda_m^{\alpha/(1+\alpha)}(\log\lambda_m)^{-\beta/(1+\alpha)}\}.$$

On the other hand
$$e^{\zeta s\lambda_m}-e^{-\zeta s\lambda_m}$$
is of the form
$$\exp\{\eta_\zeta\lambda_m^{\alpha/(1+\alpha)}(\log\lambda_m)^{-\beta/(1+\alpha)}\}.$$

We have thus

(4·541)
$$A\{(1-\zeta)\lambda_m\} < \exp\{(1+\eta_\zeta+\eta)\,B\lambda_m^{\alpha/(1+\alpha)}(\log\lambda_m)^{-\beta/(1+\alpha)}\},$$

(4·542)
$$A\{(1+\zeta)\lambda_m\} > \exp\{(1-\eta_\zeta-\eta)\,B\lambda_m^{\alpha/(1+\alpha)}(\log\lambda_m)^{-\beta/(1+\alpha)}\}.$$

Now let ν be any number such that

(4·55)
$$(1-\zeta)\lambda_m \leqslant \lambda_\nu \leqslant (1+\zeta)\lambda_m.$$

Since $\lambda_n/\lambda_{n-1}\to 1$, it is clear that *all* numbers n from a certain point onwards will fall among the numbers ν. It follows from (4·541) and (4·542) that

(4·56)
$$\exp\{(1-\eta_\zeta-\eta)(1-\eta_\zeta)\,B\lambda_\nu^{\alpha/(1+\alpha)}(\log\lambda_\nu)^{-\beta/(1+\alpha)}\} < A(\lambda_\nu) < \exp\{(1+\eta_\zeta+\eta)(1+\eta_\zeta)\,B\lambda_\nu^{\alpha/(1+\alpha)}(\log\lambda_\nu)^{-\beta/(1+\alpha)}\};$$

and therefore that, given ϵ, we can choose first ζ and then n_0 so that

(4·57)
$$\exp\{(1-\epsilon)\,B\lambda_n^{\alpha/(1+\alpha)}(\log\lambda_n)^{-\beta/(1+\alpha)}\} < A(\lambda_n) < \exp\{(1+\epsilon)\,B\lambda_n^{\alpha/(1+\alpha)}(\log\lambda_n)^{-\beta/(1+\alpha)}\},$$

for $n\geqslant n_0$. This completes the proof of the theorem.

4·6. There is of course a corresponding "Abelian" theorem, which we content ourselves with enunciating. This theorem is naturally not limited by the restriction that the coefficients a_n are positive.

THEOREM B. *Suppose that*

(1) $\lambda_1 \geqslant 0, \quad \lambda_n > \lambda_{n-1}, \quad \lambda_n \to \infty$;

(2) $\lambda_n/\lambda_{n-1} \to 1$;

(3) $A > 0, \quad 0 < \alpha < 1$;

(4) $A_n = a_1 + a_2 + \ldots + a_n = \exp[\{1 + o(1)\} A \lambda_n^{\alpha} (\log \lambda_n)^{-\beta}]$,

when $n \to \infty$. *Then the series* $\Sigma a_n e^{-\lambda_n s}$ *is convergent for* $s > 0$, *and*

$$f(s) = \Sigma a_n e^{-\lambda_n s} = \exp\left[\{1 + o(1)\} B s^{-\alpha/(1-\alpha)} \left(\log \frac{1}{s}\right)^{-\beta/(1-\alpha)}\right],$$

where $$B = A^{1/(1-\alpha)} \alpha^{\alpha/(1-\alpha)} (1-\alpha)^{1+[\beta/(1-\alpha)]},$$

when $s \to 0$.

The proof of this theorem, which is naturally easier than that of the correlative Tauberian theorem, should present no difficulty to anyone who has followed the analysis which precedes.

4·7. The simplest and most interesting cases of Theorems A and B are those in which

$$\lambda_n = n, \quad \beta = 0.$$

It is then convenient to write x for e^{-s}. We thus obtain

THEOREM C. *If* $A > 0, \ 0 < \alpha < 1$, *and*

$$\log A_n = \log(a_1 + a_2 + \ldots + a_n) \sim A n^{\alpha},$$

then the series $\Sigma a_n x^n$ *is convergent for* $|x| < 1$, *and*

$$\log f(x) = \log(\Sigma a_n x^n) \sim B(1-x)^{-\alpha/(1-\alpha)},$$

where $$B = (1-\alpha)\alpha^{\alpha/(1-\alpha)} A^{1/(1-\alpha)},$$

when $x \to 1$ *by real values.*

If the coefficients are positive the converse inference is also correct. *That is to say, if*

$$A > 0, \quad \alpha > 0,$$

and $$\log f(x) \sim A(1-x)^{-\alpha},$$

then $$\log A_n \sim B n^{\alpha/(1+\alpha)},$$

where $$B = (1+\alpha)\alpha^{-\alpha/(1+\alpha)} A^{1/(1+\alpha)}.$$

5. *Application to our problem, and to other problems in the Theory of Numbers.*

5·1. We proved in 3 that

(3·56) $$\log \mathfrak{Q}(s) \sim \frac{\pi^2}{6s \log(1/s)}.$$

In Theorem A take

$$\lambda_n = \log n, \quad A = \frac{\pi^2}{6}, \quad \alpha = 1, \quad \beta = 1.$$

Then all the conditions of the theorem are satisfied. And A_n is $Q(n)$, the number of numbers q not exceeding n. We have therefore

$$\log Q(n) \sim B \sqrt{\left(\frac{\log n}{\log \log n}\right)}, \tag{5·11}$$

where

$$B = 2^{\frac{3}{2}} \sqrt{\left(\frac{\pi^2}{6}\right)} = \frac{2\pi}{\sqrt{3}}. \tag{5·12}$$

5·2. The method which we have followed in solving this problem is one capable of many other interesting applications.

Suppose, for example, that $R_r(n)$ is the number of ways in which n can be represented as the sum of any number of rth powers of positive integers*. We shall prove that

$$\log R_r(n) \sim (r+1) \left\{\frac{1}{r} \Gamma\left(\frac{1}{r}+1\right) \zeta\left(\frac{1}{r}+1\right)\right\}^{r/(r+1)} n^{1/(r+1)}. \tag{5·21}$$

In particular, if $P(n) = R_1(n)$ is the number of partitions of n, then

$$\log P(n) \sim \pi \sqrt{\left(\frac{2n}{3}\right)}. \tag{5·22}$$

We need only sketch the proof, which is in principle similar to the main proof of this paper. We have

$$\sum_1^\infty R_r(n) e^{-ns} = \prod_1^\infty \left(\frac{1}{1-e^{-s\nu^r}}\right),$$

and so

$$f(s) = \sum_1^\infty \{R_r(n) - R_r(n-1)\} e^{-ns} = \prod_2^\infty \left(\frac{1}{1-e^{-s\nu^r}}\right)\dagger. \tag{5·23}$$

It is obvious that $R_r(n)$ increases with n and that all the coefficients in $f(s)$ are positive. Again,

$$\log f(s) = \sum_2^\infty \log\left(\frac{1}{1-e^{-s\nu^r}}\right) = \sum_2^\infty (e^{-s\nu^r} + \tfrac{1}{2} e^{-2s\nu^r} + \ldots) \tag{5·24}$$

$$= \sum_{k=1}^\infty \frac{1}{k} \phi(ks),$$

where

$$\phi(s) = \sum_2^\infty e^{-s\nu^r}. \tag{5·241}$$

* Thus $28 = 3^3 + 1^3 = 3.2^3 + 4.1^3 = 2.2^3 + 12.1^3 = 2^3 + 20.1^3 = 28.1^3$:

and $R_3(28) = 5$.

The *order* of the powers is supposed to be indifferent, so that (e.g.) $3^3 + 1^3$ and $1^3 + 3^3$ are not reckoned as separate representations.

† $R_r(0)$ is to be interpreted as zero.

But

$$(5{\cdot}25) \qquad \phi(s) \sim \Gamma\left(\frac{1}{r}+1\right) s^{-1/r},$$

when $s \to 0$; and we can deduce, by an argument similar to that of 3·5, that

$$(5{\cdot}26) \qquad \log f(s) \sim \Gamma\left(\frac{1}{r}+1\right) \zeta\left(\frac{1}{r}+1\right) s^{-1/r}.$$

We now obtain (5·21) by an application of Theorem A, taking

$$\lambda_n = n, \quad \alpha = \frac{1}{r}, \quad \beta = 0, \quad A = \Gamma\left(\frac{1}{r}+1\right) \zeta\left(\frac{1}{r}+1\right).$$

In a similar manner we can shew that, if $S(n)$ is the number of partitions of n into *different* positive integers, so that

$$\Sigma S(n) e^{-ns} = (1+e^{-s})(1+e^{-2s})(1+e^{-3s}) + \ldots$$
$$= \frac{1}{(1-e^{-s})(1-e^{-3s})(1-e^{-5s})\ldots},$$

then

$$(5{\cdot}27) \qquad \log S(n) \sim \pi \sqrt{\left(\frac{n}{3}\right)};$$

that if $T_r(n)$ is the number of representations of n as the sum of rth powers of *primes*, then

$$(5{\cdot}28) \qquad \log T_r(n) \sim (r+1) \left\{\Gamma\left(\frac{1}{r}+2\right) \zeta\left(\frac{1}{r}+1\right)\right\}^{r/(r+1)} n^{1/(r+1)} (\log n)^{-r/(r+1)};$$

and, in particular, that if $T(n) = T_1(n)$ is the number of partitions of n into primes, then

$$(5{\cdot}281) \qquad \log T(n) \sim \frac{2\pi}{\sqrt{3}} \sqrt{\left(\frac{n}{\log n}\right)}.$$

Finally, we can shew that if r and s are positive integers, $a > 0$, and $0 \leqslant b \leqslant 1$, and

$$(5{\cdot}291) \qquad \Sigma \phi(n) x^n = \frac{\{(1+ax)(1+ax^2)(1+ax^3)\ldots\}^r}{\{(1-bx)(1-bx^2)(1-bx^3)\ldots\}^s},$$

then

$$(5{\cdot}292) \qquad \log \phi(n) \sim 2\sqrt{(cn)},$$

where

$$(5{\cdot}2921) \qquad c = r\int_0^a \frac{\log(1+t)}{t}\,dt - s\int_0^b \frac{\log(1-t)}{t}\,dt.$$

In particular, if $a = 1$, $b = 1$, and $r = s$, we have

$$(5{\cdot}293) \qquad \Sigma \phi(n) x^n = (1 - 2x + 2x^4 - 2x^9 + \ldots)^{-r},$$

$$(5{\cdot}294) \qquad \log \phi(n) \sim \pi \sqrt{(rn)}.$$

[*Added March 28th*, 1917.—Since this paper was written M. G. Valiron ("Sur la croissance du module maximum des séries entières," *Bulletin de la*

Société mathématique de France, Vol. XLIV, 1916, pp. 45—64) has published a number of very interesting theorems concerning power-series which are more or less directly related to ours. M. Valiron considers power-series only, and his point of view is different from ours, in some respects more restricted and in others more general.

He proves in particular that *the necessary and sufficient conditions that*

$$\log M(r) \sim \frac{A}{(1-r)^{\alpha}},$$

where $M(r)$ *is the maximum modulus of* $f(x) = \Sigma a_n x^n$ *for* $|x| = r$, *are that*

$$\log |a_n| < (1+\epsilon)(1+\alpha) A^{1/(1+\alpha)} \left(\frac{n}{\alpha}\right)^{\alpha/(1+\alpha)}$$

for $n > n_0(\epsilon)$, *and*

$$\log |a_n| > (1-\epsilon_p)(1+\alpha) A^{1/(1+\alpha)} \left(\frac{n}{\alpha}\right)^{\alpha/(1+\alpha)}$$

for $n = n_p$ $(p = 1, 2, 3, \ldots)$, *where* $n_{p+1}/n_p \to 1$ *and* $\epsilon_p \to 0$ *as* $p \to \infty$.

M. Valiron refers to previous, but less general or less precise, results given by Borel (*Leçons sur les séries à termes positifs*, 1902, Ch. v) and by Wiman ("Über dem Zusammenhang zwischen dem Maximal-betrage einer analytischen Funktion und dem grössten Gliede der zugehörigen Taylor'schen Reihe," *Acta Mathematica*, Vol. XXXVII, 1914, pp. 305—326). We may add a reference to Le Roy, "Valeurs asymptotiques de certaines séries procédant suivant les puissances entières et positives d'une variable réelle," *Bulletin des sciences mathématiques*, Ser. 2, Vol. XXIV, 1900, pp. 245—268.

We have more recently obtained results concerning $P(n)$, the number of partitions of n, far more precise than (5·22). A preliminary account of these researches has appeared, under the title "Une formule asymptotique pour le nombre des partitions de n," in the *Comptes Rendus* of January 2nd, 1917*; and a fuller account has been presented to the Society. See *Records of Proceedings at Meetings*, March 1st, 1917 †.]

* [No. 31 of this volume.]
† [No. 33 of this volume; see also No. 36.]

35

THE NORMAL NUMBER OF PRIME FACTORS OF A NUMBER n

(*Quarterly Journal of Mathematics*, XLVIII, 1917, 76—92)

I.

Statement of the problem.

1·1. The problem with which we are concerned in this paper may be stated roughly as follows: *What is the normal degree of compositeness of a number n?* We shall prove a number of theorems the general result of which is to shew that n is, as a rule, composed of about $\log\log n$ factors.

These statements are vague, and we must define our problem more precisely before we proceed further.

1·2. There are two ways in which it is natural to measure the "degree of compositeness" of n, viz.

(1) by its number of *divisors*,

(2) by its number of *prime factors*.

In this paper we adopt the second point of view. A distinction arises according as multiple factors are or are not counted multiply. We shall denote the number of *different* prime factors by $f(n)$, and the total number of prime factors by $F(n)$, so that (e.g.)

$$f(2^3 3^2 5) = 3, \quad F(2^3 3^2 5) = 6.$$

With regard to these functions (or any other arithmetical functions of n) four questions naturally suggest themselves.

(1) In the first place we may ask what is the *minimum* order of the function considered. We wish to determine an elementary function of n, with as low a rate of increase as possible, such that (e.g.)

$$f(n) \leqslant \phi(n)$$

for an infinity of values of n. This question is, for the functions now under consideration, trivial; for it is plain that

$$f(n) = F(n) = 1$$

when n is a prime.

(2) Secondly, we may ask what is the *maximum* order of the function. This question also is trivial for $F(n)$: we have

$$F(n) = \frac{\log n}{\log 2}$$

whenever n is of the form 2^k; and there is no $\phi(n)$ of slower increase which satisfies the conditions. The answer is less obvious in the case of $f(n)$: but, supposing n to be the product of the first k primes, we can shew (by purely elementary reasoning) that, if ϵ is any positive number, we have

$$f(n) < (1+\epsilon)\frac{\log n}{\log\log n}$$

for all sufficiently large values of n, and

$$f(n) > (1-\epsilon)\frac{\log n}{\log\log n}$$

for an infinity of values; so that the maximum order of $f(n)$ is

$$\frac{\log n}{\log\log n}.$$

It is worth mentioning, for the sake of comparison, that the minimum order of $d(n)$, the number of *divisors* of n, is 2, while the maximum order lies between

$$2^{(1-\epsilon)\log n/\log\log n}, \quad 2^{(1+\epsilon)\log n/\log\log n}$$

for every positive value of ϵ*.

(3) Thirdly, we may ask what is the *average* order of the function. It is well known—to return for a moment to the theory of $d(n)$—that

$$(1{\cdot}21) \qquad d(1)+d(2)+\ldots+d(n) \sim n\log n.$$

This result, indeed, is almost trivial, and far more is known; it is known in fact that

$$(1{\cdot}22) \quad d(1)+d(2)+\ldots+d(n) = n\log n + (2\gamma-1)n + O(n^{\frac{1}{3}}\log n),$$

this result being one of the deepest in the analytic theory of numbers. It would be natural, then, to say that the average order of $d(n)$ is $\log n$.

A similar problem presents itself for $f(n)$ and $F(n)$; and it is easy to shew that, in this sense, *the average order of* $f(n)$ *and* $F(n)$ *is* $\log\log n$. In fact it may be shewn, by purely elementary methods, that

$$(1{\cdot}23) \quad f(1)+f(2)+\ldots+f(n) = n\log\log n + An + O\left(\frac{n}{\log n}\right),$$

$$(1{\cdot}24) \quad F(1)+F(2)+\ldots+F(n) = n\log\log n + Bn + O\left(\frac{n}{\log n}\right),$$

where A and B are certain constants.

This problem, however, we shall dismiss for the present, as results still more precise than (1·23) and (1·24) can be found by transcendental methods.

(4) Fourthly, we may ask what is the *normal* order of the function. This phrase requires a little more explanation.

* Wigert, "Sur l'ordre de grandeur du nombre des diviseurs d'un entier," *Arkiv för matematik*, Vol. III (1907), No. 18, pp. 1–9. See Ramanujan, "Highly Composite Numbers," *Proc. London Math. Soc.*, Ser. 2, Vol. XIV (1915), pp. 347—409 [No. 15 of this volume], for more precise results.

Suppose that $N(x)$ is the number of numbers, not exceeding x, which possess a certain property P. This property may be a function of n only, or of x only, or of both n and x: we shall be concerned only with cases in which it is a function of one variable alone. And suppose further that

$$N(x) \sim x \tag{1·25}$$

when $x \to \infty$. Then we shall say, if P is a function of n only, that *almost all numbers* possess the property, and, if P is a function of x only, that *almost all numbers less than x* possess the property. Thus, to take a trivial example, almost all numbers are composite.

If then $g(n)$ is an arithmetical function of n, and $\phi(n)$ an elementary increasing function, and if, for every positive ϵ, we have

$$(1-\epsilon)\,\phi(n) < g(n) < (1+\epsilon)\,\phi(n) \tag{1·26}$$

for almost all values of n, we shall say that *the normal order of $g(n)$ is $\phi(n)$*.

It is in no way necessary that a function should possess either a determinate average order or a determinate normal order, or that one should be determinate when the other is, or that, if both are determinate, they should be the same. But we shall find that each of the functions $f(n)$ and $F(n)$ has both the average order and the normal order $\log\log n$.

The definitions just given may be modified so as to apply only to numbers of a special class. If $Q(x)$ is the number of numbers of the class not exceeding x, and $N(x)$ the number of *these* numbers which possess the property P, and

$$N(x) \sim Q(x), \tag{1·27}$$

then we shall say that almost all of the numbers (or almost all of the numbers less than x) possess the property.

II.

"*Quadratfrei*" *numbers.*

2·1. It is most convenient to begin by confining our attention to *quadratfrei* numbers, numbers, that is to say, which contain no prime factor raised to a power higher than the first. It is well known that the number $Q(x)$ of such numbers, not exceeding x, satisfies the relation

$$Q(x) \sim \frac{6}{\pi^2}x\text{*}. \tag{2·11}$$

We shall denote by $\pi_\nu(x)$ the number of numbers which are products of just ν different prime factors and do not exceed x, so that $\pi_1(x) = \pi(x)$ is the number of primes not exceeding x, and

$$\sum_1^\infty \pi_\nu(x) = Q(x).$$

* Landau, *Handbuch*, p. 581.

It has been shewn by Landau* that

(2·12) $$\pi_\nu(x) \sim \frac{x}{\log x} \frac{(\log\log x)^{\nu-1}}{(\nu-1)!}$$

for every fixed value of ν. In what follows we shall require, not an asymptotic equality, but an inequality satisfied for all values of ν and x.

2·2. LEMMA A. *There are absolute constants C and K such that*

(2·21) $$\pi_{\nu+1}(x) < \frac{Kx}{\log x} \frac{(\log\log x + C)^\nu}{\nu!}$$

for $$\nu = 0, 1, 2, \ldots, \quad x \geqslant 2.$$

The inequality (2·21) is certainly true when $\nu = 0$, whatever the value of C. It is known (and may be proved by elementary methods†) that

(2·221) $$\sum_{p \leqslant x} \frac{1}{p} < \log\log x + B$$

and

(2·222) $$\sum_{p \leqslant x} \frac{\log p}{p} < H \log x,$$

where B and H are constants. We shall prove by induction that (2·21) is true if

(2·223) $$C > B + H.$$

Consider the numbers which do not exceed x and are comprised in the table

$$\begin{array}{l} 2 \,.\, p_1 \,.\, p_2 \ldots p_\nu, \\ 3 \,.\, p_1 \,.\, p_2 \ldots p_\nu, \\ 5 \,.\, p_1 \,.\, p_2 \ldots p_\nu, \\ \ldots\ldots\ldots\ldots\ldots\ldots, \\ P \,.\, p_1 \,.\, p_2 \ldots p_\nu, \end{array}$$

where $p_1, p_2, \ldots, p_\nu$ are, in each row, different primes arranged in ascending order of magnitude, and where $p_\nu \geqslant 2$ in the first row, $p_\nu \geqslant 3$ in the second, $p_\nu \geqslant 5$ in the third, and so on. It is plain that $P \leqslant \sqrt{x}$. The total number of numbers in the table is plainly not greater than

$$\sum_{p^2 \leqslant x} \pi_\nu\left(\frac{x}{p}\right).$$

If now $\omega_1, \omega_2, \ldots$ are primes and

$$\omega_1 < \omega_2 < \ldots < \omega_{\nu+1}, \quad \omega_1\omega_2 \ldots \omega_{\nu+1} \leqslant x,$$

the number $\omega_1\omega_2 \ldots \omega_{\nu+1}$ will occur at least ν times in the table, once in the row in which the first figure is ω_1, once in that in which it is ω_2, ..., once in that in which it is ω_ν. We have therefore

(2·23) $$\nu\pi_{\nu+1}(x) \leqslant \sum_{p^2 \leqslant x} \pi_\nu\left(\frac{x}{p}\right).$$

* Landau, *Handbuch*, pp. 203 *et seq.*

† In applying (2·21) we assume that $x/p \geqslant 2$. If $x/p < 2$, $\pi_\nu(x/p)$ is zero.

Assuming, then, that (2·21) is true when ν is replaced by $\nu-1$, we obtain*.

$$(2\cdot 24)\qquad \pi_{\nu+1}(x) < \frac{Kx}{\nu!} \sum_{p^2 \leqslant x} \frac{1}{p \log (x/p)} \left\{\log\log\left(\frac{x}{p}\right) + C\right\}^{\nu-1}$$

$$< \frac{Kx(\log\log x + C)^{\nu-1}}{\nu!} \sum_{p^2 \leqslant x} \frac{1}{p \log (x/p)}.$$

But

$$(2\cdot 25)\qquad \frac{1}{\log x - \log p} = \frac{1}{\log x} + \frac{\log p}{(\log x)^2}\left\{1 + \frac{\log p}{\log x} + \dots\right\} \leqslant \frac{1}{\log x} + \frac{2\log p}{(\log x)^2},$$

since $\log p \leqslant \frac{1}{2}\log x$; and so

$$(2\cdot 26)\qquad \sum_{p^2 \leqslant x} \frac{1}{p \log (x/p)} \leqslant \frac{1}{\log x} \sum_{p^2 \leqslant x} \frac{1}{p} + \frac{2}{(\log x)^2} \sum_{p^2 \leqslant x} \frac{\log p}{p}$$

$$< \frac{\log\log x + B}{\log x} + \frac{H}{\log x} < \frac{\log\log x + C}{\log x}.$$

Substituting in (2·24), we obtain (2·21).

2·3. We can now prove one of our main theorems.

THEOREM A. *Suppose that ϕ is a function of x which tends steadily to infinity with x. Then*

$$(2\cdot 31)\qquad \log\log x - \phi\sqrt{(\log\log x)} < f(n) < \log\log x + \phi\sqrt{(\log\log x)}$$

for almost all quadratfrei *numbers n less than x.*

It is plainly enough to prove that

$$(2\cdot 321)\qquad S_1 = \sum_{\nu < llx - \phi\sqrt{(llx)}} \pi_{\nu+1}(x) = o(x),$$

$$(2\cdot 322)\qquad S_2 = \sum_{\nu > llx + \phi\sqrt{(llx)}} \pi_{\nu+1}(x) = o(x)\dagger.$$

It will be sufficient to consider one of the two sums S_1, S_2, say the latter; the discussion of S_1 proceeds on the same lines and is a little simpler. We have

$$(2\cdot 33)\qquad S_2 < K\frac{x}{\log x} \sum_{\nu > llx + \phi\sqrt{(llx)}} \frac{(llx + C)^\nu}{\nu!}.$$

Write $$\log\log x + C = \xi.$$

Then the condition that

$$\nu > \log\log x + \phi\sqrt{(\log\log x)},$$

where ϕ is some function of x which tends steadily to infinity with x, is plainly equivalent to the condition that

$$\nu > \xi + \psi\sqrt{\xi},$$

* It may be well to observe explicitly that $\log\log 2 + C$ is positive.

† We shall sometimes write lx, llx, ..., instead of $\log x$, $\log\log x$, ..., in order to shorten our formulæ.

where ψ is some function of ξ which tends steadily to infinity with ξ; and so what we have to prove is that

(2·34) $$S = \sum_{\nu > \xi + \psi\sqrt{\xi}} \frac{\xi^\nu}{\nu!} = o(e^\xi).$$

We choose a positive number δ so small that

(2·35) $$\frac{\delta}{2.3} + \frac{\delta^2}{3.4} + \frac{\delta^3}{4.5} + \ldots < \tfrac{1}{4},$$

and write

(2·36) $$S = \sum_{\xi+\psi\sqrt{\xi} < \nu \leqslant (1+\delta)\xi} \frac{\xi^\nu}{\nu!} + \sum_{\nu > (1+\delta)\xi} \frac{\xi^\nu}{\nu!} = S' + S'',$$

say. In the first place we have

(2·371) $$S'' < \frac{\xi^{\nu_1}}{\nu_1!}\left\{1 + \frac{\xi}{\nu_1+1} + \frac{\xi^2}{(\nu_1+1)(\nu_1+2)} + \ldots\right\}$$
$$< \frac{\xi^{\nu_1}}{\nu_1!}\left\{1 + \frac{1}{1+\delta} + \frac{1}{(1+\delta)^2} + \ldots\right\} = \frac{1+\delta}{\delta}\frac{\xi^{\nu_1}}{\nu_1!},$$

where ν_1 is the smallest integer greater than $(1+\delta)\xi$. It follows that

(2·372) $$S'' < \frac{K}{\delta\sqrt{\nu_1}} e^{\nu_1(\log\xi - \log\nu_1 + 1)}$$
$$< \frac{K}{\delta\sqrt{\xi}} e^{\Delta\xi},$$

where the K's are absolute constants and

$$\Delta = (1+\delta)\log(1+\delta) - \delta.$$

And since

$$(1+\delta)\log(1+\delta) - \delta > (1+\delta)(\delta - \tfrac{1}{2}\delta^2) - \delta = \tfrac{1}{2}\delta^2(1-\delta) = \eta,$$

say, we obtain

(2·373) $$S'' < \frac{K}{\delta\sqrt{\xi}} e^{(1-\eta)\xi},$$

where η is positive: so that

(2·374) $$S'' = o(e^\xi).$$

In S', we write $\nu = \xi + \mu$, so that $\psi\sqrt{\xi} < \mu \leqslant \delta\xi$. Then

(2·381) $$\frac{\xi^\nu}{\nu!} = \frac{\xi^{\xi+\mu}}{(\xi+\mu)!}$$
$$< \frac{K}{\sqrt{\xi}} \exp\{(\xi+\mu)\log\xi - (\xi+\mu)\log(\xi+\mu) + \xi + \mu\}$$
$$= \frac{K}{\sqrt{\xi}} \exp\left\{(\xi+\mu)\left(1 - \frac{\mu}{\xi} + \frac{\mu^2}{2\xi^2} - \ldots\right)\right\}$$
$$< \frac{K}{\sqrt{\xi}} e^{\xi - (\mu^2/4\xi)} \text{*}.$$

* Since the exponent is equal to

$$\xi - \mu\left(\frac{\mu}{1.2.\xi} - \frac{\mu^2}{2.3.\xi^2} + \frac{\mu^3}{3.4.\xi^3} - \ldots\right),$$

and

$$\frac{\mu^2}{2.3.\xi^2} + \frac{\mu^3}{3.4.\xi^3} + \ldots < \frac{\mu}{4\xi}$$

in virtue of (2·35).

Thus

$$(2\cdot382)\qquad S' = O\left(\frac{e^{\xi}}{\sqrt{\xi}} \sum_{\psi\sqrt{\xi}}^{\delta\xi} e^{-\mu^2/4\xi}\right)^*$$

$$= O\left\{\frac{e^{\xi}}{\sqrt{\xi}} \int_{\psi\sqrt{\xi}}^{\infty} e^{-t^2/4\xi}\, dt\right\}$$

$$= O\left(e^{\xi} \int_{\psi}^{\infty} e^{-u^2}\, du\right) = o\left(e^{\xi}\right),$$

since $\psi \rightarrow \infty$. From (2·36), (2·374), and (2·382) follows the truth of (2·34), and so that of (2·322). As (2·321) may be proved in the same manner, the proof of Theorem A is completed.

2·4. THEOREM A′. *If ϕ is a function of n which tends steadily to infinity with n, then almost all* quadratfrei *numbers have between*

$$\log\log n - \phi\sqrt{(\log\log n)} \quad \textit{and} \quad \log\log n + \phi\sqrt{(\log\log n)}$$

prime factors.

This theorem is a simple corollary of Theorem A. Consider the numbers not exceeding x. We may plainly neglect numbers less than $\sqrt{x}$; so that

$$\tfrac{1}{2}\log x \leqslant \log n \leqslant \log x,$$

$$(2\cdot41)\qquad \log\log x - \log 2 \leqslant \log\log n \leqslant \log\log x.$$

Given ϕ, we can determine a function $\psi(x)$ such that ψ tends steadily to infinity with x and

$$(2\cdot42)\qquad \psi(x) < \tfrac{1}{2}\phi(\sqrt{x}).$$

Then
$$\psi(x) < \tfrac{1}{2}\phi(n) \quad (\sqrt{x} \leqslant n \leqslant x).$$

In the first place

$$\phi\sqrt{(\log\log n)} > \psi\sqrt{(\log\log x)}$$

if
$$\log\log n > \tfrac{1}{4}\log\log x,$$

which is certainly true, in virtue of (2·41), for sufficiently large values of x. Thus

$$(2\cdot43)\qquad \log\log n - \phi\sqrt{(\log\log n)} < \log\log x - \psi\sqrt{(\log\log x)}.$$

In the second place the corresponding inequality

$$(2\cdot44)\qquad \log\log n + \phi\sqrt{(\log\log n)} > \log\log x + \psi\sqrt{(\log\log x)}$$

is certainly true if

$$\phi\sqrt{(\log\log n)} > \psi\sqrt{(\log\log x)} + \log 2;$$

and this also is true, in virtue of (2·41) and (2·42), for sufficiently large values of x. From (2·43), (2·44), and Theorem A, Theorem A′ follows at once.

As a corollary we have

THEOREM A″. *The normal order of the number of prime factors of a* quadratfrei *number is* $\log\log n$.

* In this sum the values of μ are not, in general, integral: $\xi+\mu$ is integral.

III.

The normal order of $f(n)$.

3·1. So far we have confined our attention to numbers which have no repeated factors. When we remove this restriction, the functions $f(n)$ and $F(n)$ have to be distinguished from one another.

We shall denote by $\varpi_\nu(x)$ the number of numbers, not exceeding x, for which

$$f(n) = \nu.$$

It is obvious that
$$\varpi_\nu(x) \geqslant \pi_\nu(x).$$

We require an inequality for $\varpi_\nu(x)$ similar to that for $\pi_\nu(x)$ given by Lemma A.

Lemma B. *There are absolute constants D and L such that*

$$(3{\cdot}11) \qquad \varpi_{\nu+1}(x) < \frac{Lx}{\log x}\frac{(\log\log x + D)^\nu}{\nu!}$$

for
$$\nu = 0, 1, 2, \ldots, \quad x \geqslant 2.$$

It is plain that

$$\varpi_1(x) = \pi(x) + \pi(\sqrt{x}) + \pi(\sqrt[3]{x}) + \ldots = O\left(\frac{x}{\log x}\right).$$

The inequality is therefore true for $\nu = 0$, whatever the value of D. We shall prove by induction that it is true in general if

$$(3{\cdot}12) \qquad D > B + H + J,$$

where B and H have the same values as in 2·2, and

$$(3{\cdot}13) \qquad J = \sum_2^\infty (s+1)(2^{-s} + 3^{-s} + 5^{-s} + \ldots).$$

Consider the numbers which do not exceed x and are comprised in the table

$$\begin{array}{l} 2^a \,.\, p_1^{a_1} \,.\, p_2^{a_2} \ldots p_\nu^{a_\nu}, \\ 3^a \,.\, p_1^{a_1} \,.\, p_2^{a_2} \ldots p_\nu^{a_\nu}, \\ \ldots\ldots\ldots\ldots\ldots\ldots\ldots\ldots, \\ P^a \,.\, p_1^{a_1} \,.\, p_2^{a_2} \ldots p_\nu^{a_\nu}, \end{array}$$

where $p_1, p_2, \ldots, p_\nu$ satisfy the same conditions as in the table of 2·2. It is plain that $P \leqslant \sqrt{x}$ if $a = 1$, $P \leqslant \sqrt[3]{x}$ if $a = 2$, and so on, so that the total number of numbers in the table does not exceed

$$\sum_{p^{a+1} \leqslant x} \varpi_\nu\left(\frac{x}{p^a}\right).$$

If now $\omega_1, \omega_2, \ldots$ are primes and

$$\omega_1 < \omega_2 < \ldots < \omega_{\nu+1}, \quad \omega_1^{a_1}\omega_2^{a_2}\ldots\omega_{\nu+1}^{a_{\nu+1}} \leqslant x,$$

the number $\omega_1^{a_1}\omega_2^{a_2}\ldots\omega_{\nu+1}^{a_{\nu+1}}$ will occur at least ν times in one of the tables which correspond to different values of a. We have therefore

$$(3\cdot14) \qquad \nu\varpi_{\nu+1}(x) \leqslant \sum_{p^2 \leqslant x} \varpi_\nu\left(\frac{x}{p}\right) + \sum_{p^3 \leqslant x} \varpi_\nu\left(\frac{x}{p^2}\right) + \dots.$$

Now let us suppose that (3·11) is true when $\nu - 1$ is substituted for ν. Then it is plain* that

(3·15)

$$\varpi_{\nu+1}(x) < \frac{Lx(\log\log x + D)^{\nu-1}}{\nu!}\left\{\sum_{p^2 \leqslant x} \frac{1}{p\log(x/p)} + \sum_{p^3 \leqslant x} \frac{1}{p^2\log(x/p^2)} + \dots\right\}.$$

Now

$$(3\cdot16) \qquad \sum_{p^2 \leqslant x} \frac{1}{p\log(x/p)} < \frac{\log\log x + B + H}{\log x},$$

as we have already seen in 2·2. Also, if $p^{s+1} \leqslant x$, we have

$$\frac{x}{p^s} \geqslant x^{1/(s+1)}, \quad \log\frac{x}{p^s} \geqslant \frac{\log x}{s+1};$$

and so

$$(3\cdot17) \qquad \sum_{p^{s+1} \leqslant x} \frac{1}{p^s\log(x/p^s)} \leqslant \frac{s+1}{\log x}(2^{-s} + 3^{-s} + 5^{-s} + \dots),$$

if $s \geqslant 2$. Hence

(3·18)

$$\sum_s \sum_{p^{s+1} \leqslant x} \frac{1}{p^s\log(x/p^s)} < \frac{\log\log x + B + H + J}{\log x} < \frac{\log\log x + D}{\log x}.$$

From (3·15) and (3·18) the truth of (3·11) follows immediately.

3·2. We can now argue with $\varpi_\nu(x)$ as we argued with $\pi_\nu(x)$ in 2·3 and the paragraphs which follow; and we may, without further preface, state the following theorems.

THEOREM B. *If ϕ is a function of x which tends steadily to infinity to x, then*

$$\log\log x - \phi\sqrt{(\log\log x)} < f(n) < \log\log x + \phi\sqrt{(\log\log x)}$$

for almost all numbers n less than x.

THEOREM B′. *If ϕ is a function of n which tends steadily to infinity with n, then almost all numbers have between*

$$\log\log n - \phi\sqrt{(\log\log n)} \ \textit{and} \ \log\log n + \phi\sqrt{(\log\log n)}$$

different prime factors.

THEOREM B″. *The normal order of the number of different prime factors of a number is* $\log\log n$.

* See the footnote to p. 81. [Footnote * to p. 266.]

IV.

The normal order of $F(n)$.

4·1. We have now to consider the corresponding theorems for $F(n)$, the number of prime factors of n when multiple factors are counted multiply. These theorems are slightly more difficult. The additional difficulty occurs, however, only in the first stage of the argument, which requires the proof of some inequality analogous to those given by Lemmas A and B.

We denote by
$$\Pi_\nu(x)$$
the number of numbers, not exceeding x, for which
$$F(n) = \nu.$$
It would be natural to expect an inequality of the same form as those of Lemmas A and B, though naturally with different values of the constants. It is easy to see, however, that no such inequality can possibly be true.

For, if ν is greater than a constant multiple of $\log x$, the function
$$\frac{x}{\log x}\frac{(\log\log x + C)^\nu}{\nu\,!}$$
is—as may be seen at once by a simple approximation based upon Stirling's theorem—exceedingly small; and $\Pi_{\nu+1}(x)$, being an integer, cannot be small unless it is zero. Thus such an inequality as is suggested would shew that $F(n)$ *cannot* be of order as high as $\log x$ for any n less than x; and this is false, as we can see by taking
$$x = 2^k + 1,\quad n = 2^k,\quad F(n) = \frac{\log n}{\log 2}.$$
The inequality required must therefore be of a less simple character.

4·2. Lemma C. *Suppose that K and C have the same meaning as in Lemma A, and that*

$$\tfrac{9}{10} \leqslant \lambda < 1. \tag{4·21}$$

Then

$$\Pi_{\nu+1}(x) < \frac{Kx}{\log x} \tag{4·22}$$
$$\times\left\{\frac{(\log\log x + C)^\nu}{\nu\,!} + \lambda\frac{(\log\log x + C)^{\nu-1}}{(\nu-1)\,!} + \lambda^2\frac{(\log\log x + C)^{\nu-2}}{(\nu-2)\,!} + \dots\right\},$$

the series being continued to the term λ^ν.

It is plainly sufficient to prove this inequality when $\lambda = \frac{9}{10}$. In what follows we shall suppose that λ has this particular value.

We require a preliminary inequality analogous to (2·23) and (3·14). Consider the numbers which do not exceed x and are comprised in the table

$$2^{a} \,.\, p_1 \,.\, p_2 \ldots p_{\nu+1-a},$$
$$3^{a} \,.\, p_1 \,.\, p_2 \ldots p_{\nu+1-a},$$
$$\ldots\ldots\ldots\ldots\ldots\ldots\ldots\ldots,$$
$$P^{a} \,.\, p_1 \,.\, p_2 \ldots p_{\nu+1-a},$$

where now $p_1, p_2, \ldots, p_{\nu+1-a}$ are primes, arranged in ascending order of magnitude, but not necessarily different, and where $p_{\nu+1-a} \geqslant 2$ in the first row, $p_{\nu+1-a} \geqslant 3$ in the second, and so forth. Arguing as in 2·2 and 3·1, we now find

(4·23)
$$\nu \Pi_{\nu+1}(x) < \sum_{p^2 \leqslant x} \Pi_\nu \left(\frac{x}{p}\right) + \sum_{p^3 \leqslant x} \Pi_{\nu-1} \left(\frac{x}{p^2}\right) + \sum_{p^4 \leqslant x} \Pi_{\nu-2} \left(\frac{x}{p^3}\right) + \ldots.$$

This inequality differs formally from (3·14) in that the suffixes of the functions on the right-hand side sink by one from term to term.

We can now prove (4·22) by a process of induction similar in principle to that of 2·2 and 3·1. Let us suppose that the inequality is true when ν is replaced by $\nu - 1$, and let us write

$$\log \log x + C = \xi,$$

as in 2·3. Substituting in (4·23), and observing that

$$\log \log \frac{x}{p^s} + C \leqslant \xi,$$

we obtain

$$(4·24) \quad \Pi_{\nu+1}(x) < \frac{Kx}{\nu} \sum_{p^2 \leqslant x} \frac{1}{p \log (x/p)} \left\{ \frac{\xi^{\nu-1}}{(\nu-1)!} + \lambda \frac{\xi^{\nu-2}}{(\nu-2)!} + \ldots \right\}$$
$$+ \frac{Kx}{\nu} \sum_{p^3 \leqslant x} \frac{1}{p^2 \log (x/p^2)} \left\{ \frac{\xi^{\nu-2}}{(\nu-2)!} + \lambda \frac{\xi^{\nu-3}}{(\nu-3)!} + \ldots \right\}$$
$$+ \frac{Kx}{\nu} \sum_{p^4 \leqslant x} \frac{1}{p^3 \log (x/p^3)} \left\{ \frac{\xi^{\nu-3}}{(\nu-3)!} + \lambda \frac{\xi^{\nu-4}}{(\nu-4)!} + \ldots \right\} + \ldots.$$

Now, as we have seen already,

$$(4·251) \qquad \sum_{p^2 \leqslant x} \frac{1}{p \log (x/p)} < \frac{\log \log x + C}{\log x};$$

and

$$(4·252) \qquad \sum_{p^{s+1} \leqslant x} \frac{1}{p^s \log (x/p^s)} < \frac{s+1}{\log x} (2^{-s} + 3^{-s} + 5^{-s} + \ldots)$$

if $s \geqslant 2$. It is moreover easy to prove that

$$(4·2531) \qquad (s+1)(2^{-s} + 3^{-s} + 5^{-s} + \ldots) < 2\lambda^s = 2\left(\tfrac{9}{10}\right)^s$$

if $s = 2$, and

$$(4·2532) \qquad (s+1)(2^{-s} + 3^{-s} + 5^{-s} + \ldots) < \lambda^s = \left(\tfrac{9}{10}\right)^s$$

if $s > 2$*. From (4·24)—(4·2532) it follows that

$$(4{\cdot}25)\quad \Pi_{\nu+1}(x) < \frac{Kx}{\nu \log x}\left\{\frac{\xi^{\nu}}{(\nu-1)!} + \lambda \frac{\xi^{\nu-1}}{(\nu-2)!} + \lambda^2 \frac{\xi^{\nu-2}}{(\nu-3)!} + \ldots\right\}$$
$$+ \frac{2Kx}{\nu \log x}\lambda^2\left\{\frac{\xi^{\nu-2}}{(\nu-2)!} + \lambda \frac{\xi^{\nu-3}}{(\nu-3)!} + \ldots\right\}$$
$$+ \frac{Kx}{\nu \log x}\lambda^3\left\{\frac{\xi^{\nu-3}}{(\nu-3)!} + \lambda \frac{\xi^{\nu-4}}{(\nu-4)!} + \ldots\right\}$$
$$+ \ldots \dagger.$$

When we collect together the various terms on the right-hand side which involve the same powers of ξ, it will be found that the coefficient of $\xi^{\nu-p}$ is exactly

$$\frac{Kx}{\log x}\frac{\lambda^p}{(\nu-p)!},$$

except when $p = 1$, when it is

$$\left(1 - \frac{1}{\nu}\right)\frac{Kx}{\log x}\frac{\lambda}{(\nu-1)!}.$$

We thus obtain (4·22).

4·3. We may now argue substantially as in 2·3. We have to shew, for example, that

$$(4{\cdot}31)\qquad S_2 = \sum_{\nu > llx + \phi\sqrt{(llx)}} \Pi_{\nu+1}(x) = o(x);$$

and this is equivalent to proving that

$$(4{\cdot}32)\qquad S = \sum_{\nu > \xi + \psi\sqrt{\xi}} \Pi_{\nu+1}(x) = o(x),$$

where ψ is any function of x which tends to infinity with x.

We choose δ so that

$$(4{\cdot}33)\qquad \frac{1}{9\log(1+\frac{1}{9})} - 1 < \delta < \tfrac{1}{9}\ddagger,$$

and write

$$(4{\cdot}34)\qquad S = \sum_{\xi+\psi\sqrt{\xi} < \nu \leq (1+\delta)\xi} \Pi_{\nu+1}(x) + \sum_{\nu > (1+\delta)\xi} \Pi_{\nu+1}(x)$$
$$= S' + S'',$$

say. In S', we have

(4·351)

$$\Pi_{\nu+1}(x) < \frac{Kx}{\log x}\frac{\xi^{\nu}}{\nu!}\left\{1 + \tfrac{9}{10}\frac{\nu}{\xi} + (\tfrac{9}{10})^2\frac{\nu(\nu-1)}{\xi^2} + \ldots\right\} < \frac{K_1 x}{\log x}\frac{\xi^{\nu}}{\nu!},$$

where

$$(4{\cdot}352)\qquad K_1 = \frac{K}{1 - \frac{9}{10}(1+\delta)};$$

* The inequalities may be verified directly for $s=2$ and $s=3$, and then proved to be true generally by induction.

† The factor 2 occurs in the second line only.

‡ $\frac{1}{9\log(1+\frac{1}{9})} - 1 < \frac{1}{1-\frac{1}{18}} - 1 < \frac{1}{9}$.

and by means of this inequality we can shew, just as in 2·3, that

(4·353) $$S' = o(x).$$

In discussing S'' we must use (4·22) in a different manner. We have

(4·361)
$$\Pi_{\nu+1}(x) < \frac{Kx}{\log x}\left(\lambda^{\nu} + \lambda^{\nu-1}\frac{\xi}{1!} + \lambda^{\nu-2}\frac{\xi^2}{2!} + \dots\right) < \frac{Kx}{\log x}\lambda^{\nu} e^{\xi/\lambda};$$

and so

(4·362)
$$S'' < \frac{Kx}{\log x} e^{\xi/\lambda} \sum_{\nu > (1+\delta)\xi} \lambda^{\nu} < \frac{K}{1-\lambda}\frac{x}{\log x} e^{\xi/\lambda} \lambda^{(1+\delta)\xi} = O\left\{\frac{x}{(\log x)^{\eta}}\right\},$$

where

(4·363) $$\eta = 1 - (1/\lambda) + (1+\delta)\log(1/\lambda) = (1+\delta)\log\tfrac{10}{9} - \tfrac{1}{9} > 0,$$

by (4·33). Hence

(4·364) $$S'' = o(x).$$

From (4·34), (4·353), and (4·364), it follows that $S = o(x)$.

4·4. We have therefore the following theorems.

THEOREM C. *The result of Theorem B remains true when $F(n)$ is substituted for $f(n)$.*

THEOREM C′. *The result of Theorem B′ remains true when the word "different" is omitted.*

THEOREM C″. *The normal order of the total number of prime factors of a number is* $\log\log n$.

V.

The normal order of $d(n)$.

5·1. It is natural to ask whether similar theorems cannot be proved with regard to some of the other standard arithmetical functions of n, such as $d(n)$.

If
$$n = p_1^{a_1} p_2^{a_2} \dots p_{\nu}^{a_{\nu}},$$
we have
$$d(n) = (1+a_1)(1+a_2)\dots(1+a_{\nu}).$$

Since
$$2 \leqslant 1 + a \leqslant 2^a$$
if $a \geqslant 1$, we obtain at once
$$2^{\nu} \leqslant d(n) \leqslant 2^{a_1+a_2+\dots+a_{\nu}},$$
or

(5·11) $$2^{f(n)} \leqslant d(n) \leqslant 2^{F(n)}.$$

From (5·11), and Theorems B′ and C′, we obtain at once

THEOREM D′. *The inequalities*

(5·12) $$2^{\log\log n - \phi\sqrt{(\log\log n)}} < d(n) < 2^{\log\log n + \phi\sqrt{(\log\log n)}},$$

where ϕ *is any function of* n *which tends to infinity with* n, *are satisfied for almost all numbers* n.

The inequalities (5·12) are of a much less precise type than (1·26): we cannot say that the normal order of $d(n)$ is $2^{\log\log n}$. We can however say that (to put it roughly) the normal order of $d(n)$ is *about*

$$2^{\log\log n} = (\log n)^{\log 2} = (\log n)^{0\cdot 69\ldots}.$$

It should be observed that this order is far removed from $\log n$, the *average* order of $d(n)$. The explanation of this apparent paradox is simple. The majority of numbers have about $(\log n)^{\log 2}$ divisors. But *those which have an abnormal number may have a very much larger number indeed*: the excess of the maximum order over the normal order is so great that, when we compute the average order, it is the numbers with an abnormal number of divisors which dominate the calculation. The maximum order of the number of *prime factors* is not large enough to give rise to a similar phenomenon.

36

ASYMPTOTIC FORMULÆ IN COMBINATORY ANALYSIS*

(*Proceedings of the London Mathematical Society*, 2, XVII, 1918, 75—115)

1. Introduction and summary of results.

1·1. The present paper is the outcome of an attempt to apply to the principal problems of the theory of partitions the methods, depending upon the theory of analytic functions, which have proved so fruitful in the theory of the distribution of primes and allied branches of the analytic theory of numbers.

The most interesting functions of the theory of partitions appear as the coefficients in the power-series which represent certain elliptic modular functions. Thus $p(n)$, the number of unrestricted partitions of n, is the coefficient of x^n in the expansion of the function

(1·11) $$f(x) = 1 + \sum_1^\infty p(n)\, x^n = \frac{1}{(1-x)(1-x^2)(1-x^3)\ldots}\dagger.$$

If we write

(1·12) $$x = q^2 = e^{2\pi i\tau},$$

where the imaginary part of τ is positive, we see that $f(x)$ is substantially the reciprocal of the modular function called by Tannery and Molk‡ $h(\tau)$; that, in fact,

(1·13) $$h(\tau) = q^{\frac{1}{12}} q_0 = q^{\frac{1}{12}} \prod_1^\infty (1-q^{2n}) = \frac{x^{\frac{1}{24}}}{f(x)}.$$

The theory of partitions has, from the time of Euler onwards, been developed from an almost exclusively algebraical point of view. It consists of an assemblage of formal identities—many of them, it need hardly be said, of an exceedingly ingenious and beautiful character. Of *asymptotic* formulæ, one may fairly say, there are none§. So true is this, in fact, that we have

* A short abstract of the contents of part of this paper appeared under the title "Une formule asymptotique pour le nombre des partitions de n," in the *Comptes Rendus*, January 2nd, 1917 [No. 31 of this volume].

† P. A. MacMahon, *Combinatory Analysis*, Vol. II, 1916, p. 1.

‡ J. Tannery and J. Molk, *Fonctions elliptiques*, Vol. II, 1896, pp. 31 *et seq.* We shall follow the notation of this work whenever we have to quote formulæ from the theory of elliptic functions.

§ We should mention one exception to this statement, to which our attention was called by Major MacMahon. The number of partitions of n *into parts none of which exceed* r is the coefficient $p_r(n)$ in the series

$$1 + \sum_1^\infty p_r(n)\, x^n = \frac{1}{(1-x)(1-x^2)\ldots(1-x^r)}.$$

This function has been studied in much detail, for various special values of r, by Cayley,

been unable to discover in the literature of the subject any allusion whatever to the question of the order of magnitude of $p(n)$.

1·2. The function $p(n)$ may, of course, be expressed in the form of an integral

(1·21) $$p(n) = \frac{1}{2\pi i}\int_\Gamma \frac{f(x)}{x^{n+1}}\,dx,$$

by means of Cauchy's theorem, the path Γ enclosing the origin and lying entirely inside the unit circle. The idea which dominates this paper is that of obtaining asymptotic formulæ for $p(n)$ by a detailed study of the integral (1·21). This idea is an extremely obvious one; it is the idea which has dominated nine-tenths of modern research in the analytic theory of numbers: and it may seem very strange that it should never have been applied to this particular problem before. Of this there are no doubt two explanations. The first is that the theory of partitions has received its most important developments, since its foundation by Euler, at the hands of a series of mathematicians whose interests have lain primarily in algebra. The second and more fundamental reason is to be found in the extreme complexity of the behaviour of the generating function $f(x)$ near a point of the unit circle.

It is instructive to contrast this problem with the corresponding problems which arise for the arithmetical functions $\pi(n)$, $\vartheta(n)$, $\psi(n)$, $\mu(n)$, $d(n)$, ... which have their genesis in Riemann's Zeta-function and the functions allied

Sylvester, and Glaisher: we may refer in particular to J. J. Sylvester, "On a discovery in the theory of partitions," *Quarterly Journal*, Vol. I, 1857, pp. 81—85, and "On the partition of numbers," *ibid.*, pp. 141—152 (Sylvester's *Works*, Vol. II, pp. 86—89 and 90—99); J. W. L. Glaisher, "On the number of partitions of a number into a given number of parts," *Quarterly Journal*, Vol. XL, 1909, pp. 57—143; "Formulæ for partitions into given elements, derived from Sylvester's Theorem," *ibid.*, pp. 275—348; "Formulæ for the number of partitions of a number into the elements 1, 2, 3, ..., n up to $n=9$," *ibid.*, Vol. XLI, 1910, pp. 94—112: and further references will be found in MacMahon, *loc. cit.*, pp. 59—71, and E. Netto, *Lehrbuch der Combinatorik*, 1901, pp. 146—158. Thus, for example, the coefficient of x^n in

$$\frac{1}{(1-x)(1-x^2)(1-x^3)}$$

is $$p_3(n) = \tfrac{1}{12}(n+3)^2 - \tfrac{7}{72} + \tfrac{1}{8}(-1)^n + \tfrac{2}{9}\cos\frac{2n\pi}{3};$$

as is easily found by separating the function into partial fractions. This function may also be expressed in the forms

$$\tfrac{1}{12}(n+3)^2 + (\tfrac{1}{2}\cos\tfrac{1}{2}\pi n)^2 - (\tfrac{2}{3}\sin\tfrac{1}{3}\pi n)^2,$$

$$1 + [\tfrac{1}{12}n(n+6)], \quad \{\tfrac{1}{12}(n+3)^2\},$$

where $[n]$ and $\{n\}$ denote the greatest integer contained in n and the integer nearest to n. These formulæ do, of course, furnish incidentally asymptotic formulæ for the functions in question. But they are, from this point of view, of a very trivial character: the interest which they possess is algebraical.

to it. In the latter problems we are dealing with functions defined by Dirichlet's series. The study of such functions presents difficulties far more fundamental than any which confront us in the theory of the modular functions. These difficulties, however, relate to the distribution of the zeros of the functions and their general behaviour at infinity: no difficulties whatever are occasioned by the crude singularities of the functions in the finite part of the plane. The single finite singularity of $\zeta(s)$, for example, the pole at $s=1$, is a singularity of the simplest possible character. It is this pole which gives rise to the *dominant* terms in the asymptotic formulæ for the arithmetical functions associated with $\zeta(s)$. To prove such a formula rigorously is often exceedingly difficult; to determine precisely the order of the error which it involves is in many cases a problem which still defies the utmost resources of analysis. But to write down the dominant terms involves, as a rule, no difficulty more formidable than that of deforming a path of integration over a pole of the subject of integration and calculating the corresponding residue.

In the theory of partitions, on the other hand, we are dealing with functions which do not exist at all outside the unit circle. Every point of the circle is an essential singularity of the function, and no part of the contour of integration can be deformed in such a manner as to make its contribution obviously negligible. Every element of the contour requires special study; and there is no obvious method of writing down a "dominant term."

The difficulties of the problem appear then, at first sight, to be very serious. We possess, however, in the formulæ of the theory of the linear transformation of the elliptic functions, an extremely powerful analytical weapon by means of which we can study the behaviour of $f(x)$ near any assigned point of the unit circle*. It is to an appropriate use of these formulæ that the accuracy of our final results, an accuracy which will, we think, be found to be quite startling, is due.

1·3. It is very important, in dealing with such a problem as this, to distinguish clearly the various stages to which we can progress by arguments of a progressively "deeper" and less elementary character. The earlier results are naturally (so far as the particular problem is concerned) superseded by the later. But the more elementary methods are likely to be applicable to other problems in which the more subtle analysis is impracticable.

We have attacked this particular problem by a considerable number of different methods, and cannot profess to have reached any very precise conclusions as to the possibilities of each. A detailed comparison of the results

* See G. H. Hardy and J. E. Littlewood, "Some problems of Diophantine approximation (II: The trigonometrical series associated with the elliptic Theta-functions)," *Acta Mathematica*, Vol. XXXVII, 1914, pp. 193—238, for applications of the formulæ to different but not unrelated problems.

to which they lead would moreover expand this paper to a quite unreasonable length. But we have thought it worth while to include a short account of two of them. The first is quite elementary; it depends only on Euler's identity

(1·31)
$$\frac{1}{(1-x)(1-x^2)(1-x^3)\ldots} = 1 + \frac{x}{(1-x)^2} + \frac{x^4}{(1-x)^2(1-x^2)^2} + \ldots$$

—an identity capable of wide generalisation—and on elementary algebraical reasoning. By these means we shew, in section 2, that

(1·32)
$$e^{A\sqrt{n}} < p(n) < e^{B\sqrt{n}},$$

where A and B are positive constants, for all sufficiently large values of n.

It follows that

(1·33)
$$A\sqrt{n} < \log p(n) < B\sqrt{n};$$

and the next question which arises is the question whether a constant C exists such that

(1·34)
$$\log p(n) \sim C\sqrt{n}.$$

We prove that this is so in section 3. Our proof is still, in a sense, "elementary." It does not appeal to the theory of analytic functions, depending only on a general arithmetic theorem concerning infinite series; but this theorem is of the difficult and delicate type which Messrs Hardy and Littlewood have called "Tauberian." The actual theorem required was proved by us in a paper recently printed in these *Proceedings**. It shews that

(1·35)
$$C = \frac{2\pi}{\sqrt{6}};$$

in other words that

(1·36)
$$p(n) = \exp\left\{\pi\sqrt{\left(\frac{2n}{3}\right)}(1+\epsilon)\right\},$$

where ϵ is small when n is large. This method is one of very wide application. It may be used, for example, to prove that, if $p^{(s)}(n)$ denotes the number of partitions of n into perfect s-th powers, then

$$\log p^{(s)}(n) \sim (s+1)\left\{\frac{1}{s}\Gamma\left(1+\frac{1}{s}\right)\zeta\left(1+\frac{1}{s}\right)\right\}^{s/(s+1)} n^{1/(s+1)}.$$

It is certainly possible to obtain, by means of arguments of this general character, information about $p(n)$ more precise than that furnished by the formula (1·36). And it is equally possible to prove (1·36) by reasoning of a more elementary, though more special, character: we have a proof, for example, based on the identity

$$np(n) = \sum_{\nu=1}^{n} \sigma(\nu)\, p(n-\nu),$$

* G. H. Hardy and S. Ramanujan, "Asymptotic formulæ for the distribution of integers of various types," *Proc. London Math. Soc.*, Ser. 2, Vol. XVI, 1917, pp. 112—132 [No. 34 of this volume].

where $\sigma(\nu)$ is the sum of the divisors of ν, and a process of induction. But we are at present unable to obtain, by any method which does not depend upon Cauchy's theorem, a result as precise as that which we state in the next paragraph, a result, that is to say, which is "vraiment asymptotique."

1·4. Our next step was to replace (1·36) by the much more precise formula

$$p(n) \sim \frac{1}{4n\sqrt{3}} \exp\left\{\pi \sqrt{\left(\frac{2n}{3}\right)}\right\}. \tag{1·41}$$

The proof of this formula appears to necessitate the use of much more powerful machinery, Cauchy's integral (1·21) and the functional relation

$$f(x) = \frac{x^{\frac{1}{24}}}{\sqrt{(2\pi)}} \sqrt{\left(\log \frac{1}{x}\right)} \exp\left\{\frac{\pi^2}{6 \log (1/x)}\right\} f(x'), \tag{1·42}$$

where

$$x' = \exp\left\{-\frac{4\pi^2}{\log (1/x)}\right\}. \tag{1·43}$$

This formula is merely a statement in a different notation of the relation between $h(\tau)$ and $h(\mathrm{T})$, where

$$\mathrm{T} = \frac{c + d\tau}{a + b\tau}, \quad a = d = 0, \quad b = 1, \quad c = -1;$$

viz.

$$h(\tau) = \sqrt{\left(\frac{i}{\tau}\right)}\, h(\mathrm{T}) *.$$

It is interesting to observe the correspondence between (1·41) and the results of numerical computation. Numerical data furnished to us by Major MacMahon gave the following results: we denote the right-hand side of (1·41) by $\varpi(n)$.

n	$p(n)$	$\varpi(n)$	ϖ/p
10	42	48·104	1·145
20	627	692·385	1·104
50	204226	217590·499	1·065
80	15796476	16606781·567	1·051

It will be observed that the progress of ϖ/p towards its limit unity is not very rapid, and that $\varpi - p$ is always positive and appears to tend rapidly to infinity.

* Tannery and Molk, *loc. cit.*, p. 265 (Table XLV, 5).

1·5. In order to obtain more satisfactory results it is necessary to construct some auxiliary function $F(x)$ which is regular at all points of the unit circle save $x=1$, and has there a singularity of a type as near as possible to that of the singularity of $f(x)$. We may then hope to obtain a much more precise approximation by applying Cauchy's theorem to $f-F$ instead of to f. For although every point of the circle is a singular point of f, $x=1$ is, to put it roughly, much the *heaviest* singularity. When $x \to 1$ by real values, $f(x)$ tends to infinity like an exponential

$$\exp\left\{\frac{\pi^2}{6(1-x)}\right\};$$

when

$$x = re^{2p\pi i/q},$$

p and q being co-prime integers, and $r \to 1$, $|f(x)|$ tends to infinity like an exponential

$$\exp\left\{\frac{\pi^2}{6q^2(1-r)}\right\};$$

while, if

$$x = re^{2\theta\pi i},$$

where θ is irrational, $|f(x)|$ can become infinite at most like an exponential of the type

$$\exp\left\{o\left(\frac{1}{1-r}\right)\right\}^*.$$

The function required is

$$(1{\cdot}51) \qquad F(x) = \frac{1}{\pi\sqrt{2}}\sum_1^\infty \psi(n)\, x^n,$$

where

$$(1{\cdot}52) \qquad \psi(n) = \frac{d}{dn}\left\{\frac{\cosh C\lambda_n - 1}{\lambda_n}\right\},$$

$$(1{\cdot}53) \qquad C = 2\pi/\sqrt{6} = \pi\sqrt{(\tfrac{2}{3})}, \quad \lambda_n = \sqrt{(n-\tfrac{1}{24})}.$$

This function may be transformed into an integral by means of a general formula given by Lindelöf†; and it is then easy to prove that the "principal branch" of $F(x)$ is regular all over the plane except at $x=1$‡; and that

$$F(x) - \chi(x),$$

* The statements concerning the "rational" points are corollaries of the formulæ of the transformation theory, and proofs of them are contained in the body of the paper. The proposition concerning "irrational" points may be proved by arguments similar to those used by Hardy and Littlewood in their memoir already quoted. It is not needed for our present purpose. As a matter of fact it is *generally* true that $f(x) \to 0$ when θ is irrational, and very nearly as rapidly as $\sqrt[4]{(1-r)}$. It is in reality owing to this that our final method is so successful.

† E. Lindelöf, *Le calcul des résidus et ses applications à la théorie des fonctions* (Gauthier-Villars, Collection Borel, 1905), p. 111.

‡ We speak, of course, of the principal branch of the function, viz. that represented by the series (1·51) when x is small. The other branches are singular at the origin.

where

$$(1{\cdot}54)\qquad \chi(x) = \frac{x^{\frac{1}{24}}}{\sqrt{(2\pi)}} \sqrt{\left(\log\frac{1}{x}\right)} \left[\exp\left\{\frac{\pi^2}{6\log(1/x)}\right\} - 1\right]$$

is regular for $x = 1$. If we compare (1·42) and (1·54), and observe that $f(x')$ tends to unity with extreme rapidity when x tends to 1 along any regular path which does not touch the circle of convergence, we can see at once the very close similarity between the behaviour of f and F inside the unit circle and in the neighbourhood of $x = 1$.

It should be observed that the term -1 in (1·52) and (1·54) is—so far as our present assertions are concerned—otiose: all that we have said remains true if it is omitted; the resemblance between the singularities of f and F becomes indeed even closer. The term is inserted merely in order to facilitate some of our preliminary analysis, and will prove to be without influence on the final result.

Applying Cauchy's theorem to $f - F$, we obtain

$$(1{\cdot}55)\qquad p(n) = \frac{1}{2\pi\sqrt{2}}\frac{d}{dn}\left(\frac{e^{C\lambda_n}}{\lambda_n}\right) + O(e^{D\sqrt{n}}),$$

where D is any number greater than

$$\tfrac{1}{2}C = \tfrac{1}{2}\pi\sqrt{(\tfrac{2}{3})}.$$

1·6. The formula (1·55) is an asymptotic formula of a type far more precise than that of (1·41). The error term is, however, of an exponential type, and may be expected ultimately to increase with very great rapidity. It was therefore with considerable surprise that we found what exceedingly good results the formula gives for fairly large values of n. For $n = 61$, 62, 63 it gives

1121538·972, 1300121·359, 1505535·606,

while the correct values are

1121505, 1300156, 1505499.

The errors 33·972, $-$34·641, 36·606

are relatively very small, and alternate in sign.

The next step is naturally to direct our attention to the singular point of $f(x)$ next in importance after that at $x = 1$, viz., that at $x = -1$; and to subtract from $f(x)$ a second auxiliary function, related to this point as $F(x)$ is to $x = 1$. No new difficulty of principle is involved, and we find that

$$(1{\cdot}61)\qquad p(n) = \frac{1}{2\pi\sqrt{2}}\frac{d}{dn}\left(\frac{e^{C\lambda_n}}{\lambda_n}\right) + \frac{(-1)^n}{2\pi}\frac{d}{dn}\left(\frac{e^{\frac{1}{2}C\lambda_n}}{\lambda_n}\right) + O(e^{D\sqrt{n}}),$$

where D is now any number greater than $\frac{1}{3}C$. It now becomes obvious why our earlier approximation gave errors alternately of excess and of defect.

It is obvious that this process may be repeated indefinitely. The singularities next in importance are those at $x=e^{\frac{2}{3}\pi i}$ and $x=e^{\frac{4}{3}\pi i}$; the next those at $x=i$ and $x=-i$; and so on. The next two terms in the approximate formula are found to be

$$\frac{\sqrt{3}}{\pi\sqrt{2}}\cos\left(\tfrac{2}{3}n\pi-\tfrac{1}{18}\pi\right)\frac{d}{dn}\left(\frac{e^{\frac{1}{3}C\lambda_n}}{\lambda_n}\right)$$

and

$$\frac{\sqrt{2}}{\pi}\cos\left(\tfrac{1}{2}n\pi-\tfrac{1}{8}\pi\right)\frac{d}{dn}\left(\frac{e^{\frac{1}{4}C\lambda_n}}{\lambda_n}\right).$$

As we proceed further, the complexity of the calculations increases. The auxiliary function associated with the point $x=e^{2p\pi i/q}$ involves a certain $24q$-th root of unity, connected with the linear transformation which must be used in order to elucidate the behaviour of $f(x)$ near the point; and the explicit expression of this root in terms of p and q, though known, is somewhat complex. But it is plain that, by taking a sufficient number of terms, we can find a formula in which the error is

$$O\left(e^{C\lambda_n/\nu}\right),$$

where ν is a fixed but arbitrarily large integer.

1·7. A final question remains. We have still the resource of making ν a function of n, that is to say of making the number of terms in our approximate formula itself a function of n. In this way we may reasonably hope, at any rate, to find a formula in which the error is of order less than that of any exponential of the type e^{an}; of the order of a power of n, for example, or even bounded.

When, however, we proceeded to test this hypothesis by means of the numerical data most kindly provided for us by Major MacMahon, we found a correspondence between the real and the approximate values of such astonishing accuracy as to lead us to hope for even more. Taking $n=100$, we found that the first six terms of our formula gave

$$\begin{array}{r} 190568944{\cdot}783 \\ +348{\cdot}872 \\ -2{\cdot}598 \\ +{\cdot}685 \\ +{\cdot}318 \\ -{\cdot}064 \\ \hline 190569291{\cdot}996, \end{array}$$

while

$$p(100)=190569292;$$

so that the error after six terms is only ·004. We then proceeded to calculate $p(200)$, and found

$$\begin{array}{r} 3,\ 972,\ 998,\ 993,\ 185{\cdot}896 \\ +36,\ 282{\cdot}978 \\ -87{\cdot}555 \\ +\ 5{\cdot}147 \\ +\ 1{\cdot}424 \\ +\ 0{\cdot}071 \\ +\ 0{\cdot}000^{*} \\ +\ 0{\cdot}043 \\ \hline 3,\ 972,\ 999,\ 029,\ 388{\cdot}004, \end{array}$$

and Major MacMahon's subsequent calculations shewed that $p(200)$ is, in fact,

3, 972, 999, 029, 388.

These results suggest very forcibly that it is possible to obtain a formula for $p(n)$, which not only exhibits its order of magnitude and structure, but may be used to calculate its *exact* value for any value of n. That this is in fact so is shewn by the following theorem.

Statement of the main theorem.

THEOREM. *Suppose that*

(1·71) $$\phi_q(n) = \frac{\sqrt{q}}{2\pi\sqrt{2}}\frac{d}{dn}\left(\frac{e^{C\lambda_n/q}}{\lambda_n}\right),$$

where C and λ_n are defined by the equations (1·53), *for all positive integral values of q; that p is a positive integer less than and prime to q; that $\omega_{p,q}$ is a $24q$-th root of unity, defined when p is odd by the formula*

(1·721)

$$\omega_{p,q} = \left(\frac{-q}{p}\right)\exp\left[-\left\{\tfrac{1}{4}(2-pq-p)+\tfrac{1}{12}\left(q-\frac{1}{q}\right)(2p-p'+p^2p')\right\}\pi i\right],$$

and when q is odd by the formula

(1·722)

$$\omega_{p,q} = \left(\frac{-p}{q}\right)\exp\left[-\left\{\tfrac{1}{4}(q-1)+\tfrac{1}{12}\left(q-\frac{1}{q}\right)(2p-p'+p^2p')\right\}\pi i\right],$$

where (a/b) is the symbol of Legendre and Jacobi†, *and p' is any positive integer such that $1+pp'$ is divisible by q; that*

(1·73) $$A_q(n) = \sum_{(p)} \omega_{p,q}\, e^{-2np\pi i/q};$$

and that α is any positive constant, and ν the integral part of $\alpha\sqrt{n}$.

* This term vanishes identically.

† See Tannery and Molk, *loc. cit.*, pp. 104—106, for a complete set of rules for the calculation of the value of (a/b), which is, of course, always 1 or -1. When *both* p and q are odd it is indifferent which formula is adopted.

Then

(1·74) $$p(n) = \sum_1^{\nu} A_q \phi_q + O(n^{-\frac{1}{4}}),$$

so that $p(n)$ *is, for all sufficiently large values of* n, *the integer nearest to*

(1·75) $$\sum_1^{\nu} A_q \phi_q.$$

It should be observed that all the numbers A_q are real. A table of A_q from $q = 1$ to $q = 18$ is given at the end of the paper (Table II).

The proof of this theorem is given in section 5; section 4 being devoted to a number of preliminary lemmas. The proof is naturally somewhat intricate; and we trust that we have arranged it in such a form as to be readily intelligible. In section 6 we draw attention to one or two questions which our theorem, in spite of its apparent completeness, still leaves open. In section 7 we indicate some other problems in combinatory analysis and the analytic theory of numbers to which our method may be applied; and we conclude by giving some functional and numerical tables: for the latter we are indebted to Major MacMahon and Mr H. B. C. Darling. To Major MacMahon in particular we owe many thanks for the amount of trouble he has taken over very tedious calculations. It is certain that, without the encouragement given by the results of these calculations, we should never have attempted to prove theoretical results at all comparable in precision with those which we have enunciated.

2. Elementary proof that $e^{A\sqrt{n}} < p(n) < e^{B\sqrt{n}}$ for sufficiently large values of n.

2·1. In this section we give the elementary proof of the inequalities (1·32). We prove, in fact, rather more, viz., that positive constants H and K exist such that

(2·11) $$\frac{H}{n} e^{2\sqrt{n}} < p(n) < \frac{K}{n} e^{2\sqrt{(2n)}}$$

for $n \geqslant 1$*. We shall use in our proof only Euler's formula (1·31) and a debased form of Stirling's theorem, easily demonstrable by quite elementary methods: the proposition that

$$n!\, e^n / n^{n+\frac{1}{2}}$$

lies between two positive constants for all positive integral values of n.

* Somewhat inferior inequalities, of the type

$$2^{A[\sqrt{n}]} < p(n) < n^{B[\sqrt{n}]},$$

may be proved by *entirely* elementary reasoning; by reasoning, that is to say, which depends only on the arithmetical definition of $p(n)$ and on elementary finite algebra, and does not presuppose the notion of a limit or the definitions of the logarithmic or exponential functions.

2·2. The proof of the first of the two inequalities is slightly the simpler. It is obvious that if

$$\Sigma p_r(n)\, x^n = \frac{1}{(1-x)(1-x^2)\dots(1-x^r)}$$

so that $p_r(n)$ is the number of partitions of n into parts not exceeding r, then

$$(2·21) \qquad p_r(n) = p_{r-1}(n) + p_{r-1}(n-r) + p_{r-1}(n-2r) + \dots.$$

We shall use this equation to prove, by induction, that

$$(2·22) \qquad p_r(n) \geqslant \frac{rn^{r-1}}{(r!)^2}.$$

It is obvious that (2·22) is true for $r=1$. Assuming it to be true for $r=s$, and using (2·21), we obtain

$$p_{s+1}(n) \geqslant \frac{s}{(s!)^2}\{n^{s-1} + (n-s-1)^{s-1} + (n-2s-2)^{s-1} + \dots\}$$

$$\geqslant \frac{s}{(s!)^2}\left\{\frac{n^s-(n-s-1)^s}{s(s+1)} + \frac{(n-s-1)^s-(n-2s-2)^s}{s(s+1)} + \dots\right\}$$

$$= \frac{n^s}{(s+1)(s!)^2} = \frac{(s+1)\,n^s}{\{(s+1)!\}^2}.$$

This proves (2·22). Now $p(n)$ is obviously not less than $p_r(n)$, whatever the value of r. Take $r=[\sqrt{n}]$: then

$$p(n) \geqslant p_{[\sqrt{n}]}(n) \geqslant \frac{[\sqrt{n}]}{n}\frac{n^{[\sqrt{n}]}}{\{[\sqrt{n}]!\}^2} > \frac{H}{n}e^{2\sqrt{n}},$$

by a simple application of the degenerate form of Stirling's theorem mentioned above.

2·3. The proof of the second inequality depends upon Euler's identity. If we write

$$\Sigma q_r(n)\, x^n = \frac{1}{(1-x)^2(1-x^2)^2\dots(1-x^r)^2},$$

we have

$$(2·31) \qquad q_r(n) = q_{r-1}(n) + 2q_{r-1}(n-r) + 3q_{r-1}(n-2r) + \dots,$$

and

$$(2·32) \qquad p(n) = q_1(n-1) + q_2(n-4) + q_3(n-9) + \dots.$$

We shall first prove by induction that

$$(2·33) \qquad q_r(n) \leqslant \frac{(n+r^2)^{2r-1}}{(2r-1)!\,(r!)^2}.$$

This is obviously true for $r=1$. Assuming it to be true for $r=s$, and using (2·31), we obtain

$$q_{s+1}(n) \leqslant \frac{1}{(2s-1)!\,(s!)^2}\{(n+s^2)^{2s-1} + 2(n+s^2-s-1)^{2s-1} + 3(n+s^2-2s-2)^{2s-1} + \dots\}.$$

Now $$m(m-1)\,a^{m-2}b^2 \leqslant (a+b)^m - 2a^m + (a-b)^m,$$
if m is a positive integer, and a, b, and $a-b$ are positive, while if $a-b \leqslant 0$, and m is odd, the term $(a-b)^m$ may be omitted. In this inequality write
$$m = 2s+1, \quad a = n + s^2 - ks - k \ (k = 0, 1, 2, \ldots), \quad b = s+1,$$
and sum with respect to k. We find that
$$(2s+1)\,2s\,(s+1)^2\{(n+s^2)^{2s-1} + 2(n+s^2-s-1)^{2s-1} + \ldots\} \leqslant (n+s^2+s+1)^{2s+1};$$
and so
$$q_{s+1}(n) \leqslant \frac{(n+s^2+s+1)^{2s+1}}{(2s+1)\,2s\,(s+1)^2(2s-1)!\,(s!)^2} \leqslant \frac{\{n+(s+1)^2\}^{2s+1}}{(2s+1)!\,\{(s+1)!\}^2}.$$
Hence (2·33) is true generally.

It follows from (2·32) that
$$p(n) = q_1(n-1) + q_2(n-4) + \ldots \leqslant \sum_1^\infty \frac{n^{2r-1}}{(2r-1)!\,(r!)^2}.$$
But, using the degenerate form of Stirling's theorem once more, we find without difficulty that
$$\frac{1}{(2r-1)!\,(r!)^2} < \frac{2^{6r}K}{4r!},$$
where K is a constant. Hence
$$p(n) < 8K \sum_1^\infty \frac{(8n)^{2r-1}}{4r!} < 8K \sum_1^\infty \frac{(8n)^{\frac{1}{2}r-1}}{r!} < \frac{K}{n} e^{2\sqrt{(2n)}}.$$
This is the second of the inequalities (2·11).

3. Application of a Tauberian theorem to the determination of the constant C.

3·1. The value of the constant
$$C = \lim \frac{\log p(n)}{\sqrt{n}},$$
is most naturally determined by the use of the following theorem.

If $g(x) = \Sigma a_n x^n$ is a power-series with positive coefficients, and
$$\log g(x) \sim \frac{A}{1-x}$$
when $x \to 1$, then
$$\log s_n = \log(a_0 + a_1 + \ldots + a_n) \sim 2\sqrt{(An)}$$
when $n \to \infty$.

This theorem is a special case* of Theorem C in our paper already referred to.

Now suppose that
$$g(x) = (1-x)f(x) = \Sigma\{p(n) - p(n-1)\}x^n = \frac{1}{(1-x^2)(1-x^3)(1-x^4)\ldots}.$$

* *Loc. cit.*, p. 129 (with $a = 1$) [p. 258 of this volume].

Then
$$a_n = p(n) - p(n-1)$$
is plainly positive. And

(3·11)
$$\log g(x) = \sum_2^\infty \log \frac{1}{1-x^\mu} = \sum_1^\infty \frac{1}{\nu} \frac{x^{2\nu}}{1-x^\nu} \sim \frac{1}{1-x} \sum_1^\infty \frac{1}{\nu^2} = \frac{\pi^2}{6(1-x)},$$
when $x \to 1$*. Hence

(3·12)
$$\log p(n) = a_0 + a_1 + \ldots + a_n \sim C\sqrt{n},$$
where $C = 2\pi/\sqrt{6} = \pi\sqrt{(\frac{2}{3})}$, as in (1·53).

3·2. There is no doubt that it is possible, by "Tauberian" arguments, to prove a good deal more about $p(n)$ than is asserted by (3·12). The functional equation satisfied by $f(x)$ shews, for example, that
$$g(x) \sim \frac{(1-x)^{\frac{3}{2}}}{\sqrt{(2\pi)}} \exp\left\{\frac{\pi^2}{6(1-x)}\right\},$$
a relation far more precise than (3·11). From this relation, and the fact that the coefficients in $g(x)$ are positive, it is certainly possible to deduce more than (3·12). But it hardly seems likely that arguments of this character will lead us to a proof of (1·41). It would be exceedingly interesting to know exactly how far they will carry us, since the method is comparatively elementary, and has a much wider range of application than the more powerful methods employed later in this paper. We must, however, reserve the discussion of this question for some future occasion.

4. LEMMAS PRELIMINARY TO THE PROOF OF THE MAIN THEOREM.

4·1. We proceed now to the proof of our main theorem. The proof is somewhat intricate, and depends on a number of subsidiary theorems which we shall state as lemmas.

Lemmas concerning Farey's series.

4·21. The *Farey's series of order m* is the aggregate of irreducible rational fractions
$$p/q \quad (0 \leqslant p \leqslant q \leqslant m),$$

* This is a special case of much more general theorems: see K. Knopp, "Grenzwerte von Reihen bei der Annäherung an die Konvergenzgrenze," *Inaugural-Dissertation*, Berlin, 1907, pp. 25 *et seq.*; K. Knopp, "Über Lambertsche Reihen," *Journal für Math.*, Vol. CXLII, 1913, pp. 283—315; G. H. Hardy, "Theorems connected with Abel's Theorem on the continuity of power series," *Proc. London Math. Soc.*, Ser. 2, Vol. IV, 1906, pp. 247—265 (pp. 252, 253); G. H. Hardy, "Some theorems concerning infinite series," *Math. Ann.*, Vol. LXIV, 1907, pp. 77—94; G. H. Hardy, "Note on Lambert's series," *Proc. London Math. Soc.*, Ser. 2, Vol. XIII, 1913, pp. 192—198.

A direct proof is very easy: for
$$\nu x^{\nu-1}(1-x) < 1 - x^\nu < \nu(1-x),$$
$$\frac{1}{1-x} \Sigma \frac{x^{2\nu}}{\nu^2} < \log g(x) < \frac{1}{1-x} \Sigma \frac{x^{\nu+1}}{\nu^2}.$$

arranged in ascending order of magnitude. Thus

$$\tfrac{0}{1},\ \tfrac{1}{7},\ \tfrac{1}{6},\ \tfrac{1}{5},\ \tfrac{1}{4},\ \tfrac{2}{7},\ \tfrac{1}{3},\ \tfrac{2}{5},\ \tfrac{3}{7},\ \tfrac{1}{2},\ \tfrac{4}{7},\ \tfrac{3}{5},\ \tfrac{2}{3},\ \tfrac{5}{7},\ \tfrac{3}{4},\ \tfrac{4}{5},\ \tfrac{5}{6},\ \tfrac{6}{7},\ \tfrac{1}{1}$$

is the Farey's series of order 7.

LEMMA 4·21. *If p/q, p'/q' are two successive terms of a Farey's series, then*

$$(4{\cdot}211) \qquad p'q - pq' = 1.$$

This is, of course, a well-known theorem, first observed by Farey and first proved by Cauchy*. The following exceedingly simple proof is due to Hurwitz†.

The result is plainly true when $m = 1$. Let us suppose it true for $m = k$; and let p/q, p'/q' be two consecutive terms in the series of order k.

Suppose now that p''/q'' is a term of the series of order $k+1$ which falls between p/q and p'/q'. Let

$$p''q - pq'' = \lambda > 0, \qquad p'q'' - p''q' = \mu > 0.$$

Solving these equations for p'', q'', and observing that $p'q - pq' = 1$, we obtain

$$p'' = \mu p + \lambda p', \qquad q'' = \mu q + \lambda q'.$$

Consider now the aggregate of fractions

$$(\mu p + \lambda p')/(\mu q + \lambda q'),$$

where λ and μ are positive integers without common factor. All of these fractions lie between p/q and p'/q'; and all are in their lowest terms, since a factor common to numerator and denominator would divide

$$\lambda = q(\mu p + \lambda p') - p(\mu q + \lambda q'),$$

and

$$\mu = p'(\mu q + \lambda q') - q'(\mu p + \lambda p').$$

Each of them first makes its appearance in the Farey's series of order $\mu q + \lambda q'$, and the *first* of them to make its appearance must be that for which $\lambda = 1$, $\mu = 1$. Hence

$$p'' = p + p', \qquad q'' = q + q',$$
$$p''q - pq'' = p'q'' - p''q' = 1.$$

The lemma is consequently proved by induction.

LEMMA 4·22. *Suppose that p/q is a term of the Farey's series of order m, and p''/q'', p'/q' the adjacent terms on the left and the right: and let $j_{p,q}$ denote the interval*

$$\frac{p}{q} - \frac{1}{q(q+q'')}, \qquad \frac{p}{q} + \frac{1}{q(q+q')}\ddagger.$$

* J. Farey, "On a curious property of vulgar fractions," *Phil. Mag.*, Ser. 1, Vol. XLVII, 1816, pp. 385, 386; A. L. Cauchy, "Démonstration d'un théorème curieux sur les nombres," *Exerciçes de mathématiques*, Vol. I, 1826, pp. 114—116. Cauchy's proof was first published in the *Bulletin de la Société Philomatique* in 1816.

† A. Hurwitz, "Ueber die angenäherte Darstellung der Zahlen durch rationale Brüche," *Math. Ann.*, Vol. XLIV, 1894, pp. 417—436.

‡ When p/q is 0/1 or 1/1, only the part of this interval inside (0, 1) is to be taken; thus $j_{0,1}$ is 0, $1/(m+1)$ and $j_{1,1}$ is $1 - 1/(m+1)$, 1.

Then (i) *the intervals* $j_{p,q}$ *exactly fill up the continuum* (0, 1), *and* (ii) *the length of each of the parts into which* $j_{p,q}$ *is divided by* p/q* *is greater than* $1/2mq$ *and less than* $1/mq$.

(i) Since

$$\frac{1}{q(q+q')}+\frac{1}{q'(q'+q)}=\frac{1}{qq'}=\frac{p'q-pq'}{qq'}=\frac{p'}{q'}-\frac{p}{q},$$

the intervals just fill up the continuum.

(ii) Since neither q nor q' exceeds m, and one at least must be less than m, we have

$$\frac{1}{q(q+q')}>\frac{1}{2mq}.$$

Also $q+q'>m$, since otherwise $(p+p')/(q+q')$ would be a term in the series between p/q and p'/q'. Hence

$$\frac{1}{q(q+q')}<\frac{1}{mq}.$$

Standard dissection of a circle.

4·23. The following mode of dissection of a circle, based upon Lemma 4·22, is of fundamental importance for our analysis.

Suppose that the circle is defined by

$$x=Re^{2\pi i\theta}\quad(0\leqslant\theta\leqslant 1).$$

Construct the Farey's series of order m, and the corresponding intervals $j_{p,q}$. When these intervals are considered as intervals of variation of θ, and the two extreme intervals, which correspond to abutting arcs on the circle, are regarded as constituting a single interval $\xi_{1,1}$, the circle is divided into a number of arcs

$$\xi_{p,q},$$

where q ranges from 1 to m and p through the numbers not exceeding and prime to q†. We call this dissection of the circle *the dissection* Ξ_m.

Lemmas from the theory of the linear transformation of the elliptic modular functions.

4·3. Lemma 4·31. *Suppose that* q *is a positive integer; that* p *is a positive integer not exceeding and prime to* q*; that* p' *is a positive integer such that* $1+pp'$ *is divisible by* q*; that* $\omega_{p,q}$ *is defined by the formulæ* (1·721) *or* (1·722)*; that*

$$x=\exp\left(-\frac{2\pi z}{q}+\frac{2p\pi i}{q}\right),\quad x'=\exp\left(-\frac{2\pi}{qz}+\frac{2p'\pi i}{q}\right),$$

* See the preceding footnote [footnote ‡ of p. 289].

† $p=0$ occurring with $q=1$ only.

where the real part of z is positive; and that

$$f(x) = \frac{1}{(1-x)(1-x^2)(1-x^3)\dots}.$$

Then $$f(x) = \omega_{p,q} \sqrt{z} \exp\left(\frac{\pi}{12qz} - \frac{\pi z}{12q}\right) f(x').$$

This lemma is merely a restatement in a different notation of well-known formulæ in the transformation theory.

Suppose, for example, that p is odd. If we take

$$a = p, \quad b = -q, \quad c = \frac{1+pp'}{q}, \quad d = -p',$$

so that $ad - bc = 1$; and write

$$x = q^2 = e^{2\pi i\tau}, \quad x' = Q^2 = e^{2\pi i T},$$

so that $$\tau = \frac{p}{q} + \frac{iz}{q}, \quad T = \frac{p'}{q} + \frac{i}{qz};$$

then we can easily verify that

$$T = \frac{c + d\tau}{a + b\tau}.$$

Also, in the notation of Tannery and Molk, we have

$$f(x) = \frac{q^{\frac{1}{12}}}{h(\tau)}, \quad f(x') = \frac{Q^{\frac{1}{12}}}{h(T)};$$

and the formula for the linear transformation of $h(\tau)$ is

$$h(T) = \left(\frac{b}{a}\right) \exp\left[\{\tfrac{1}{4}(a-1) - \tfrac{1}{12}[a(b-c) + bd(a^2-1)]\}\pi i\right] \sqrt{(a+b\tau)}\, h(\tau),$$

where $\sqrt{(a+b\tau)}$ has its real part positive*. A little elementary algebra will shew the equivalence of this result and ours.

The other formula for $\omega_{p,q}$ may be verified similarly, but in this case we must take

$$a = -p, \quad b = q, \quad c = -\frac{1+pp'}{q}, \quad d = p'.$$

We have included in the Appendix (Table I) a short table of some values of $\omega_{p,q}$, or rather of $(\log \omega_{p,q})/\pi i$.

LEMMA 4·32. *The function $f(x)$ satisfies the equation*

(4·321)

$$f(x) = \omega_{p,q} \sqrt{\left\{\frac{q}{2\pi} \log\left(\frac{1}{x_{p,q}}\right)\right\}}\, x_{p,q}^{\frac{1}{24}} \exp\left\{\frac{\pi^2}{6q^2 \log(1/x_{p,q})}\right\} f(x'_{p,q}),$$

where

(4·322)

$$x_{p,q} = xe^{-2p\pi i/q}, \quad x'_{p,q} = \exp\left\{-\frac{4\pi^2}{q^2 \log(1/x_{p,q})} + \frac{2p'\pi i}{q}\right\}.$$

* Tannery and Molk, *loc. cit.*, pp. 113, 267.

This is an immediate corollary from Lemma 4·31, since

$$z = \frac{q}{2\pi} \log\left(\frac{1}{x_{p,q}}\right), \quad e^{-\pi z/12q} = x_{p,q}^{\frac{1}{24}},$$

$$\frac{\pi}{12qz} = \frac{\pi^2}{6q^2 \log(1/x_{p,q})}, \quad x' = \exp\left(-\frac{2\pi}{qz} + \frac{2p'\pi i}{q}\right) = x'_{p,q}.$$

If we observe that

$$f(x'_{p,q}) = 1 + p(1)\, x'_{p,q} + \ldots,$$

we see that, if x tends to $e^{2p\pi i/q}$ along a radius vector, or indeed any regular path which does not touch the circle of convergence, the difference

$$f(x) - \omega_{p,q} \sqrt{\left\{\frac{q}{2\pi} \log\left(\frac{1}{x_{p,q}}\right)\right\}}\, x_{p,q}^{\frac{1}{24}} \exp\left\{\frac{\pi^2}{6q^2 \log(1/x_{p,q})}\right\}$$

tends to zero with great rapidity. It is on this fact that our analysis is based.

Lemmas concerning the auxiliary function $F_a(x)$.

4·41. The auxiliary function $F_a(x)$ is defined by the equation

$$F_a(x) = \sum_1^\infty \psi_a(n)\, x^n,$$

where

$$\psi_a(n) = \frac{d}{dn} \frac{\cosh a\lambda_n - 1}{\lambda_n},$$

$$\lambda_n = \sqrt{(n - \tfrac{1}{24})}, \quad a > 0.$$

LEMMA 4·41. *Suppose that a cut is made along the segment* (1, ∞) *in the plane of x. Then $F_a(x)$ is regular at all points inside the region thus defined.*

This lemma is an immediate corollary of a general theorem proved by Lindelöf on pp. 109 *et seq.* of his *Calcul des résidus**.

The function

$$\psi_a(z) = \frac{d}{dz} \frac{\cosh a\sqrt{(z - \frac{1}{24})} - 1}{\sqrt{(z - \frac{1}{24})}}$$

satisfies the conditions imposed upon it by Lindelöf, if the number which he calls α is greater than $\frac{1}{24}$; and

$$(4·411) \qquad F_a(x) = \int_{\alpha - i\infty}^{\alpha + i\infty} \frac{x^z}{1 - e^{2\pi i z}} \phi(z)\, dz,$$

if

$$x = re^{i\theta}, \quad 0 < \theta < 2\pi, \quad x^z = \exp\{z(\log r + i\theta)\}.$$

4·42. LEMMA 4·42. *Suppose that D is the region defined by the inequalities*

$$-\pi < -\theta_0 < \theta < \theta_0 < \pi, \quad r_0 < r, \quad 0 < r_0 < 1,$$

* Lindelöf gives references to Mellin and Le Roy, who had previously established the theorem in less general forms.

and that $\log(1/x)$ *has its principal value, so that* $\log(1/x)$ *is one-valued, and its square root two-valued, in* D. *Further, let*

$$\chi_a(x) = \sqrt{\{\pi \log(1/x)\}}\, x^{\frac{1}{24}} \left[\exp\left\{\frac{a^2}{4\log(1/x)}\right\} - 1\right],$$

that value of the square root being chosen which is positive when $0 < x < 1$. *Then*

$$F_a(x) - \chi_a(x)$$

is regular inside D*.

We observe first that, when θ has a fixed value between 0 and 2π, the integral on the right-hand side of (4·411) is uniformly convergent for $\frac{1}{24} \leqslant \alpha \leqslant \alpha_0$. Hence we may take $\alpha = \frac{1}{24}$ in (4·411). We thus obtain

$$F_a(x) = ix^{\frac{1}{24}} \int_0^\infty \frac{x^{it}}{1 - e^{\frac{1}{12}\pi i - 2\pi t}} \psi_a(\tfrac{1}{24} + it)\, dt + ix^{\frac{1}{24}} \int_0^\infty \frac{x^{-it}}{1 - e^{\frac{1}{12}\pi i + 2\pi t}} \psi_a(\tfrac{1}{24} - it)\, dt,$$

where the $\sqrt{(it)}$ and $\sqrt{(-it)}$ which occur in $\psi_a(\frac{1}{24} + it)$ and $\psi_a(\frac{1}{24} - it)$ are to be interpreted as $e^{\frac{1}{4}\pi i}\sqrt{t}$ and $e^{-\frac{1}{4}\pi i}\sqrt{t}$ respectively. We write this in the form

$$(4\cdot421) \quad F_a(x) = X_a(x) + ix^{\frac{1}{24}} \int_0^\infty \frac{x^{it}}{e^{-\frac{1}{12}\pi i + 2\pi t} - 1} \psi_a(\tfrac{1}{24} + it)\, dt$$

$$+ ix^{\frac{1}{24}} \int_0^\infty \frac{x^{-it}}{1 - e^{\frac{1}{12}\pi i + 2\pi t}} \psi_a(\tfrac{1}{24} - it)\, dt$$

$$= X_a(x) + \Theta_1(x) + \Theta_2(x),$$

say, where

$$X_a(x) = ix^{\frac{1}{24}} \int_0^\infty x^{it} \psi_a(\tfrac{1}{24} + it)\, dt.$$

Now, since

$$|x^{it}| = e^{-\theta t}, \quad |x^{-it}| = e^{\theta t},$$

the functions Θ are regular throughout the angle of Lemma 4·42. And

$$X_a(x) = \frac{x^{\frac{1}{24}}}{\sqrt{i}} \int_0^\infty e^{-\lambda t} \frac{d}{dt}\left(\frac{\cosh \mu\sqrt{t} - 1}{\sqrt{t}}\right) dt,$$

where

$$\lambda = i \log \frac{1}{x}, \quad \mu = a\sqrt{i}.$$

The form of this integral may be calculated by supposing λ and μ positive, when we obtain

$$\int_0^\infty e^{-\lambda w^2} \frac{d}{dw}\left(\frac{\cosh \mu w - 1}{w}\right) dw = 2\lambda \int_0^\infty e^{-\lambda w^2} (\cosh \mu w - 1)\, dw$$

$$= \sqrt{(\lambda\pi)}\,(e^{\mu^2/4\lambda} - 1).$$

Hence

$$(4\cdot422) \quad X_a(x) = \sqrt{\{\pi \log(1/x)\}}\, x^{\frac{1}{24}} \left[\exp\left\{\frac{a^2}{4\log(1/x)}\right\} - 1\right] = \chi_a(x),$$

and the proof of the lemma is completed.

* Both $F_a(x)$ and $\chi_a(x)$ are two-valued in D. The value of $F_a(x)$ contemplated is naturally that represented by the power-series.

Lemmas 4·41 and 4·42 shew that $x=1$ is the sole finite singularity of the principal branch of $F_a(x)$.

4·43. Lemma 4·43. *Suppose that* P, θ_1, *and* A *are positive constants,* θ_1 *being less than* π. *Then*

$$|F_a(x)| < K = K(\mathrm{P}, \theta_1, A),$$

for
$$0 \leqslant r \leqslant \mathrm{P}, \quad \theta_1 \leqslant \theta \leqslant 2\pi - \theta_1, \quad 0 < a \leqslant A.$$

We use K generally to denote a positive number independent of x and of a. We may employ the formula (4·411). It is plain that

$$\left|\frac{x^z}{1-e^{2\pi i z}}\right| < Ke^{-\theta_1|\eta|},$$

$$|\psi_a(z)| = \left|\frac{d}{dz}\left\{\frac{\cosh a\sqrt{(z-\frac{1}{24})}-1}{\sqrt{(z-\frac{1}{24})}}\right\}\right| < Ke^{K\sqrt{|\eta|}},$$

where η is the imaginary part of z. Hence

$$|F_a(x)| < K\int_{-\infty}^{\infty} e^{K\sqrt{|\eta|}-\theta_1|\eta|}\,d\eta < K.$$

4·44. Lemma 4·44. *Let* c *be a circle whose centre is* $x=1$, *and whose radius* δ *is less than unity. Then*

$$|F_a(x) - \chi_a(x)| < Ka^2,$$

if x *lies in* c *and* $0 < a \leqslant A$, $K = K(\delta, A)$ *being as before independent of* x *and of* a.

If we refer back to (4·421) and (4·422), we see that it is sufficient to prove that

$$|\Theta_1(x)| < Ka^2, \quad |\Theta_2(x)| < Ka^2;$$

and we may plainly confine ourselves to the first of these inequalities. We have

$$\Theta_1(x) = \frac{x^{\frac{1}{24}}}{\sqrt{i}}\int_0^\infty \frac{x^{it}}{e^{-\frac{1}{12}\pi i+2\pi t}-1}\frac{d}{dt}\left\{\frac{\cosh a\sqrt{(it)}-1}{\sqrt{t}}\right\}dt.$$

Rejecting the extraneous factor, which is plainly without importance, and integrating by parts, we obtain

$$\Theta(x) = \int_0^\infty \Phi(t)\frac{\cosh a\sqrt{(it)}-1}{\sqrt{t}}\,dt,$$

where
$$\Phi(t) = -\frac{ix^{it}\log x}{e^{-\frac{1}{12}\pi i+2\pi t}-1} + \frac{2\pi x^{it}e^{-\frac{1}{12}\pi i+2\pi t}}{(e^{-\frac{1}{12}\pi i+2\pi t}-1)^2}.$$

Now $|\theta| < \frac{1}{2}\pi$ and $|x^{it}| < Ke^{\frac{1}{2}\pi t}$. It follows that

$$|\Phi(t)| < Ke^{-\pi t};$$

and
$$|\Theta(x)| < K\int_0^\infty \frac{e^{-\pi t}}{\sqrt{t}} |\sinh^2 \tfrac{1}{2}a\sqrt{(it)}|\, dt$$
$$< K\int_0^\infty \frac{e^{-\pi t}}{\sqrt{t}} \{\cosh a\sqrt{(\tfrac{1}{2}t)} - \cos a\sqrt{(\tfrac{1}{2}t)}\}\, dt$$
$$< K\int_0^\infty e^{-\pi w^2}\left(\cosh\frac{aw}{\sqrt{2}} - \cos\frac{aw}{\sqrt{2}}\right) dw$$
$$= K(e^{a^2/8\pi} - e^{-a^2/8\pi}) < Ka^2.$$

5. Proof of the main theorem.

5·1. We write

(5·11)
$$F_{p,q}(x) = \omega_{p,q}\frac{\sqrt{q}}{\pi\sqrt{2}}F_{C/q}(x_{p,q}),$$
where $C = \pi\sqrt{\tfrac{2}{3}}$, $x_{p,q} = xe^{-2p\pi i/q}$; and

(5·12)
$$\Phi(x) = f(x) - \sum_q\sum_p F_{p,q}(x),$$
where the summation applies to all values of p not exceeding q and prime to q, and to all values of q such that

(5·13)
$$1 \leqslant q \leqslant \nu = [\alpha\sqrt{n}],$$
α being positive and independent of n. If then

(5·14)
$$F_{p,q}(x) = \sum c_{p,q,n}x^n,$$
we have

(5·15)
$$p(n) - \sum_q\sum_p c_{p,q,n} = \frac{1}{2\pi i}\int_\Gamma \frac{\Phi(x)}{x^{n+1}}dx,$$
where Γ is a circle whose centre is the origin and whose radius R is less than unity. We take

(5·16)
$$R = 1 - \frac{\beta}{n},$$
where β also is positive and independent of n.

Our object is to shew that the integral on the right-hand side of (5·15) is of the form $O(n^{-\frac{1}{4}})$; the constant implied in the O will of course be a function of α and β. It is to be understood throughout that O's are used in this sense; $O(1)$, for instance, stands for *a function of x, n, p, q, α, and β* (or of some only of these variables) *which is less in absolute value than a number $K = K(\alpha, \beta)$ independent of x, n, p, and q.*

We divide up the circle Γ, by means of the dissection Ξ_ν of 4·23, into arcs $\xi_{p,q}$ each associated with a point $Re^{2p\pi i/q}$; and we denote by $\eta_{p,q}$ the arc of Γ complementary to $\xi_{p,q}$. This being so, we have

(5·17)
$$\int_\Gamma \frac{\Phi(x)}{x^{n+1}}dx = \sum\int_{\xi_{p,q}} \frac{f(x) - F_{p,q}(x)}{x^{n+1}}dx - \sum\int_{\eta_{p,q}} \frac{F_{p,q}(x)}{x^{n+1}}dx$$
$$= \sum J_{p,q} - \sum j_{p,q},$$

say. We shall prove that each of these sums is of the form $O(n^{-\frac{1}{4}})$; and we shall begin with the second sum, which only involves the auxiliary functions F.

Proof that $\Sigma j_{p,q} = O(n^{-\frac{1}{4}})$.

5·21. We have, by Cauchy's theorem,

(5·211) $$j_{p,q} = \int_{\eta_{p,q}} \frac{F_{p,q}(x)}{x^{n+1}}\,dx = \int_{\zeta_{p,q}} \frac{F_{p,q}(x)}{x^{n+1}}\,dx,$$

where $\zeta_{p,q}$ consists of the contour $LMNM'L'$ shewn in the figure. Here L and L' are the ends of $\xi_{p,q}$, LM and $M'L'$ are radii vectores, and MNM' is part of a circle Γ_1 whose radius R_1 is greater than 1. P is the point $e^{2p\pi i/q}$; and we suppose that R_1 is small enough to ensure that all points of LM and $M'L'$ are at a distance from P less than $\frac{1}{2}$. The other circle c shewn in the figure has P as its centre and radius $\frac{1}{2}$. We denote LM by $\varpi_{p,q}$, $M'L'$ by $\varpi'_{p,q}$, and MNM' by $\gamma_{p,q}$: and we write

(5·212) $$j_{p,q} = \int_{\zeta_{p,q}} = \int_{\gamma_{p,q}} + \int_{\varpi_{p,q}} + \int_{\varpi'_{p,q}} = j^1_{p,q} + j^2_{p,q} + j^3_{p,q}.$$

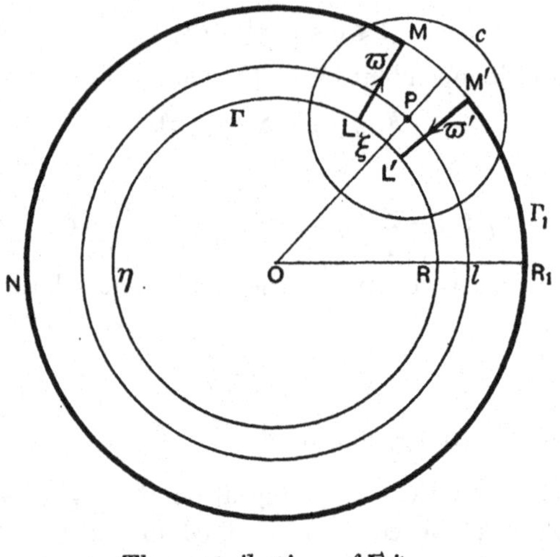

The contribution of $\Sigma j^1_{p,q}$.

5·22. Suppose first that x lies on $\gamma_{p,q}$ and outside c. Then, in virtue of (5·11) and Lemma 4·43, we have

(5·221) $$F_{p,q}(x) = O(\sqrt{q}).$$

If on the other hand x lies on $\gamma_{p,q}$, but inside c, we have, by (5·11) and Lemma 4·44,

(5·222) $$F_{p,q}(x) - \chi_{p,q}(x) = O(q^{-\frac{3}{2}}),$$

where

(5·2221) $$\chi_{p,q}(x) = \omega_{p,q}\frac{\sqrt{q}}{\pi\sqrt{2}}\chi_{C/q}(x_{p,q}).$$

But, if we recur to the definition of $\chi_a(x)$ in Lemma 4·42, and observe that

$$\left| \exp \frac{a^2}{4 \log (1/x)} \right| = \exp \frac{a^2 \log (1/r)}{4 [\{\log (1/r)\}^2 + \theta^2]} < 1$$

if $x = re^{i\theta}$ and $r > 1$, we see that

(5·223) $$\chi_{p,q}(x) = O(\sqrt{q})$$

on the part of $\gamma_{p,q}$ in question. Hence (5·221) holds for all $\gamma_{p,q}$. It follows that

$$j^1_{p,q} = O(R_1^{-n} \sqrt{q}),$$

(5·224) $$\Sigma j_{p,q} = O(R_1^{-n} \sum_q q^{\frac{3}{2}}) = O(n^{\frac{5}{4}} R_1^{-n})^*.$$

This sum tends to zero more rapidly than any power of n, and is therefore completely trivial.

The contributions of $\Sigma j^2_{p,q}$ *and* $\Sigma j^3_{p,q}$.

5·231. We must now consider the sums which arise from the integrals along $\varpi_{p,q}$ and $\varpi'_{p,q}$; and it is evident that we need consider in detail only the first of these two lines. We write

(5·2311)
$$j^2_{p,q} = \int_{\varpi_{p,q}} \frac{F_{p,q}(x) - \chi_{p,q}(x)}{x^{n+1}} dx + \int_{\varpi_{p,q}} \frac{\chi_{p,q}(x)}{x^{n+1}} dx = j'_{p,q} + j''_{p,q},$$

say.

In the first place we have, from (5·222),

$$j'_{p,q} = O\left(q^{-\frac{3}{2}} \int_R^{R_1} \frac{dr}{r^{n+1}}\right) = O(q^{-\frac{3}{2}} n^{-1}),$$

since

(5·2312) $$R^{-n} = \left(1 - \frac{\beta}{n}\right)^{-n} = O(1).$$

Thus

(5·2313) $$\Sigma j'_{p,q} = O\{n^{-1} \sum_{q < O(\sqrt{n})} q^{-\frac{1}{2}}\} = O(n^{-\frac{3}{4}}).$$

5·232. In the second place we have

$$j''_{p,q} = \omega_{p,q} \frac{\sqrt{q}}{\pi \sqrt{2}} \int_{\varpi_{p,q}} \frac{\chi_{C/q}(x_{p,q})}{x^{n+1}} dx.$$

It is plain that, if we substitute y for $xe^{-2p\pi i/q}$, then write x again for y, and finally substitute for $\chi_{C/q}$ its explicit expression as an elementary function, given in Lemma 4·42, we obtain

(5·2321)
$$j''_{p,q} = O(\sqrt{q}) \int \{E(x) - 1\} \sqrt{\left(\log \frac{1}{x}\right)}\, x^{-n-\frac{23}{24}} dx = O(\sqrt{q})\, J,$$

* Here, and in many passages in our subsequent argument, it is to be remembered that the number of values of p, corresponding to a given q, is less than q, and that the number of values of q is of order $\sqrt{n}$. Thus we have generally

$$\Sigma O(q^s) = O(\sum_{q < O(\sqrt{n})} q^{s+1}) = O(n^{\frac{1}{2}s+1}).$$

say, where

(5·23211) $$E(x) = \exp\left\{\frac{\pi^2}{6q^2 \log(1/x)}\right\},$$

and the path of integration is now a line related to $x = 1$ as $\varpi_{p,q}$ is to $x = e^{2p\pi i/q}$: the line defined by $x = re^{i\theta}$, where $R \leqslant r \leqslant R_1$, and θ is fixed and (by Lemma 4·22) lies between $1/2q\nu$ and $1/q\nu$.

Integrating J by parts, we find

(5·2322)

$$(n - \tfrac{1}{24}) J = -\left[\{E(x) - 1\} \sqrt{\left(\log\frac{1}{x}\right)} x^{-n+\frac{1}{24}}\right]_{r=R}^{r=R_1}$$

$$- \tfrac{1}{2}\int \{E(x) - 1\} \left(\log\frac{1}{x}\right)^{-\frac{1}{2}} x^{-n-\frac{23}{24}} dx$$

$$+ \frac{\pi^2}{6q^2}\int E(x) \left(\log\frac{1}{x}\right)^{-\frac{3}{2}} x^{-n-\frac{23}{24}} dx = J_1 + J_2 + J_3,$$

say.

5·233. In estimating J_1, J_2, and J_3, we must bear the following facts in mind.

(1) Since $|x| \geqslant R$, it follows from (5·2312) that $|x|^{-n} = O(1)$ throughout the range of integration.

(2) Since $1 - R = \beta/n$ and $1/2q\nu < \theta < 1/q\nu$, where $\nu = [\alpha\sqrt{n}]$, we have

$$\log\left(\frac{1}{x}\right) = O\left(\frac{1}{q\sqrt{n}}\right),$$

when $r = R$, and $$\frac{1}{\log(1/x)} = O(q\sqrt{n}),$$

throughout the range of integration.

(3) Since $$|E(x)| = \exp\frac{\pi^2 \log(1/r)}{6q^2[\{\log(1/r)\}^2 + \theta^2]},$$

$E(x)$ is less than 1 in absolute value when $r > 1$. And, on the part of the path for which $r < 1$, it is of the form

$$\exp O\left(\frac{1}{q^2 n\theta^2}\right) = \exp O(1) = O(1).$$

It is accordingly of the form $O(1)$ throughout the range of integration.

5·234. Thus we have, first

(5·2341)

$$J_1 = O(1)\, O(1)\, O(R_1^{-n}) + O(1)\, O(q^{-\frac{1}{2}} n^{-\frac{1}{4}})\, O(1) = O(q^{-\frac{1}{2}} n^{-\frac{1}{4}}),$$

secondly

(5·2342) $$J_2 = O(1)\, O(q^{\frac{1}{2}} n^{\frac{1}{4}}) \int_R^{R_1} \frac{dr}{r^{n+\frac{23}{24}}} = O(q^{\frac{1}{2}} n^{-\frac{3}{4}}),$$

and thirdly

$$(5{\cdot}2343)\qquad J_3 = O(q^{-2})\,O(1)\,O(q^{\frac{3}{2}}n^{\frac{3}{4}})\int_R^{R_1}\frac{dr}{r^{n+\frac{23}{24}}} = O(q^{-\frac{1}{2}}n^{-\frac{1}{4}}).$$

From (5·2341), (5·2342), (5·2343), and (5·2322), we obtain

$$J = O(q^{-\frac{1}{2}}n^{-\frac{5}{4}}) + O(q^{\frac{1}{2}}n^{-\frac{7}{4}});$$

and, from (5·2321), $\qquad j''_{p,q} = O(n^{-\frac{5}{4}}) + O(qn^{-\frac{7}{4}}).$

Summing, we obtain

$$(5{\cdot}2344)\qquad \Sigma j''_{p,q} = O\Big(n^{-\frac{5}{4}}\sum_{q<O(\surd n)} q\Big) + O\Big(n^{-\frac{7}{4}}\sum_{q<O(\surd n)} q^2\Big)$$
$$= O(n^{-\frac{1}{4}}) + O(n^{-\frac{1}{4}}) = O(n^{-\frac{1}{4}}).$$

5·235. From (5·2311), (5·2313), and (5·2344), we obtain

$$(5{\cdot}2351)\qquad \Sigma j^2_{p,q} = O(n^{-\frac{1}{4}});$$

and in exactly the same way we can prove

$$(5{\cdot}2352)\qquad \Sigma j^3_{p,q} = O(n^{-\frac{1}{4}}).$$

And from (5·212), (5·224), (5·2351), and (5·2352), we obtain, finally,

$$(5{\cdot}2353)\qquad \Sigma j_{p,q} = O(n^{-\frac{1}{4}}).$$

Proof that $\Sigma J_{p,q} = O(n^{-\frac{1}{4}})$.

5·31. We turn now to the discussion of

$$(5{\cdot}311)\qquad J_{p,q} = \int_{\xi_{p,q}} \frac{f(x) - F_{p,q}(x)}{x^{n+1}}\,dx$$
$$= \int_{\xi_{p,q}} \frac{f(x) - X_{p,q}(x)}{x^{n+1}}\,dx - \int_{\xi_{p,q}} \frac{F_{p,q}(x) - \chi_{p,q}(x)}{x^{n+1}}\,dx + \int_{\xi_{p,q}} \frac{\rho_{p,q}(x)}{x^{n+1}}\,dx$$
$$= J^1_{p,q} + J^2_{p,q} + J^3_{p,q},$$

say, where $\qquad \rho_{p,q}(x) = \omega_{p,q}\sqrt{\left(\dfrac{q}{2\pi}\log\dfrac{1}{x_{p,q}}\right)}\,x_{p,q}^{\frac{1}{24}},$

$$X_{p,q}(x) = \chi_{p,q}(x) + \rho_{p,q}(x) = \rho_{p,q}(x)\,E(x_{p,q}),$$

$E(x)$ being defined as in (5·23211).

Discussion of $\Sigma J^2_{p,q}$ and $\Sigma J^3_{p,q}$.

5·32. The discussion of these two sums is, after the analysis which precedes, a simple matter. The arc $\xi_{p,q}$ is less than a constant multiple of $1/q\surd n$; and $x^{-n} = O(1)$ on $\xi_{p,q}$. Also

$$|F_{p,q}(x) - \chi_{p,q}(x)| = O(q^{-\frac{3}{2}}),$$

by (5·222); and

$$(5{\cdot}321)\qquad \sqrt{\left(\log\frac{1}{x_{p,q}}\right)} = O(q^{-\frac{1}{2}}n^{-\frac{1}{4}}),$$

since $\qquad |x_{p,q}| = R = 1 - (\beta/n), \quad |am\ x_{p,q}| < 1/q\nu.$

Hence
$$J^2_{p,q} = O(q^{-\frac{5}{2}} n^{-\frac{1}{2}}),$$

(5·322)
$$\Sigma J^2_{p,q} = O(n^{-\frac{1}{2}} \sum_{q < O(\sqrt{n})} q^{-\frac{3}{2}}) = O(n^{-\frac{1}{2}});$$

and
$$J^3_{p,q} = O(q^{-1} n^{-\frac{3}{4}}),$$

(5·323)
$$\Sigma J^3_{p,q} = O(n^{-\frac{3}{4}} \sum_{q < O(\sqrt{n})} 1) = O(n^{-\frac{1}{4}}).$$

Discussion of $\Sigma J^1_{p,q}$.

5·33. From (4·321) and (5·2221), we have

(5·331)
$$f(x) - X_{p,q}(x) = \omega_{p,q} \sqrt{\left\{\frac{q}{2\pi} \log\left(\frac{1}{x_{p,q}}\right)\right\}}\, x_{p,q}^{\frac{1}{24}} E(x_{p,q})\, \Omega(x'_{p,q}),$$

where
$$\Omega(z) = f(z) - 1 = \prod_1^\infty \left(\frac{1}{1 - z^\nu}\right) - 1 = \sum_1^\infty p(\nu) z^\nu,$$

if $|z| < 1$, and
$$x'_{p,q} = \exp\left\{-\frac{4\pi^2}{q^2 \log(1/x_{p,q})} + \frac{2\pi i p'}{q}\right\}.$$

Now
$$|x'_{p,q}| = \exp\left[-\frac{4\pi^2 \log(1/R)}{q^2 \{[\log(1/R)]^2 + \theta^2\}}\right],$$

where θ is the amplitude of $x_{p,q}$. Also
$$q^2 \{[\log(1/R)]^2 + \theta^2\} = O\left\{q^2 \left(\frac{1}{n^2} + \frac{1}{q^2 n}\right)\right\} = O\left(\frac{1}{n}\right),$$

while $\log(1/R)$ is greater than a constant multiple of $1/n$. There is therefore a positive number δ, less than unity and independent of n and of q, such that
$$|x'_{p,q}| < \delta;$$

and we may write
$$\Omega(x'_{p,q}) = O(|x'_{p,q}|).$$

We have therefore
$$E(x_{p,q})\, \Omega(x'_{p,q}) = O(|x'_{p,q}|^{-\frac{1}{24}})\, O(|x'_{p,q}|) = O(|x'_{p,q}|^{\frac{23}{24}}) = O(1);$$

and so, by (5·321),
$$f(x) - \chi_{p,q}(x) = O(\sqrt{q})\, O\left(\sqrt{\left|\log \frac{1}{x_{p,q}}\right|}\right) O(1) = O(n^{-\frac{1}{4}}).$$

And hence, as the length of $\xi_{p,q}$ is of the form $O(1/q\sqrt{n})$, we obtain
$$J^1_{p,q} = O(q^{-1} n^{-\frac{3}{4}}),$$

(5·332)
$$\Sigma J^1_{p,q} = O(n^{-\frac{3}{4}} \sum_{q < O(\sqrt{n})} 1) = O(n^{-\frac{1}{4}}).$$

5·34. From (5·311), (5·322), (5·323), and (5·332), we obtain

(5·341)
$$\Sigma J_{p,q} = O(n^{-\frac{1}{4}}).$$

Completion of the proof.

5·4. From (5·15), (5·17), (5·2353), and (5·341), we obtain

(5·41) $$p(n) - \sum_q \sum_p c_{p,q,n} = O(n^{-\frac{1}{4}}).$$

But $$\sum_p c_{p,q,n} = \frac{\sqrt{q}}{\pi\sqrt{2}} A_q \frac{d}{dn} \frac{\cosh(C\lambda_n/q) - 1}{\lambda_n},$$

where $$A_q = \sum_p \omega_{p,q} e^{-2np\pi i/q}.$$

All that remains, in order to complete the proof of the theorem, is to shew that

$$\cosh(C\lambda_n/q) - 1$$

may be replaced by $$\tfrac{1}{2} e^{C\lambda_n/q};$$

and in order to prove this it is only necessary to shew that

$$\sum_{q<O(\sqrt{n})} q^{\frac{3}{2}} \frac{d}{dn} \frac{\frac{1}{2}e^{C\lambda_n/q} - \cosh(C\lambda_n/q) + 1}{\lambda_n} = O(n^{-\frac{1}{4}}).$$

On differentiating we find that the sum is of the form

$$\sum_{q<O(\sqrt{n})} q^{\frac{3}{2}} \left\{ O\left(\frac{1}{qn}\right) + O\left(\frac{1}{n^{\frac{3}{2}}}\right) \right\} = O\left\{ \frac{1}{n} \sum_{q<O(\sqrt{n})} q^{\frac{1}{2}} \right\} = O(n^{-\frac{1}{4}}).$$

Thus the theorem is proved.

6. Additional remarks on the theorem.

6·1. The theorem which we have proved gives information about $p(n)$ which is in some ways extraordinarily exact. We are for this reason the more anxious to point out explicitly two respects in which the results of our analysis are incomplete.

6·21. We have proved that

$$p(n) = \sum A_q \phi_q + O(n^{-\frac{1}{4}}),$$

where the summation extends over the values of q specified in the theorem, for every fixed value of α; that is to say that, when α is given, a number $K = K(\alpha)$ can be found such that

$$|p(n) - \sum A_q \phi_q| < Kn^{-\frac{1}{4}}$$

for every value of n. It follows that

(6·211) $$p(n) = \{\sum A_q \phi_q\},$$

where $\{x\}$ denotes the integer nearest to x, for $n \geqslant n_0$, where $n_0 = n_0(\alpha)$ is a certain function of α.

The question remains whether we can, by an appropriate choice of α, secure the truth of (6·211) for *all* values of n, and not merely for all sufficiently large values. Our opinion is that this is possible, and that it could be proved to be possible without any fundamental change in our analysis. Such a proof would however involve a very careful revision of our argument. It would be necessary to replace all formulæ involving O's by inequalities, containing only numbers expressed explicitly as functions of the various parameters employed. This process would certainly add very considerably to the length and the complexity of our argument. It is, as it stands, sufficient to prove what is, from our point of view, of the greatest interest; and we have not thought it worth while to elaborate it further.

6·22. The second point of incompleteness of our results is of much greater interest and importance. We have not proved either that the series

$$\sum_1^\infty A_q \phi_q$$

is convergent, or that, if it is convergent, it represents $p(n)$. Nor does it seem likely that our method is one intrinsically capable of proving these results, if they are true—a point on which we are not prepared to express any definite opinion.

It should be observed in this connection that we have not even discovered anything definite concerning the order of magnitude of A_q for large values of q. We can prove nothing better than the absolutely trivial equation $A_q = O(q)$. On the other hand we cannot assert that A_q is, for an infinity of values of q, effectively of an order as great as q, or indeed even that it does not tend to zero (though of course this is most unlikely).

6·3. Our formula directs us, if we wish to obtain the exact value of $p(n)$ for a large value of n, to take a number of terms of order $\sqrt{n}$. The numerical data suggest that a considerably smaller number of terms will be equally effective; and it is easy to see that this conjecture is correct.

Let us write $\beta = 4\pi\sqrt{(\frac{2}{3})} = 4C$, $\mu = \left[\dfrac{\beta\sqrt{n}}{\log n}\right]$,

and let us suppose that $\alpha < 2$. Then

$$\sum_{\mu+1}^{\nu} A_q\phi_q = \sum_{\mu+1}^{\nu} O(q^{\frac{3}{2}})\, O\left(\frac{1}{qn}\right) O\left(e^{C\sqrt{n}/q}\right) = O\left(\frac{1}{n}\sum_{\mu+1}^{\nu}\sqrt{q}\, e^{C\sqrt{n}/q}\right)$$

$$= O\left(\frac{1}{n}\int_\mu^\nu \sqrt{x}\, e^{C\sqrt{n}/x}\, dx\right),$$

since $\sqrt{q}e^{C\sqrt{n}/q}$ decreases steadily throughout the range of summation*.

Writing $\sqrt{n}/y$ for x, we obtain

$$O\left(n^{-\frac{1}{4}}\int_{1/\alpha}^{\sqrt{n}/\mu} y^{-\frac{5}{2}}\, e^{Cy}\, dy\right) = O\left\{n^{-\frac{1}{4}}\left(\frac{\sqrt{n}}{\mu}\right)^{-\frac{5}{2}} e^{C\sqrt{n}/\mu}\right\} = O\left\{n^{-\frac{1}{4}}(\log n)^{-\frac{5}{2}} e^{\frac{1}{4}\log n}\right\}$$

$$= O(\log n)^{-\frac{5}{2}} = o(1).$$

* The minimum occurs when q is about equal to $2C\sqrt{n}$.

It follows that it is enough, when n is sufficiently large, to take

$$\left[\frac{\beta\sqrt{n}}{\log n}\right]$$

terms of the series. It is probably also *necessary* to take a number of terms of order $\sqrt{n}/(\log n)$; but it is not possible to prove this rigorously without a more exact knowledge of the properties of A_q than we possess.

6·4. We add a word on certain simple approximate formulæ for $\log p(n)$ found empirically by Major MacMahon and by ourselves. Major MacMahon found that if

$$\log_{10} p(n) = \sqrt{(n+4)} - a_n, \tag{6·41}$$

then a_n is approximately equal to 2 within the limits of his table of values of $p(n)$ (Table IV). This suggested to us that we should endeavour to find more accurate formulæ of the same type. The most striking that we have found is

$$\log_{10} p(n) = \tfrac{10}{9}\{\sqrt{(n+10)} - a_n\}; \tag{6·42}$$

the mode of variation of a_n is shewn in Table III.

In this connection it is interesting to observe that the function

$$13^{-\sqrt{n}}p(n)$$

(which ultimately tends to infinity with exponential rapidity) is equal to ·973 for $n = 30000000000$.

7. Further applications of the method.

7·1. We shall conclude with a few remarks concerning actual or possible applications of our method to other problems in Combinatory Analysis or the Analytic Theory of Numbers.

The class of problems in which the method gives the most striking results may be defined as follows. Suppose that $q(n)$ is the coefficient of x^n in the expansion of $F(x)$, where $F(x)$ is a function of the form

$$\frac{\{f(\pm x^a)\}^\alpha \{f(\pm x^{a'})\}^{\alpha'} \dots}{\{f(\pm x^b)\}^\beta \{f(\pm x^{b'})\}^{\beta'} \dots} \text{*}; \tag{7·11}$$

$f(x)$ being the function considered in this paper, the a's, b's, α's, and β's being positive integers, and the number of factors in numerator and denominator being finite; and suppose that $|F(x)|$ tends exponentially to infinity when x tends in an appropriate manner to some or all of the points $e^{2p\pi i/q}$. Then our method may be applied in its full power to the asymptotic study of $q(n)$, and yields results very similar to those which we have found concerning $p(n)$.

* Since
$$f(-x) = \frac{\{f(x^2)\}^3}{f(x)f(x^4)},$$
the arguments with a negative sign may be eliminated if this is desired.

Thus, if

$$F(x) = \frac{f(x)}{f(x^2)} = (1+x)(1+x^2)(1+x^3)\ldots = \frac{1}{(1-x)(1-x^3)(1-x^5)\ldots},$$

so that $q(n)$ is the number of partitions of n into odd parts, or into unequal parts*, we find that

$$q(n) = \frac{1}{\sqrt{2}}\frac{d}{dn} J_0[i\pi\sqrt{\{\tfrac{1}{3}(n+\tfrac{1}{24})\}}]$$
$$+\sqrt{2}\cos(\tfrac{2}{3}n\pi - \tfrac{1}{9}\pi)\frac{d}{dn} J_0[\tfrac{1}{3}i\pi\sqrt{\{\tfrac{1}{3}(n+\tfrac{1}{24})\}}]+\ldots.$$

The error after $[\alpha\sqrt{n}]$ terms is of the form $O(1)$. We are not in a position to assert that the *exact* value of $q(n)$ can always be obtained from the formula (though this is probable); but the error is certainly bounded.

If $$F(x) = \frac{f(x^2)}{f(-x)} = \frac{f(x)f(x^4)}{\{f(x^2)\}^2} = (1+x)(1+x^3)(1+x^5)\ldots,$$

so that $q(n)$ is the number of partitions of n into parts which are both odd and unequal, then

$$q(n) = \frac{d}{dn} J_0[i\pi\sqrt{\{\tfrac{1}{6}(n-\tfrac{1}{24})\}}]$$
$$+2\cos(\tfrac{2}{3}n\pi - \tfrac{2}{9}\pi)\frac{d}{dn} J_0[\tfrac{1}{3}i\pi\sqrt{\{\tfrac{1}{6}(n-\tfrac{1}{24})\}}]+\ldots.$$

The error is again bounded (and probably tends to zero).

If $$F(x) = \frac{\{f(x)\}^2}{f(x^2)} = \frac{1}{1-2x+2x^4-2x^9+\ldots},$$

$q(n)$ has no very simple arithmetical interpretation; but the series is none the less, as the direct reciprocal of a simple ϑ-function, of particular interest. In this case we find

$$q(n) = \frac{1}{4\pi}\frac{d}{dn}\frac{e^{\pi\sqrt{n}}}{\sqrt{n}} + \frac{\sqrt{3}}{2\pi}\cos(\tfrac{2}{3}n\pi - \tfrac{1}{6}\pi)\frac{d}{dn}\frac{e^{\frac{1}{3}\pi\sqrt{n}}}{\sqrt{n}}+\ldots.$$

The error here is (as in the partition problem) of order $O(n^{-\frac{1}{4}})$, and the exact value can always be found from the formula.

7·2. The method may also be applied to products of the form (7·11) which have (to put the matter roughly) no exponential infinities. In such cases the approximation is of a much less exact character. On the other hand the problems of this character are of even greater arithmetical interest.

The standard problem of this category is that of the representation of a number as the sum of s squares, s being any positive integer odd or even†. We must reserve the application of our method to this problem for another occasion; but we can indicate the character of our main result as follows.

* Cf. MacMahon, *loc. cit.*, p. 11. We give at the end of the paper a table (Table V) of the values of $q(n)$ up to $n=100$. This table was calculated by Mr Darling.

† As is well known, the arithmetical difficulties of the problem are much greater when s is odd.

If $r_s(n)$ is the number of representations of n as the sum of s squares, we have

$$F(x) = \Sigma r_s(n) x^n = (1 + 2x + 2x^4 + \ldots)^s = \frac{\{f(x^2)\}^s}{\{f(-x)\}^{2s}} = \frac{\{f(x)\}^{2s}\{f(x^4)\}^{2s}}{\{f(x^2)\}^{5s}}.$$

We find that

$$(7{\cdot}21) \qquad r_s(n) = \frac{\pi^{\frac{1}{2}s}}{\Gamma(\frac{1}{2}s)} n^{\frac{1}{2}s-1} \Sigma \frac{c_q}{q^{\frac{1}{2}s}} + O(n^{\frac{1}{4}s}),$$

where c_q is a function of q and of n of the same general type as the function A_q of this paper. The series

$$(7{\cdot}22) \qquad \Sigma \frac{c_q}{q^{\frac{1}{2}s}}$$

is absolutely convergent for sufficiently large values of s, and the summation in (7·21) may be regarded indifferently as extended over all values of q or only over a range $1 \leqslant q \leqslant \alpha\sqrt{n}$. It should be observed that the series (7·22) is quite different in form from any of the infinite series which are already known to occur in connection with this problem.

7·3. There is also a wide range of problems to which our methods are *partly* applicable. Suppose, for example, that

$$F(x) = \Sigma p^2(n) x^n = \frac{1}{(1-x)(1-x^4)(1-x^9)\ldots},$$

so that $p^2(n)$ is the number of partitions of n into *squares*. Then $F(x)$ is not an elliptic modular function; it possesses no general transformation theory: and the full force of our method cannot be applied. We can still, however, apply some of our preliminary methods. Thus the "Tauberian" argument shews that

$$\log p^2(n) \sim 2^{-\frac{4}{3}} 3\pi^{\frac{1}{3}} \{\zeta(\tfrac{3}{2})\}^{\frac{2}{3}} n^{\frac{1}{3}}.$$

And although there is no general transformation theory, there is a formula which enables us to specify the nature of the singularity at $x = 1$. This formula is

$$\frac{1}{f(e^{-\pi z})} = 2\sqrt{\left(\frac{\pi}{z}\right)} \exp\left\{\frac{2\pi}{\sqrt{z}}\zeta(-\tfrac{1}{2})\right\}$$
$$\times \prod_{1}^{\infty} \{1 - 2e^{-2\pi\sqrt{(n/z)}} \cos 2\pi\sqrt{(n/z)} + e^{-4\pi\sqrt{(n/z)}}\}.$$

By the use of this formula, in conjunction with Cauchy's theorem, it is certainly possible to obtain much more precise information about $p^2(n)$, and in particular the formula

$$p^2(n) \sim 3^{-\frac{1}{2}} (4\pi n)^{-\frac{7}{6}} \{\zeta(\tfrac{3}{2})\}^{\frac{2}{3}} e^{2^{-\frac{4}{3}} 3\pi^{\frac{1}{3}} \{\zeta(\frac{3}{2})\}^{\frac{2}{3}} n^{\frac{1}{3}}}.$$

The corresponding formula for $p^s(n)$, the number of partitions of n into perfect s-th powers, is

$$p^s(n) \sim (2\pi)^{-\frac{1}{2}(s+1)} \sqrt{\left(\frac{s}{s+1}\right)} k n^{\frac{1}{s+1} - \frac{3}{2}} e^{(s+1) k n^{1/(s+1)}},$$

where

$$k = \left\{\frac{1}{s}\Gamma\left(1 + \frac{1}{s}\right)\zeta\left(1 + \frac{1}{s}\right)\right\}^{\frac{s}{s+1}}.$$

The series (7·21) may be written in the form

$$\frac{\pi^{\frac{1}{2}s}}{\Gamma(\frac{1}{2}s)} n^{\frac{1}{2}s-1} \sum_{p,q} \frac{\omega^s_{p,q}}{q^{\frac{1}{2}s}} e^{-np\pi i/q},$$

where $\omega_{p,q}$ is always one of the five numbers 0, $e^{\frac{1}{4}\pi i}$, $e^{-\frac{1}{4}\pi i}$, $-e^{\frac{1}{4}\pi i}$, $-e^{-\frac{1}{4}\pi i}$. When s is even it begins

$$\frac{\pi^{\frac{1}{2}s}}{\Gamma(\frac{1}{2}s)} n^{\frac{1}{2}s-1} \{1^{-\frac{1}{2}s} + 2\cos(\tfrac{1}{2}n\pi - \tfrac{1}{4}s\pi)\, 2^{-\frac{1}{2}s} + 2\cos(\tfrac{2}{3}n\pi - \tfrac{1}{2}s\pi)\, 3^{-\frac{1}{2}s} + \ldots\}.$$

It has been proved by Ramanujan that the series gives an *exact* representation of $r_s(n)$ when $s = 4, 6, 8$; and by Hardy that this is also true when $s = 3, 5, 7$. See Ramanujan, "On certain trigonometrical sums and their applications in the Theory of Numbers"; Hardy, "On the expression of a number as the sum of any number of squares, and in particular of five or seven*."

TABLE I $\omega_{p,q}$.

p	q	$\log \omega_{p,q}/\pi i$	p	q	$\log \omega_{p,q}/\pi i$	p	q	$\log \omega_{p,q}/\pi i$
1	1	0	3	11	3/22	8	15	7/18
1	2	0	4	"	3/22	11	"	−19/90
1	3	1/18	5	"	−5/22	13	"	−7/18
2	"	−1/18	6	"	5/22	14	"	−1/90
1	4	1/8	7	"	−3/22	1	16	−29/32
3	"	−1/8	8	"	−3/22	3	"	−27/32
1	5	1/5	9	"	−5/22	5	"	−5/32
2	"	0	10	"	−15/22	7	"	−3/32
3	"	0	1	12	55/72	9	"	3/32
4	"	−1/5	5	"	−1/72	11	"	5/32
1	6	5/18	7	"	1/72	13	"	27/32
5	"	−5/18	11	"	−55/72	15	"	29/32
1	7	5/14	1	13	11/13	1	17	−14/17
2	"	1/14	2	"	4/13	2	"	8/17
3	"	−1/14	3	"	1/13	3	"	5/17
4	"	1/14	4	"	−1/13	4	"	0
5	"	−1/14	5	"	0	5	"	1/17
6	"	−5/14	6	"	−4/13	6	"	5/17
1	8	7/16	7	"	4/13	7	"	1/17
3	"	1/16	8	"	0	8	"	−8/17
5	"	−1/16	9	"	1/13	9	"	8/17
7	"	−7/16	10	"	−1/13	10	"	−1/17
1	9	14/27	11	"	−4/13	11	"	−5/17
2	"	4/27	12	"	−11/13	12	"	−1/17
4	"	−4/27	1	14	13/14	13	"	0
5	"	4/27	3	"	3/14	14	"	−5/17
7	"	−4/27	5	"	3/14	15	"	−8/17
8	"	−14/27	9	"	−3/14	16	"	14/17
1	10	3/5	11	"	−3/14	1	18	−20/27
3	"	0	13	"	−13/14	5	"	2/27
7	"	0	1	15	1/90	7	"	−2/27
9	"	−3/5	2	"	7/18	11	"	2/27
1	11	15/22	4	"	19/90	13	"	−2/27
2	"	5/22	7	"	−7/18	17	"	20/27

* [Ramanujan's paper referred to is No. 21 of this volume. That of Hardy was published, in the first instance, in *Proc. National Acad. of Sciences* (Washington), Vol. IV, 1918, pp. 189—193, and later (in fuller form and with a slightly different title) in *Trans. American Math. Soc.*, Vol. XXI, 1920, pp. 255—284.]

TABLE II: A_q.

$A_1 = 1.$

$A_2 = \cos n\pi.$

$A_3 = 2\cos(\frac{2}{3}n\pi - \frac{1}{18}\pi).$

$A_4 = 2\cos(\frac{1}{2}n\pi - \frac{1}{8}\pi).$

$A_5 = 2\cos(\frac{2}{5}n\pi - \frac{1}{5}\pi) + 2\cos\frac{4}{5}n\pi.$

$A_6 = 2\cos(\frac{1}{3}n\pi - \frac{5}{18}\pi).$

$A_7 = 2\cos(\frac{2}{7}n\pi - \frac{5}{14}\pi) + 2\cos(\frac{4}{7}n\pi - \frac{1}{14}\pi) + 2\cos(\frac{6}{7}n\pi + \frac{1}{14}\pi).$

$A_8 = 2\cos(\frac{1}{4}n\pi - \frac{7}{16}\pi) + 2\cos(\frac{3}{4}n\pi - \frac{1}{16}\pi).$

$A_9 = 2\cos(\frac{2}{9}n\pi - \frac{14}{27}\pi) + 2\cos(\frac{4}{9}n\pi - \frac{4}{27}\pi) + 2\cos(\frac{8}{9}n\pi + \frac{4}{27}\pi).$

$A_{10} = 2\cos(\frac{1}{5}n\pi - \frac{3}{5}\pi) + 2\cos\frac{3}{5}n\pi.$

$A_{11} = 2\cos(\frac{2}{11}n\pi - \frac{15}{22}\pi) + 2\cos(\frac{4}{11}n\pi - \frac{5}{22}\pi) + 2\cos(\frac{6}{11}n\pi - \frac{3}{22}\pi) + 2\cos(\frac{8}{11}n\pi - \frac{3}{22}\pi)$
$+ 2\cos(\frac{10}{11}n\pi + \frac{5}{22}\pi).$

$A_{12} = 2\cos(\frac{1}{6}n\pi - \frac{55}{72}\pi) + 2\cos(\frac{5}{6}n\pi + \frac{1}{72}\pi).$

$A_{13} = 2\cos(\frac{2}{13}n\pi - \frac{11}{13}\pi) + 2\cos(\frac{4}{13}n\pi - \frac{4}{13}\pi) + 2\cos(\frac{6}{13}n\pi - \frac{1}{13}\pi) + 2\cos(\frac{8}{13}n\pi + \frac{1}{13}\pi)$
$+ 2\cos\frac{10}{13}n + 2\cos(\frac{12}{13}n\pi + \frac{4}{13}\pi).$

$A_{14} = 2\cos(\frac{1}{7}n\pi - \frac{13}{14}\pi) + 2\cos(\frac{3}{7}n\pi - \frac{3}{14}\pi) + 2\cos(\frac{5}{7}n\pi - \frac{3}{14}\pi).$

$A_{15} = 2\cos(\frac{2}{15}n\pi - \frac{1}{90}\pi) + 2\cos(\frac{4}{15}n\pi - \frac{7}{18}\pi) + 2\cos(\frac{8}{15}n\pi - \frac{19}{90}\pi) + 2\cos(\frac{14}{15}n\pi + \frac{7}{18}\pi).$

$A_{16} = 2\cos(\frac{1}{8}n\pi + \frac{29}{32}\pi) + 2\cos(\frac{3}{8}n\pi + \frac{27}{32}\pi) + 2\cos(\frac{5}{8}n\pi + \frac{5}{32}\pi) + 2\cos(\frac{7}{8}n\pi + \frac{3}{32}\pi).$

$A_{17} = 2\cos(\frac{2}{17}n\pi + \frac{14}{17}\pi) + 2\cos(\frac{4}{17}n\pi - \frac{8}{17}\pi) + 2\cos(\frac{6}{17}n\pi - \frac{5}{17}\pi) + 2\cos\frac{8}{17}n\pi$
$+ 2\cos(\frac{10}{17}n\pi - \frac{1}{17}\pi) + 2\cos(\frac{12}{17}n\pi - \frac{5}{17}n\pi) + 2\cos(\frac{14}{17}n\pi - \frac{1}{17}\pi) + 2\cos(\frac{16}{17}n\pi + \frac{8}{17}\pi).$

$A_{18} = 2\cos(\frac{1}{9}n\pi + \frac{20}{27}\pi) + 2\cos(\frac{5}{9}n\pi - \frac{2}{27}\pi) + 2\cos(\frac{7}{9}n\pi + \frac{2}{27}\pi).$

It may be observed that

$A_5 = 0\ (n \equiv 1, 2 \pmod 5)$, $A_7 = 0\ (n \equiv 1, 3, 4 \pmod 7)$,
$A_{10} = 0\ (n \equiv 1, 2 \pmod 5)$, $A_{11} = 0\ (n \equiv 1, 2, 3, 5, 7 \pmod{11})$,
$A_{13} = 0\ (n \equiv 2, 3, 5, 7, 9, 10 \pmod{13})$, $A_{14} = 0\ (n \equiv 1, 3, 4 \pmod 7)$,
$A_{16} = 0\ (n \equiv 0 \pmod 2)$, $A_{17} = 0\ (n \equiv 1, 3, 4, 6, 7, 9, 13, 14 \pmod{17})$;

while A_1, A_2, A_3, A_4, A_6, A_8, A_9, A_{12}, A_{15}, and A_{18} never vanish.

TABLE III: $\log_{10} p(n) = \frac{10}{9}\{\sqrt{(n+10)} - a_n\}$.

n	a_n	n	a_n
1	3·317	10000	4·148
3	3·176	30000	4·364
10	3·011	100000	4·448
30	2·951	300000	4·267
100	3·036	1000000	3·554
300	3·237	3000000	2·072
1000	3·537	10000000	−1·188
3000	3·838	30000000	−6·796
		∞	$-\infty$

TABLE IV*: $p(n)$.

1...	1	51...	239943	101...	214481126	151...	45060624582
2...	2	52...	281589	102...	241265379	152...	49686288421
3...	3	53...	329931	103...	271248950	153...	54770336324
4...	5	54...	386155	104...	304801365	154...	60356673280
5...	7	55...	451276	105...	342325709	155...	66493182097
6...	11	56...	526823	106...	384276336	156...	73232243759
7...	15	57...	614154	107...	431149389	157...	80630964769
8...	22	58...	715220	108...	483502844	158...	88751778802
9...	30	59...	831820	109...	541946240	159...	97662728555
10...	42	60...	966467	110...	607163746	160...	107438159466
11...	56	61...	1121505	111...	679903203	161...	118159068427
12...	77	62...	1300156	112...	761002156	162...	129913904637
13...	101	63...	1505499	113...	851376628	163...	142798995930
14...	135	64...	1741630	114...	952050665	164...	156919475295
15...	176	65...	2012558	115...	1064144451	165...	172389800255
16...	231	66...	2323520	116...	1188908248	166...	189334822579
17...	297	67...	2679689	117...	1327710076	167...	207890420102
18...	385	68...	3087735	118...	1482074143	168...	228204732751
19...	490	69...	3554345	119...	1653668665	169...	250438925115
20...	627	70...	4087968	120...	1844349560	170...	274768617130
21...	792	71...	4697205	121...	2056148051	171...	301384802048
22...	1002	72...	5392783	122...	2291320912	172...	330495499613
23...	1255	73...	6185689	123...	2552338241	173...	362326859895
24...	1575	74...	7089500	124...	2841940500	174...	397125074750
25...	1958	75...	8118264	125...	3163127352	175...	435157697830
26...	2436	76...	9289091	126...	3519222692	176...	476715857290
27...	3010	77...	10619863	127...	3913864295	177...	522115831195
28...	3718	78...	12132164	128...	4351078600	178...	571701605655
29...	4565	79...	13848650	129...	4835271870	179...	625846753120
30...	5604	80...	15796476	130...	5371315400	180...	684957390936
31...	6842	81...	18004327	131...	5964539504	181...	749474411781
32...	8349	82...	20506255	132...	6620830889	182...	819876908323
33...	10143	83...	23338469	133...	7346629512	183...	896684817527
34...	12310	84...	26543660	134...	8149040695	184...	980462880430
35...	14883	85...	30167357	135...	9035836076	185...	1071823774337
36...	17977	86...	34262962	136...	10015581680	186...	1171432692373
37...	21637	87...	38887673	137...	11097645016	187...	1280011042268
38...	26015	88...	44108109	138...	12292341831	188...	1398341745571
39...	31185	89...	49995925	139...	13610949895	189...	1527273599625
40...	37338	90...	56634173	140...	15065878135	190...	1667727404093
41...	44583	91...	64112359	141...	16670689208	191...	1820701100652
42...	53174	92...	72533807	142...	18440293320	192...	1987276856363
43...	63261	93...	82010177	143...	20390982757	193...	2168627105469
44...	75175	94...	92669720	144...	22540654445	194...	2366022741845
45...	89134	95...	104651419	145...	24908858009	195...	2580840212973
46...	105558	96...	118114304	146...	27517052599	196...	2814570987591
47...	124754	97...	133230930	147...	30388671978	197...	3068829878530
48...	147273	98...	150198136	148...	33549419497	198...	3345365983698
49...	173525	99...	169229875	149...	37027355200	199...	3646072432125
50...	204226	100...	190569292	150...	40853235313	200...	3972999029388

* The numbers in this table were calculated by Major MacMahon, by means of the recurrence formulæ obtained by equating coefficients in the identity

$$(1-x-x^2+x^5+x^7-x^{12}-x^{15}+\dots)\sum_0^\infty p(n)\,x^n=1.$$

We have verified the table by direct calculation up to $n=158$. Our calculation of $p(200)$ from the asymptotic formula then seemed to render further verification unnecessary.

TABLE V*: $q(n)$.

n	c_n	n	c_n	n	c_n	n	c_n
1...	1	26...	165	51...	4097	76...	53250
2...	1	27...	192	52...	4582	77...	58499
3...	2	28...	222	53...	5120	78...	64234
4...	2	29...	256	54...	5718	79...	70488
5...	3	30...	296	55...	6378	80...	77312
6...	4	31...	340	56...	7108	81...	84756
7...	5	32...	390	57...	7917	82...	92864
8...	6	33...	448	58...	8808	83...	101698
9...	8	34...	512	59...	9792	84...	111322
10...	10	35...	585	60...	10880	85...	121792
11...	12	36...	668	61...	12076	86...	133184
12...	15	37...	760	62...	13394	87...	145578
13...	18	38...	864	63...	14848	88...	159046
14...	22	39...	982	64...	16444	89...	173682
15...	27	40...	1113	65...	18200	90...	189586
16...	32	41...	1260	66...	20132	91...	206848
17...	38	42...	1426	67...	22250	92...	225585
18...	46	43...	1610	68...	24576	93...	245920
19...	54	44...	1816	69...	27130	94...	267968
20...	64	45...	2048	70...	29927	95...	291874
21...	76	46...	2304	71...	32992	96...	317788
22...	89	47...	2590	72...	36352	97...	345856
23...	104	48...	2910	73...	40026	98...	376256
24...	122	49...	3264	74...	44046	99...	409174
25...	142	50...	3658	75...	48446	100...	444793

* We are indebted to Mr Darling for this table.

37

ON THE COEFFICIENTS IN THE EXPANSIONS OF CERTAIN MODULAR FUNCTIONS

(*Proceedings of the Royal Society*, A, xcv, 1919, 144—155)

1. A very large proportion of the most interesting arithmetical functions —of the functions, for example, which occur in the theory of partitions, the theory of the divisors of numbers, or the theory of the representation of numbers by sums of squares—occur as the coefficients in the expansions of elliptic modular functions in powers of the variable $q = e^{\pi i \tau}$. All of these functions have a restricted region of existence, the unit circle $|q| = 1$ being a "natural boundary" or line of essential singularities. The most important of them, such as the functions*

$$(1{\cdot}1) \qquad (\omega_1/\pi)^{12}\Delta = q^2\{(1-q^2)(1-q^4)\ldots\}^{24},$$

$$(1{\cdot}2) \qquad \vartheta_3(0) = 1 + 2q + 2q^4 + 2q^9 + \ldots,$$

$$(1{\cdot}3) \qquad 12\left(\frac{\omega_1}{\pi}\right)^4 g_2 = 1 + 240\left(\frac{1^3q^2}{1-q^2} + \frac{2^3q^4}{1-q^4} + \ldots\right),$$

$$(1{\cdot}4) \qquad 216\left(\frac{\omega_1}{\pi}\right)^6 g_3 = 1 - 504\left(\frac{1^5q^2}{1-q^2} + \frac{2^5q^4}{1-q^4} + \ldots\right),$$

are regular inside the unit circle; and many, such as the functions (1·1) and (1·2), have the additional property of having no zeros inside the circle, so that their reciprocals are also regular.

In a series of recent papers† we have applied a new method to the study of these arithmetical functions. Our aim has been to express them as series which exhibit explicitly their order of magnitude, and the genesis of their irregular variations as n increases. We find, for example, for $p(n)$, the number

* We follow, in general, the notation of Tannery and Molk's *Éléments de la théorie des fonctions elliptiques*. Tannery and Molk, however, write $16G$ in place of the more usual Δ.

† (1) G. H. Hardy and S. Ramanujan, "Une formule asymptotique pour le nombre des partitions de n," *Comptes Rendus*, January 2, 1917 [No. 31 of this volume]; (2) G. H. Hardy and S. Ramanujan, "Asymptotic Formulæ in Combinatory Analysis," *Proc. London Math. Soc.*, Ser. 2, Vol. XVII, 1918, pp. 75—115 [No. 36 of this volume]; (3) S. Ramanujan, "On Certain Trigonometrical Sums and their Applications in the Theory of Numbers," *Trans. Camb. Phil. Soc.*, Vol. XXII, 1918, pp. 259—276 [No. 21 of this volume]; (4) G. H. Hardy, "On the Expression of a Number as the Sum of any Number of Squares, and in Particular of Five or Seven," *Proc. National Acad. of Sciences*, Vol. IV, 1918, pp. 189—193: [and G. H. Hardy, "On the expression of a number as the sum of any number of squares, and in particular of five," *Trans. American Math. Soc.*, Vol. XXI, 1920, pp. 255—284].

of unrestricted partitions of n, and for $r_s(n)$, the number of representations of n as the sum of an even number s of squares, the series

(1·5)

$$\frac{1}{2\pi\sqrt{2}}\frac{d}{dn}\left(\frac{e^{C\lambda_n}}{\lambda_n}\right)+\frac{(-1)^n}{2\pi}\frac{d}{dn}\left(\frac{e^{\frac{1}{2}C\lambda_n}}{\lambda_n}\right)$$
$$+\pi\sqrt{\left(\frac{3}{2}\right)}\cos\left(\frac{2}{3}n\pi-\frac{1}{18}\pi\right)\frac{d}{dn}\left(\frac{e^{\frac{1}{3}C\lambda_n}}{\lambda_n}\right)+\ldots,$$

where $\lambda_n=\sqrt{\left(n-\frac{1}{24}\right)}$ and $C=\pi\sqrt{\left(\frac{2}{3}\right)}$, and

(1·6)

$$\frac{\pi^{\frac{1}{2}s}}{\Gamma(\frac{1}{2}s)}n^{\frac{1}{2}s-1}\{1^{-\frac{1}{2}s}+2\cos(\tfrac{1}{2}n\pi-\tfrac{1}{4}s\pi)\,2^{-\frac{1}{2}s}+2\cos(\tfrac{2}{3}n\pi-\tfrac{1}{2}s\pi)\,3^{-\frac{1}{2}s}+\ldots\};$$

and our methods enable us to write down similar formulæ for a very large variety of other arithmetical functions.

The study of series such as (1·5) and (1·6) raises a number of interesting problems, some of which appear to be exceedingly difficult. The first purpose for which they are intended is that of obtaining approximations to the functions with which they are associated. Sometimes they give also an exact representation of the functions, and sometimes they do not. Thus the sum of the series (1·6) is equal to $r_s(n)$ if s is 4, 6, or 8, but not in any other case. The series (1·5) enables us, by stopping after an appropriate number of terms, to find approximations to $p(n)$ of quite startling accuracy; thus six terms of the series give $p(200)=3972999029388$, a number of 13 figures, with an error of 0·004. But we have never been able to prove that the sum of the series is $p(n)$ exactly, nor even that it is convergent.

There is one class of series, of the same general character as (1·5) or (1·6), which lends itself to comparatively simple treatment. These series arise when the generating modular function $f(q)$ or $\phi(\tau)$ satisfies an equation

$$\phi(\tau)=(a+b\tau)^n\phi\left(\frac{c+d\tau}{a+b\tau}\right),$$

where n is a positive integer, and behaves, inside the unit circle, like a rational function; that is to say, possesses no singularities but poles. The simplest examples of such functions are the reciprocals of the functions (1·3) and (1·4). The coefficients in their expansions are integral, but possess otherwise no particular arithmetical interest. The results, however, are very remarkable from the point of view of approximation; and it is, in any case, well worth while, in view of the many arithmetical applications of this type of series, to study in detail any example in which the results can be obtained by comparatively simple analysis.

We begin by proving a general theorem (Theorem 1) concerning the expression of a modular function with poles as a series of partial fractions.

This series is (as appears in Theorem 2) a "Poincaré's series": what our theorem asserts is, in effect, that the sum of a certain Poincaré's series is the only function which satisfies certain conditions. It would, no doubt, be possible to obtain this result as a corollary from propositions in the general theory of automorphic functions; but we thought it best to give an independent proof, which is interesting in itself and demands no knowledge of this theory.

2. THEOREM 1. *Suppose that*

$$f(q) = f(e^{\pi i \tau}) = \phi(\tau) \tag{2·1}$$

is regular for $q=0$, *has no singularities save poles within the unit circle, and satisfies the functional equation*

$$\phi(\tau) = (a+b\tau)^n \phi\left(\frac{c+d\tau}{a+b\tau}\right) = (a+b\tau)^n \phi(T), \tag{2·2}$$

n being a positive integer and a, b, c, d *any integers such that* $ad-bc=1$. *Then*

$$f(q) = \Sigma R, \tag{2·3}$$

where R is a residue of $\quad f(x)/(q-x)$

at a pole of $f(x)$, *if* $|q|<1$; *while if* $|q|>1$ *the sum of the series on the right-hand side of* (2·3) *is zero.*

The proof requires certain geometrical preliminaries.

3. The half-plane $\mathbf{I}(\tau)>0$, which corresponds to the inside of the unit circle in the plane of q, is divided up, by the substitutions of the modular group, into a series of triangles whose sides are arcs of circles and whose angles are $\frac{1}{3}\pi$, $\frac{1}{3}\pi$, and 0*. One of these, which is called the *fundamental polygon* (P)†, has its vertices at the points ρ, ρ^2, and $i\infty$, where $\rho = e^{\frac{1}{3}\pi i}$, and its sides are parts of the unit circle $|\tau|=1$ and the lines $\mathbf{R}(\tau) = \pm\frac{1}{2}$.

Suppose that F_m is the "Farey's series" of order m, that is to say the aggregate of the rational fractions between 0 and 1, whose denominators are not greater than m, arranged in order of magnitude‡, and that h'/k' and h/k, where $0<h'/k'<h/k<1$, are two adjacent terms in the series. We shall consider what regions in the τ-plane correspond to P in the T-plane, when

(3·1) $\quad T = -\dfrac{h'-k'\tau}{h-k\tau}, \qquad$ (3·2) $\quad T = \dfrac{h-k\tau}{h'-k'\tau}.$

Both of these substitutions belong to the modular group, since $hk'-h'k=1$. The points $i\infty$, $\frac{1}{2}$, $-\frac{1}{2}$, in the T-plane correspond to h/k, $(h+2h')/(k+2k')$,

* It is for many purposes necessary to divide each triangle into two, whose angles are $\frac{1}{2}\pi$, $\frac{1}{3}\pi$, and 0; but this further subdivision is not required for our present purpose. For the detailed theory of the modular group, see Klein-Fricke, *Vorlesungen über die Theorie der Elliptischen Modulfunktionen*, 1890—1892.

† See Fig. 1.

‡ The first and last terms are 0/1 and 1/1. A brief account of the properties of Farey's series is given in § 4·2 of our paper (2) [pp. 288—290 of this volume].

$(h-2h')/(k-2k')$ in the τ-plane. Thus the lines $\mathbf{R}(T)=\frac{1}{2}$, $\mathbf{R}(T)=-\frac{1}{2}$ correspond to semicircles described on the segments

$$\left(\frac{h}{k},\ \frac{h+2h'}{k+2k'}\right),\qquad \left(\frac{h}{k},\ \frac{h-2h'}{k-2k'}\right)$$

respectively as diameters. Similarly the upper half of the unit circle corresponds to a semicircle on the segment

$$\left(\frac{h+h'}{k+k'},\ \frac{h-h'}{k-k'}\right).$$

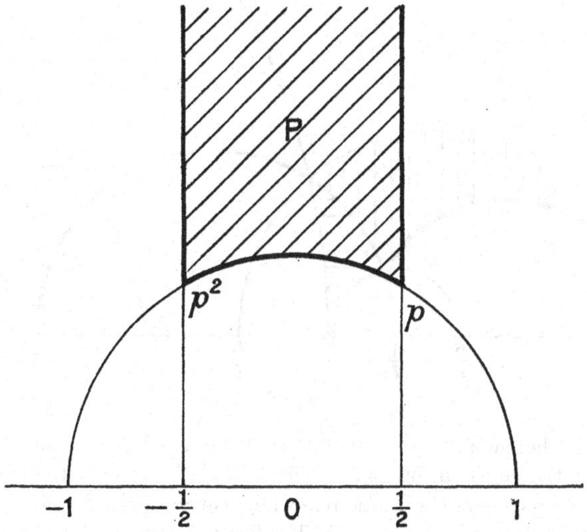

Fig. 1.

The polygon P corresponds to the region bounded by these three semicircles. In particular, the right-hand edge of P corresponds to a circular arc stretching from h/k (where it cuts the real axis at right angles) to the point

$$\text{(3·3)}\qquad \frac{h'k'+hk+\frac{1}{2}(hk'+h'k)+\frac{1}{2}i\sqrt{3}}{k^2+kk'+k'^2}$$

corresponding to $\tau=\rho$.

Similarly we find that the substitution (3·2) correlates to P a triangle bounded by semicircles on the segments

$$\left(\frac{h'}{k'},\ \frac{h'-2h}{k'-2k}\right),\qquad \left(\frac{h'}{k'},\ \frac{h'+2h}{k'+2k}\right),\qquad \left(\frac{h'-h}{k'-k},\ \frac{h'+h}{k'+k}\right).$$

In particular, the left-hand edge of P corresponds to a circular arc from h'/k' to the point (3·3). These two arcs, meeting at the point (3·3), form a continuous path ω, connecting h/k and h'/k', every point of which corresponds, in virtue of one or other of the substitutions (3·1) and (3·2), to a point on one of the rectilinear boundaries of P*.

* Fig. 2 illustrates the case in which $h/k=\frac{3}{5}$, $h'/k'=\frac{1}{2}$. These fractions are adjacent in F_5 and F_6, but not in F_7.

Performing a similar construction for every pair of adjacent fractions of F_m, we obtain a continuous path from $\tau=0$ to $\tau=1$. This path, and its reflexion in the imaginary axis, give a continuous path from $\tau=-1$ to $\tau=1$, which we shall denote by Ω_m. To Ω_m corresponds a path in the q-plane, which we call H_m; H_m is a closed path, formed entirely by arcs of circles which cut the unit circle at right angles.

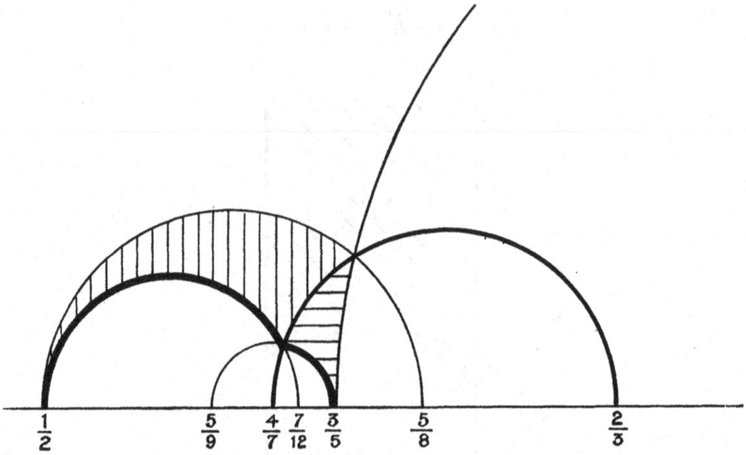

Fig. 2.

The region shaded horizontally corresponds to P for the substitution (3·1), that shaded vertically for the substitution (3·2). The thickest lines shew the path ω; the line of medium thickness shews the semicircle which corresponds (for either substitution) to the unit semicircle in the plane of T. The large incomplete semicircle passes through $\tau=1$.

Since
$$\frac{h'}{k'} < \frac{h'+2h}{k'+2k}, \quad \frac{h+2h'}{k+2k'} < \frac{h}{k},$$
the path ω from h'/k' to h/k is always passing from left to right, and its length is less than twice that of the semicircle on $(h'/k', h/k)$, i.e., than π/kk'. The total length of Ω_m is less than 2π; and, since
$$\left|\frac{dq}{d\tau}\right| = \left|\pi i e^{\pi i \tau}\right| \leqslant \pi,$$
the length of H_m is less than $2\pi^2$. Finally, we observe that the maximum distance of Ω_m from the real axis is less than half the maximum distance between two adjacent terms of F_m, and so less than $1/2m$*. Hence Ω_m tends uniformly to the real axis, and H_m to the unit circle, when $m \to \infty$.

4. The function $\phi(\tau)$ can have but a finite number of poles in P; we suppose, for simplicity, that none of them lie on the boundary. There is then a constant K such that $|f(q)| < K$ on the boundary of P.

* See Lemma 4·22 of our paper (2) [pp. 289—290 of this volume].

We now consider the integral

(4·1) $$\frac{1}{2\pi i}\int \frac{f(x)}{x-q}\,dx,$$

where $|q|<1$ and the contour of integration is H_m*. By Cauchy's Theorem, the integral is equal to

$$f(q)-\Sigma R,$$

where R is a residue of $f(x)/(q-x)$ at a pole of $f(x)$ inside H_m†. To prove our theorem, then, we have merely to shew that the integral (4·1) tends to zero when $m\to\infty$.

Let ω_1' and ω_1 be the left- and right-hand parts of ω, and ζ_1', ζ_1 and ζ the corresponding arcs of H_m. The length of ω_1 is, as we have seen, less than $\frac{1}{2}\pi/kk'$, and that of ζ_1 than $\frac{1}{2}\pi^2/kk'$. Further, we have, on ζ_1,

$$\left|f(x)\right| = \left|\phi(\tau)\right| = \left|h-k\tau\right|^n \left|\phi(T)\right| < K\left\{k\left(\frac{h}{k}-\frac{h'}{k'}\right)\right\}^n = \frac{K}{k'^n}.$$

Thus the contribution of ζ_1 to the integral is numerically less than $C/(kk'^{n+1})$, where C is independent of m; and the whole integral (4·1) is numerically less than

(4·2) $$2C\Sigma\frac{1}{kk'}\left(\frac{1}{k^n}+\frac{1}{k'^n}\right),$$

where the summation extends to all pairs of adjacent terms of F_m.

When ν is fixed and $m>\nu$, the number of terms of F_m whose denominators are less than ν is a function of ν only, say $N(\nu)$. If h/k is one of these, and h'/k' is adjacent to it, $k+k'>m$‡, and so $k'>m-\nu$. Thus the terms of (4·2) in which either k or k' is less than ν contribute less than $8CN(\nu)/(m-\nu)$. The remaining terms contribute less than

$$\frac{4C}{\nu^n}\Sigma\frac{1}{kk'}=\frac{4C}{\nu^n}.$$

Hence the sum (4·3) is less than

$$\frac{8CN(\nu)}{m-\nu}+\frac{4C}{\nu^n},$$

and it is plain that, by choice of first ν and then m, this may be made as small as we please. Thus (4·1) tends to zero and the theorem is proved. It should be observed that ΣR must, for the present at any rate, be interpreted as meaning the limit of the sum of terms corresponding to poles inside H_m; we have not established the absolute convergence of the series.

* Strictly speaking, $f(x)$ is not defined at the points where H_m meets the unit circle, and we should integrate round a path just inside H_m and proceed to the limit. The point is trivial, as $f(x)$, in virtue of the functional equation, tends to zero when we approach a cusp of H_m from inside.

† We suppose m large enough to ensure that $x=q$ lies inside H_m.

‡ See our paper (2), *loc. cit.* [p. 290].

We supposed that no pole of $\phi(\tau)$ lies on the boundary of P. This restriction, however, is in no way essential; if it is not satisfied, we have only to select our "fundamental polygon" somewhat differently. The theorem is consequently true independently of any such restriction.

So far we have supposed $|q|<1$. It is plain that, if $|q|>1$, the same reasoning proves that

$$\Sigma R = 0. \tag{4·3}$$

5. Suppose in particular that $\phi(\tau)$ has one pole only, and that a simple pole at $\tau=\alpha$, with residue A. The complete system of poles is then given by

$$\tau = \mathfrak{a} = \frac{c+d\alpha}{a+b\alpha} \quad (ad-bc=1). \tag{5·1}$$

If a and b are fixed, and (c, d) is one pair of solutions of $ad-bc=1$, the complete system of solutions is $(c+ma, d+mb)$, where m is an integer. To each pair (a, b) correspond an infinity of poles in the plane of τ; but these poles correspond to two different poles only in the plane of q, viz.,

$$q = \pm\mathbf{q} = \pm e^{\pi i \mathfrak{a}}, \tag{5·2}$$

the positive and negative signs corresponding to even and odd values of m respectively. It is to be observed, moreover, that different pairs (a, b) may give rise to the same pole $\mathbf{q}$.

The residue of $\phi(\tau)$ for $\tau=\mathfrak{a}$ is, in virtue of the functional equation (2·2),

$$\frac{A}{(a+b\alpha)^{n+2}};$$

and the residue of $f(q)$ for $q=\mathbf{q}$ is

$$\frac{A}{(a+b\alpha)^{n+2}}\left(\frac{dq}{d\tau}\right)_{\tau=\mathfrak{a}} = \frac{\pi i A\mathbf{q}}{(a+b\alpha)^{n+2}}.$$

Thus the sum of the terms of our series which correspond to the poles (5·2) is

$$\frac{\pi i A}{(a+b\alpha)^{n+2}}\left(\frac{\mathbf{q}}{q-\mathbf{q}} - \frac{\mathbf{q}}{q+\mathbf{q}}\right) = \frac{2\pi i A}{(a+b\alpha)^{n+2}}\frac{\mathbf{q}^2}{q^2-\mathbf{q}^2}.$$

We thus obtain:

THEOREM 2. *If $\phi(\tau)$ has one pole only in P, viz., a simple pole at $\tau=\alpha$, with residue A, and $|q|<1$, then*

$$f(q) = 2\pi i A\,\Sigma\,\frac{1}{(a+b\alpha)^{n+2}}\frac{\mathbf{q}^2}{q^2-\mathbf{q}^2}, \tag{5·3}$$

where

$$\mathbf{q} = \exp\left(\frac{c+d\alpha}{a+b\alpha}\right)\pi i;$$

c, d being any pair of solutions of $ad-bc=1$, and the summation extending over all pairs a, b, which give rise to distinct values of $\mathbf{q}$. If $|q|>1$, the sum of the series on the right-hand side of (5·3) *is zero.*

If $\phi(\tau)$ has several poles in P, $f(q)$, of course, will be the sum of a number of series such as (5·3). Incidentally, we may observe that it now appears that the series in question are absolutely convergent.

6. As an example, we select the function

$$(6{\cdot}1)\qquad f(q)=\frac{\pi^6}{216\omega_1{}^6 g_3}=1\Big/\left(1-504\sum_1^\infty\frac{r^5q^{2r}}{1-q^{2r}}\right)=\sum_0^\infty p_n x^n,$$

say, where $x=q^2$. It is evident that p_n is always an integer; the values of the first 13 coefficients are

$$p_0=1,\quad p_1=504,\quad p_2=270648,\quad p_3=144912096,$$
$$p_4=77599626552,\quad p_5=41553943041744,\quad p_6=22251789971649504,$$
$$p_7=11915647845248387520,\quad p_8=6380729991419236488504,$$
$$p_9=3416827666558895485479576,$$
$$p_{10}=1829682703808504464920468048,$$
$$p_{11}=979779820147442370107345764512,$$
$$p_{12}=524663917940510191509934144603104;$$

so that p_{12} is a number of 33 figures.

By means of the formulæ*

$$g_3=\tfrac{8}{27}(e_1-e_3)^2(1+k^2)(1-\tfrac{1}{2}k^2)(1-2k^2),$$
$$e_1-e_3=\left(\frac{\pi}{2\omega_1}\right)^2\{\vartheta_3(0)\}^4,\qquad \frac{2K}{\pi}=\{\vartheta_3(0)\}^2,$$

we find that
$$\frac{1}{f(q)}=\left(\frac{2K}{\pi}\right)^6(1+k^2)(1-\tfrac{1}{2}k^2)(1-2k^2).$$

The value of n is 6. The poles of $f(q)$ correspond to the values of τ which make $K=k^2$ equal to -1, 2, or $\frac{1}{2}$. It is easily verified† that these values are given by the general formula

$$\tau=\frac{c+di}{a+bi}\qquad (ad-bc=1),$$

so that

$$(6{\cdot}2)\qquad \mathrm{q}=\exp\left(\frac{c+di}{a+bi}\,\pi i\right)=\exp\left(\frac{ac+bd}{a^2+b^2}\,\pi i-\frac{\pi}{a^2+b^2}\right).$$

The value of α is i‡. In order to determine A we observe that

$$-504\frac{d}{dq}\left(\frac{1^5q^2}{1-q^2}+\frac{2^5q^4}{1-q^4}+\dots\right)=-\frac{1008}{q}\left\{\frac{1^6q^2}{(1-q^2)^2}+\frac{2^6q^4}{(1-q^4)^2}+\dots\right\}.$$

* All the formulæ which we quote are given in Tannery and Molk's Tables; see in particular Tables XXXVI (3), LXXI (3), XCVI, CX (3).

† A full account of the problem of finding τ when κ is given will be found in Tannery and Molk, *loc. cit.*, Vol. III, ch. 7 ("On donne k^2 ou g_2, g_3; trouver τ ou ω_1, ω_3").

‡ It will be observed that in this case α is on the boundary of P; see the concluding remarks of § 4. As it happens, $\tau=i$ lies on that edge of P (the circular edge) which was not used in the construction of H_m, so that our analysis is applicable as it stands.

The series in curly brackets is the function called by Ramanujan* $\Phi_{1,6}$ and†

$$1008\Phi_{1,6} = Q^2 - PR,$$

where $$P = \frac{12\eta_1\omega_1}{\pi^2}, \quad Q = 12g_2\left(\frac{\omega_1}{\pi}\right)^4, \quad R = 216g_3\left(\frac{\omega_1}{\pi}\right)^6.$$

Here $R = 0$, so that

$$1008\Phi_{1,6} = Q^2 = 1 + 480\Phi_{0,7}‡ = 1 + 480\left(\frac{1^7q^2}{1-q^2} + \frac{2^7q^4}{1-q^4} + \dots\right).$$

Hence we find that $A = i/\pi C, \quad 2\pi i A = -2/C,$

where

(6·3) $$C = 1 + 480\left(\frac{1^7}{e^{2\pi}-1} + \frac{2^7}{e^{4\pi}-1} + \dots\right).$$

Another expression for C is

(6·4) $$C = 144\left(\frac{K_0}{\pi}\right)^8,$$

where

(6·41) $$K_0 = \int_0^{\frac{1}{2}\pi} \frac{d\theta}{\sqrt{(1-\frac{1}{2}\sin^2\theta)}} = \frac{\{\Gamma(\frac{1}{4})\}^2}{4\sqrt{\pi}}.$$

We have still to consider more closely the values of a and b, over which the summation is effected. Let us fix k, and suppose that (a, b) is a pair of positive solutions of the equation $a^2 + b^2 = k$. This pair gives rise to a system of eight solutions, viz.,

$$(\pm a, \pm b), \quad (\pm b, \pm a).$$

But it is obvious that, if we change the signs of both a and b, we do not affect the aggregate of values of $\mathbf{a}$. Thus we need only consider the four pairs

$$(a, b), \quad (a, -b), \quad (b, a), \quad (b, -a).$$

If a or b is zero, or if $a = b$, these four pairs reduce to two.

It is easily verified that, if (a, b) leads to the pair of poles

$$q = \pm \mathbf{q} = \pm \exp\left(\frac{ac+bd}{a^2+b^2}\pi i - \frac{\pi}{a^2+b^2}\right),$$

then $(a, -b)$ and (b, a) each lead to $q = \pm \bar{\mathbf{q}}$, where $\bar{\mathbf{q}}$ is the conjugate of $\mathbf{q}$. Thus, in general (a, b) and the solutions derived from it lead to four distinct poles, viz., $\pm \mathbf{q}$ and $\pm \bar{\mathbf{q}}$. These four reduce to two in two cases, when $\mathbf{q}$ is real, so that $\mathbf{q} = \bar{\mathbf{q}}$, and when $\mathbf{q}$ is purely imaginary, so that $\mathbf{q} = -\bar{\mathbf{q}}$. It is

* S. Ramanujan, "On Certain Arithmetical Functions," *Trans. Camb. Phil. Soc.*, Vol. XXII, pp. 159—184 (p. 163) [No. 18 of this volume, p. 140].

† Ramanujan, *loc. cit.*, p. 164 [p. 142].

‡ Ramanujan, *loc. cit.*, p. 163 [p. 141].

easy to see that the first case can occur only when $k=1$, and the second when $k=2$*.

If $k=1$ we take $a=1$, $b=0$, $c=0$, $d=1$; and $\mathsf{q}=\bar{\mathsf{q}}=e^{-\pi}$. If $k=2$ we take $a=1$, $b=1$, $c=0$, $d=1$; and $\mathsf{q}=-\bar{\mathsf{q}}=ie^{-\frac{1}{2}\pi}$. The corresponding terms in our series are

$$\frac{1}{1-q^2e^{2\pi}},\quad \frac{1}{2^4(1+qe^{\pi})}.$$

If k is greater than 2, and is the sum of two coprime squares a^2 and b^2, it gives rise to terms

$$\frac{1}{(a+bi)^8}\frac{1}{1-(q/\mathsf{q})^2}+\frac{1}{(a-bi)^8}\frac{1}{1-(q/\bar{\mathsf{q}})^2}.$$

There is, of course, a similar pair of terms corresponding to every other distinct representation of k as a sum of coprime squares. Thus finally we obtain the following result:

THEOREM 3. *If*

$$f(q)=\frac{\pi^6}{216\omega_1^6 g_3}=1\Big/\left(1-504\sum_1^\infty\frac{r^5q^{2r}}{1-q^{2r}}\right)=\sum_0^\infty p_n q^{2n},$$

and $|q|<1$, *then*

$$(6\cdot5)\qquad \tfrac{1}{2}Cf(q)=\frac{1}{1-q^2e^{2\pi}}+\frac{1}{2^4(1+q^2e^{\pi})}$$
$$+\sum\left\{\frac{1}{(a+bi)^8}\frac{1}{1-(q/\mathsf{q})^2}+\frac{1}{(a-bi)^8}\frac{1}{1-(q/\bar{\mathsf{q}})^2}\right\};$$

where

$$C=1+480\left(\frac{1^7}{e^{2\pi}-1}+\frac{2^7}{e^{4\pi}-1}+\dots\right)=\frac{9\pi^4}{16\{\Gamma(\frac{3}{4})\}^{16}},$$

$$\mathsf{q}=\exp\left(\frac{c+di}{a+bi}\pi i\right)=\exp\left(\frac{ac+bd}{a^2+b^2}\pi i-\frac{\pi}{a^2+b^2}\right),$$

and $\bar{\mathsf{q}}$ *is the conjugate of* q. *The summation applies to every pair of coprime positive numbers* a *and* b, *such that* $k=a^2+b^2\geqslant 5$, *such pairs, however, only being counted as distinct if they correspond to independent representations of* k *as a sum of squares. If* $|q|>1$, *then the sum of the series on the right-hand side of* (6·5) *is zero.*

* When a and b are given, we can always choose c and d so that $|ac+bd|\leqslant\frac{1}{2}(a^2+b^2)$. If q is real, we have $ad-bc=1$ and $ac+bd=0$ simultaneously: whence

$$(a^2+b^2)(c^2+d^2)=1.$$

If q is purely imaginary, we have

$$ad-bc=1,\quad 2|ac+bd|=a^2+b^2,$$

whence

$$(c^2+d^2)^2=(|ac+bd|-c^2-d^2)^2+1.$$

This is possible only if $c^2+d^2=1$ and $|ac+bd|=1$, whence $a^2+b^2=2$.

7. It follows that

$$(7{\cdot}1)\quad \tfrac{1}{2}Cp_n = e^{2n\pi} + \frac{(-1)^n}{2^4}e^{n\pi} + \Sigma\left\{\frac{1}{(a+bi)^8}\mathfrak{q}^{-2n} + \frac{1}{(a-bi)^8}\bar{\mathfrak{q}}^{-2n}\right\} = \sum_{(\lambda)}\frac{c_\lambda(n)}{\lambda^4}e^{2n\pi/\lambda},$$

say. Here λ is the sum of two coprime squares, so that

$$\lambda = 2^{a_2}5^{a_5}13^{a_{13}}17^{a_{17}}\ldots,$$

where a_2 is 0 or 1 and 5, 13, 17, ... are the primes of the form $4k+1$; and the first few values of $c_\lambda(n)$ are

$$c_1(n) = 1,\quad c_2(n) = (-1)^n,\quad c_5(n) = 2\cos(\tfrac{4}{5}n\pi + 8\arctan 2),$$

$$c_{10}(n) = 2\cos(\tfrac{3}{5}n\pi - 8\arctan 2),\quad c_{13}(n) = 2\cos(\tfrac{10}{13}n\pi + 8\arctan 5).$$

The approximations to the coefficients given by the formula (7·1) are exceedingly remarkable. Dividing by $\tfrac{1}{2}C$, and taking $n = 0, 1, 2, 3, 6$, and 12, we find the following results:

(0)	(1)	(2)
0·944	505·361	270616·406
+0·059	−1·365	+31·585
−0·003	+0·004	+0·009
$p_0 = 1{\cdot}000$	$p_1 = 504{\cdot}000$	$p_2 = 270648{\cdot}000$

(3)	(6)
144912827·002	22251789962592450·237
−730·900	+9057051·688
−0·101	+2·081
−0·001	−0·006
$p_3 = 144912096{\cdot}000$	$p_6 = 22251789971649504{\cdot}000$

(12)
524663917940510190119197271938395·329
+1390736872662028·140
+2680·418
+0·130
−0·014
−0·003
$p_{12} = 524663917940510191509934144603104{\cdot}000$

An alternative expression for C is

$$C = 96^2 e^{-8\pi/3}\{(1-e^{-4\pi})(1-e^{-8\pi})\ldots\}^{16},$$

by means of which C may be calculated with great accuracy*. To five places we have $2/C = 0{\cdot}94373$, which is very nearly equal to $352/373 = 0{\cdot}94370$.

* Gauss, *Werke*, Vol. III, pp. 418—419, gives the values of various powers of $e^{-\pi}$ to a large number of figures.

It is easy to see directly that p_n lies between the coefficients of x^n in the expansions of

$$\frac{1}{(1-535x)(1+31x)}, \quad \frac{1-7\cdot 5x}{(1-535\cdot 5x)(1+24x)},$$

and so that

$$\frac{(535)^{n+1}-(-31)^{n+1}}{566} \leqslant p_n \leqslant \frac{352\,(535\cdot 5)^n + 21\,(-24)^n}{373}.$$

The function

$$\Omega(x) = \sum_{(\lambda)} \frac{c_\lambda(x)}{\lambda^4} e^{2x\pi/\lambda}$$

has very remarkable properties. It is an integral function of x, whose maximum modulus is less than a constant multiple of $e^{2\pi|x|}$. It is equal to p_n, an integer, when $x = n$, a positive integer; and to zero when $x = -n$. But we must reserve the discussion of these peculiarities for some other occasion.

QUESTIONS AND SOLUTIONS

(published by SRINIVASA RAMANUJAN in the *Journal of the Indian Mathematical Society*)

Question 260 (III, 43; *communicated by* P. V. Seshu Aiyar):

Shew, without using calculus, that

$$\tfrac{3}{2}\log 2 = 1 + \frac{2}{4^3-4} + \frac{2}{8^3-8} + \frac{2}{12^3-12} + \dots \textit{ ad inf.}$$

[*Solution by* K. J. Sanjana and S. Narayanan, III, 86—87.]

Question 261 (III, 43; *communicated by* P. V. Seshu Aiyar):

Shew that

$$(a)\quad \left(1+\frac{1}{1^3}\right)\left(1+\frac{1}{2^3}\right)\left(1+\frac{1}{3^3}\right)\dots = \frac{1}{\pi}\cosh\left(\tfrac{1}{2}\pi\sqrt{3}\right),$$

$$(b)\quad \left(1-\frac{1}{2^3}\right)\left(1-\frac{1}{3^3}\right)\left(1-\frac{1}{4^3}\right)\dots = \frac{1}{3\pi}\cosh\left(\tfrac{1}{2}\pi\sqrt{3}\right);$$

and prove from first principles that $(b) = \frac{1}{3}(a)$.

[*Solution by* N. B. Pendse, III, 124—125.]

Question 283 (III, 89, and IV, 106):

Shew that it is possible to solve the equations

$$x+y+z=a, \qquad px+qy+rz=b,$$

$$p^2x+q^2y+r^2z=c, \quad p^3x+q^3y+r^3z=d,$$

$$p^4x+q^4y+r^4z=e, \quad p^5x+q^5y+r^5z=f,$$

where x, y, z, p, q, r are the unknowns. Solve the above when $a=2$, $b=3$, $c=4$, $d=6$, $e=12$, and $f=32$.

[*Solutions* (1) *by* M. K. Sadasiva Aiyar, and (2) by Sankara Aiyar, III, 198—200. See also "Note on a Set of Simultaneous Equations," *Journal of the Indian Mathematical Society*, IV (1912), 94—96, and pp. 18—19 of this volume.]

Question 284 (III, 89):

Solve
$$\frac{x^5-6}{x^2-y} = \frac{y^5-9}{y^2-x} = 5(xy-1).$$

Solution by Srinivasa Ramanujan, IV, 183.

Let
$$x = \alpha+\beta+\gamma, \quad y = \alpha\beta+\beta\gamma+\gamma\alpha,$$

where α, β, γ may have any values we choose. By substituting the supposed values of x and y in the given equations we find that $\alpha\beta\gamma$ should have the value 1. Now

$$(\alpha+\beta+\gamma)^5 - 6 = 5\,\{(\alpha+\beta+\gamma)^2 - (\alpha\beta+\beta\gamma+\gamma\alpha)\}\{(\alpha+\beta+\gamma)(\alpha\beta+\beta\gamma+\gamma\alpha) - \alpha\beta\gamma\};$$

$$(\alpha\beta+\beta\gamma+\gamma\alpha)^5 - 9 = 5\,\{(\alpha\beta+\beta\gamma+\gamma\alpha)^2 - \alpha\beta\gamma(\alpha+\beta+\gamma)\}\{(\alpha+\beta+\gamma)(\alpha\beta+\beta\gamma+\gamma\alpha)\,\alpha\beta\gamma - \alpha^2\beta^2\gamma^2\}.$$

Simplifying,
$$\alpha^5+\beta^5+\gamma^5 = 6,$$

$$(\alpha\beta)^5+(\beta\gamma)^5+(\gamma\alpha)^5 = 9,$$

$$(\alpha\beta\gamma)^5 = 1.$$

Thus we see that α^5, β^5, γ^5 are the roots of the equation

$$t^3-6t^2+9t-1=0.$$

But it is easily seen that the roots of this equation are

$$4\cos^2 20^\circ, \quad 4\cos^2 40^\circ, \quad 4\cos^2 80^\circ.$$

Hence

$$\alpha=(2\cos 20^\circ)^{\frac{2}{5}}, \quad \beta=(2\cos 40^\circ)^{\frac{2}{5}}, \quad \gamma=(2\cos 80^\circ)^{\frac{2}{5}},$$

$$x=(2\cos 20^\circ)^{\frac{2}{5}}+(2\cos 40^\circ)^{\frac{2}{5}}+(2\cos 80^\circ)^{\frac{2}{5}},$$

$$y=(\tfrac{1}{2}\sec 20^\circ)^{\frac{2}{5}}+(\tfrac{1}{2}\sec 40^\circ)^{\frac{2}{5}}+(\tfrac{1}{2}\sec 80^\circ)^{\frac{2}{5}}.$$

All the values of x are

$$\alpha+\beta+\gamma, \quad \alpha+\beta\rho+\gamma\rho^4, \quad \alpha+\beta\rho^2+\gamma\rho^3, \quad \alpha\rho+\beta\rho+\gamma\rho^3,$$

$$\alpha\rho+\beta\rho^2+\gamma\rho^2, \quad \alpha\rho^2+\beta\rho^4+\gamma\rho^4, \quad \alpha\rho^3+\beta\rho^3+\gamma\rho^4,$$

where ρ is an imaginary fifth root of unity. Similarly for y.

Question 289 (III, 90):

Find the value of

(i) $\sqrt{[1+2\sqrt{\{1+3\sqrt{(1+\ldots)}\}}]}$,

(ii) $\sqrt{[6+2\sqrt{\{7+3\sqrt{(8+\ldots)}\}}]}$.

Solution by Srinivasa Ramanujan, IV, 226.

(i) $n(n+2)=n\sqrt{\{1+(n+1)(n+3)\}}$.

Let

$$n(n+2)=f(n);$$

then we see that

$$\begin{aligned} f(n)&=n\sqrt{\{1+f(n+1)\}} \\ &=n\sqrt{\{1+(n+1)\sqrt{(1+f(n+2))}\}} \\ &=n\sqrt{[1+(n+1)\sqrt{\{1+(n+2)\sqrt{(1+f(n+3))}\}}]} \\ &=\ldots\ldots\ldots; \end{aligned}$$

that is, $n(n+2)=n\sqrt{[1+(n+1)\sqrt{\{1+(n+2)\sqrt{(1+(n+3)\sqrt{1+\ldots})}\}}]}$.

Putting $n=1$, we have

$$\sqrt{[1+2\sqrt{\{1+3\sqrt{(1+\ldots)}\}}]}=3.$$

(ii) In a similar manner,

$$n(n+3)=n\sqrt{\{n+5+(n+1)(n+4)\}}.$$

Supposing

$$f(n)=n(n+3),$$

we have

$$\begin{aligned} f(n)&=n\sqrt{\{n+5+f(n+1)\}} \\ &=n\sqrt{[n+5+(n+1)\sqrt{\{n+6+f(n+2)\}}]} \\ &=\ldots\ldots. \end{aligned}$$

Thus $n(n+3)=n\sqrt{[n+5+(n+1)\sqrt{\{n+6+(n+2)\sqrt{(n+7+\ldots)}\}}]}$.

Hence

$$\sqrt{[6+2\sqrt{\{7+3\sqrt{(8+4\sqrt{(9+\ldots)})}\}}]}=4.$$

Question 294 (III, 128):

Shew that [if x is a positive integer]

$$\tfrac{1}{2}e^x=1+\frac{x}{1!}+\frac{x^2}{2!}+\ldots+\frac{x^x}{x!}\Theta,$$

where Θ lies between $\frac{1}{2}$ and $\frac{1}{3}$.

Partial solution by Srinivasa Ramanujan, IV, 151—152.

$$\int_0^\infty e^{-z}\left(1+\frac{z}{x}\right)^x dz=\int_0^\infty e^{-z}\left\{\left(\frac{z}{x}\right)^x+\frac{x}{1\,!}\left(\frac{z}{x}\right)^{x-1}+\frac{x(x-1)}{2\,!}\left(\frac{z}{x}\right)^{x-2}+\ldots\right\}dz$$

$$=\frac{x\,!}{x^x}\left(1+\frac{x}{1\,!}+\frac{x^2}{2\,!}+\ldots\frac{x^x}{x\,!}\right).$$

Now, if we suppose that y is a function of x such that

$$\tfrac{1}{2}e^x=1+\frac{x}{1\,!}+\frac{x^2}{2\,!}+\ldots+\frac{x^{x-1}}{(x-1)\,!}+\frac{x^x}{x\,!}y,$$

we see that

$$y=\tfrac{1}{2}\frac{e^x x\,!}{x^x}-\frac{x\,!}{x^x}\left\{1+\frac{x}{1\,!}+\frac{x^2}{2\,!}+\ldots+\frac{x^{x-1}}{(x-1)\,!}\right\}$$

$$=\tfrac{1}{2}\frac{e^x x\,!}{x^x}+1-\int_0^\infty e^{-z}\left(1+\frac{z}{x}\right)^x dz$$

$$=\tfrac{1}{2}\frac{e^x x\,!}{x^x}+1-\int_0^\infty e^{-z}e^{\{z-z^2/(2x)+z^3/(3x^2)-\ldots\}}dz$$

$$=\tfrac{1}{2}\frac{e^x x\,!}{x^x}+1-\int_0^\infty e^{\{-z^2/(2x)+z^3/(3x^2)-\ldots\}}dz$$

$$=\tfrac{1}{2}\frac{e^x x\,!}{x^x}+1-\sqrt{(2x)}\int_0^\infty e^{\{-z^2+2\sqrt{2}z^3/(3\sqrt{x})-\ldots\}}dz$$

$$=\tfrac{1}{2}\frac{e^x x\,!}{x^x}+1-\int_0^\infty e^{-z^2}\{\sqrt{(2x)}+\tfrac{4}{3}z^3-\ldots\}\,dz.$$

Therefore, when $x=\infty$,

$$y=\lim_{x\to\infty}\left\{\frac{1}{2}\frac{e^x x\,!}{x^x}-\sqrt{(\tfrac{1}{2}\pi x)}+1-\int_0^\infty e^{-z^2}\left(\tfrac{4}{3}z^3+\text{terms containing } x^{-\frac{1}{2}},\ x^{-1},\text{ etc.}\right)dz\right\}$$

$$=1-\tfrac{2}{3}=\tfrac{1}{3},$$

since, by Stirling's theorem,

$$\lim_{x\to\infty}\left\{\frac{e^x x\,!}{x^x}-\sqrt{(\tfrac{1}{2}\pi x)}\right\}=0.$$

Also $y=\frac{1}{2}$ when $x=0$.

Though the limits of y are known to be $\frac{1}{2}$ and $\frac{1}{3}$ when x is 0 and ∞ respectively, yet it is difficult to prove that y lies between $\frac{1}{2}$ and $\frac{1}{3}$.

The expansion of y in ascending powers of x^{-1} (which can be found from the above integral) is

$$\frac{1}{3}+\frac{4}{135x}-\frac{8}{2835x^2}-\frac{16}{8505x^3}+\ldots.$$

This result is important as it helps to express e^x between rational limits.

Question 295 (III, 128):

If $\alpha\beta=\pi$, shew that

$$\sqrt{\alpha}\int_0^\infty\frac{e^{-x^2}dx}{\cosh\alpha x}=\sqrt{\beta}\int_0^\infty\frac{e^{-x^2}dx}{\cosh\beta x}.$$

Solution by Srinivasa Ramanujan, v, 65.

Since
$$\int_0^\infty \frac{\cos 2nz}{\cosh \pi z}\,dz = \frac{1}{2\cosh n},$$

$$\sqrt{a}\int_0^\infty \frac{e^{-x^2}dx}{\cosh ax} = 2\sqrt{a}\int_0^\infty\int_0^\infty \frac{e^{-x^2}\cos 2axz}{\cosh \pi z}\,dx\,dz$$
$$= \sqrt{(a\pi)}\int_0^\infty \frac{e^{-a^2z^2}}{\cosh \pi z}\,dz = \sqrt{\left(\frac{\pi}{a}\right)}\int_0^\infty \frac{e^{-z^2}}{\cosh(\pi z/a)}\,dz$$
$$= \sqrt{\beta}\int_0^\infty \frac{e^{-x^2}dx}{\cosh \beta x},$$

since $a\beta = \pi$.

[See also "Some definite integrals," *Messenger of Mathematics*, XLIV (1915), 10—18, and pp. 53—58 of this volume, equation (12).]

Question 308 (III, 168):

Shew that

$$\text{(i)}\quad \int_0^{\frac{1}{2}\pi} \theta\cot\theta\log\sin\theta\,d\theta = -\frac{\pi^3}{48} - \frac{\pi}{4}(\log 2)^2,$$

$$\text{(ii)}\quad \int_0^{1/\sqrt{2}} \frac{\sin^{-1}x}{x}\,dx - \tfrac{1}{2}\int_0^1 \frac{\tan^{-1}x}{x}\,dx = \frac{\pi}{8}\log 2.$$

[*Solution by* K. J. Sanjana, III, 248.]

Question 327 (III, 209):

Shew that Euler's constant, namely [the limit of]

$$1 + \tfrac{1}{2} + \tfrac{1}{3} + \ldots + \frac{1}{n} - \log n$$

when n is infinite, is equal to

$$\log 2 - 1\left(\frac{2}{3^3-3}\right) - 2\left(\frac{2}{6^3-6} + \frac{2}{9^3-9} + \frac{2}{12^3-12}\right)$$
$$-3\left(\frac{2}{15^3-15} + \frac{2}{18^3-18} + \ldots + \frac{2}{39^3-39}\right) - \ldots,$$

the first term in the nth group being

$$\frac{2}{\{\frac{1}{2}(3^n+3)\}^3 - \frac{1}{2}(3^n+3)}.$$

[See "A series for Euler's constant γ," *Messenger of Mathematics*, XLVI (1917), 73—80, and pp. 163—168 of this volume.]

Question 352 (IV, 40):

Shew that

$$\text{(i)}\quad \frac{1}{1-}\,\frac{e^{-2\pi}}{1+}\,\frac{e^{-4\pi}}{1+}\,\frac{e^{-6\pi}}{1+}\ldots = [\sqrt{\{\tfrac{1}{2}(5+\sqrt{5})\}} - \tfrac{1}{2}(\sqrt{5}+1)]\,e^{\frac{2}{5}\pi},$$

$$\text{(ii)}\quad \frac{1}{1-}\,\frac{e^{-\pi}}{1+}\,\frac{e^{-2\pi}}{1-}\,\frac{e^{-3\pi}}{1+}\ldots = [\sqrt{\{\tfrac{1}{2}(5-\sqrt{5})\}} - \tfrac{1}{2}(\sqrt{5}-1)]\,e^{\frac{1}{5}\pi}.$$

Question 353 (IV, 40):

If n is any positive odd integer, shew that

$$\int_0^\infty \frac{\sin nx}{\cosh x + \cos x}\,\frac{dx}{x} = \tfrac{1}{4}\pi;$$

and hence prove that

$$\tfrac{1}{8}\pi = \left(\cosh\frac{\pi}{2n}+\cos\frac{\pi}{2n}\right)^{-1} - \tfrac{1}{3}\left(\cosh\frac{3\pi}{2n}+\cos\frac{3\pi}{2n}\right)^{-1} + \tfrac{1}{5}\left(\cosh\frac{5\pi}{2n}+\cos\frac{5\pi}{2n}\right)^{-1} - \dots$$

for all odd values of n.

[*Solution by* A. C. L. Wilkinson, VIII, 106—110; also *Remarks by* S. D. S. Chowla, XVI, 119—120.]

Question 358 (IV, 78):

If n is a multiple of 4, excluding zero, shew that

$$1^{n-1}\operatorname{sech}(\tfrac{1}{2}\pi) - 3^{n-1}\operatorname{sech}(\tfrac{3}{2}\pi) + 5^{n-1}\operatorname{sech}(\tfrac{5}{2}\pi) - \dots = 0.$$

[*Solution by* M. Bhimasena Rao, VII, 99—101.]

Question 359 (IV, 78):

If $\sin(x+y) = 2\sin(\tfrac{1}{2}(x-y)), \quad \sin(y+z) = 2\sin(\tfrac{1}{2}(y-z)),$

prove that $(\tfrac{1}{2}\sin x\cos z)^{\frac{1}{4}} + (\tfrac{1}{2}\cos x\sin z)^{\frac{1}{4}} = (\sin 2y)^{\frac{1}{12}},$

and verify the result when

$$\sin 2x = (\sqrt{5}-2)^3(4+\sqrt{15})^2, \quad \sin 2y = \sqrt{5}-2, \quad \sin 2z = (\sqrt{5}-2)^3(4-\sqrt{15})^2.$$

[*Solution by* T. R. Srinivasa Iyer, XV, 114—117.]

Question 386 (IV, 120):

Shew that

$$\int_0^\infty \frac{dx}{(1+x^2)(1+n^2x^2)(1+n^4x^2)(1+n^6x^2)\dots} = \frac{\pi}{2(1+n+n^3+n^6+n^{10}+\dots)},$$

where the indices in the denominator are the sums of the natural numbers.

[*Solution by* N. Durai Rajan, VII, 143—144. See also "Some definite integrals," *Messenger of Mathematics*, XLIV (1915), 10—18, and pp. 53—58 of this volume, equation (24).]

Question 387 (IV, 120):

Shew that $\dfrac{1}{e^{2\pi}-1} + \dfrac{2}{e^{4\pi}-1} + \dfrac{3}{e^{6\pi}-1} + \dots = \dfrac{1}{24} - \dfrac{1}{8\pi}.$

Question 427 (IV, 238):

Express $(Ax^2+Bxy+Cy^2)(Ap^2+Bpq+Cq^2)$

in the form $Au^2+Buv+Cv^2$; and hence shew that, if

$$(2x^2+3xy+5y^2)(2p^2+3pq+5q^2) = 2u^2+3uv+5v^2,$$

then one set of the values of u and v is

$$u = \tfrac{5}{2}(x+y)(p+q) - 2xp, \quad v = 2qy-(x+y)(p+q).$$

[*Solutions by* (1) "Zero," and (2) S. Narayanan, X, 320—321.]

Question 441 (V, 39):

Shew that

$$(6a^2-4ab+4b^2)^3 = (3a^2+5ab-5b^2)^3 + (4a^2-4ab+6b^2)^3 + (5a^2-5ab-3b^2)^3,$$

and find other quadratic expressions satisfying similar relations.

[*Solution by* S. Narayanan, VI, 226.]

Question 463 (v, 120):

If
$$\int_0^\infty \frac{\cos nx}{e^{2\pi\sqrt{x}}-1}\,dx=\phi(n),$$
then
$$\int_0^\infty \frac{\sin nx}{e^{2\pi\sqrt{x}}-1}\,dx=\phi(n)-\frac{1}{2n}+\phi\left(\frac{\pi^2}{n}\right)\sqrt{\left(\frac{2\pi^3}{n^3}\right)}.$$
Find $\phi(n)$, and hence shew that
$$\phi(0)=\tfrac{1}{12},\quad \phi(\tfrac{1}{2}\pi)=\frac{1}{4\pi},\quad \phi(\pi)=\frac{2-\sqrt{2}}{8},$$
$$\phi(2\pi)=\tfrac{1}{16},\quad \phi(\infty)=0.$$

[See "Some definite integrals connected with Gauss's sums," *Messenger of Mathematics*, XLIV (1915), 75—85, and pp. 59—67 of this volume.]

Question 464 (v, 120):

2^n-7 is a perfect square for the values 3, 4, 5, 7, 15 of n. Find other values.

[*Solution by* K. J. Sanjana and T. P. Trivedi, v, 227—228.]

Question 469 (v, 159):

The number $1+n!$ is a perfect square for the values 4, 5, 7 of n. Find other values.

[*Comment by* M. Bhimasena Rao, XV, 97.]

Question 489 (v, 200):

Shew that
$$(1+e^{-\pi\sqrt{55}})(1+e^{-3\pi\sqrt{55}})(1+e^{-5\pi\sqrt{55}})\ldots=\frac{1+\sqrt{(3+2\sqrt{5})}}{\sqrt{2}}e^{-\frac{1}{24}\pi\sqrt{55}}.$$

[*Solution by* A. C. L. Wilkinson, VII, 104. See also "Modular equations and approximations to π," *Quarterly Journal of Mathematics*, XLV (1914), 350—372, and pp. 23—39 of this volume.]

Question 507 (v, 240):

Solve completely $\quad x^2=y+a,\quad y^2=z+a,\quad z^2=x+a;$
and hence shew that

(a) $\sqrt{[8-\sqrt{\{8+\sqrt{(8-\ldots)}\}}]}=1+2\sqrt{3}\sin 20°,$

(b) $\sqrt{[11-2\sqrt{\{11+2\sqrt{(11-\ldots)}\}}]}=1+4\sin 10°,$

(c) $\sqrt{[23-2\sqrt{\{23+2\sqrt{(23+2\sqrt{23-\ldots)}}\}}]}=1+4\sqrt{3}\sin 20°.$

Solution by Srinivasa Ramanujan, VI, 74—77.

We have
$$x=z^2-a=(y^2-a)^2-a=\{(x^2-a)^2-a\}^2-a,$$
or
$$x^8-4x^6a+2x^4(3a^2-a)-4x^2(a^3-a^2)-x+a^4-2a^3+a^2-a=0. \qquad (1)$$
Evidently y and z also are roots of this equation. But x may be written as
$$\sqrt{(a+y)}=\sqrt{\{a+\sqrt{(a+z)}\}}=\sqrt{[a+\sqrt{\{a+\sqrt{(a+x)}\}}]}$$
$$=\sqrt{[a+\sqrt{\{a+\sqrt{(a+\ldots)}\}}]}=\sqrt{(a+x)}.$$
Hence we have
$$x^2-x-a=0; \qquad (2)$$
therefore x^2-x-a must be a factor of the expression in (1). Now, dividing (1) by x^2-x-a, we have
$$x^6+x^5+x^4(1-3a)+x^3(1-2a)+x^2(1-3a+3a^2)+x(1-2a+a^2)$$
$$+(1-a+2a^2-a^3)=0. \qquad (3)$$

Let $\alpha, \beta, \gamma, \alpha', \beta', \gamma'$ be the roots of this equation. Then, since y and z are also roots, we may suppose that

$$\alpha^2=a+\beta, \quad \alpha'^2=a+\beta',$$
$$\beta^2=a+\gamma, \quad \beta'^2=a+\gamma',$$
$$\gamma^2=a+\alpha, \quad \gamma'^2=a+\alpha'.$$

Let $$\alpha+\beta+\gamma=u, \quad \alpha'+\beta'+\gamma'=v;$$

then we see that $$\alpha^2+\beta^2+\gamma^2=3a+u,$$

that is $$\alpha\beta+\beta\gamma+\gamma\alpha=\tfrac{1}{2}(u^2-u-3a).$$

Again we have $$\alpha^2\beta=a\beta+\beta^2, \quad \gamma\alpha^2=a\gamma+\beta\gamma,$$
$$\beta^2\gamma=a\gamma+\gamma^2, \quad \alpha\beta^2=a\alpha+\gamma\alpha,$$
$$\gamma^2\alpha=a\alpha+\alpha^2, \quad \beta\gamma^2=a\beta+\alpha\beta;$$

adding up all the six results we have

$$\Sigma\alpha^2(\beta+\gamma)=2a(\alpha+\beta+\gamma)+\alpha^2+\beta^2+\gamma^2+\alpha\beta+\beta\gamma+\gamma\alpha;$$

i.e. $$(\alpha+\beta+\gamma)(\alpha\beta+\beta\gamma+\gamma\alpha)-3\alpha\beta\gamma=2a(\alpha+\beta+\gamma)+(\alpha^2+\beta^2+\gamma^2)+(\alpha\beta+\beta\gamma+\gamma\alpha),$$

i.e. $$\tfrac{1}{2}u(u^2-u-3a)-3\alpha\beta\gamma=2au+(3a+u)+\tfrac{1}{2}(u^2-u-3a).$$

Hence we have $$6\alpha\beta\gamma=u^3-2u^2-7au-u-3a.$$

Similarly we have $$\alpha'+\beta'+\gamma'=v, \quad \alpha'\beta'+\beta'\gamma'+\gamma'\alpha'=\tfrac{1}{2}(v^2-v-3a),$$
$$6\alpha'\beta'\gamma'=v^3-2v^2-7av-v-3a.$$

Hence the sextic in (3) is identical with

$$\{x^3-x^2u+\tfrac{1}{2}x(u^2-u-3a)-\tfrac{1}{6}(u^3-2u^2-7au-u-3a)\}$$
$$\{x^3-x^2v+\tfrac{1}{2}x(v^2-v-3a)-\tfrac{1}{6}(v^3-2v^2-7av-v-3a)\}. \quad \text{.........(4)}$$

Now, equating the coefficients of x^5 and x^3 in (3) and (4), we have

$$u+v=-1,$$
$$u^3+v^3-2(u^2+v^2)-(7a+1)(u+v)-6a+3\{uv(u+v)-2uv-3a(u+v)\}=6(2a-1)$$

Substituting for $u+v$ in the above result, we have

$$uv=2-a;$$

hence $$u=-\tfrac{1}{2}\{1+\sqrt{(4a-7)}\}, \quad v=-\tfrac{1}{2}\{1-\sqrt{(4a-7)}\}. \quad \text{.....................(5)}$$

These values of u and v, when substituted in (4), reduce the sextic equation (3) to the two cubics:

$$x^3+\tfrac{1}{2}x^2\{1+\sqrt{(4a-7)}\}-\tfrac{1}{2}x\{2a+1+\sqrt{(4a-7)}\}+\tfrac{1}{2}\{a-2+a\sqrt{(4a-7)}\}=0,$$
$$x^3+\tfrac{1}{2}x^2\{1-\sqrt{(4a-7)}\}-\tfrac{1}{2}x\{2a+1-\sqrt{(4a-7)}\}+\tfrac{1}{2}\{a-2-a\sqrt{(4a-7)}\}=0,$$

which can be solved by the usual methods.

In the numerical examples proposed, the combinations of the signs plus and minus may be determined by proceeding as follows:

$$\begin{aligned}1+2\sqrt{3}\sin 20^\circ &= \sqrt{(1+4\sqrt{3}\sin 20^\circ+12\sin^2 20^\circ)}\\
&=\sqrt{(7+4\sqrt{3}\sin 20^\circ-6\cos 40^\circ)}\\
&=\sqrt{(7+4\sqrt{3}\sin 20^\circ-4\sqrt{3}\cos 30^\circ\cos 40^\circ)}\\
&=\sqrt{(7+4\sqrt{3}\sin 20^\circ-2\sqrt{3}\cos 70^\circ-2\sqrt{3}\cos 10^\circ)}\\
&=\sqrt{(7+2\sqrt{3}\cos 70^\circ-2\sqrt{3}\cos 10^\circ)}\\
&=\sqrt{(7-4\sqrt{3}\sin 30^\circ\sin 40^\circ)}\\
&=\sqrt{\{8-(1+2\sqrt{3}\sin 40^\circ)\}}\\
&=\sqrt{[8-\sqrt{\{8+(2\sqrt{3}\sin 80^\circ-1)\}}]}\\
&=\sqrt{[8-\sqrt{\{8+\sqrt{(8-(1+2\sqrt{3}\sin 20^\circ))}\}}]}.\end{aligned}$$

In a similar manner we have

$$1+4\sin 10^\circ = \surd\{11-2(1+4\sin 50^\circ)\}$$
$$= \surd[11-2\surd\{11+2(4\sin 70^\circ - 1)\}]$$
$$= \surd[11-2\surd\{11+2\surd(11-2(1+4\sin 10^\circ))\}];$$

and also

$$1+4\surd 3\sin 20^\circ = \surd\{23-2(4\surd 3\sin 80^\circ - 1)\}$$
$$= \surd[23-2\surd\{23+2(1+4\surd 3\sin 40^\circ)\}]$$
$$= \surd[23-2\surd\{23+2\surd(23+2(1+4\surd 3\sin 20^\circ))\}].$$

Question 524 (VI, 39):

Shew that (i) $\sqrt[3]{(\cos\frac{2}{7}\pi)}+\sqrt[3]{(\cos\frac{4}{7}\pi)}+\sqrt[3]{(\cos\frac{8}{7}\pi)}=\sqrt[3]{\{\frac{1}{2}(5-3\surd 7)\}}$,

(ii) $\sqrt[3]{(\cos\frac{2}{9}\pi)}+\sqrt[3]{(\cos\frac{4}{9}\pi)}+\sqrt[3]{(\cos\frac{8}{9}\pi)}=\sqrt[3]{\{\frac{1}{2}(3\sqrt[3]{9}-6)\}}$.

[*Solution by* N. Sankara Aiyar, VI, 190—191.]

Question 525 (VI, 39):

Shew how to find the square roots of surds of the form $\sqrt[3]{A}+\sqrt[3]{B}$, and hence prove that

(i) $\surd(\sqrt[3]{5}-\sqrt[3]{4})=\frac{1}{3}(\sqrt[3]{2}+\sqrt[3]{20}-\sqrt[3]{25})$,

(ii) $\surd(\sqrt[3]{28}-\sqrt[3]{27})=\frac{1}{3}(\sqrt[3]{98}-\sqrt[3]{28}-1)$.

[*Solution by* N. Sankara Aiyar, VI, 191—192.]

Question 526 (VI, 39):

If n is positive shew that

$$\frac{1}{n}>\frac{1}{n+1}+\frac{1}{(n+2)^2}+\frac{3}{(n+3)^3}+\frac{4^2}{(n+4)^4}+\frac{5^3}{(n+5)^5}+\dots;$$

and find approximately the difference when n is great. Hence shew that

$$\frac{1}{1001}+\frac{1}{1002^2}+\frac{3}{1003^3}+\frac{4^2}{1004^4}+\frac{5^3}{1005^5}+\dots$$

is less than $\frac{1}{1000}$ by approximately 10^{-440}.

Question 541 (VI, 79):

Prove that

$$1+\frac{1}{1.3}+\frac{1}{1.3.5}+\frac{1}{1.3.5.7}+\dots+\frac{1}{1+}\,\frac{1}{1+}\,\frac{2}{1+}\,\frac{3}{1+}\,\frac{4}{1+\dots}=\surd(\tfrac{1}{2}\pi e).$$

[*Solutions, with remarks, by* K. B. Madhava, VIII, 17—20.]

Question 546 (VI, 80):

Shew that

$$\text{(i)}\quad \left(\frac{1}{3}-\frac{1}{4}\right)+\frac{2}{3^2}\left(\frac{1}{3}-\frac{1}{4^2}\right)+\frac{2.4}{3.5^2}\left(\frac{1}{3}-\frac{1}{4^3}\right)+\dots=\tfrac{1}{12}\pi\log(2+\surd 3),$$

$$\text{(ii)}\quad 1-\frac{2}{3^2}+\frac{2.4}{3.5^2}-\frac{2.4.6}{3.5.7^2}+\dots=\tfrac{1}{8}\pi^2-\tfrac{1}{2}\{\log(1+\surd 2)\}^2.$$

[*Solution by* M. Bhimasena Rao, VII, 107—109; also K. J. Sanjana, "Notes on Questions 546, 572, 573, 583, 585, 606," VII, 136—141.]

Question 571 (VI, 160):

If

$$\tfrac{1}{2}\pi\alpha=\log\tan\{\tfrac{1}{4}\pi(1+\beta)\}$$

shew that

$$\left(\frac{1^2+\alpha^2}{1^2-\beta^2}\right)\left(\frac{3^2-\beta^2}{3^2+\alpha^2}\right)^3\left(\frac{5^2+\alpha^2}{5^2-\beta^2}\right)^5\dots=e^{\frac{1}{2}\pi\alpha\beta}.$$

[*Solution by* J. C. Swaminarayan, R. J. Pocock, A. Narasinga Rao and K. B. Madhava, VII, 32. See also "On the Integral $\int_0^x \frac{\tan^{-1}t}{t}\,dt$," *Journal of the Indian Mathematical Society*, VII (1915), 93—96, and pp. 40—43 of this volume, equation (17).]

Question 584 (VI, 199):

Examine the correctness of the following results:

$$\text{(i)}\quad 1+\frac{x}{1-x}+\frac{x^4}{(1-x)(1-x^2)}+\frac{x^9}{(1-x)(1-x^2)(1-x^3)}+\dots$$
$$=\frac{1}{(1-x)(1-x^6)(1-x^{11})(1-x^{16})\dots}\times\frac{1}{(1-x^4)(1-x^9)(1-x^{14})\dots};$$

here 1, 4, 9, ... on the left are the squares of natural numbers, while 1, 6, 11, 16, ... and 4, 9, 14, ... on the right are numbers in arithmetical progression, with 5 for common difference.

$$\text{(ii)}\quad 1+\frac{x^2}{1-x}+\frac{x^6}{(1-x)(1-x^2)}+\frac{x^{12}}{(1-x)(1-x^2)(1-x^3)}+\dots$$
$$=\frac{1}{(1-x^2)(1-x^7)(1-x^{12})\dots}\times\frac{1}{(1-x^3)(1-x^8)(1-x^{13})\dots};$$

here the nth term of the sequence 2, 6, 12, ... is $n(n+1)$, and 2, 7, 12, ..., 3, 8, 13, ... increase by 5.

[See "Proof of certain identities in Combinatory Analysis," *Proc. Camb. Phil. Soc.*, XIX (1919), 214—216, and pp. 214—215 of this volume.]

Question 605 (VI, 239):

Shew that, when $x=\infty$,

$$\frac{(x+a-b)!\,(8x+2b)!\,(9x+a+b)!}{(3x+a-c)!\,(3x+a-b+c)!\,(12x+3b)!}=\sqrt{\tfrac{2}{3}}.$$

[*Solution by* K. J. Sanjana and A. Narasinga Rao, VII, 191—192.]

Question 606 (VI, 239):

Shew that

$$\sum_0^\infty\frac{(\sqrt{5}-2)^{2n+1}}{(2n+1)^2}=\tfrac{1}{24}\pi^2-\tfrac{1}{12}\{\log(2+\sqrt{5})\}^2.$$

[*Solution by* N. Sankara Aiyar, VII, 192; also the note by K. J. Sanjana quoted with reference to Question 546.]

Question 629 (VII, 40):

Prove that

$$\tfrac{1}{2}+\sum_{n=1}^\infty e^{-\pi n^2 x}\cos\{\pi n^2\sqrt{(1-x^2)}\}=\frac{\sqrt{2}+\sqrt{(1+x)}}{\sqrt{(1-x)}}\sum_{n=1}^\infty e^{-\pi n^2 x}\sin\{\pi n^2\sqrt{(1-x^2)}\};$$

and deduce the following:

$$\text{(i)}\quad \tfrac{1}{2}+\sum_{n=1}^\infty e^{-\pi n^2}=\sqrt{(5\sqrt{5}-10)}\left[\tfrac{1}{2}+\sum_{n=1}^\infty e^{-5\pi n^2}\right],$$

$$\text{(ii)}\quad \sum_{n=1}^\infty e^{-\pi n^2}(\pi n^2-\tfrac{1}{4})=\tfrac{1}{8}.$$

[*Solutions by* (1) K. B. Madhava, (2) N. Durai Rajan, and (3) M. Bhimasena Rao, VIII, 25—30.]

Question 642 (VII, 80):

Shew that

$$\text{(i)}\quad \sum_{n=0}^\infty\left(1+\tfrac{1}{3}+\tfrac{1}{5}+\dots+\frac{1}{2n+1}\right)\frac{5^{-n}}{2n+1}=\frac{\pi^2}{4\sqrt{5}},$$

$$\text{(ii)}\quad \sum_{n=0}^\infty\left(1+\tfrac{1}{3}+\tfrac{1}{5}+\dots+\frac{1}{2n+1}\right)\frac{9^{-n}}{2n+1}=\tfrac{1}{8}\pi^2-\tfrac{3}{8}(\log 2)^2.$$

[*Solution, with correction, by* M. Bhimasena Rao, VII, 232—233.]

Question 661 (VII, 119):

Solve in integers $$x^3+y^3+z^3=u^6,$$
and deduce the following:

$$6^3-5^3-3^3=2^6, \qquad 8^3+6^3+1^3=3^6,$$
$$12^3-10^3+1^3=3^6, \qquad 46^3-37^3-3^3=6^6,$$
$$174^3+133^3-45^3=14^6, \qquad 1188^3-509^3-3^3=34^6.$$

[*Solutions by* N. B. Mitra, XIII, 15—17. Additional solution and remarks by N. B. Mitra, XIV, 73—77.]

Question 662 (VII, 119—120):

Let AB be a diameter and BC be a chord of a circle ABC. Bisect the minor arc BC at M; and draw a chord BN equal to half of the chord BC. Join AM. Describe two circles with A and B as centres and AM and BN as radii, cutting each other at S and S', and cutting the given circle again at the points M' and N' respectively. Join AN and BM intersecting at R, and also join AN' and BM' intersecting at R'. Through B draw a tangent to the given circle, meeting AM and AM' produced at Q and Q' respectively. Produce AN and $M'B$ to meet at P, and also produce AN' and MB to meet at P'. Shew that the eight points P, Q, R, S, S', R', Q', P' are cyclic, and that the circle passing through these eight points is orthogonal to the given circle ABC.

Question 666 (VII, 120):

Solve in positive rational numbers
$$x^y=y^x.$$
For example: $$x=4,\ y=2;\quad x=3\tfrac{3}{8},\ y=2\tfrac{1}{4}.$$

[*Solution by* J. C. Swaminarayan and R. Vythynathaswamy, VIII, 31.]

Question 681 (VII, 160):

Solve in integers $$x^3+y^3+z^3=1,$$
and deduce the following:

$$6^3+8^3=9^3-1, \qquad 9^3+10^3=12^3+1, \qquad 135^3+138^3=172^3-1,$$
$$791^3+812^3=1010^3-1, \qquad 11161^3+11468^3=14258^3+1, \qquad 65601^3+67402^3=83802^3+1.$$

[*Partial solution by* N. B. Mitra, XIII, 17. See also N. B. Mitra, XIV, 73—77 (76—77).]

Question 682 (VII, 160):

Shew how to find the cube roots of surds of the form $A+\sqrt[3]{B}$; and deduce that
$$\sqrt[3]{(\sqrt[3]{2}-1)}=\sqrt[3]{(\tfrac{1}{9})}-\sqrt[3]{(\tfrac{2}{9})}+\sqrt[3]{(\tfrac{4}{9})}.$$

[*Solution by* "Zero," X, 325.]

Question 699 (VII, 199):

Shew that the roots of the equations
$$\text{(i)}\ x^6-x^3+x^2+2x-1=0,$$
$$\text{(ii)}\ x^6+x^5-x^3-x^2-x+1=0$$
can be expressed in terms of radicals.

Question 700 (VII, 199):

Sum the series
$$(a+b+1)\left(\frac{a}{b}\right)^2+(a+b+3)\left\{\frac{a\,(a+1)}{b\,(b+1)}\right\}^2+(a+b+5)\left\{\frac{a\,(a+1)\,(a+2)}{b\,(b+1)\,(b+2)}\right\}^2+\ldots \text{ to } n \text{ terms}.$$

[*Solutions by* (1) K. R. Rama Aiyar, (2) K. Appukuttan Erady, VIII, 152.]

Question 722 (VII, 240):

Solve completely

$$x^2=a+y,\quad y^2=a+z,\quad z^2=a+u,\quad u^2=a+x;$$

and deduce that, if $\quad x=\sqrt{[5+\sqrt{\{5+\sqrt{(5-\sqrt{(5+x)})}\}}]}$,

then $\quad x=\frac{1}{2}\{2+\sqrt{5}+\sqrt{(15-6\sqrt{5})}\}$;

and that, if $\quad x=\sqrt{[5+\sqrt{\{5-\sqrt{(5-\sqrt{(5+x)})}\}}]}$,

then $\quad x=\frac{1}{4}[\sqrt{5}-2+\sqrt{(13-4\sqrt{5})}+\sqrt{\{50+12\sqrt{5}-2\sqrt{(65-20\sqrt{5})}\}}]$.

Question 723 (VII, 240):

If $[x]$ denotes the greatest integer in x, and n is any positive integer, shew that

(i) $\left[\frac{n}{3}\right]+\left[\frac{n+2}{6}\right]+\left[\frac{n+4}{6}\right]=\left[\frac{n}{2}\right]+\left[\frac{n+3}{6}\right]$,

(ii) $[\frac{1}{2}+\sqrt{(n+\frac{1}{2})}]=[\frac{1}{2}+\sqrt{(n+\frac{1}{4})}]$,

(iii) $[\sqrt{n}+\sqrt{(n+1)}]=[\sqrt{(4n+2)}]$.

[*Solution by* H. Br., X, 357—358.]

Question 724 (VII, 240):

Shew that

(i) $\tan^{-1}\frac{1}{2n+1}+\tan^{-1}\frac{1}{2n+3}+\tan^{-1}\frac{1}{2n+5}+\ldots$ to n terms

$=\tan^{-1}\frac{1}{1+2.1^2}+\tan^{-1}\frac{1}{3(1+2.3^2)}+\tan^{-1}\frac{1}{5(1+2.5^2)}+\ldots$ to n terms,

(ii) $\tan^{-1}\frac{1}{(2n+1)\sqrt{3}}+\tan^{-1}\frac{1}{(2n+3)\sqrt{3}}+\tan^{-1}\frac{1}{(2n+5)\sqrt{3}}+\ldots$ to n terms

$=\tan^{-1}\frac{1}{(\sqrt{3})^3}+\tan^{-1}\frac{1}{(3\sqrt{3})^3}+\tan^{-1}\frac{1}{(5\sqrt{3})^3}+\ldots$ to n terms.

[*Solution, with correction, by* K. B. Madhava, VIII, 191—192; also *Remarks by* Mehr Chand Suri, XVI, 121.]

Question 738 (VIII, 40):

If $\quad \phi(x)=e^{-x}+\frac{x}{1!}e^{-2x}+\frac{3x^2}{2!}e^{-3x}+\frac{4^2x^3}{3!}e^{-4x}+\frac{5^3x^4}{4!}e^{-5x}+\ldots,$

shew that $\phi(x)=1$ when x lies between 0 and 1; and that $\phi(x)\neq 1$ when $x>1$. Find the limit of

$$\{\phi(1+\epsilon)-\phi(1)\}/\epsilon$$

as $\epsilon\to 0$ through positive values.

Question 739 (VIII, 40):

Shew that $\quad \int_0^\infty e^{-nx}(\cot x+\coth x)\sin nx\,dx=\frac{1}{2}\pi\left(\frac{1+e^{-n\pi}}{1-e^{-n\pi}}\right)^{(-1)^n}$

for all positive integral values of n.

[*Solution by* A. C. L. Wilkinson, VIII, 218—219.]

Question 740 (VIII, 40):

If $\quad \phi(x)=\left\{\frac{e^x[x]!}{x^{[x]}}\right\}^2-2\pi x,$

where $[x]$ denotes the greatest integer in x, shew that $\phi(x)$ is a continuous function of x for all positive values of x, and oscillates from $\frac{1}{3}\pi$ to $-\frac{1}{6}\pi$ when x becomes infinite. Also differentiate $\phi(x)$.

[*Solution by* A. C. L. Wilkinson, VIII, 220—221.]

Question 753 (VIII, 80):

If
$$\phi(x)=\tfrac{1}{2}\log 2\pi x-x+\int_1^x\frac{[t]}{t}\,dt,$$
where $[t]$ denotes the greatest integer in t, shew that
$$\overline{\lim_{x\to\infty}}\,x\phi(x)=\tfrac{1}{24},\quad \underline{\lim_{x\to\infty}}\,x\phi(x)=-\tfrac{1}{12}.$$

[*Solutions by* (1) N. Durairajan, (2) K. B. Madhava, X, 395—397.]

Question 754 (VIII, 80):

Shew that
$$e^x x^{-x}\pi^{-\frac{1}{2}}\Gamma(1+x)=(8x^3+4x^2+x+E)^{\frac{1}{6}},$$
where E lies between $\frac{1}{100}$ and $\frac{1}{30}$ for all positive values of x.

[*Partial solution by* K. B. Madhava, XII, 101. See also remarks by E. H. Neville and C. Krishnamachari, XIII, 151.]

Question 755 (VIII, 80):

Let p be the perimeter and e the eccentricity of an ellipse whose centre is C, and let CA and CB be a semi-major and a semi-minor axis. From CA cut off CQ equal to CB, and also produce AC to P making CP equal to CB. From A draw AN perpendicular to CA (in the direction of CB). From Q draw QM making with QA an angle equal to ϕ (which is to be determined) and meeting AN at M. Join PM and draw PN making with PM an angle equal to half of the angle APM, and meeting AN at N. With P as centre and PA as radius describe a circle, cutting PN at K, and meeting PB produced at L. Then, if
$$\frac{\text{arc } AL}{\text{arc } AK}=\frac{p}{4AN},$$
trace the changes in ϕ when e varies from 0 to 1. In particular, shew that $\phi=30°$ when $e=0$; $\phi\to 30°$ when $e\to 1$; $\phi=30°$ when $e=0{\cdot}99948$ nearly; ϕ assumes the minimum value of about $29°\ 58\tfrac{3}{4}'$ when e is about $0{\cdot}999886$; and ϕ assumes the maximum value of about $30°\ 44\tfrac{1}{4}'$ when e is about $0{\cdot}9589$.

Question 768 (VIII, 119):

If
$$\psi(x)=(x+2)/(x^2+x+1),$$
shew that (i) $\tfrac{1}{3}\psi(x^{\frac{1}{3}})+\tfrac{1}{9}\psi(x^{\frac{1}{9}})+\tfrac{1}{27}\psi(x^{\frac{1}{27}})+\ldots=(\log x)^{-1}+(1-x)^{-1}$

for all positive values of x; and that

(ii) $\tfrac{1}{3}\psi(x^{\frac{1}{3}})+\tfrac{1}{9}\psi(x^{\frac{1}{9}})+\tfrac{1}{27}\psi(x^{\frac{1}{27}})+\ldots=(1-x)^{-1}$

for all negative values of x.

[*Solution, with correction, by* K. J. Sanjana and N. Durairajan, VIII, 227.]

Question 769 (VIII, 120):

Shew that
$$\log 2\left(\frac{1}{2\log 2\log 4}+\frac{1}{3\log 3\log 6}+\frac{1}{4\log 4\log 8}+\ldots\right)$$
$$+\frac{1}{2\log 2}-\frac{1}{3\log 3}+\frac{1}{4\log 4}-\frac{1}{5\log 5}+\ldots=\frac{1}{\log 2}.$$

[*Solution, with correction, by* K. B. Madhava, M. K. Kewalramani, N. Durairajan, and S. V. Venkatachala Aiyar, IX, 120—121.]

Question 770 (VIII, 120):

If δ_n denote the number of divisors of n (e.g., $\delta_1=1$, $\delta_2=2$, $\delta_3=2$, $\delta_4=3$, ...) shew that

(i) $\delta_1-\tfrac{1}{3}\delta_3+\tfrac{1}{5}\delta_5-\tfrac{1}{7}\delta_7+\tfrac{1}{9}\delta_9-\ldots$

is a convergent series; and that

(ii) $\delta_1-\tfrac{1}{2}\delta_2+\tfrac{1}{3}\delta_3-\tfrac{1}{4}\delta_4+\tfrac{1}{5}\delta_5-\ldots$

is a divergent series in the *strict* sense (i.e. not oscillating).

Question 783 (VIII, 159):

If
$$x=y^n-y^{n-1},$$
$$J_n=\int_0^1 \frac{\log y}{x}\,dx,$$
shew that (i) $J_0=\frac{1}{6}\pi^2,\quad J_{\frac{1}{2}}=\frac{1}{10}\pi^2,\quad J_1=\frac{1}{12}\pi^2,\quad J_2=\frac{1}{15}\pi^2;$

(ii) $$J_n+J_{1/n}=\tfrac{1}{6}\pi^2.$$

[*Solution by* N. Durai Rajan and "Zero," x, 397—399.]

Question 784 (VIII, 159):

If $F(x)$ denotes the fractional part of x (e.g. $F(\pi)=0\cdot14159\ldots$), and if N is a positive integer, shew that

(i) $$\lim_{N\to\infty} NF(N\sqrt{2})=\frac{1}{2\sqrt{2}},\quad \lim_{N\to\infty} NF(N\sqrt{3})=\frac{1}{\sqrt{3}},$$
$$\lim_{N\to\infty} NF(N\sqrt{5})=\frac{1}{2\sqrt{5}},\quad \lim_{N\to\infty} NF(N\sqrt{6})=\frac{1}{\sqrt{6}},$$
$$\lim_{N\to\infty} NF(N\sqrt{7})=\frac{3}{2\sqrt{7}};$$

(ii) $$\lim_{N\to\infty} N(\log N)^{1-p}F(Ne^{2/n})=0,$$

where n is any integer and p is any positive number; shew further that in (ii) p cannot be zero.

Question 785 (VIII, 159—160):

Shew that $$[3\{(a^3+b^3)^{\frac{1}{3}}-a\}\{(a^3+b^3)^{\frac{1}{3}}-b\}]^{\frac{1}{3}}=(a+b)^{\frac{2}{3}}-(a^2-ab+b^2)^{\frac{1}{3}}.$$
This is analogous to
$$[2\{(a^2+b^2)^{\frac{1}{2}}-a\}\{(a^2+b^2)^{\frac{1}{2}}-b\}]^{\frac{1}{2}}=a+b-(a^2+b^2)^{\frac{1}{2}}.$$

[*Solution by* K. K. Ranganatha Aiyar, R. D. Karve, G. A. Kamtekar, L. N. Datta and L. N. Subramanyam, VIII, 232.]

Question 1049 (XI, 120):

Shew that (i) $$\int_0^\infty \frac{\sin nx\,dx}{x+\dfrac{1}{x+}\,\dfrac{2}{x+}\,\dfrac{3}{x+}\cdots}=\frac{\sqrt{(\frac{1}{2}\pi)}}{n+\dfrac{1}{n+}\,\dfrac{2}{n+}\,\dfrac{3}{n+}\cdots};$$

(ii) $$\int_0^\infty \frac{\sin\frac{1}{2}\pi nx\,dx}{x+\dfrac{1^2}{x+}\,\dfrac{2^2}{x+}\,\dfrac{3^2}{x+}\cdots}=\frac{1}{n+}\,\frac{1^2}{n+}\,\frac{2^2}{n+}\,\frac{3^2}{n+}\cdots.$$

Question 1070 (XI, 160):

Shew that

(i) $$(\sqrt[5]{\tfrac{1}{5}}+\sqrt[5]{\tfrac{4}{5}})^{\frac{1}{2}}=(1+\sqrt[5]{2}+\sqrt[5]{8})^{\frac{1}{5}}=\sqrt[5]{\tfrac{16}{125}}+\sqrt[5]{\tfrac{8}{125}}+\sqrt[5]{\tfrac{2}{125}}-\sqrt[5]{\tfrac{1}{125}};$$

(ii) $$(\sqrt[5]{\tfrac{32}{5}}-\sqrt[5]{\tfrac{27}{5}})^{\frac{1}{3}}=\sqrt[5]{\tfrac{1}{25}}+\sqrt[5]{\tfrac{3}{25}}-\sqrt[5]{\tfrac{9}{25}};$$

(iii) $$\left(\frac{3+2\sqrt[4]{5}}{3-2\sqrt[4]{5}}\right)^{\frac{1}{4}}=\frac{\sqrt[4]{5}+1}{\sqrt[4]{5}-1}.$$

[*Solution by* S. D. Chowla, N. B. Mitra, and S. V. Venkataraya Sastri, XVI, 122—123.]

Question 1076 (XI, 199):

Shew that (i) $$(7\sqrt[3]{20}-19)^{\frac{1}{6}}=\sqrt[3]{\tfrac{5}{3}}-\sqrt[3]{\tfrac{2}{3}};$$

(ii) $$(4\sqrt[3]{\tfrac{2}{3}}-5\sqrt[3]{\tfrac{1}{3}})^{\frac{1}{6}}=\sqrt[3]{\tfrac{4}{9}}-\sqrt[3]{\tfrac{2}{9}}+\sqrt[3]{\tfrac{1}{9}}.$$

APPENDIX I: NOTES ON THE PAPERS

1.

Page 1, (1). Ramanujan writes B_{2n} where B_n is usual.

P. 1, (2). $$c_k = \binom{n}{k} = \frac{n!}{k!\,(n-k)!}.$$

P. 4, l. 4. (12) is true if n is odd and greater than 1, and the last terms are given in the three cases $n \equiv 1, \; -1, \; 3 \pmod{6}$.

P. 8, ll. 7—9. The proof is invalid as it stands, since the series in the integrand is divergent when $x \geqq n$. The same remark applies to p. 9, ll. 6—8.

P. 9, l. 2 *from below*. The coefficients, if integral, are necessarily odd.

P. 10, (32). Apply (31) to the well-known formula

$$\psi(z+1) - \psi(\tfrac{1}{2}z+1) = \log 2 - \sum_{m=1}^{\infty} \frac{(-1)^{m-1}}{z+m},$$

where $\psi(z)$ is the derivative of $\log \Gamma(z)$.

P. 11, l. 2 *from below*. N and D are the numerator and denominator of B_{444}.

The paper is interesting as a specimen of Ramanujan's earliest manner, but the principal results are well known and the proofs are incomplete. Thus (20) of § 8 is simply the theorem of von Staudt and Clausen, and (16), which is a consequence of (20), was known to Euler: see P. Bachmann, *Niedere Zahlentheorie*, vol. II, pp. 43—51 and p. 54; or N. Nielsen, *Traité élémentaire des nombres de Bernoulli*, pp. 244—245 and 258.

Ramanujan states eight theorems, viz. (14)—(21), embodying arithmetical properties of the B's. Of these, proofs are indicated for three, viz. (16), (20), and (21); but the theorems on which these proofs would depend, viz. (28) and the corresponding propositions about the series (30) and (32), are never proved. Two other theorems, (17) and (18), are stated to be corollaries of (16) and (14); and (14), (15), and (19) are stated merely as conjectures.

4.

All the results of this note are naturally familiar in the elements of the analytic theory of numbers: see Landau's *Handbuch der Lehre von der Verteilung der Primzahlen*, especially pp. 567 *et seq.*, dealing with Mobius' function $\mu(n)$.

6.

This paper is of special interest because it embodies so much of Ramanujan's early Indian work. It was rewritten in England, the references to Weber inserted, and many results given by Weber deleted, but all the work was done quite independently.

Professor L. J. Mordell writes:

"It would be extremely interesting to know if and how much Ramanujan is indebted to other writers. Some results are easily accessible in English books. See, for example, Greenhill's *Elliptic Functions*, chap. X, in particular pp. 330—339, for results and references concerning complex multiplication; and Cayley's *Elliptic Functions*, chap. VII. It is of course possible that these books were known to Ramanujan.

"As Ramanujan says, his functions G_n and g_n are practically Weber's $f\{\sqrt{(-n)}\}$ and $f_1\{\sqrt{(-n)}\}$. The properties of $f(w)$ and $f_1(w)$ are given in Weber's *Lehrbuch der Algebra*, vol. III (1908), § 34, pp. 112—116. The modular equations, which are merely the algebraic equations connecting $f(nw)$ and $f(w)$, etc., are developed in §§ 73—76, pp. 256—280. All the other roots of the modular equation for $f(w)$ are expressible very simply by means of a finite number of the values $f\left(\frac{aw+b}{cw+d}\right)$, where a, b, c, d are integers such that $ad-bc=n$. If we put $f(nw)=f(w)$ in the equation it splits up into factors, the roots of each factor being of the form $f_D^{\lambda}(\Omega)$, where now D is a function of n and Ω is a root, with positive imaginary part, of one of the system of non-equivalent quadratic forms of determinant $-D$. Each of these factors in turn splits up into simpler factors, whose roots are now restricted to belong to a genus of quadratic forms of determinant $-D$. All this is developed in chapters XVII—XX of Weber's book. It appears from Ramanujan's § 4 that he was not aware that the character of the surds for G_n and g_n depends on the number of classes of quadratic forms.

"Ramanujan's method of approximating to π by means of equations

$$e^{\pi\sqrt{n}}=m,$$

where m is nearly an integer, is due to Hermite. See Ch. Hermite, 'Sur la theorie des equations modulaires' [*Comptes Rendus*, 16 May, 13 June, 20 June, 4 July, 18 July, and 25 July, 1859 (20 June), and *Œuvres*, vol. II, pp. 38—82 (pp. 60—61)]; L. Kronecker, *Berliner Monatsberichte*, 1863, pp. 340—345; and H. J. S. Smith, 'Report on the theory of numbers' [*Report of the British Association*, 1865, pp. 322—375, and *Collected Papers*, vol. I, pp. 289—358 (p. 357)]. Hermite and Kronecker give some of Ramanujan's approximations, or results equivalent to them. On the other hand Ramanujan's method of obtaining purely algebraical approximations appears to be new.

"It is unfortunate that Ramanujan has not developed in detail the corresponding theories referred to in § 14. His K_1 is

$$1+\frac{1.3}{4^2}k^2+\frac{1.3.5.7}{4^2.8^2}k^4+\ldots=\frac{2}{\pi}\int_0^{\frac{1}{2}\pi}\frac{d\theta}{\sqrt{1-k^2\sin^4\theta}},$$

which is reducible to an elliptic integral. The result in § 15, that K can be expressed in terms of Γ-functions when q is of the forms $e^{-\pi n}$, $e^{-\pi n\sqrt{2}}$, $e^{-\pi n\sqrt{3}}$ and n is rational, also seems to be new. It may be deduced from the equation

$$q^{\frac{1}{12}}(1-q^2)(1-q^4)\ldots=(2kk')^{\frac{1}{6}}\sqrt{\left(\frac{K}{\pi}\right)}.$$

For the left-hand side is Weber's $\eta(w)$ (Weber, *l.c.*, p. 85), and there is an algebraic equation connecting $\eta(w)$ and $\eta(nw)$ when n is an integer. There is also an algebraic equation connecting kk' and its transformed value, and so one connecting K and its transformed value. Finally, K is known in terms of Γ-functions when q is $e^{-\pi}$, $e^{-\pi\sqrt{2}}$, or $e^{-\pi\sqrt{3}}$: for these classic results see Whittaker and Watson, *Modern Analysis* (ed. 3, 1920), pp. 524—526.

"Weber's $\eta(w)$ is practically the modular invariant $\Delta(w_1, w_2)$ of Klein-Fricke. For the theory of the algebraic equations connecting $\Delta(w_1, w_2)$ and $\Delta(nw_1, w_2)$, see their *Elliptische Modulfunktionen*, vol. II, pp. 62—82. See also pp. 117—159 for the theory of the modular equation and for further references."

7.

P. 40, (4). If $R(x)>0$, we have

$$\phi(x)=\int_{1/x}^{\infty}\tan^{-1}\frac{1}{t}\frac{dt}{t}=\lim_{X\to\infty}\int_{1/x}^{X}\frac{\frac{1}{2}\pi-\tan^{-1}t}{t}dt$$

$$=\tfrac{1}{2}\pi\log x+\phi\left(\frac{1}{x}\right)+\lim_{X\to\infty}\{\tfrac{1}{2}\pi\log X-\phi(X)\}.$$

If we write A for the last limit, then

$$\phi(x) = \tfrac{1}{2}\pi \log x + \phi\left(\frac{1}{x}\right) + A,$$

and A may be determined by putting $x=1$.

P. 40, (6). This equation should read

$$\frac{1}{1^2}+\frac{1}{3^2}-\frac{1}{5^2}-\frac{1}{7^2}+\frac{1}{9^2}+\ldots=\sqrt{2}\,\phi(\sqrt{2}-1)+\frac{\pi}{4\sqrt{2}}\log(1+\sqrt{2}).$$

P. 42, § 4. Since $-1 < I(x) < 1$ in (15), the x of this section must in the first instance lie (when real) between $-\frac{1}{2}$ and $\frac{1}{2}$. The results must then be extended by analytic continuation, in which some care is necessary with the many-valued functions.

P. 43, (26). The real parts of α and β must be positive.

The transcendent considered in this note is closely connected with the integrals

$$\int \frac{\log(1-x)}{x}\,dx, \quad \int \log\left(\frac{1+x}{1-x}\right)\frac{dx}{x}, \quad \int x \cot x\, dx, \quad \int \log\cos x\, dx, \ \ldots,$$

whose properties have been investigated by a number of writers. See, for example Bertrand's *Calcul intégral*, §§ 270 *et seq.*; Lobatschewsky's memoirs "Imaginäre Geometrie" and "Anwendung der imaginären Geometrie auf einige Integrale" (German edition by H. Liebmann, Leipzig, 1904); and L. J. Rogers, "On function sum theorems connected with the series $\Sigma n^{-2}x^n$", *Proc. London Math. Soc.* (2), vol. IV (1906), pp. 169—189. It appears from Ramanujan's notebooks that he had found the values of

$$L(x) = \int_0^x \frac{\log(1-t)}{t}\,dt$$

for the special values $\frac{1}{2}$, 2, $\frac{1}{2}(\sqrt{5}-1), \ldots$, given by Bertrand, and Rogers' result that

$$L(x)+L(y)-L(xy)-L\left\{\frac{x(1-y)}{1-xy}\right\}-L\left\{\frac{y(1-x)}{1-xy}\right\}$$

is an elementary function, without knowledge of the work of these writers.

8.

This is merely a preliminary account of a small part of No. 15.

9.

P. 48, l. 17. The constant is of course the constant of the Euler-Maclaurin sum-formula, which is in this case $\zeta(1-r)$.

11.

P. 57, *bottom*. The reference is to G. H. Hardy, "Proof of a formula of Mr Ramanujan *Messenger of Math.*, vol. XLIV (1915), pp. 18—21. The proof given by Hardy depends on the application of Cauchy's Theorem to the integral

$$\frac{1}{2\pi i}\int \frac{F(abz)}{F(z)}(-z)^{n-1}dz,$$

where

$$F(z) = (1+z)(1+bz)(1+b^2z)\ldots.$$

P. 58, formulæ after (24). See No. 6, text and comments in Appendix. The special cases are naturally more difficult than the general formula.

12.

P. 61, *bottom*. The theorem referred to is due to Lerch. See M. Lerch, "Sur un point de la théorie des fonctions génératrices d'Abel", *Acta Math.*, vol. XXVII (1903), pp. 339—351; E. W. Hobson, "The fundamental lemma of the calculus of variations", *Proc. London Math. Soc.* (2), vol. XI (1911), pp. 17—28. Lerch proves that if $f(x)$ is continuous and

$$\int_0^\infty e^{-ax} f(x)\, dx = 0$$

for $a = \alpha + n\beta$ $(n = 0, 1, 2, \ldots)$, then $f(x) = 0$. A substantially equivalent theorem is that

$$\int_0^1 x^n f(x)\, dx = 0 \qquad (n = 0, 1, 2, \ldots)$$

implies $f(x) = 0$. Hobson gives more general forms of this result. Ramanujan requires only the special case of Lerch's theorem in which the integral vanishes for all (sufficiently large) positive values of a; this indeed was one of his favourite weapons.

P. 63, § 3. See A. L. Cauchy, *Exercices de mathématiques*, vol. II (1827), pp. 141—156.

A great deal of calculation is suppressed in this paper, and the verification of particular formulæ is sometimes laborious. More general results were found independently by L. J. Mordell, "The value of the definite integral $\int_{-\infty}^{\infty} \frac{e^{at^2+bt}}{e^{ct}+d}\, dt$", *Quarterly Journal of Math.*, vol. XLVIII (1920), pp. 329—342.

14.

Ramanujan does not attempt to use his identities to obtain approximations to $\zeta(s)$, but the idea of splitting up $\zeta(s)$ into two parts is the same as that which underlies Hardy and Littlewood's "approximate functional equation". See G. H. Hardy and J. E. Littlewood, "The zeros of Riemann's Zeta-function on the critical line", *Math. Zeitschrift*, vol. X (1921), pp. 283—317; and "The approximate functional equation in the theory of the Zeta-function, with applications to the divisor problems of Dirichlet and Piltz", *Proc. London Math. Soc.* (2), vol. XXI (1923), pp. 39—74.

This paper was followed in the *Quarterly Journal* by a short note by Hardy which we do not reproduce.

15.

P. 79, l. 6 *from below*. The identity attributed to Hardy was first proved by Voronoï: see G. H. Hardy, "On Dirichlet's divisor problem", *Proc. London Math. Soc.* (2), vol. XV (1916), pp. 1—25 (p. 21). The result referred to immediately above, viz. that (in the notation of **17**) $\Delta_1(n)$ is of order not lower than $n^{\frac{1}{4}}$, was proved independently and almost simultaneously by Hardy [*l.c. supra*, and "Sur le problème des diviseurs de Dirichlet", *Comptes Rendus*, 10 May 1915] and Landau ["Über die Gitterpunkte in einem Kreise (Zweite Mitteilung)", *Göttinger Nachrichten*, 1915, pp. 161—171: see also "Über die Heckesche Funktionalgleichung", *ibid.*, 1917, pp. 102—111 (footnote 3 to p. 102)]. For further developments, see the notes to **17**.

P. 80, l. 5. See also E. Landau, *Handbuch der Lehre von der Verteilung der Primzahlen*, § 60, pp. 219—222.

P. 85, l. 8 *from below*. The argument beginning "Now, if N is a function of t,..." is correct in principle but expressed in a puzzling manner. Write

$$d(N) \leqq \left(1 + \frac{\log N}{t}\right)^{\frac{t}{\log t} + \frac{At}{(\log t)^2}},$$

where A is constant, and allow t to vary in a range $(T, \log N)$, where T is sufficiently large but fixed. It is then easily verified that the right-hand side is greatest when $t = \log N$.

P. 88. The number

$$293{,}318{,}625{,}600 = 2^6 . 3^4 . 5^2 . 7^2 . 11 . 13 . 17 . 19,$$

with $5040 = 7 . 5 . 3^2 . 2^4$ divisors, was omitted by Ramanujan. The omission was pointed out by T. Vijayaraghavan, *Journal London Math. Soc.*, vol. I (1926), p. 192.

P. 107, l. 5. n is of course chosen to be a divisor of N.

P. 119, § 40. The ρ's in series and products, when these are not absolutely convergent, are arranged in order of increasing moduli. For a full discussion of the convergence of the series $\Sigma\,(x^\rho/\rho)$ see E. Landau, "Über die Nullstellen der Zetafunktion", *Math. Annalen*, vol. LXXI (1912), pp. 548—564.

P. 121, (229). We have deleted a superfluous $(\log t)^2$ and $(\log x)^2$ in the O terms.

This paper, long as it is, is not complete. The London Mathematical Society was in some financial difficulty at the time, and Ramanujan suppressed part of what he had written in order to save expense.

16.

P. 129. The deduction of (5) from (4) depends upon the well-known formula

$$\int_0^\infty x^{-\frac{3}{2}} e^{-ax-\frac{b}{x}}\,dx = \sqrt{\left(\frac{\pi}{b}\right)}\, e^{-2\sqrt{(ab)}},$$

where the real parts of a and b are positive. The same remark applies to (11) and (16).

Pp. 130--132. The equations (9), (14), and (19) should be

$$\frac{\sin(\frac{1}{2}\pi/a)}{\cosh\frac{1}{2}\pi} - \frac{3\sin(\frac{9}{2}\pi/a)}{\cosh\frac{3}{2}\pi} + \dots$$

$$= \tfrac{1}{2} a\sqrt{a}\left\{\frac{\cosh\frac{1}{4}\pi a}{\cosh\frac{1}{2}\pi a} + \frac{3\cosh\frac{3}{4}\pi a}{\cosh\frac{3}{2}\pi a} - \frac{5\cosh\frac{5}{4}\pi a}{\cosh\frac{5}{2}\pi a} - \frac{7\cosh\frac{7}{4}\pi a}{\cosh\frac{7}{2}\pi a} + \dots\right\}, \quad \dots(9')$$

$$\frac{1}{4\pi} - \frac{\cosh(2\pi/a)}{\sinh\pi} + \frac{2\cosh(8\pi/a)}{\sinh 2\pi} - \dots$$

$$= \tfrac{1}{4} a\sqrt{a}\left\{\frac{\sinh\frac{1}{4}\pi a}{\cosh\frac{1}{2}\pi a} - \frac{3\sinh\frac{3}{4}\pi a}{\cosh\frac{3}{2}\pi a} - \frac{5\sinh\frac{5}{4}\pi a}{\cosh\frac{5}{2}\pi a} + \frac{7\sinh\frac{7}{4}\pi a}{\cosh\frac{7}{2}\pi a} + \dots\right\} \quad \dots(14')$$

and

$$\frac{1}{8\pi} + \sum_{n=1}^\infty \frac{n\cos(2n^2\pi/a)}{e^{2n\pi}-1} = \int_0^\infty \frac{x\cos(2\pi x^2/a)}{e^{2\pi x}-1}\,dx + \tfrac{1}{4} a\sqrt{a}\,S,$$

where

$$S = \sum_{n=1}^\infty \frac{(-1)^{n-1} n}{e^{n\pi a} + (-1)^n}, \quad S = \sum_{n=1}^\infty \frac{n}{e^{n\pi a} + (-1)^{n-1}}, \quad \dots\dots\dots\dots(19')$$

according as $a \equiv 1$ or $a \equiv 3 \pmod 4$.

17.

Proofs of the formulæ contained in this paper were given by B. M. Wilson, "Proofs of some formulæ enunciated by Ramanujan", *Proc. London Math. Soc.* (2), vol. XXI (1922), pp. 235—255. In some cases the orders of Ramanujan's error terms are improved upon.

It has since been proved by T. Estermann that if

$$f_{k,r}(s) = \Sigma\,\frac{\{d_k(n)\}^r}{n^s} \qquad (k = 2, 3, \dots;\ r = 1, 2, \dots;\ \sigma > 1)$$

then $f_{k,r}(s)$ is meromorphic in the half plane $\sigma > 0$ and, except when $r = 1$ or $r = 2$, $k = 2$, has the line $\sigma = 0$ as a natural boundary.

E. C. Titchmarsh has proved the formula

$$f_{k,2}(s) = \{\zeta(s)\}^k \prod_p P_{k-1}\left(\frac{1+p^{-s}}{1-p^{-s}}\right),$$

where P_{k-1} is Legendre's polynomial.

P. 134, footnote †. See G. H. Hardy, "On Dirichlet's divisor problem", *Proc. London Math. Soc.* (2), vol. XV (1916), pp. 1—25. We take this opportunity of observing that (as was pointed out to us by Professor E. Landau) Hardy states there rather more than is justified. His argument proves that $\Delta_1(n)$ is sometimes greater than $K(n \log n)^{\frac{1}{4}} \log \log n$, and so justifies Ramanujan's statement, but it fails to prove that $\Delta_1(n)$ is sometimes less than $-K(n \log n)^{\frac{1}{4}} \log \log n$. For further results in this direction see G. Szegö und A. Walfisz, "Über das Piltzsche Teilerproblem in algebraischen Zahlkörpern", *Math. Zeitschrift*, vol. XXVI (1927), pp. 138—156 and 467—486.

18.

Many of the results conjectured by Ramanujan in the latter part of this paper were afterwards proved by Mordell: see L. J. Mordell, "On Mr Ramanujan's empirical expansions of modular functions", *Proc. Camb. Phil. Soc.*, vol. XIX (1919), pp. 117—124. In particular Mordell proves Ramanujan's results (101), (103), (118), (123), (127), (128), and (159), and indicates how, in a similar way, proofs of (155) and (162) may be obtained.

A considerable number of problems remain open for future research. Thus Ramanujan's conjecture that

$$|\tau(n)| \leq n^{\frac{11}{2}} d(n)$$

(p. 153, equation (105)) is still unproved; it is not even known that $\tau(n) = O(n^{\frac{11}{2}+\epsilon})$ for every positive ϵ, though Hardy has proved that $\tau(n) = O(n^6)$ and

$$\{\tau(1)\}^2 + \{\tau(2)\}^2 + \ldots + \{\tau(n)\}^2 = O(n^{12}).$$

Similar questions remain open concerning the functions $e_{2s}(n)$.

Other interesting problems suggested by Ramanujan's work are those of finding asymptotic formulæ for

$$d(1)\,d(n-1) + d(2)\,d(n-2) + \ldots + d(n-1)\,d(1)$$

and

$$\sum_{n \leq x} d(n)\,d(n+k), \quad \sum_{n \leq x} r(n)\,r(n+k),$$

where

$$r(n) = r_2(n).$$

It has been proved by A. E. Ingham ["Some asymptotic formulæ in the theory of numbers", *Journal London Math. Soc.*, vol. II (1927), pp. 202—208] that

$$\sum_{n \leq x} d(n)\,d(n+k) = \frac{6}{\pi^2} x (\log x)^2 \sum_{d|k} \frac{1}{d} + O(x \log x),$$

and

$$\sum_{\nu < n} d(\nu)\,d(n-\nu) = \frac{6}{\pi^2} n \sum_{d|n} \frac{1}{d} \left(\log \frac{n}{2d^2}\right)^2 + O\left(n \log n \sum_{d|n} \frac{1}{d}\right);$$

more generally, Ingham obtains asymptotic formulæ for

$$\sum_{n \leq x} \sigma_r(n)\,\sigma_s(n+k), \quad \sum_{\nu < n} \sigma_r(\nu)\,\sigma_s(n-\nu)$$

valid for all real positive values of r and s.

Among other recent memoirs dealing with similar questions we may mention

L. J. Mordell, "On the representations of numbers as a sum of $2r$ squares", *Quarterly Journal of Math.*, vol. XLVIII (1920), pp. 93—104;

L. J. Mordell, "On the representations of a number as a sum of an odd number of squares", *Trans. Camb. Phil. Soc.*, vol. XXII (1923), pp. 361—372;

G. H. Hardy, "On the expression of a number as the sum of any number of squares, and in particular of five or seven", *Proc. National Acad. of Sciences*, vol. IV (1918), pp. 189—193;

G. H. Hardy, "On the representation of a number as the sum of any number of squares, and in particular of five", *Trans. American Math. Soc.*, vol. XXI (1920), pp. 255—284;

G. H. Hardy, "Note on Ramanujan's arithmetical function $\tau(n)$", *Proc. Camb. Phil. Soc.*, vol. XXIII (1927), pp. 675—680;

G. K. Stanley, "On the representation of a number as the sum of seven squares", *Journal London Math. Soc.*, vol. II (1927), pp. 91—96;

E. Landau, "Über Gitterpunkte in mehrdimensionalen Ellipsoiden", *Math. Zeitschrift*, vol. XXI (1924), pp. 126—132, and vol. XXIV (1926), pp. 299—310;

A. Walfisz, "Über Gitterpunkte in mehrdimensionalen Ellipsoiden", *Math. Zeitschrift*, vol. XIX (1924), pp. 300—307, and vol. XXVI (1927), pp. 106—124;

H. Petersson, "Über die Anzahl der Gitterpunkte in mehrdimensionalen Ellipsoiden", *Hamburg Math. Abhandlungen*, vol. V (1926), pp. 116—150.

These memoirs give further references to the literature of the subject.

20.

P. 170. It has been pointed out by L. E. Dickson that Ramanujan overlooks the fact that (1, 2, 5, 5) will not represent 15: see his note in the *Bulletin Amer. Math. Soc.* referred to below. This reduces the 55 forms to 54.

P. 171, l. 6. The forms 1, 1, 3, 3 and 1, 2, 3, 6 were considered also by H. J. S. Smith, *Collecteana Mathematica in memoriam Dominici Chelini* (1881), pp. 117—143 (*Collected Math. Papers*, vol. II, pp. 287—311, particularly pp. 309—311).

Pp. 171—172. A number of the results concerning ternary forms stated by Ramanujan seem to be new. In this connection Prof. Mordell writes: "Suppose that

$$f(x, y, z) = ax^2 + by^2 + cz^2 + 2fyz + 2gzx + 2hxy$$

is a properly primitive positive ternary form, so that a, b, c, f, g, h have no common factor and a, b, c are not all even. This form has two arithmetical invariants for linear transformations of determinant 1, namely the invariants Ω, Δ defined by

$$\begin{vmatrix} a & h & g \\ h & b & f \\ g & f & c \end{vmatrix} = \Omega^2\Delta, \qquad (1)$$

(2) Ω is the highest common factor of the coefficients of the adjoint form

$$F(x, y, z) = (bc - f^2)\,x^2 + \ldots + 2\,(gh - af)\,yz + \ldots.$$

There are only a finite number of classes of forms with given invariants Ω, Δ.

Consider the simplest case, in which Ω and Δ are both odd. If f is any form of a given class, ω is any prime factor of Ω, and δ any prime factor of Δ, the symbols of quadratic reciprocity

$$\left(\frac{f(x, y, z)}{\omega}\right), \quad \left(\frac{F(x, y, z)}{\delta}\right) \qquad (3)$$

have constant values for all x, y, z which make f prime to ω and F to δ. In fact, given N, we can find a ϕ equivalent to f, with adjoint Φ, such that

$$\phi \equiv ax^2 + \beta\Omega y^2 + \gamma\Omega\Delta z^2, \quad \Phi \equiv \beta\gamma\Omega\Delta x^2 + \gamma a\Delta y^2 + a\beta z^2,$$

where a, β, γ are odd and $a\beta\gamma \equiv 1$, the congruences being to modulus N: see P. Bachmann, *Die Arithmetik der quadratischen Formen*, vol. I (Leipzig, 1898), p. 54. Hence, if ω is any prime factor of Ω, and $N = \omega$,

$$\left(\frac{\phi}{\omega}\right) = \left(\frac{ax^2}{\omega}\right) = \left(\frac{a}{\omega}\right),$$

and is independent of x, y, z.

Hence the classes may be subdivided into genera corresponding to the values of the symbols (3). Given a genus, we can calculate

$$E = \left(\frac{f}{\Omega}\right)\left(\frac{F}{\Delta}\right)(-1)^{\frac{1}{4}(\Omega+1)(\Delta+1)},$$

and it may be shewn that

$$E = (-1)^{\frac{1}{4}(\Delta f+1)(\Omega F+1)};$$

see Bachmann, *loc. cit.*, pp. 68—69.

If $E=-1$ for a particular genus, we deduce from this that $f(x, y, z)$ can take odd values congruent to Δ, 3Δ, 5Δ (mod 8), but not those congruent to 7Δ. If *e.g.*

$$f=x^2+y^2+z^2,$$

then $\Delta=\Omega=1$, so that there is only one form in the genus, and $E=-1$. Hence f takes odd values congruent to 1, 3, 5 (mod 8), but not those congruent to 7. Similarly $x^2+y^2+5z^2$ takes odd values congruent to 1, 5, 7, but not those congruent to 3. When $E=1$, the form can take values congruent to 1, 3, 5, or 7.

It is thus a simple matter to write down the conditions that a given odd number m, which must of course give the symbols $\left(\frac{m}{\omega}\right)$ their appropriate values, should be representable by a given genus of forms; and when there is only one form in the genus, the problem of representation of an odd number by that form is solved. The results for even numbers are sometimes corollaries but sometimes require an independent discussion. Thus $x^2+y^2+z^2$ and $x^2+y^2+5z^2$ is divisible by 4 only if x, y, z are all even, so that the forms cannot assume the values $4^a(8n+7)$ or $4^a(8n+3)$ respectively; but the residues 2 and 6 (mod 8) cannot be disposed of so simply.

The foundations of the theory were laid by Eisenstein [*Journal für Math.*, 35 (1847), 117—136; or *Math. Abhandlungen*, 177—206], and it is likely enough that he knew most of Ramanujan's results. He mentions this case ($\Delta=\Omega=1$), and states that $x^2+y^2+3z^2$ will represent all odd numbers, and that he will not delay to discuss even numbers also. I may add that some of Ramanujan's results are easily deduced from the fundamental result about the form (3·1). Thus $m=x^2+y^2+2z^2$ gives

$$2m=(x+y)^2+(x-y)^2+(2z)^2=\xi^2+\eta^2+\zeta^2,$$

say, and ξ, η have the same parity. If both are even, m is even and

$$\tfrac{1}{2}m=(\tfrac{1}{2}\xi)^2+(\tfrac{1}{2}\eta)^2+z^2,$$

which is possible unless $m=2.4^r(8\mu+7)$. If both are odd, m is odd. In this case $2m$ is representable by three squares, say

$$2m=X^2+Y^2+Z^2,$$

where we may suppose X, Y odd and Z even; and

$$m=\{\tfrac{1}{2}(X+Y)\}^2+\{\tfrac{1}{2}(X-Y)\}^2+2(\tfrac{1}{2}Z)^2=x^2+y^2+2z^2.$$

In a recent note in the *Bulletin of the American Math. Soc.* [vol. XXXIII (1927), pp. 63—70] L. E. Dickson gives proofs of all Ramanujan's results about ternary forms, and others. To show, for example, that any integer not of the form $25^\lambda(25\mu\pm10)$ is representable by $f(x, y, z)=x^2+2y^2+5z^2$, he uses the fact that the classes of forms with $\Omega^2\Delta=10$ are given by this form and three others. He writes down a particular ternary form with $\Omega^2\Delta=10$ and first coefficient n and shews that it must be equivalent to $f(x, y, z)$ and so that n is representable by $f(x, y, z)$. A continuation of this note will appear in the *Annals of Mathematics*. There are two other recent papers by Dickson relevant to Ramanujan's work, viz. (1) 'Quadratic forms which represent all integers', *Proc. Nat. Acad. of Sciences*, vol. XII (1926), pp. 756—757, and (2) 'Quaternary quadratic forms representing all integers', *American Journal of Math.*, vol. XLIX (1927), pp. 39—56."

See also H. D. Kloosterman: (1) "Over het splitsen van gehule positive getallen in een som van kwadraten", *Dissertation*, Groningen, 1924; (2) "On the representation of numbers in the form $ax^2+by^2+cz^2+dt^2$", *Proc. London Math. Soc.* (2), vol. XXV (1926), pp. 143—173; (3) "Über Gitterpunkte in vierdimensionalen Ellipsoiden", *Math. Zeitschrift*, vol. XXIV (1926), pp. 519—529; (4) "On the representation of numbers in the form $ax^2+by^2+cz^2+dt^2$", *Acta Math.*, vol. XLIX (1927), pp. 407—464.

Kloosterman solves the problem "will $ax^2+by^2+cz^2+dt^2$ represent all numbers from a certain point onwards?" for all but a finite number of forms, viz. the forms

(1, 2, 11, 38), (1, 2, 19, 38), (1, 2, 19, 22), (1, 2, 17, 34)

and certain other forms derived from them.

21.

In connection with this paper see G. H. Hardy, "Note on Ramanujan's trigonometrical function $c_q(n)$, and certain series of arithmetical functions", *Proc. Camb. Phil. Soc.*, vol. XX (1921), pp. 263—271. Hardy there proves, in a different manner, Ramanujan's formulæ (2·7), (6·1), and (9·6), and others of a similar character. There is a misprint in his statement of (1·3) (Ramanujan's (6·1)).

The formula (2·7) was given by Kluyver [J. C. Kluyver, "Eenige formules aangaande de getallen kleiner dan n en ondeelbaar met n", *Versl. Kon. Akad. van Wetenschappen te Amsterdam*, vol. 15 (1906), pp. 423—429]. See also Landau, *Handbuch*, p. 572, and J. L. W. V. Jensen, "Et nyt Udtryk for den talteoretiske Funktion $\Sigma\mu(n)=M(n)$", *Beretning om den 3 Skandinaviske Matematiker-Kongres*, Kristiania, 1915. The deduction of the general formula (2·7) from the special case given by Landau is trivial. But Ramanujan's main results are new and his point of view is quite individual.

P. 194, § 15. Series such as (11·11) are what Hardy and Littlewood, in their researches on Waring's Problem, call "singular series", and are in fact the simplest of such series.

P. 196, (16·9). See also A. Walfisz, "Über die summatorischen Funktionen einiger Dirichletscher Reihen", *Dissertation*, Göttingen (1922); A. Oppenheim, "Some identities in the theory of numbers", *Proc. London Math. Soc.* (2), vol. XXIV (1926), *Records* for 15 Jan. 1925, pp. xxiii—xxvi, and vol. XXVI (1927), pp. 295—350.

P. 198, (17·5). See also B. M. Wilson, "An asymptotic relation between the arithmetic sums $\sum_{n\leqslant x}\sigma_r(n)$ and $x^r\sum_{n\leqslant x}\sigma_{-r}(n)$", *Proc. Camb. Phil. Soc.*, vol. XXI (1922), pp. 140—149; and E. Landau, "Über einige zahlentheoretische Funktionen", *Göttinger Nachrichten*, 1924, pp. 116—134.

23.

P. 203, (10). Some errors of sign have been corrected.

24.

P. 208, l. 5. $\nu(x)$ is what is usually denoted by $\vartheta(x)$.

25.

Alternative proofs of (1) and (2), and a proof of (3), appear in No. 30. The results (5), (7), and (8) are corollaries. The manuscript from which No. 30 is extracted also contains alternative proofs of (4), (6), and a proof of (9), which we hope may ultimately be published.

For other proofs of (1) and (2) see

H. B. C. Darling: (1) "On Mr Ramanujan's congruence properties of $p(n)$", *Proc. Camb. Phil. Soc.*, vol. XIX (1919), pp. 217—218; (2) "Proofs of certain identities and congruences enunciated by S. Ramanujan", *Proc. London Math. Soc.* (2), vol. XIX (1921), pp. 350—372;

L. J. Rogers, "On a type of modular relation", *Proc. London Math. Soc.* (2), vol. XIX (1921), pp. 387—397;

L. J. Mordell, "Note on certain modular relations considered by Messrs Ramanujan, Darling, and Rogers", *Proc. London Math. Soc.* (2), vol. XX (1922), pp. 408—416.

26.

This paper appeared in the *Proceedings of the Cambridge Philosophical Society* accompanied by an explanatory note by Hardy and a new proof of the identities by Rogers, to whom the identities are originally due. Both these notes are reprinted below.

1. *By G. H. Hardy.*

The identities in question are those numbered (10) and (11) in each of the two following notes, viz.,

$$1+\frac{q}{1-q}+\frac{q^4}{(1-q)(1-q^2)}+\frac{q^9}{(1-q)(1-q^2)(1-q^3)}+\ldots$$
$$=\frac{1}{(1-q)(1-q^4)(1-q^6)(1-q^9)(1-q^{11})(1-q^{14})\ldots}, \quad \ldots(1)$$

and

$$1+\frac{q^2}{1-q}+\frac{q^6}{(1-q)(1-q^2)}+\frac{q^{12}}{(1-q)(1-q^2)(1-q^3)}+\ldots$$
$$=\frac{1}{(1-q^2)(1-q^3)(1-q^7)(1-q^8)(1-q^{12})(1-q^{13})\ldots}. \quad \ldots(2)$$

On the left-hand sides the indices of the powers of q in the numerators are n^2 and $n(n+1)$, while in the products on the right-hand sides the indices of the powers of q form two arithmetical progressions with difference 5.

The formulæ (1) and (2) were first discovered by Prof. Rogers, and are contained in a paper published by him in 1894*. In this paper they appear as corollaries of a series of general theorems, and, possibly for this reason, they seem to have escaped notice, in spite of their obvious interest and beauty. They were rediscovered nearly 20 years later by Mr Ramanujan, who communicated them to me in a letter from India in February 1913. Mr Ramanujan had then no proof of the formulæ, which he had found by a process of induction. I communicated them in turn to Major MacMahon and to Prof. O. Perron of Tübingen; but none of us was able to suggest a proof; and they appear, unproved, in ch. III, vol. II, 1916, of Major MacMahon's *Combinatory Analysis*†.

Since 1916 three further proofs have been published, one by Prof. Rogers‡ and two by Prof. I. Schur of Strassburg, who appears to have rediscovered the formulæ once more§.

The proofs which follow are very much simpler than any published hitherto. The first is extracted from a letter written by Prof. Rogers to Major MacMahon in October 1917; the second from a letter written by Mr Ramanujan to me in April of this year. They are in principle the same, though the details differ‖. It seemed to me most desirable that the simplest and most elegant proofs of such very beautiful formulæ should be made public without delay, and I have therefore obtained the consent of the authors to their insertion here.

* L. J. Rogers, "Second memoir on the expansion of certain infinite products", *Proc. London Math. Soc.* (1), vol. XXV (1894), pp. 318—343 (§ 5, pp. 328—329, formulæ (1) and (2)).

† Pp. 33, 35.

‡ L. J. Rogers, "On two theorems of Combinatory Analysis and some allied identities", *Proc. London Math. Soc.* (2), vol. XVI (1917), pp. 315—336 (pp. 315—317).

§ I. Schur, "Ein Beitrag zur additiven Zahlentheorie und zur Theorie der Kettenbrüche", *Berliner Sitzungsberichte*, 1917, No. 23, pp. 301—321. [Prof. Schur's work was entirely independent; but it was naturally impossible to communicate with him at the time.]

‖ I have altered the notation of Mr Ramanujan's letter so as to agree with that of Prof. Rogers.

It should be observed that the transformation of the infinite products on the right-hand sides of (1) and (2) into quotients of Theta-series, and the expression of the quotient of the series on the left-hand sides as a continued fraction, exhibited explicitly in Prof. Rogers' original paper and in Mr Ramanujan's present note, offer no serious difficulty. All the difficulty lies in the expression of these series as products, or as quotients of Theta-series.

2. *By L. J. Rogers.*

Suppose that $|q|<1$, and let V_m denote the convergent series

$$(1-x^m)-x^n q^{n+1-m}(1-x^m q^{2m})C_1+x^{2n}q^{4n+3-2m}(1-x^m q^{4m})C_2-\ldots,\quad\ldots\ldots(1)$$

where

$$C_r=\frac{(1-x)(1-xq)(1-xq^2)\ldots(1-xq^{r-1})}{(1-q)(1-q^2)(1-q^3)\ldots(1-q^r)};$$

the general term being

$$(-1)^r x^{nr}q^{nr^2+\frac{1}{2}r(r+1)-mr}(1-x^m q^{2mr})C_r.$$

Then

$$V_m-V_{m-1}=x^{m-1}(1-x)-x^n q^{n+1-m}\{(1-q)+x^{m-1}q^{2m-1}(1-xq)\}C_1$$
$$+x^{2n}q^{4n+3-2m}\{(1-q^2)+x^{m-1}q^{4m-2}(1-xq^2)\}C_2+\ldots.\quad\ldots\ldots(2)$$

Suppose now that the symbol η is defined by the equation

$$\eta f(x)=f(xq).$$

Then

$$(1-q^r)C_r=(1-x)\eta C_{r-1},\quad (1-xq^r)C_r=(1-x)\eta C_r.$$

Hence, arranging (2) in terms of ηC_1, ηC_2, ..., we obtain

$$\frac{V_m-V_{m-1}}{1-x}=(x^{m-1}-x^n q^{n-m+1})-x^{n+m-1}q^{n+m}(1-x^{n-m+1}q^{3n-3m+3})\eta C_1+\ldots$$
$$=x^{m-1}\{(1-x^{n-m+1}q^{n-m+1})-x^n q^{n+m}(1-x^{n-m+1}q^{3n-3m+3})\eta C_1+\ldots\}$$
$$=x^{m-1}\eta V_{n-m+1}.\quad\ldots\ldots(3)$$

If we write

$$v_m\prod_{r=0}^{\infty}(1-xq^r)=V_m,\quad\ldots\ldots(4)$$

then (3) becomes

$$v_m-v_{m-1}=x^{m-1}\eta v_{n-m+1}.\quad\ldots\ldots(5)$$

It should be observed that V_0 and v_0 vanish identically.

In particular take $n=2$, $m=1$, and $n=2$, $m=2$. We then obtain

$$v_1=\eta v_2,\quad v_2-v_1=x\eta v_1;$$

and so

$$v_1-\eta v_1=xq\eta^2 v_1.\quad\ldots\ldots(6)$$

Now let

$$v_1=1+a_1x+a_2x^2+\ldots.\quad\ldots\ldots(7)$$

Then from (5)

$$1+a_1x+a_2x^2+\ldots-(1+a_1xq+a_2x^2q^2+\ldots)=xq(1+a_1xq^2+a_2x^2q^4+\ldots);$$

and so

$$a_1=\frac{q}{1-q},\quad a_2=\frac{q^4}{(1-q)(1-q^2)}.\quad\ldots\ldots(8)$$

But when $x=q$, $C_r=1$; and so

$$V_1=(1-q)-q^4(1-q^3)+q^{13}(1-q^5)-\ldots.\quad\ldots\ldots(9)$$

From (4), (6), (7), and (8) it follows that

$$1+\frac{q^2}{1-q}+\frac{q^6}{(1-q)(1-q^2)}+\ldots=\frac{(1-q)-q^4(1-q^3)+q^{13}(1-q^5)-\ldots}{(1-q)(1-q^2)(1-q^3)\ldots}$$
$$=\frac{1}{(1-q^2)(1-q^3)(1-q^7)(1-q^8)(1-q^{12})(1-q^{13})\ldots}.\quad\ldots(10)$$

Similarly we have $$v_2=\frac{v_1}{\eta}=1+\frac{x}{1-q}+\frac{x^2q^2}{(1-q)(1-q^2)}+\ldots;$$

and, when $x=q$, $$v_2=1+\frac{q}{1-q}+\frac{q^4}{(1-q)(1-q^2)}+\ldots$$

and $$V_2=(1-q^2)-q^3(1-q^6)+q^{11}(1-q^{10})-\ldots.$$

Thus $$1+\frac{q}{1-q}+\frac{q^4}{(1-q)(1-q^2)}+\ldots=\frac{(1-q^2)-q^3(1-q^6)+q^{11}(1-q^{10})-\ldots}{(1-q)(1-q^2)(1-q^3)}$$
$$=\frac{1}{(1-q)(1-q^4)(1-q^6)(1-q^9)(1-q^{11})(1-q^{14})\ldots}. \qquad \ldots(11)$$

27.

P. 224. An alternative proof of (5·4) may be found in Watson's *Bessel Functions*, p. 449.

28.

For a proof that $p(11n+6)\equiv 0 \pmod{11}$ see No. 30. The manuscript from which No. 30 was extracted contains a proof that $\tau(5n)\equiv 0 \pmod 5$; others were given by Darling (2) and Mordell in the papers referred to under **25**.

29.

Proofs of the two formulæ stated here were given later by Rogers and Mordell: see the papers referred to under **25**.

30.

The footnote by Hardy explains the history of this paper. It is impossible to include more of it without very elaborate editorial work, but we hope that the manuscript may ultimately be published in full. In connection with the question of the parity of $p(n)$, see P. A. MacMahon, "Note on the parity of the number which enumerates the partitions of a number", *Proc. Camb. Phil. Soc.*, vol. XX, 1921, pp. 281—283, and "The parity of $p(n)$, the number of partitions of n, when $n\leqq 1000$", *Journal London Math. Soc.*, vol. I, 1926, pp. 224—225.

31.

A number of misprints in the original have been corrected.

34.

In connection with this paper, and the more elementary parts of No. 36, the reader should consult

K. Knopp, "Asymptotische Formeln der additiven Zahlentheorie", *Schriften der Königsberger Gelehrten Gesellschaft, Naturwiss. Klasse*, vol. II, 1925, pp. 45–74;

K. Knopp and I. Schur, "Elementarer Beweis einiger asymptotischer Formeln der additiven Zahlentheorie", *Math. Zeitschrift*, vol. XXIV (1926), pp. 559—574.

Knopp and Schur give simpler proofs of a number of results deduced here from the general Tauberian theorems. They base the results on the following general theorem: *if $P(x)=\Sigma p_n x^n$ is a power-series with positive increasing coefficients, and*

$$P(x)=\exp\left\{\int_0^x (1-t)\,S(t)\,dt\right\},$$

where $$S(x)=\Sigma s_n x^{n-1},\quad s_n\sim\tfrac{1}{2}an^2 \qquad (a>0),$$

then $$\log p_n\sim 2\sqrt{(an)}.$$

Knopp (whose memoir is a development of the joint memoir, though published earlier) proves as his principal theorem: *if* λ_n *is an integer and*

$$1 \leqq \lambda_1 < \lambda_2 < \lambda_3 < \ldots, \qquad \lambda_n \to \infty,$$

and $L(z)$, *the number of* λ_ν*'s which do not exceed* z, *satisfies*

$$L(z) \sim \gamma z^\rho \qquad (\gamma > 0,\ \rho > 0);$$

and if

$$P(x) = \Pi \frac{1}{1 - x^{\lambda_\nu}} = \Sigma p_n x^n, \quad Q(x) = \Pi (1 + x^{\lambda_\nu}) = \Sigma q_n x^n$$

(*so that* p_n *is the number of partitions of* n *into* λ*'s*, q_n *the number of partitions into unequal* λ*'s*), *then*

$$\log p_n \sim \frac{1+\rho}{\rho} a^{\frac{1}{1+\rho}} n^{\frac{\rho}{\rho+1}},$$

$$\log (q_1 + q_2 + \ldots + q_n) \sim \frac{1+\rho}{\rho} \beta^{\frac{1}{1+\rho}} n^{\frac{\rho}{1+\rho}},$$

where $$a = \gamma\rho\Gamma(1+\rho)\,\zeta(1+\rho), \quad \beta = \gamma\rho\Gamma(1+\rho)(1-2^{-\rho})\,\zeta(1+\rho).$$

P. 246. The assertion that $\mathfrak{Q}(s)$ has the imaginary axis as a line of singularities requires justification.

35.

P. 263, (1·22). More than this is now known, van der Corput having proved that

$$d(1) + d(2) + \ldots + d(n) = n \log n + (2\gamma - 1)n + O(n^\Theta),$$

where $\Theta < \frac{33}{100}$. See J. G. van der Corput, "Verschärfung der Abschätzung beim Teilerproblem", *Math. Annalen*, vol. LXXXVII (1922), pp. 39—65; and other recent memoirs of van der Corput, Landau, Littlewood, and Walfisz. Van der Corput's method is extremely difficult. Littlewood and Walfisz ["The lattice-points of a circle", *Proc. Royal Soc.* (A), vol. CVI, 1924, pp. 478—488] treat the corresponding problem for $r(n)$, the number of representations of n as a sum of two squares, and prove, by somewhat simpler methods, that the error term in this problem is $O(n^{\frac{37}{112}+\epsilon})$ for every positive ϵ; while Landau, in a note added to this paper, obtains a slightly more precise result. It is to be observed that $\frac{33}{100} < \frac{37}{112} < \frac{1}{3}$. Still more recently Landau ["Über das Konvergenzgebiet einer mit der Riemannschen Zetafunktion zusammenhängenden Reihe", *Math. Annalen*, vol. XCVII (1927), pp. 251—290] arrives at the same exponent $\frac{37}{112}$ in the divisor problem by methods also simpler than van der Corput's. There is however no easy proof, in either problem, of any result which goes beyond the exponent $\frac{1}{3}$. See for all this theory vol. II of Landau's *Vorlesungen über Zahlentheorie*, Leipzig, 1927.

P. 263, (1·23) and (1·24). The Dirichlet's series associated with $f(n)$ and $F(n)$ are

$$g(s) = \zeta(s) \Sigma \frac{1}{p^s} = \zeta(s) \sum_1^\infty \frac{\mu(n)}{n} \log \zeta(ns),$$

and $$G(s) = \zeta(s) \Sigma \frac{1}{p^s - 1}.$$

Also $$f(1) + f(2) + \ldots + f(n) = \sum_{\nu=1}^{n} \pi\left(\frac{n}{\nu}\right),$$

$$F(1) + F(2) + \ldots + F(n) = \sum_{\nu=1}^{n} \left\{\pi\left(\frac{n}{\nu}\right) + \pi\left(\sqrt{\frac{n}{\nu}}\right) + \pi\left(\sqrt[3]{\frac{n}{\nu}}\right) + \ldots\right\}.$$

The results stated are readily deducible from these.

36.

P. 287, § 3·1. See also the memoirs of Knopp and Schur referred to under **34**.

P. 288, § 3·2. Mr T. Vijayaraghavan has proved that, if $c_n \geqq 0$, $f(x)=\Sigma c_n x^n$, and

$$f(x) \sim A\sqrt{(1-x)} \exp \frac{\pi^2}{6(1-x)},$$

then

$$C_n = c_0 + c_1 + \ldots + c_n = O\left\{n^{-\frac{1}{4}} \exp \pi \sqrt{\left(\frac{2n}{3}\right)}\right\},$$

and

$$C_n > \exp\left\{\pi \sqrt{\left(\frac{2n}{3}\right)} - Bn^{\frac{1}{4}}\right\}$$

for a constant B; and that these results are substantially the best of their kind.

Questions and solutions.

We have made no systematic attempt to verify the questions of which no solutions are given, but we add a note on Question 289, p. 323. Mr T. Vijayaraghavan has pointed out that in order to justify Ramanujan's formal processes one has only to observe that *if*

$$a_n \geqslant 0, \quad T_n = \sqrt{[a_1 + \sqrt{\{a_2 + \sqrt{(a_3 + \ldots + \sqrt{a_n})}\}}]},$$

then a necessary and sufficient condition for the existence of the limit $\lim_{n\to\infty} T_n$ *is that*

$$\overline{\lim_{n\to\infty}} \frac{\log a_n}{2^n} < \infty.$$

For, if the condition is satisfied, there is a constant H such that

$$T_n \leqslant \sqrt{[H^2 + \sqrt{\{H^4 + \sqrt{(H^8 + \ldots + \sqrt{H^{2^n}})}\}}]} < H\sqrt{[1 + \sqrt{\{1 + \sqrt{(1 + \ldots)}\}}]} = HP = \tfrac{1}{2}H(1+\sqrt{5});$$

and the sufficiency of the condition follows, since T_n increases steadily with n. On the other hand, if

$$a_n = H_n^{2^n},$$

then $T_n \geqslant H_n$, so that the condition is also necessary. For the evaluation of P, and a less precise form of the convergence criterion, see G. Pólya and G. Szegö, *Aufgabe und Lehrsätze aus der Analysis*, vol. I, pp. 29—30 and 184.

In the first of Ramanujan's examples,

$$a_n = 2^{2^{n-1}} 3^{2^{n-2}} 4^{2^{n-3}} \ldots (n-1)^{2^2} n^2,$$

$$\frac{\log a_n}{2^n} \to \frac{\log 2}{2} + \frac{\log 3}{4} + \frac{\log 4}{8} + \ldots,$$

so that the formula is convergent. The second example differs only trivially.

APPENDIX II. FURTHER EXTRACTS FROM RAMANUJAN'S LETTERS TO G. H. HARDY

We give here the theorems stated by Ramanujan in his letters from India, apart from those already quoted in Hardy's notice, and those given on one sheet of the first letter which has been lost. The gaps in the numeration correspond to the formulæ printed in the notice. Thus the formulæ (20), (i) and (v), missing on p. 353, will be found on p. xxix, and what is printed under (21) on p. 353 is a continuation of what appears under the same number on p. xxix.

16 *Jan.* 1913.

I. I have found a function which exactly represents the number of prime numbers less than x, "exactly" in the sense that the difference between the function and the actual number of primes is generally 0 or some small finite value even when x becomes infinite. I have got the function in the form of infinite series and have expressed it in two ways.

(1) In terms of Bernoullian numbers*. From this we can easily calculate the number of prime numbers up to 100 millions, with generally no error and in some cases with an error of 1 or 2.

(2) As a definite integral from which we can calculate for all values [see formula 1, p. xxvii, bottom].

II. I have also got expressions to find the actual number of prime numbers of the form $An+B$....

III. I have found out expressions for finding not only irregularly increasing functions but also irregular functions without increase (*e.g.* the number of divisors of natural numbers), not merely the order but the exact form.

IV. ......

(2)
$$\int_0^\infty \frac{1}{\left\{1+\left(\frac{x}{a}\right)^2\right\}\left\{1+\left(\frac{x}{a+1}\right)^2\right\}\cdots}\,\frac{dx}{\left\{1+\left(\frac{x}{b}\right)^2\right\}\left\{1+\left(\frac{x}{b+1}\right)^2\right\}\cdots} = \tfrac{1}{2}\sqrt{\pi}\,\frac{\Gamma(a+\frac{1}{2})}{\Gamma(a)}\,\frac{\Gamma(b+\frac{1}{2})}{\Gamma(b)}\,\frac{\Gamma(a+b)}{\Gamma(a+b+\frac{1}{2})}.$$

......

(6)
$$\int_0^\infty \tan^{-1}\frac{2nz}{n^2+x^2-z^2}\,\frac{dz}{e^{2\pi z}-1}$$

can be exactly found if $2n$ is any integer.

V. ......

(7)
$$\frac{1}{(1^2+2^2)(\sinh 3\pi-\sinh\pi)}+\frac{1}{(2^2+3^2)(\sinh 5\pi-\sinh\pi)}+\frac{1}{(3^2+4^2)(\sinh 7\pi-\sinh\pi)}+\dots=\frac{1}{2\sinh\pi}\left(\frac{1}{\pi}+\coth\pi-\frac{\pi}{2}\tanh^2\frac{\pi}{2}\right).$$

(8)
$$\frac{1}{\left(25+\frac{1^4}{100}\right)(e^{\pi}+1)}+\frac{3}{\left(25+\frac{3^4}{100}\right)(e^{3\pi}+1)}+\frac{5}{\left(25+\frac{5^4}{100}\right)(e^{5\pi}+1)}+\dots=\frac{\pi}{8}\coth^2\frac{5\pi}{2}-\frac{4689}{11890}.$$

* [See formula 2, p. xxvii, bottom; this formula was given by J. P. Gram, *K. Danske Vidensk. Selsk. Skrifter* (6), vol. II (1881—1886), pp. 185—308.]

(9) $$\frac{1}{1^7 \cosh \frac{1}{2}\pi\sqrt{3}} - \frac{1}{3^7 \cosh \frac{3}{2}\pi\sqrt{3}} + \ldots = \frac{\pi^7}{23040}.$$

(10) $$\left\{1+\left(\frac{n}{1}\right)^3\right\}\left\{1+\left(\frac{n}{2}\right)^3\right\}\left\{1+\left(\frac{n}{3}\right)^3\right\}\ldots$$

can always be exactly found if n is any integer positive or negative.

(11) $$\frac{2}{3}\int_0^1 \frac{\tan^{-1}x}{x}\,dx - \int_0^{2-\sqrt{3}}\frac{\tan^{-1}x}{x}\,dx = \frac{\pi}{12}\log(2+\sqrt{3}).$$

VI. ......

(2) $$\frac{\log 1}{\sqrt{1}} - \frac{\log 3}{\sqrt{3}} + \frac{\log 5}{\sqrt{5}} - \ldots = (\tfrac{1}{4}\pi - \tfrac{1}{2}\gamma - \tfrac{1}{2}\log 2\pi)\left(\frac{1}{\sqrt{1}} - \frac{1}{\sqrt{3}} + \frac{1}{\sqrt{5}} - \ldots\right),$$

where $\gamma = \cdot5772\ldots$, the Eulerian constant.

......

(4) If $$\int_0^a \phi(p,x)\cos nx\,dx = \psi(p,n),$$

then $$\frac{\pi}{2}\int_0^a \phi(p,x)\,\phi(q,nx)\,dx = \int_0^\infty \psi(q,x)\,\psi(p,nx)\,dx.$$

(5) If $\alpha\beta = \pi$ then $$\sqrt{\alpha}\int_0^\infty \frac{e^{-x^2}}{\cosh \alpha x}\,dx = \sqrt{\beta}\int_0^\infty \frac{e^{-x^2}}{\cosh \beta x}\,dx.$$

......

VII. (1) $1^2\log 1 + 2^2\log 2 + 3^2\log 3 + \ldots + x^2\log x$

$$= \tfrac{1}{6}x(x+1)(2x+1)\log x - \tfrac{1}{9}x^3 + \frac{1}{4\pi^2}\left(\frac{1}{1^3}+\frac{1}{2^3}+\ldots\right) + \frac{x}{12} - \frac{1}{360x} + \ldots.$$

......[The missing page included the formulæ headed VIII.]

IX. ......

(2) If

$$P = \frac{\Gamma\{\tfrac{1}{4}(x+m+n+1)\}\,\Gamma\{\tfrac{1}{4}(x+m-n+1)\}\,\Gamma\{\tfrac{1}{4}(x-m+n+3)\}\,\Gamma\{\tfrac{1}{4}(x-m-n+3)\}}{\Gamma\{\tfrac{1}{4}(x-m+n+1)\}\,\Gamma\{\tfrac{1}{4}(x-m-n+1)\}\,\Gamma\{\tfrac{1}{4}(x+m+n+3)\}\,\Gamma\{\tfrac{1}{4}(x+m-n+3)\}},$$

then $$\frac{1-P}{1+P} = \frac{m}{x+}\,\frac{1^2-n^2}{x+}\,\frac{2^2-m^2}{x+}\,\frac{3^2-n^2}{x+}\,\frac{4^2-m^2}{x+\ldots}.$$

(3) If $$z = 1 + \left(\frac{1}{2}\right)^2 x + \left(\frac{1.3}{2.4}\right)^2 x^2 + \ldots,$$

and $$y = \frac{\pi}{2}\,\frac{1+\left(\frac{1}{2}\right)^2(1-x)+\left(\frac{1.3}{2.4}\right)^2(1-x)^2+\ldots}{1+\left(\frac{1}{2}\right)^2 x+\left(\frac{1..3}{2.4}\right)^2 x^2+\ldots},$$

then $$\frac{1}{(1+a^2)\cosh y} + \frac{1}{(1+9a^2)\cosh 3y} + \frac{1}{(1+25a^2)\cosh 5y} + \ldots$$

$$= \frac{1}{2}\left\{\frac{z\sqrt{x}}{1+}\,\frac{(az)^2}{1+}\,\frac{(2az)^2 x}{1+}\,\frac{(3az)^2}{1+}\,\frac{(4az)^2 x}{1+\ldots}\right\},$$

a being any quantity.

......

I have got theorems on divergent series, theorems to calculate the convergent values corresponding to the divergent series, viz.

$$1-2+3-4+\dots \quad =\tfrac{1}{4},$$
$$1-1!+2!-3!+\dots=\cdot 596\dots,$$
$$1+2+3+4+\dots \quad =-\tfrac{1}{12},$$
$$1^3+2^3+3^3+4^3+\dots=\tfrac{1}{120},\ \dots$$

and theorems to calculate such values for any given series (say $1-1^1+2^2-3^3+4^4-\dots$)....

I have also given meanings to the fractional and negative number of terms in a series and can calculate such values exactly and approximately. Many remarkable results have been got from such theorems; *e.g.*

$$\frac{1}{n}+\left(\frac{1}{2}\right)^2\frac{1}{n+1}+\left(\frac{1.3}{2.4}\right)^2\frac{1}{n+2}+\dots=\left\{\frac{\Gamma(n)}{\Gamma(n+\frac{1}{2})}\right\}^2\left\{1+\left(\frac{1}{2}\right)^2+\left(\frac{1.3}{2.4}\right)^2+\dots\text{to } n \text{ terms}\right\}.$$

......

27 *Feb.* 1913.

......

3. The number of prime numbers less than n is

$$\int_\mu^n \frac{dx}{\log x}-\frac{1}{2}\int_\mu^{\sqrt{n}}\frac{dx}{\log x}-\frac{1}{3}\int_\mu^{\sqrt[3]{n}}\frac{dx}{\log x}-\frac{1}{5}\int_\mu^{\sqrt[5]{n}}\frac{dx}{\log x}\,\Bigg|+\frac{1}{6}\int_\mu^{\sqrt[6]{n}}\frac{dx}{\log x}$$
$$-\frac{1}{7}\int_\mu^{\sqrt[7]{n}}\frac{dx}{\log x}\,\Bigg|+\frac{1}{10}\int_\mu^{\sqrt[10]{n}}\frac{dx}{\log x}-\frac{1}{11}\int_\mu^{\sqrt[11]{n}}\frac{dx}{\log x}\,\Bigg|-\frac{1}{13}\int_\mu^{\sqrt[13]{n}}\frac{dx}{\log x}$$
$$+\frac{1}{14}\int_\mu^{\sqrt[14]{n}}\frac{dx}{\log x}\,\Bigg|+\frac{1}{15}\int_\mu^{\sqrt[15]{n}}\frac{dx}{\log x}-\frac{1}{17}\int_\mu^{\sqrt[17]{n}}\frac{dx}{\log x}\,\Bigg|-\frac{1}{19}\int_\mu^{\sqrt[19]{n}}\frac{dx}{\log x}+\dots,$$

where $\mu=1{\cdot}45136380$ nearly. The numbers 1, 2, 3, 5, 6, 7, 10, 11, 13, ... above are numbers containing dissimilar prime divisors; hence 4, 8, 9, 12, ... are excluded: plus sign for even number of prime divisors and minus sign for odd number of prime divisors. As soon as a term becomes less than unity in practical calculation we should stop at the term before any vertical line marked above and not anywhere; hence the first four terms are necessary even when n is very small. Prime numbers begin with 2 and not with 1.

For practical calculations

$$\int_\mu^n \frac{dx}{\log x}=n\left(\frac{1}{\log n}+\frac{1!}{(\log n)^2}+\dots+\frac{(k-1)!}{(\log n)^k}\theta\right),$$

where
$$\theta=\tfrac{2}{3}-\delta+\frac{1}{\log n}\left\{\frac{4}{135}-\frac{\delta^2(1-\delta)}{3}\right\}$$
$$+\frac{1}{(\log n)^2}\left\{\frac{8}{2835}+\frac{2\delta(1-\delta)}{135}-\frac{\delta(1-\delta^2)(2-3\delta^2)}{45}\right\}+\dots,$$

and $\delta=k-\log n$.

It is better to choose k to be the integer just greater than $\log n$.

The number of primes less than 50 = 15 and by my formula 14·9,

" " 300 = 62 " " 61·9,

" " 1000 = 168 " " 168·2,

and so on.

I have also found expressions for the number of prime numbers of a given form (say of the form $24n+17$) less than any given number.

Primes of the form $4n+1$ = primes of the form $6n+1$,

" " $4n-1$ = " " $6n-1$,

" " $8n+1$ = " " $12n+1$.

Those of the forms $8n+3$, $8n+5$, $8n+7$, $12n+5$, $12n+7$ and $12n+11$ are all equal.

But

(primes of the form $4n-1$) − (those of the form $4n+1$) $\to \infty$,
(„ „ $6n-1$) − („ „ $6n+1$) $\to \infty$,
(„ „ $8n+3$) − („ „ $8n+1$) $\to \infty$,
(„ „ $12n+5$) − („ „ $12n+1$) $\to \infty$.

I have not merely shown that the difference tends to infinity, but found out expressions (like those for prime numbers) for the difference, within any given number....

The sum of the number of divisors of n natural numbers $= n(2\gamma - 1 + \log n) + \frac{1}{2}t_n + E$, where t_n is the number of divisors of n and E is of lower order than $\log n$. In practical calculations you will find E is very small*.

The coefficient of x^n in $(1-2x+2x^4-\ldots)^{-1}$ is

$$\frac{1}{4n}\left(\cosh \pi\sqrt{n} - \frac{\sinh \pi\sqrt{n}}{\pi\sqrt{n}}\right) + F(\cos \pi\sqrt{n}) + f(\sin \pi\sqrt{n}).$$

I have not written here the forms of F and f as they are very irregular and complicated, and their values are very small. Hence the coefficient is an integer very near to

$$\frac{1}{4n}\left(\cosh \pi\sqrt{n} - \frac{\sinh \pi\sqrt{n}}{\pi\sqrt{n}}\right),$$

and not always the nearest integer, as I hastily wrote to you before.... At present we may be contented with the result, the ratio of the coefficient of x^n in the above function, divided by the approximation, is very nearly equal to 1 and very rapidly approaches 1 when n becomes infinite... †.

......

(4) $$\frac{1}{1^3}\left(\coth \pi x + x^2 \coth\frac{\pi}{x}\right) + \frac{1}{2^3}\left(\coth 2\pi x + x^2 \coth\frac{2\pi}{x}\right) + \frac{1}{3^3}\left(\coth 3\pi x + x^2 \coth\frac{3\pi}{x}\right) + \ldots = \frac{\pi^3}{90x^3}(x^4+5x^2+1).$$

(5) $$\frac{1^5}{e^{2\pi}-1}\,\frac{1}{2500+1^4} + \frac{2^5}{e^{4\pi}-1}\,\frac{1}{2500+2^4} + \ldots = \frac{123826979}{6306456} - \frac{25\pi}{4}\coth^2 5\pi.$$

......

(8) $$\int_0^\infty \frac{(1+ab^2x^2)(1+ab^4x^2)(1+ab^6x^2)\ldots}{(1+x^2)(1+b^2x^2)(1+b^4x^2)\ldots}\,dx = \frac{\pi}{2(1+b+b^3+b^6+b^{10}+\ldots)}\,\frac{(1-ab^2)(1-ab^4)(1-ab^6)\ldots}{(1-ab)(1-ab^3)(1-ab^5)\ldots}.$$

(9) $$\int_0^a \frac{\sin z}{z}\,dz = \tfrac{1}{2}\pi - r\cos(a-\theta),$$

where $r = \sqrt{\left\{\int_0^\infty \frac{e^{-z}}{z}\log\left(1+\frac{z^2}{a^2}\right)dz\right\}}$, $\tan\theta = \left(\int_0^\infty \frac{ze^{-z}}{a^2+z^2}\,dz\right) \Big/ \left(\int_0^\infty \frac{ae^{-z}}{a^2+z^2}\,dz\right)$.

......

(11) When x is not great

$$xe^{\frac{1}{2}}\{e^{-\frac{1}{2}(1+x)^2} + e^{-\frac{1}{2}(1+2x)^2} + e^{-\frac{1}{2}(1+3x)^2} + \ldots\} = \frac{1}{1+}\,\frac{1}{1+}\,\frac{2}{1+}\,\frac{3}{1+}\,\frac{4}{1+\ldots} - \frac{x}{2} + \frac{x^2}{12} + \frac{x^4}{360} + \frac{x^6}{5040} + \frac{x^8}{60480} + \frac{x^{10}}{1710720} + \ldots.$$

* [Actually E is at least of the order $n^{\frac{1}{4}}$.]

† [See p. xxvii, formula VII (7), for Ramanujan's first statement; and No. 36, p. 304, for the correct result. The error is roughly of order $e^{\frac{1}{2}\pi\sqrt{n}}$.]

(12) If $\frac{1}{2}\pi a = \log\tan\{\frac{1}{4}\pi(1+\beta)\}$, then

$$\frac{1^2+a^2}{1^2-\beta^2}\left(\frac{3^2-\beta^2}{3^2+a^2}\right)^3\left(\frac{5^2+a^2}{5^2-\beta^2}\right)^5\left(\frac{7^2-\beta^2}{7^2+a^2}\right)^7\ldots = e^{\frac{1}{2}\pi a\beta}.$$

(13)...... which is a particular case of the continued fraction

$$\frac{a}{p+n+}\;\frac{1\,.\,p\,.\,a^2}{p+n+2+}\;\frac{2\,(p+1)\,a^2}{p+n+3+\ldots},$$

which is a particular case of a corollary to a theorem on transformation of integrals and continued fractions.

......

(16) If $$F(a,\beta) = \tan^{-1}\left\{\frac{a}{x+}\;\frac{\beta^2+k^2}{x+}\;\frac{a^2+(2k)^2}{x+}\;\frac{\beta^2+(3k)^2}{x+\ldots}\right\},$$

then $$F(a,\beta)+F(\beta,a) = 2F\{\tfrac{1}{2}(a+\beta),\ \tfrac{1}{2}(a+\beta)\}.$$

......

(18) If $$\frac{1+\frac{1\,.\,2}{3^2}(1-a)+\frac{1\,.\,2\,.\,4\,.\,5}{3^2\,.\,6^2}(1-a)^2+\ldots}{1+\frac{1\,.\,2}{3^2}a+\frac{1\,.\,2\,.\,4\,.\,5}{3^2\,.\,6^2}a^2+\ldots} = 5\,\frac{1+\frac{1\,.\,2}{3^2}(1-\beta)+\ldots}{1+\frac{1\,.\,2}{3^2}\beta+\ldots},$$

then $$(a\beta)^{\frac{1}{3}}+\{(1-a)(1-\beta)\}^{\frac{1}{3}}+3\{a\beta(1-a)(1-\beta)\}^{\frac{1}{6}} = 1.$$

(19) If $$\frac{1+\frac{1\,.\,3}{4^2}(1-a)+\frac{1\,.\,3\,.\,5\,.\,7}{4^2\,.\,8^2}(1-a)^2+\ldots}{1+\frac{1\,.\,3}{4^2}a+\frac{1\,.\,3\,.\,5\,.\,7}{4^2\,.\,8^2}a^2+\ldots} = 5\,\frac{1+\frac{1\,.\,3}{4^2}(1-\beta)+\ldots}{1+\frac{1\,.\,3}{4^2}\beta+\ldots},$$

then $$(a\beta)^{\frac{1}{2}}+\{(1-a)(1-\beta)\}^{\frac{1}{2}}+8\{a\beta(1-a)(1-\beta)\}^{\frac{1}{6}}\left[(a\beta)^{\frac{1}{6}}+\{(1-a)(1-\beta)\}^{\frac{1}{6}}\right] = 1.$$

(20)...... (ii) $$(a\delta)^{\frac{1}{8}}-\{(1-a)(1-\delta)\}^{\frac{1}{8}} = (\beta\gamma)^{\frac{1}{8}}-\{(1-\beta)(1-\gamma)\}^{\frac{1}{8}}.$$

(iii) $$\left[1+(\beta\gamma)^{\frac{1}{8}}+\{(1-\beta)(1-\gamma)\}^{\frac{1}{8}}\right]\left[1-(a\delta)^{\frac{1}{8}}-\{(1-a)(1-\delta)\}^{\frac{1}{8}}\right] = 2\{16a\beta(1-a)(1-\beta)(1-\gamma)(1-\delta)\}^{\frac{1}{24}}.$$

(iv) $$\frac{1+(\beta\gamma)^{\frac{1}{8}}+\{(1-\beta)(1-\gamma)\}^{\frac{1}{8}}}{1-(a\delta)^{\frac{1}{8}}-\{(1-a)(1-\delta)\}^{\frac{1}{8}}} = \left\{\frac{\beta\gamma(1-\beta)(1-\gamma)}{a\delta(1-a)(1-\delta)}\right\}^{\frac{1}{8}}.$$

......

(vi) $$Q+\frac{1}{Q} = \left(P+\frac{1}{P}\right)\sqrt{2},$$

where $$P = \{256a\beta\gamma\delta(1-a)(1-\beta)(1-\gamma)(1-\delta)\}^{\frac{1}{48}},\quad Q = \left\{\frac{a\delta(1-a)(1-\delta)}{\beta\gamma(1-\beta)(1-\gamma)}\right\}^{\frac{1}{16}}.$$

(21)...... [each of the expressions printed on p. xxvii is also equal to]

$$\frac{\left(\frac{1}{2}\{1+(a\delta)^{\frac{1}{2}}+[(1-a)(1-\delta)]^{\frac{1}{2}}\}\right)^{\frac{1}{2}}-\{(1-a)(1-\delta)\}^{\frac{1}{8}}}{(\beta\gamma)^{\frac{1}{8}}+\{(1-\beta)(1-\gamma)\}^{\frac{1}{8}}} = \left\{\frac{1+(\frac{1}{2})^2\beta+\ldots}{1+(\frac{1}{2})^2a+\ldots}\;\frac{1+(\frac{1}{2})^2\gamma+\ldots}{1+(\frac{1}{2})^2\delta+\ldots}\right\}^{\frac{1}{2}}.$$

(22) If
$$\begin{aligned} & F(a) = 3F(\beta) = 29F(\gamma) = 87F(\delta)\\ \text{or}\quad & F(a) = 5F(\beta) = 27F(\gamma) = 135F(\delta)\\ \text{or}\quad & F(a) = 7F(\beta) = 25F(\gamma) = 175F(\delta)\\ \text{or}\quad & F(a) = 9F(\beta) = 23F(\gamma) = 207F(\delta)\\ \text{or}\quad & F(a) = 11F(\beta) = 21F(\gamma) = 231F(\delta)\\ \text{or}\quad & F(a) = 13F(\beta) = 19F(\gamma) = 247F(\delta)\\ \text{or}\quad & F(a) = 15F(\beta) = 17F(\gamma) = 255F(\delta), \end{aligned}$$

then $$\left(\tfrac{1}{2}\{1+(\beta\gamma)^{\frac{1}{2}}+[(1-\beta)(1-\gamma)]^{\frac{1}{2}}\}\right)^{\frac{1}{2}}+(\beta\gamma)^{\frac{1}{8}}+\{(1-\beta)(1-\gamma)\}^{\frac{1}{8}}+\{\beta\gamma(1-\beta)(1-\gamma)\}^{\frac{1}{8}}$$
$$= \left(1+(a\delta)^{\frac{1}{4}}+\{(1-a)(1-\delta)\}^{\frac{1}{4}}\right)\left\{\frac{1+(\frac{1}{2})^2a+\ldots}{1+(\frac{1}{2})^2\beta+\ldots}\;\frac{1+(\frac{1}{2})^2\gamma+\ldots}{1+(\frac{1}{2})^2\delta+\ldots}\right\}^{\frac{1}{2}}.$$

12 *Jan.* 1920.

[This letter was written under difficulties and is in places very obscure. Ramanujan however makes it clear that what he means by a "mock ϑ-function" is a function, defined by a q-series convergent for $|q|<1$, for which we can calculate asymptotic formulæ, when q tends to a "rational point" $e^{2r\pi i/s}$, of the same degree of precision as those furnished, for the ordinary ϑ-functions, by the theory of linear transformation. Thus he asserts, for example, that if

$$f(q)=1+\frac{q}{(1+q)^2}+\frac{q^4}{(1+q)^2(1+q^2)^2}+\dots$$

and $q=e^{-t}\to 1$ by positive values, then

$$f(q)+\sqrt{\left(\frac{\pi}{t}\right)}\exp\left(\frac{\pi^2}{24t^2}-\frac{t}{24}\right)\to 4.$$

His list of functions is]

Mock ϑ-functions.

$$\phi(q)=1+\frac{q}{1+q^2}+\frac{q^4}{(1+q^2)(1+q^4)}+\dots,$$

$$\psi(q)=\frac{q}{1-q}+\frac{q^4}{(1-q)(1-q^3)}+\frac{q^9}{(1-q)(1-q^3)(1-q^5)}+\dots$$

$$\chi(q)=1+\frac{q}{1-q+q^2}+\frac{q^4}{(1-q+q^2)(1-q^2+q^4)}+\dots,$$

$$2\phi(-q)-f(q)=f(q)+4\psi(-q)=\frac{1-2q+2q^4-2q^9+\dots}{(1+q)(1+q^2)(1+q^3)\dots},$$

$$4\chi(q)-f(q)=3\frac{(1-2q^3+2q^{12}-\dots)^2}{(1-q)(1-q^2)(1-q^3)\dots},$$

[$f(q)$ being the function defined above].

Mock ϑ-functions (of 5th order).

$$f(q)=1+\frac{q}{1+q}+\frac{q^4}{(1+q)(1+q^2)}+\dots,$$

$$\phi(q)=1+q(1+q)+q^4(1+q)(1+q^3)+q^9(1+q)(1+q^3)(1+q^5)+\dots,$$

$$\psi(q)=q+q^3(1+q)+q^6(1+q)(1+q^2)+q^{10}(1+q)(1+q^2)(1+q^3)+\dots,$$

$$\chi(q)=1+\frac{q}{1-q^2}+\frac{q^2}{(1-q^3)(1-q^4)}+\frac{q^3}{(1-q^4)(1-q^5)(1-q^6)}+\dots$$

$$=1+\frac{q}{1-q}+\frac{q^3}{(1-q^2)(1-q^3)}+\frac{q^5}{(1-q^3)(1-q^4)(1-q^5)}+\dots,$$

$$F(q)=1+\frac{q^2}{1-q}+\frac{q^8}{(1-q)(1-q^3)}+\dots,$$

$$\phi(-q)+\chi(q)=2F(q),$$

$$f(-q)+2F(q^2)-2=\phi(-q^2)+\psi(-q)$$

$$=2\phi(-q^2)-f(q)=\frac{1-2q+2q^4-2q^9+\dots}{(1-q)(1-q^4)(1-q^6)(1-q^9)\dots},$$

$$\psi(q)-F(q^2)+1=q\frac{1+q^2+q^6+q^{12}+\dots}{(1-q^8)(1-q^{12})(1-q^{28})\dots}.$$

Mock ϑ-functions (of 5th order).

$$f(q)=1+\frac{q^2}{1+q}+\frac{q^6}{(1+q)(1+q^2)}+\frac{q^{12}}{(1+q)(1+q^2)(1+q^3)}+\ldots,$$

$$\phi(q)=q+q^4(1+q)+q^9(1+q)(1+q^2)+\ldots,$$

$$\psi(q)=1+q(1+q)+q^3(1+q)(1+q^2)+q^6(1+q)(1+q^2)(1+q^3)+\ldots,$$

$$\chi(q)=\frac{1}{1-q}+\frac{q}{(1-q^2)(1-q^3)}+\frac{q^2}{(1-q^3)(1-q^4)(1-q^5)}+\ldots,$$

$$F(q)=\frac{1}{1-q}+\frac{q^4}{(1-q)(1-q^3)}+\frac{q^{12}}{(1-q)(1-q^3)(1-q^5)}+\ldots$$

satisfy similar relations.

Mock ϑ-functions (of 7th order).

$$1+\frac{q}{1-q^2}+\frac{q^4}{(1-q^3)(1-q^4)}+\frac{q^9}{(1-q^4)(1-q^5)(1-q^6)}+\ldots,$$

$$\frac{q}{1-q}+\frac{q^4}{(1-q^2)(1-q^3)}+\frac{q^9}{(1-q^3)(1-q^4)(1-q^5)}+\ldots,$$

$$\frac{1}{1-q}+\frac{q^2}{(1-q^2)(1-q^3)}+\frac{q^6}{(1-q^3)(1-q^4)(1-q^5)}+\ldots.$$

These are not related to each other.

Printed in the United States
By Bookmasters